ATTENTION IN NEUROPHYSIOLOGY

Lord Adrian
Guest of Honour at the Conference

ATTENTION
IN NEUROPHYSIOLOGY

An International Conference

Edited by

C. R. EVANS, Ph.D.

Division of Computer Science
National Physical Laboratory
Teddington, England

T. B. MULHOLLAND, Ph.D.

Perception Laboratory
Veterans Administration Hospital
Bedford, Mass., U.S.A.

NEW YORK
APPLETON-CENTURY-CROFTS
Educational Division
MEREDITH CORPORATION

LONDON
BUTTERWORTHS

Suggested U.D.C. No.: 612.821.2
Suggested additional No. 159.952.3

Standard Book Number: 408 11900 4

MADE AND PRINTED IN GREAT BRITAIN BY
WILLIAM CLOWES AND SONS, LIMITED, LONDON AND BECCLES

Dedication

To

LORD ADRIAN, O.M., F.R.S.

Acknowledgements

Very many people in one way or another have contributed to the success of this Symposium, and have helped in the various stages prior to its publication. In particular we should like to express our grateful appreciation to the Carnegie Institution of Washington (U.S.A.), to the National Institute of Mental Health (U.S.A.) and to the Veterans Administration Central Office (U.S.A.) for financial and other assistance. Thanks are also due to the Director of the National Physical Laboratory in England for providing such excellent Conference facilities. We would also like to say how much we owe to the hard work and enthusiasm of Miss Margie Rosenman (now Mrs. J. M. Clegg) who helped with the organization of the Conference itself, and to our respective secretaries, Miss J. Wilson and Mrs. K. Tanguay. On Jackie Wilson in particular has fallen the heavy burden of the preparation of the MS for press and it is hard to see how the work could have been completed without her.

Contents

* *Nature*, *Lond.* 1967, **216**, 1124

Introduction

C. R. Evans
N.P.L., Teddington

T. B. Mulholland
Veterans Administration Hospital, Bedford, Mass.

The concept of 'Attention' in neurophysiology and in the behavioural sciences provided a controversial but productive theme for this Conference which was held at the National Physical Laboratory in October 1967. The proceedings extended over three days, sessions being chaired by Professor Herbert Jasper, Dr. W. Grey Walter and Dr. John Kennedy. Present for the three days, and also an active participant, was the Guest of Honour, Lord Adrian, whose distinguished work in this area has laid the foundations for much of the current research reported at the Conference. To many of the scientists present, the opportunity to meet or to renew acquaintanceship with this great English physiologist was the highlight of the occasion, and on the final day, Dr. C. R. Evans, one of the organizers of the Conference, announced that participants had expressed the wish to dedicate the Proceedings to Lord Adrian, who kindly agreed to this.

In many ways the Conference was interdisciplinary in its approach, contributions coming from psychologists, physiologists, anatomists, physicians, and even mathematicians and physicists. This posed certain problems in communication, though these were surprisingly minor—a suggestion that the role of interdisciplinary exchange (particularly in such areas as brain research) will probably grow steadily in the next few years.

The opening paper is by Professor D. Berlyne of Toronto, and this appropriately gives an historical account of the development of the concept of attention in psychology, and indicates the complexity of the process of attention which traditionally includes a number of quite different phenomena lumped together under a single heading.

No fewer than five participants raised the issue of the relation of the visual control systems to the electroencephalographic (EEG) occipital alpha rhythm. In his studies on the effect of eye deviation on alpha rhythm, Dr. P. Fenwick (London) confirms and extends the original results by Mulholland and Evans that occipital alpha can be facilitated by elevation of the eyes. Alpha was consistently increased in these subjects by upward, lateral, and downward deviations of the eye in 30–50 per cent of untrained subjects. This effect has been dubbed the 'improper alpha rhythm' by Dr. C. R. Evans (Teddington) who showed the Conference a film illustrating the facilitation of alpha by eye elevation—despite the extremely bright lights required for filming.

Co-organizer of the Conference, Dr. T. B. Mulholland (Bedford) presents results concerning eye position and alpha, and interprets these and other experiments in terms of the hypothesis that the occipital alpha-activation cycle reflects the behaviour of the visual control systems. These systems, grouped under the term 'triad of accommodation', include pupillary control, lens-accommodation control and eye-position control. When the triad of accommodation is unstable, alpha occurs; when it is stable, alpha occurs much less or not at all.

In his brief contribution, Professor Morris Bender (New York) points out that when the brain stem is stimulated many systems are activated including the oculomotor, and this invariably causes alterations in eye movements. Oculomotor pathways and reticular-activation systems are located in the same zone of the brain stem tegmentum. He concludes 'since it is claimed that alpha rhythm becomes more evident during upward rolling of the eyes with partial eyelid closure and since such an ocular syndrome could be activated in the brain stem it is conceivable that some of the electrical activity in the occipital lobe may originate in the brain stem'.

One of the Czechoslovakian participants, Dr. V. Zikmund (Bratislava), discusses the oculomotor component of the orienting response, showing that orienting to auditory stimuli includes a series of rapid eye movements supplanted by slower undulatory movements during intervals of no stimulation.

Electroencephalography is used to determine the level of attention for diagnosis and the evaluation of the alertness of astronauts in orbital flight. Professor Alan Mirsky (Boston) reports on a study with humans suffering from *petit mal* epilepsy who looked at a recurring visual pattern. Every now and then a different pattern was presented. Mirsky finds that the probability of *not detecting* the different pattern was closely correlated with the occurrence of a change in the EEG, characteristic of *petit mal*—the so-called spike-and-wave episode. The episodic impairment of performance began and ended a brief time before the recorded spike-and-wave episode began and ended. In other studies of background EEG in relation to performance, monkeys were trained to respond to one colour and to withhold response to another. Performance was best with theta (4–7 Hz) low and beta (21–32 Hz) high, worst with theta high and beta low. Mirsky does not find rhythms of 8–13 Hz a useful index in these studies.

Dr. O. Creutzfeldt (Munich) describes EEG recordings obtained by radio telemetry while subjects were performing mental and visuomotor tasks. Occipital alpha was present in all subjects with eyes open, though less than when eyes were closed. During various tasks alpha increased for some and decreased for others. Creutzfeldt's results offer little support to the idea that alpha occurrence is a useful measure of attention during performance of complex tasks or a measure of the rated difficulty or complexity of the tasks.

Despite wide individual differences in basic EEG patterns, Dr. Ross Adey (Los Angeles) reports that spectral-density distributions have characteristic features in a wide gamut of waking and sleeping states. These measures have proved valuable not only in establishing norms but have stimulated the use of multivariate pattern recognition of the EEG signatures that characterize, in an individual way, the performance of subjects on a standard set of tasks graded in difficulty. His paper includes remarkable data gathered from U.S. astronaut subjects.

In a paper, which included a motion picture illustration of a subject producing voluntary control of EEG, Dr. E. Dewan of the U.S. Air Force discusses 'entrainment' and 'adaptive control' relating these to electrophysiological activity associated with activity of the central nervous system that in turn relates to attention. In addition, the Wiener optimum filter as extended by Kalman and Bucy (for use in computerized space technology) is compared with models for attention, orientation, and habituation. In the past, Grey Walter and Norbert Wiener noted that 'mutual entrainment', whereby linked oscillations converge to a common frequency, may play a role in normal and abnormal EEG rhythms. 'Adaptive control' is an extension of the feedback control to increase the range of situations in which a control system can continue to perform adequately. This is achieved by regulating the control system itself by means of an *additional* control system. This additional control system constantly monitors input and output of the main control system and alters the latter's internal structure so that it will be best suited for ongoing tasks. A crucial aspect of such a system is the procedure of 'identification' of the main control system's potential response. This is often done by means of 'testing signals' which can consist of sinusoidal waves and, often, pure noise. Since both forms of electrical activity are found in large amounts in the central nervous system, and since adaptive control is to be expected in biology, Dewan suggests that some of the observed signals—for example noise and 10 Hz rhythms—might serve the purpose of testing signals. Finally, he compares optimum filtering to attention processes and provides a suggestive computer analogy to some important aspects of attention.

In five papers the relation between evoked responses and attention is directly considered. Dr. W. A. Cobb (London) describes the effect on the visual-evoked potential of binocular rivalry considered as a special case of alternations in attention. Flicker presented to an eye when it is dominant or when it is suppressed does not produce any differences in the waveform of the resulting evoked response. When patterns are viewed, there is a difference depending on whether the dominant or suppressed eye is stimulated. Dr. E. García-Austt (Montevideo) reports changes in visual evoked responses which co-vary with the processing of visual information. The results suggest that the critical determinant of the waveform of a visual-evoked response is the attitude of the subject toward the stimuli presented. In his interesting contribution, Dr. Grey Walter of Bristol asks whether attention can be defined physiologically. The results of evoked-response recording from multiple electrodes implanted on the human cortex in the frontal area suggest that this supposedly non-specific area of the brain is an integral part of the sensory system that receives signals in all modalities and produces many evoked-response patterns to incoming signals. Even closely adjacent points on the cortex show markedly different evoked-response waveforms and individual specificity to different stimulus parameters.

Dr. E. C. Beck (Salt Lake City) reviews experiments on the function of the reticular activating system in the acquisition of conditioned responses. Abrupt, acute interruption of the function of this system disrupts a conditioned response in the behaving animal. However, if the animal is subjected to long-term deprivation of the reticular system, conditioned responses can still be acquired. Resolution of this seeming paradox may lie in the dynamic nature of brain function which permits the development and utilization of auxiliary functional systems. Another series of experiments was designed to test the possibility that the reticular system's role in attention is to stabilize selectively the ongoing electrical activity in certain areas of the brain thereby improving the signal-to-noise ratio and facilitating sensory perception. By a final series of experiments, changes in human evoked responses were investigated during changes in attentiveness, and the apparent importance of the right parietal area to the neurophysiology of attention was indicated. Dr. W. Goff (West Haven) evaluates the concept of attention in evoked-response studies. He reviews selected recent experiments and suggests the implication of the results for methodology and theory in future attention experiments. Recent evidence suggesting the necessity for revision of previous theories as to the neural pathways mediating the late components of evoked responses, which are most often studied in attention experiments, is described.

The work of the Bristol group on the contingent negative variation (CNV) in normal children, adults and psychiatric patients is reviewed by Dr. Cheyne McCallum. He concludes that various distractions reduce the CNV, and that this reduction is related to the degree of attention focused on the distracting stimuli.

Dr. C. Lombroso (Boston) raises questions concerning the hypotheses that the CNV (i) is coincidental with psychological and behavioural processes; (ii) forms an integral part of the bioelectric activity of the underlying cortex; and (iii) measurement provides information on the functional state of the cortex. Lombroso's results prove not to be consistent with the hypothesis that the CNV reflects a cortical 'priming' in frontal cortex for discharge of motor neurones.

In a discussion of the habituation of response from single cells in brain and behavioural reactions, Dr. Gabriel Horn (Cambridge) speculates that some of these may function as 'novelty detectors', exerting an influence on command systems underlying perceptual and behavioural activity. With reference to the elevation of visual threshold occurring in conjunction with all saccadic eye movements, Professor L. Stark (California) postulates that saccadic suppression is a functionless consequence of the recomputation of the visual frame of reference initiated by a saccadic anticipatory signal sent to the visual control system by a brain-stem controller for eye movements. Professor R. Jung and Dr. J. Dichgans (Freiburg) discuss optokinetic nystagmus in relation to attention and the constancy of space perception during the rapid phase of optokinetic nystagmus.

Reporting on experimental research on neurochemical mediation of neural processes related to sleep and waking Professor H. H. Jasper (Montreal) concludes that significant changes do occur in the chemistry of the cerebral cortex during different stages of behavioural responsiveness in sleep and wakefulness, and that similar changes might occur—though perhaps more localized—during the process of focal vigilance called attention or attentive behaviour.

A relation of attention and arousal to cardiac function is proposed by Dr. I. Ruttkay-Nedecký (Bratislava), who points out that while there is evidence for a reflex facilitation of sensory intake during attention arousal via central and peripheral threshold variations, autonomic changes occurring much later seem to be instrumental in preparing the subject for action, following as a consequence of sensory intake. He also argues that the conceptualization of the orienting reaction should be in terms of the probability of occurrence of its components.

Professor H. P. Koepchen (Munich) warns against over-simplified identification of brain-stem neurones, showing that criteria for discriminating between these classes of neurones are not sufficient. The multisensory connections commonly ascribed to neurones of the unspecific activating brain-stem system also exist in neurones responding to the respiratory and circulatory changes and vice versa.

The final paper of the conference was presented by Dr. Raúl Hernández-Peón whose tragic death in a motor car accident occurred only a few months after his visit to Teddington.

In his interesting and controversial paper, he concludes that attention has a superreptilian origin, evolving in birds and reaching higher development in mammals; in *Homo sapiens* it reaches a glorified culmination.

Welcoming Speech by the Director of the N.P.L., Dr. J. V. Dunworth

Lord Adrian, Ladies and Gentlemen: It is a great pleasure to welcome you all to the National Physical Laboratory as delegates to the Conference on 'The Concept of Attention in Neurophysiology'. At first thought it may seem a little strange that an organization such as the NPL, with its particular orientation towards the Physical Sciences, should act as hosts to a Conference devoted to the Life Sciences. However, the link is not as tenuous as it first appears; the new building in which we are now meeting houses the Division of Computer Science, and it is named after a distinguished Englishman, Charles Babbage who lived a century or so ago and who built the first working calculating machine. Much more recently, the NPL took a pioneering interest in the development and design of the modern digital computer, and provided a base for the immensely fruitful work of the great mathematician, the late A. M. Turing. Turing's is a name which is equally well-known in the Physical as in the Life Sciences, and it is in this area that the link between the NPL and the present Conference becomes understandable. Today we are actively engaged in the exciting field of computer research, and it is the development of automatic pattern recognition and speech recognition devices amongst other things that claims our interest. And so in planning the next and the next-but-one generation of computers we also keep a watchful eye on developments in brain research which is one other reason why your presence here today is welcome.

While you are at Teddington I hope you will find the time in between Conference proceedings to see something of the NPL—not only of the work of the computer scientists, but also of some of the other Divisions. For this reason I thought it might be appropriate to tell you a little of the background of the Laboratory, which is one of the major Government Research Stations, and something of its structure and function.

The NPL was founded in 1900 with the principal role of acting as the national standardizing laboratory—it was in fact one of the first such establishments in the world. Its objectives were defined in 1902 as being 'to bring scientific knowledge to bear practically on our everyday industrial and commercial life; to break down the barrier between the theory and practice; to effect a union between science and commerce'. Although the Laboratory is now greatly increased in size and with a considerably wider scope, these objectives still hold exceedingly well for today. The close ties between Government science and industry are rigorously maintained, and at the NPL research is in progress which in some cases parallels, complements and quite frequently antedates industrial research. A particularly good example can be seen in the Computer Science Division, where advice and experience on the most recent developments in computer usage is immediately available to both national policy-makers and to industry. In the kind of environment which is provided at the NPL, the Government can keep in touch with this and other fast-moving subjects and at the same time can undertake research into promising novel concepts.

The NPL is itself divided into three groups. The first is 'Measurement'—which incorporates the Division of Quantum Metrology, the Division of Radiation Science, and the Division of Optical Metrology. A fourth Division, Electrical Science, has also recently been formed, and a special group, the Advanced Instrumentation Unit, also forms part of the main body. The second group, 'Materials' contains four Divisions—Chemical Standards, Molecular Science, Inorganic and Metallic Structure, and the Division of Materials Applications. The third group, is known as 'Engineering Sciences'. In it we find the Aerodynamics Division, the Ship Division, the Hovercraft Unit, the Mathematics Division and the Division of Computer Science who are in fact your special hosts today.

This gives an account—of necessity a condensed one—of the major topics covered by the Laboratory's substantial research programme. I should like to think that it gives at least some idea of the very considerable scope afforded to scientific research at the NPL, and that it will stimulate your interest sufficiently to find out a little more about us in the few days you are here. May I say once again, on behalf of myself and my staff, that we hope your stay here will be pleasant and profitable. I wish this Conference every success, and am very pleased to declare it open.

Opening Remarks

H. H. Jasper
University of Montreal

I would like to say how pleased I am to have this opportunity to sit beside Lord Adrian at this table. I am delighted, as we all are, that he is with us. We are very grateful to you, Sir, for coming to this Conference, and to those who arranged it. It is a great pleasure indeed to see so many friends here interested in this difficult subject, a subject that Lord Adrian has been interested in for as long as any of us, and who did some of the first real electrophysiological work on the mechanisms of attention. We are expecting your valuable guidance in our discussions, Lord Adrian; and we hope you won't be disappointed in what has come of some of the ideas that you have launched in our minds over the years. It takes a little courage, or foolhardiness perhaps, to discuss the physiological mechanisms of attention in their full complexity. Some believe that we should *first* learn more of the mechanisms of simple sensation and perception before attempting to study how they may be modified by attention. Perhaps this is so but like so many other fields in science, if we wait until we understand some of the elementary processes we find that we don't ever get to the more complicated ones. This is because the understanding of the simpler ones carries us deeper and deeper into elemental problems until we are at the molecular level, and the more complex problems get left by the way. So I think these different approaches have to be made in parallel, not in sequence. Certainly the approach to this problem of attention today is extraordinarily different from what it was when Lord Adrian and Berger began their classical studies over thirty years ago. Progress in our understanding of sensory and perceptual mechanisms has been considerable over this period and we now have many new techniques at our disposal some of which will no doubt be described in the present symposium.

(1) Dr. T. B. Mulholland
(2) Mrs. M. Clegg (*née* Rosenman)
(3) Dr. E. García-Austt
(4) Prof. A. F. Mirsky
(5) Dr. R. A. Chase
(6) Prof. H. P. Koepchen
(7) Dr. J. L. Kennedy
(8) Dr. P. B. C. Fenwick
(9) Dr. V. Zikmund
(10) Prof. R. Jung
(11) Dr. W. R. Goff
(12) Dr. I. Ruttkay-Nedecký
(13) Dr. R. Hernández-Peón
(14) Dr. J. C. Lilly
(15) Dr. G. H. Byford
(16) Lord Adrian
(17) Prof. L. Stark
(18) Prof. D. E. Berlyne
(19) Prof. H. H. Jasper
(20) Dr. I. J. Foley
(21) Dr. G. Horn
(22) Dr. W. Grey Walter
(23) Dr. W. C. McCallum
(24) Prof. M. B. Bender
(25) Dr. O. C. Creutzfeldt
(26) Dr. W. A. Cobb
(27) Dr. W. Ross Adey
(28) Dr. C. T. Lombroso
(29) Dr. C. R. Evans

The Development of the Concept of Attention in Psychology

D. E. Berlyne

Department of Psychology, University of Toronto

The history of the dealings that psychologists have had with the concept of 'attention' can be summed up very briefly. Attention was a topic of intense interest during the early days of experimental psychology, when the primary concern was the study of conscious experience using the introspective method. Then it suffered neglect for several decades. Finally, the word 'attention' has come to be used more and more in recent years, but there is considerable ambiguity about its meaning (10). Let me elaborate a little on this course of development.

The starting-point was perhaps Leibnitz's (44) observation that many external stimuli (as we should call them now) excite sense organs without being represented in consciousness. Nevertheless, they are not without their effect on us. Leibnitz gave several examples, which have often been quoted since. If we live for some time in the vicinity of a waterfall, we cease to notice the noise, but if someone draws our attention to it, 'we remember it, and are conscious of having had at the time some feeling of it'. We are aware of the sound of the sea breaking on the shore but not of the sound made by a single wave. But the sound of one wave must be affecting us, since otherwise the sound of many such waves would not be audible; 'a hundred thousand nothings cannot make something'. According to Leibnitz, a process called 'apperception' must be added to perception to produce a conscious sensation, and he repeatedly mentions 'attention' as a factor determining what will and will not be 'apperceived'. After its introduction by Leibnitz, the notion of 'apperception' was adopted and elaborated by later continental European psychologists, notably by Herbart (29) and by Wundt (70).

The cognate notions of 'apperception' and 'attention' were naturally more congenial to the continental philosophical psychologists of the rationalist–spiritualist tradition, who were prone to view the mind as a system of active, selective processes, than to the more mechanistically inclined British empiricists and associationists. So attention and similar topics were, on the whole, neglected by English-speaking writers until the advent of American functionalism (22, 34) at the end of the nineteenth century. Among the early introspective experimentalists, the most celebrated treatments are those of

Pillsbury (53) and Titchener (64). These pioneers offered minute descriptions of the conscious experiences that accompany attention and lack of attention. They recognized and classified the various forms that attention took, enumerated its determinants, and offered theories of attention over which the characteristic disputes of the period raged. Much of this early work was reviewed in a useful summary by Paschal (51).

Then came the behaviourist revolution, which led to the eclipse of attention for several reasons, of which the following seem to be the principal ones:

1. The study of attention had been treated as inseparable from the study of consciousness. Attention had been discussed either as a process that determines what we are conscious of or else (e.g. in Titchener's writings) as an attribute ('clearness') distinguishing some experiences from others. Consequently, those who wanted psychology to concentrate on publicly observable behaviour and to eschew introspection were induced to throw attention out with the bath water.

2. The notion of attention seemed to imply a homunculus. Attention was apparently conceived as something inside us that 'decided' whether or not we would accept the influence of a particular stimulus. In fact, one variety of attention that had long been recognized was 'voluntary' attention, which seemed to stir up all the recalcitrant problems surrounding volition and perhaps even free will. This seemed quite unacceptable to those who were seeking laws that related responses to stimuli.

3. Above all, the concept of 'attention' seemed unnecessary. Objectively, we manifest attention to a particular external stimulus by performing a motor response that corresponds to it and can be attributed to its causal influence. Surely the familiar laws of behaviour fully suffice to determine how powerfully a stimulus determines the response (33). It may evoke an unlearned response because of some innate reflex or other inherited behaviour pattern. Alternatively, and more frequently in higher animals, it may evoke a response because that response has become associated with it through prior learning.

The conclusion that the concept of 'attention' is dispensable was, however, a mistake for at least two interrelated reasons:

1. The early experimental investigators of behaviour began, naturally enough, with unrepresentatively simple experimental situations. These were situations in which one stimulus was presented virtually alone or was, at any rate, so much more intense than any other that might be present that the subject's response could safely be ascribed to its influence. In the everyday life of higher animals and human beings, however, this is not how things are. Our sense organs are generally flooded with stimuli, all of which are capable of evoking unlearned and, especially, learned responses. For example, there are objects all around us that we are capable, through previous learning, of approaching and handling as well as naming and describing. However, we can hardly perform more than a very small proportion of the responses associated with the objects that are stimulating our receptors. There is severe competition among stimuli for control over behaviour. Laws of attention are therefore needed to supplement the laws of learning and the laws of unlearned behaviour if we are to be able to predict what responses will actually occur in realistic—which means complex—stimulus situations.

2. When we explain behaviour by reference to the laws of learning, these laws generally invoke such factors as the nature of the response, and of its consequences, and the circumstances in which the stimulus and the response have been paired in the past. But whether a response occurs or not, and if so how strongly, depends on properties of the stimulus also. In other words, stimuli possessing certain properties seem better able than other stimuli to evoke any response that may be associated with them, regardless of the form that response takes or of its consequences. So for this reason also, we need not only laws to tell us how responses become strong but also laws that tell us how stimuli come to be effective in evoking responses.

REASONS FOR RENEWED INTEREST IN ATTENTION

Let us then have a look at some developments that have encouraged a revival of interest in attention among psychologists.

1. Psychologists concerned with perception have long recognized that the form in which something is perceived can be affected by 'set', 'attitude', 'expectancy', and '*Einstellung*'. Words like these usually designate intra-organismic processes set up either by instructions or other utterances of the experimenter or by previous experiences to which the subject has been exposed in comparable situations. Several experiments have shown that these internal processes can determine not only how an object or event is classified or interpreted when perceived but whether it is perceived at all. Hebb (28) connected this phenomenon with neurophysiological evidence that a neurone will not normally discharge an impulse unless it is excited by two or more presynaptic neurones. Accordingly, the evocation of a response requires not only excitation of peripheral origin, transmitted along sensory nerves, but also the 'central reinforcement of a sensory process', which he identified with 'attention'. Shortly after Hebb was writing, there was a spate of neurophysiological findings indicating the importance for cortical functioning of a convergence between excitation from specific and non-specific pathways, and so the case for Hebb's point could be made even stronger.

2. During the 1930s and 1940s, it was common for psychologists to hold themselves aloof from neurophysiology, claiming that the investigation of behaviour can proceed independently of the investigation of what goes on in the brain. Many have, however, been converted away from this position by the torrent of information about neural processes that has appeared over the last 20 years. The experiments of Hernández-Peón (30) and others suggesting that corticofugal influences may attenuate or block sensory processes have aroused wide interest and been frequently cited. The same applies for work on the reticular arousal system, which has so often been singled out as the structure responsible for the bodily changes characterizing an organism that is becoming 'alert' or 'attentive'.

3. Then there is the contribution of human engineering, which, among other things, has devoted a great deal of study to 'watch-keeping' or 'vigilance' tasks. These are tasks in which a subject has to react to certain intermittent and relatively infrequent events, such as movements of pointers on dials or

'blips' on a radar screen. Performance in such situations commonly declines over time. The steepness of the decline apparently depends on several factors, and there has been a fair amount of controversy in the field over theoretical interpretations. However, since the subject's efficiency is affected by the temporal distribution of crucial stimuli and by the temporal distribution and nature of non-crucial stimuli interspersed with them, it seems that watch-keeping performance is a matter, at least partially, of responsiveness to a particular kind of stimulus and not solely of the strength of the tendency to make a particular kind of response. In particular, a signal of an infrequent kind mingled with others is particularly likely to be detected. Facts like these have encouraged the use of 'attention' or similar terms in attempts to analyse watch-keeping behaviour and the use of watch-keeping performance as a measure of 'attention'.

Other kinds of experiments on 'attention' and 'distraction' have also been inspired, at least partially, by practical problems encountered in industrial and military situations. In these experiments (19, 31), a subject is exposed simultaneously to a sequence of stimuli on which his verbal or non-verbal responses are primarily to depend, and to another, subsidiary sequence of stimuli addressed either to the same sensory modality or to a different one. Measurements are made of the extent to which he responds ('attends') to the subsidiary stimulus sequence (e.g. how far he can describe its components verbally afterwards) or of the extent to which the simultaneous presentation of the subsidiary sequence impairs performance on the primary task ('distraction'). The dichotic-listening experiments to be discussed later are cases in point.

4. In the Soviet Union, psychologists, as well as 'physiologists of higher nervous activity', have been much affected by Pavlov's (52) description of the 'orientation reflex' and the high value he placed on its contribution to biological adaptation. The extended conception of the 'orientation reaction' that resulted from the exploitation of psychophysiological recording techniques in the late 1940s (60) has compelled widespread interest among them, as it has among many of their Western colleagues. They have been impressed with experimental evidence (65) indicative of a vital and all but indispensable role for the orientation reaction in the establishment of 'temporary connections', both in providing excitation to supplement that coming from receptors and in intensifying the intake and analysis of information from the external conditions to which the response is to become attached. The writings of Sechenov (57) and Vygotski (66) have encouraged Soviet psychologists to view 'orienting-investigatory activities' as essential links in the integration of more complex processes. According to Sechenov, the internal acts that move thought along, are derivatives of the eye movements and other receptor-adjusting responses that connect one act of perception with the next. For Vygotski, intellectual functioning depends on interaction between the external stimuli of the moment and 'internalized' linguistic and perceptual processes through which past experience is given a say. Under these influences, the importance of learning to attend to relevant cues at the right time has been recognized. The ways in which 'orienting' activities come to be spontaneously organized as a child grows up have been investigated, and new teach-

ing techniques, in which training to pay attention to relevant cues plays a large part, have been devised with apparent success.

Merging Vygotski's ideas with those of Sechenov and Pavlov, Zaporozhets (71) and his associates have pointed out how the orienting responses with which we inspect one portion of a stimulus object after another must take on a form corresponding to the shape and other attributes of the object. Consequently, internalized versions of these responses can serve as bases for the imagery through which objects are represented in their absence. The processes by which we search for ideas in thinking can be traced back to the orienting activities with which we hunt for objects in the physical world. Thus, in recent Soviet psychology, perception tends, through the intermediary of the 'orienting response', to be identified with attention—and so, to some extent, does the inner contemplation or representation that constitutes thinking.

5. In Western learning theory (which has meant until recently *American* learning theory), interest in attention arose out of bitter controversy. In the learning situations that are most commonly used by experimenters, the response is performed in temporal proximity to a stimulus and a reinforcing event. The stimulus, known variously as the 'conditioned stimulus', 'cue stimulus', or 'discriminative stimulus', possesses a number of properties. If visual, for example, it will have a certain shape, colour, size, intensity, and location. In discrimination learning, the experimenter arranges for the positive stimuli (i.e. those indicating the availability of reinforcement) to be characterized by some of these properties and for the remaining properties to vary from stimulus to stimulus in a way that is uncorrelated with the availability of reinforcement. As training proceeds, the response acquires an increment of association with certain properties of the stimulus. In other words, other stimuli sharing these properties will, as a result of the learning process, become more likely to evoke that response. But which properties of the stimulus, and how many, will become implicated in the association?

During the 1930s and 1940s, two diametrically opposite answers to this question were offered. Learning theorists identified with the 'S–R' current, notably Spence (62) and Hull (32), put forward the 'continuity theory'. They held that a reinforced response acquires an increment of association with all the stimulus properties that are at the time affecting receptors. Those leaning toward the 'cognitive' current, notably Krechevsky (38) and Lashley (41), favoured 'non-continuity theory', according to which an animal undergoing discrimination training adopts in turn a number of 'hypotheses' and changes relatively abruptly from one to the other. This means that he goes through a series of states corresponding to different stimulus dimensions (shape, size, location, etc.), during each of which he 'pays attention' solely to properties belonging to the dimension in question and is impervious to the others.

There are three principal experimental methods by which these opposing views can be tested:

(*a*) If the stimulus is a compound stimulus, with components belonging to different modalities or to different portions of the visual field, the components can be presented separately. The strength with which the response then occurs is an indication of how far each of them was involved in the learning process.

(*b*) If the stimulus correlated with reinforcement is characterized by several attributes, new stimuli can be presented that resemble it in some of these attributes and differ from it in others. Then, if the unchanged attributes have become associated with the response, the response will be performed, and otherwise it will not. If the changed attributes are associated with the response, there will be a generalization decrement.

(*c*) After being put through one discrimination-learning situation, subjects can go through a second one, in which the attributes distinguishing positive from negative stimuli are changed. When their performance on the second task is compared with that of subjects that did not go through the first task beforehand, it is possible to ascertain what they learned to respond to as a consequence of training on the first task. It is customary to classify the changes from the first to the second task as 'reversal shifts' or 'non-reversal shifts'. In a reversal shift, the relevant variable or dimension remains the same, but the values characterizing positive and negative stimuli are interchanged. For example, if response to a black stimulus was formerly reinforced and a in response to a white stimulus was not, reinforcement is now forthcoming response to a white, and not to a black, stimulus. In a non-reversal shift, on the other hand, a different variable or dimension becomes relevant, so that the positive and negative stimuli may now differ in size, both being black and white equally often.

The atmosphere of the 1930s and 1940s was such that learning theorists felt obliged to align themselves with one or other of the two main contending currents and to take up either one extreme position or the other. The dispute between continuity and non-continuity theories started a flurry of experimental activity, with findings see-sawing towards one side and the other in turn. When viewed with the relative detachment that the contemporary psychologist can afford to assume—as it has recently been by Mackintosh (45)—the weight of evidence seems to favour a position midway between those taken up by the embattled contestants of the past. It seems that, during discrimination learning, the response generally acquires some degree of association with several stimulus properties at once but not with all of them to the same extent. Terrace (63) reviews a number of recent experiments suggesting that a particular stimulus dimension will not gain control over behaviour (i.e. behaviour will not vary along that dimension) unless that dimension has been made 'distinctive' by association of different values along it with differential reinforcement. He acknowledges, however, that some dimensions may acquire prepotency more readily than others because of innate factors.

While the fray was at its height, those who leaned towards the cognitive side were partial to expressions like 'pay attention' (39) or 'direction of attention' (42) in their discussions of learning. Such expressions tally with their inclination, inherited from the Gestalt psychologists and their rationalist forebears of earlier centuries, to see behaviour as the product of a perceptual or 'cognitive' organization that the nervous system imposes on incoming stimulation. The S–R theorists, maintaining the empiricist and early behaviourist bias towards peripheralism, were willing to concede the importance of 'looking at' a stimulus. Spence (62) spoke of 'preparatory responses', by which he meant 'responses which lead to the reception of the appropriate aspects of the total environmental complex on the animal's sensorium, e.g. the orientation and fixation of the head and eyes towards the actual stimulus. However, evidence accumulated that, for a stimulus property to become capable of eliciting a response, it is not enough for it simply to be present and exciting sense organs during learning. The paradoxical fact that reversal shifts are often followed by better performance than non-reversal shifts was particularly hard to ignore, since it indicated that an animal or human subject can learn to place his behaviour under the control of a particular stimulus dimension, even though the locations of positive and negative segments along that dimension may alter from time to time. So even S–R theorists found themselves eventually talking about 'mediating perceptual responses', producing 'acquired distinctiveness of cues' (43), or 'implicit orienting responses', corresponding to particular stimulus variables (36), or 'observing responses', or even 'attention'. Recent mathematical models of discrimination learning commonly find room for such processes, obviously related to attention, as habituation of irrelevant cues (55) or random selection of cues followed by retention of those providing guidance towards successful performance (18).

Unequal degrees of control over a response can be acquired by different components of a compound stimulus as well as by different properties of a single stimulus. This was noted by Pavlov in his experiments on 'overshadowing'. He found that, if two conditioned stimuli were simultaneously applied before food was presented and the two were later tested singly, one of them, generally the 'stronger' one, would show itself capable of eliciting conditioned salivation while the other would not. Razran (54) carried out similar experiments with human subjects and likewise found that components of a compound stimulus acquired unequal degrees of association with the conditioned salivary response. This line of research has recently been pursued with particularly suggestive results by Kamin (35). This experimenter uses a conditioned emotional response, consisting of suppression of bar-pressing for food while a stimulus that has been paired with electric shock is present. If rats are first trained with one conditioned stimulus, say a noise, and then as training continues, this stimulus is paired with another stimulus, say a light, it turns out that the new stimulus, when presented alone, does not evoke the conditioned response. It is as if the rat did not 'notice' the second stimulus after having learned to make its response depend on the first conditioned stimulus, apart from minor transient signs of disturbance when the new stimulus first appears. On the other hand, the new stimulus may acquire an association with the conditioned response if its introduction heralds some change in conditions that is in some way 'surprising', e.g. an intensification of the shock.

6. Perhaps the most powerful fillip to the recognition and study of the problems of attention has come from the advent of information theory. Thanks to the concepts of information theory, it is possible to see the limited information-transmitting capacity of the nervous system as the main reason for the existence of attentive processes. Sherrington (59) noted that sensory fibres are about five times as numerous as motor fibres and introduced his famous 'funnel' analogy. There have been several attempts to compare the

amount of information that enters sense organs with the amount that is given out by effectors. Different methods of calculation have produced somewhat different results, but it seems clear enough that far less than 1 per cent of the information taken in by the nervous system is able to exit through motor processes. So the bottleneck is even more severe than Sherrington suggested. The concepts of information theory enable us, moreover, to handle attention *quantitatively*. The degree to which an organism is responding to the environment as a whole, or to a particular feature of it, can be analysed in terms of information transmission, as can the distribution of attention among different sources of stimulation. In fact, the choice between relatively concentrated and relatively diffuse distributions of attention, which the early introspectionists recognized, can now be recognized as a phenomenon familiar to information theorists. Whenever the total influx of information from several inputs oversteps channel capacity, the limits to conduction can be maintained either by transmitting everything that is coming in from one source and none of what is coming in from others or by partial transmission of information (implying residual uncertainty and error) from *several* sources.

The upshot of all these developments is that psychologists are now, on the whole, more willing than they were, say, 10 or 20 years ago to use the word 'attention' and to acknowledge that some serious and ineluctable problems are bound up with it. Unfortunately, there is surprisingly little willingness to differentiate the several distinct phenomena that are commonly designated by the term or even to recognize that it does not have one single clear-cut meaning for everybody. There is even a tendency for the word 'attention' to be used at times as a cloak for ignorance. We probably all first came across the word 'attention' as a name for something that we got into trouble in school for not paying. Ever since schools began, teachers have, no doubt, accused pupils of failing to 'attend' as a means of explaining away the shortcomings of their own instructional methods. Similarly, psychologists occasionally yield to the temptation to invoke variations in supposed attentive processes whenever a stimulus fails to have some effect on behaviour that was expected of it.

CLASSIFICATION OF ATTENTIONAL PHENOMENA

I should now like to distinguish the principal phenomena to which the term 'attention' has been applied by psychologists and neurophysiologists, and I shall, if I may, mention some of the work done in our own laboratory with a view to illustrating some of the problems rather than claiming that we have their solutions.

First, we must draw a distinction between 'intensive' and 'selective' phenomena. I am deliberately refraining from speaking, as some writers have, of 'intensive' and 'selective' *aspects* of attention, since the use of the word 'aspects' suggests a unitary entity to which the various aspects belong, and we are not entitled to conclude that such exists. The intensive phenomena concern attention to the external environment as a whole, whereas the selective phenomena concern distribution of attention among competing stimuli.

Among the intensive phenomena, the following variables need to be distinguished:

1. *Attentiveness*—This is best conceptualized as the total informational flux (47)—in bits per second—transmitted from the external environment to the effectors. The term 'vigilance', introduced by Head (27), might seem an appropriate name for this variable, but there are two arguments against it. One is that the word *vigilance* has come to be widely used in French-speaking countries to denote 'arousal', which, I shall argue in a moment, should not be confused with attentiveness. The second reason is that the word 'vigilance' frequently refers to performance in watch-keeping tasks, and, as will be pointed out later, it is by no means clear that what is fluctuating in the situations in question is attentiveness to the entire environment.

2. *Arousal*—This is the concept that has had such profound influence on psychological and neurophysiological theories of motivation during the last 15 years or so, the concept of a psychophysiological dimension whose fluctuations are manifested by a wide variety of electrocortical, autonomic, somatic, and verbal indices. The justifiability of such a concept has recently been questioned (40), on the grounds that the so-called indices of arousal are not particularly highly correlated with one another and that the links among cerebral, autonomic, and behavioural manifestations of arousal can be dissolved by certain surgical and pharmacological procedures—as they apparently can be by boredom (56). Nevertheless, these observations, while they are undoubtedly important and call for caution, do not necessarily imply that the concept of general arousal is useless, as I have argued elsewhere (9).

Several writers have made the particularly pregnant suggestion that what is now commonly called 'arousal' (or 'activation') is the same phenomenon that psychologists have for a long time been calling 'drive'. Actually, this suggestion raises some difficult questions. Psychologists have used three concepts of 'drive' rather than just one (6). They have conceived of drive as an energizing factor raising the overall level of activity, as a directive factor that selectively strengthens or brings to the fore certain classes of activity, and as a factor determining what events will reinforce learned responses and how powerfully. The idea of a general arousal level most clearly approximates the first of these three. The second would seem to have more to do with activation of neural structures connected with specific biological functions than with activation of a diffuse arousal system. As for the third, the possible relations between arousal and reinforcement have been examined elsewhere (9).

To revert to the notion of drive as a general energizing factor, it has long been contended that an increase in drive usually means an increase in motor activity, whether in the form of unco-ordinated restlessness or in the form of a strengthening of any co-ordinated response that happens to be prepotent. Some writers (48, 69) have held that it means an increase in central excitation. The merger of drive with arousal implies that it also means a heightening of the organism's capacities for absorbing and processing sensory information.

Words like 'attention', 'attentiveness', and 'alertness' have not infrequently been used as synonyms for arousal. I propose, however, that attentiveness and arousal should not be equated. This is for two reasons. First, we might find it useful to have a word to designate the sensory aspects of changes in arousal when we wish to concentrate on them rather than other processes accompanying them. Secondly, it seems likely that attentiveness, in the sense outlined above, is curvilinearly related to arousal. When arousal is extremely high—as in states of overwhelming emotion or mania—we should not expect behaviour to be most finely geared to minute variations in environmental conditions. On the contrary, we would expect a person in such a state to be 'blind' or 'oblivious' to what is going on in his vicinity—to bump into furniture and take no notice of what anybody is saying to him.

3. *The orientation reaction*—Since Pavlov introduced the term 'orientational reflex', it and several equivalent terms have wavered in connotation between the intensive and the selective. Pavlov himself stated that, when this reflex is in operation, animals 'immediately orientate their appropriate receptor organ in accordance with the quality in the agent bringing about the change, making full investigation of it'. This clearly refers to responses like eye movements and head movements that aim sense organs at certain environmental features and away from others. However, according to Sokolov (61), Pavlov later ceased to identify 'adjusting' and 'orienting' reactions, reserving the former term for receptor-directing changes in posture and the latter for a general increase in excitation and overall mobilization of the sensory apparatus.

Since the 'orientation reaction' became associated with polygraphically recorded psychophysiological measures in the late 1940s, the term has certainly come to connote an intensive phenomenon, an alerting of the entire organism or 'orientation' to the external environment as a whole. This is implied by the frequently stressed point, which Sokolov (61) has introduced into the very definition of the orientation reaction, that the component processes of the orientation reaction are evoked in a similar form by any of a wide variety of changes in stimulation, regardless of modality or quality. However, Russian and, occasionally, Western psychologists have taken to speaking of 'orienting responses' or 'orienting-investigatory activities' as the objective counterparts of acts of attention or even perception. To take note of a stimulus, to let it influence performance or learning, means, according to this usage, to execute a corresponding act of orientation. This clearly signifies that orienting responses will vary in form with the stimuli evoking them, which makes them eminently selective. Receptor-adjusting responses reflect the shape, size and location of at least visual stimulus objects,which, Zaporozhets notes, means that they can function as internal equivalents to represent these objects in their absence. But whether orienting responses can be said to vary with all distinguishable stimulus properties is another matter.

What the term 'orientation reaction' designates for most neurophysiologists today is a complex of processes, mainly short-lasting, that occur in response to virtually any kind of stimulus change that has not been repeated several times without ecologically significant accompaniments within the last few minutes. These processes seem to include most of the recognized indices of heightened arousal. I have to say 'most', rather than 'all', because of evidence that a reduction in heart rate may (6, 7, 26, 40) form part of the complex. Respiration rate also may change in either direction or undergo other vicissitudes (7). However, the evidence would not seem to be distorted unduly if we concluded provisionally that the orientation reaction entails a transient rise in arousal. If we accept the suggestion that arousal is the same as drive, the inference is that the orientation reaction entails a rise in drive, which would have all sorts of interesting consequences for psychology.

To illustrate some of the far-reaching ramifications of orientation-reaction measures, I should like to mention some experiments of our own. *Figure 1*

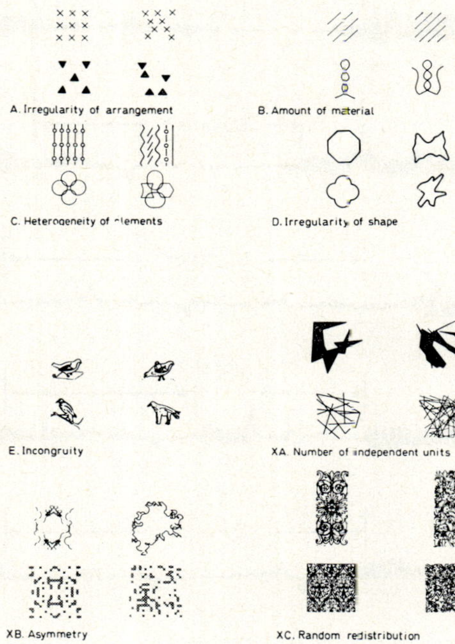

Figure 1. Stimulus patterns used in experiment by Berlyne and McDonnell
(From Berlyne and McDonnell, 1965, by courtesy of the Editor, *E.E.G. Journal*)

presents some visual patterns that we have been using in experiments for a number of years. The right-hand member of each pattern is the more 'irregular' or 'complex' member, but the actual variable distinguishing it from its partner varies from one group of pairs to another, as the captions indicate. The right-hand member can also be argued to have a higher information content. Using a number of techniques, we have found these variables to affect the duration of exploratory inspection (4, 5, 14, 15), choice of self-exposure (7), and paired-associate verbal learning (12). Latterly, we have had subjects rate or rank-order them for 'complexity', 'interestingness', 'pleasingness', and other qualities (7, 14, 17, 20, 21). Altogether, these studies, and others carried out by other investigators using different material, point to a network of important motivational phenomena dependent on stimulus complexity.

One experiment (13) supplied some evidence that one familiar component of the orientation reaction, the galvanic skin response, increases in amplitude with these complexity variables, but only when subjects are given special reasons to attend to the patterns. So we resorted to an EEG measure (16), which in this case is duration of desynchronization, measured according to the criteria introduced by Wilson and Wilson (68). (*See Figure 2.*) This is

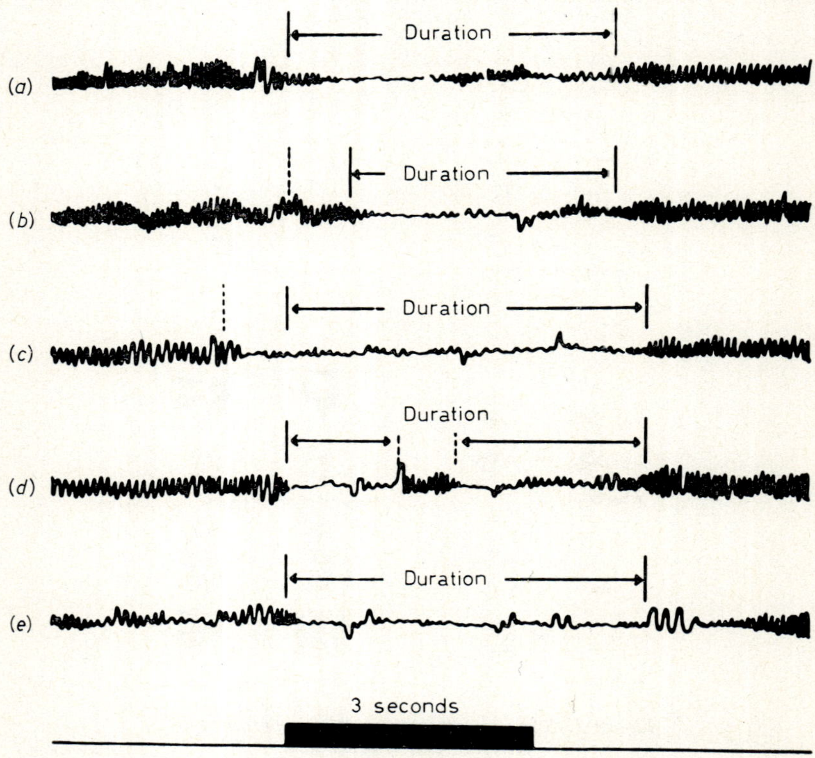

Figure 2. Different forms of the arousal response to patterned visual stimuli. The measured durations of the responses are indicated by the brackets

(From Berlyne and McDonnell, by courtesy of the Editor, *E.E.G. Journal*)

(*a*) Normal response to stimulus onset
(*b*) Delayed response to stimulus onset
(*c*) Anticipatory response conditioned by preceding stimulations
(*d*) Bursting response during stimulus
(*e*) Alpha onset with three peaks

admittedly a rather primitive way of analysing EEG data, especially when one compares it with the automated and computerized procedures that are now in vogue. Since duration of the arousal pattern declines with repetition of a stimulus (49, 58, 68) and appears to be affected in much the same way by stimuli of differing modalities and qualities, it has two of Sokolov's (61) main qualifications for recognition as an index of the orientation reaction. Furthermore, although the scoring procedure is subject to error, the practice of having tracings read 'blind' ensures that any errors are variable and not constant errors invalidating the conclusion. As Table 1 shows, there was a significant tendency over all eight categories of patterns, and in four of the categories examined separately, for the more irregular or complex patterns to evoke longer desynchronization. In a more recent experiment, using the same recording and measuring techniques, we found that 4 s exposures of blurred pictures will evoke desynchronization of significantly longer duration than exposures of clear versions of the same pictures (11). A second experiment confirmed that this effect was due to subjective uncertainty induced by the blurred picture rather than to blur as such. The blurred pictures still evoked longer-lasting desynchronization when *followed* by the corresponding clear picture but not when *preceded* by the corresponding clear picture (in which case uncertainty with regard to the content of the blurred picture would have been eliminated).

Table 1. Mean Duration of EEG Desynchronization (in seconds)
(From Berlyne and McDonnell, 1965, by courtesy of the Editor of the *EEG Journal*)

	Category	*Less complex*	*More complex*	*F* (1,258 df)	*P*
A	Irregularity of arrangement	5·8	6·5	6·08	<0·05
B	Amount of information	5·7	6·6	10·17	<0·01
C	Heterogeneity of material	6·0	6·4	—	NS
D	Irregularity of shape	6·3	6·0	—	NS
E	Incongruity	5·4	6·5	6·28	<0·05
XA	Number of independent units	5·7	6·1	—	NS
XB	Asymmetry	6·0	6·3	—	NS
XC	Random redistribution	5·6	6·2	4·98	<0·05
	All categories	5·8	6·3	24·94	<0·001

A doctoral-thesis project by Nicki (50) had already shown, with an adaptation for human subjects of the T-maze situation so often used with rats, that the replacement of a blurred picture by its clear counterpart is more reinforcing (rewarding) than its replacement by an unrelated clear picture. That the reinforcing effect depended on removal of subjective uncertainty was indicated by two pieces of evidence. First, the subjective uncertainty associated with different degrees of blur was measured by noting the number of guesses with respect to the object depicted that were voiced within a 10 s period and by obtaining a rating of the confidence placed in each. The difference in reinforcement value was found only when the degree of blur coinciding with maximum uncertainty was used. Secondly, the effect did not appear when each blurred picture was robbed of its uncertainty-inducing power by exposing the corresponding clear picture immediately before it.

These findings are consistent with the view that stimulus patterns making greater demands on the information-processing capacity of the nervous system will produce more marked orientation reactions. This presumably means that they provoke greater mobilization of the psychophysiological processes that are associated, from one point of view, with attentiveness and, from another point of view, with drive (or arousal). A more intense or more prolonged heightening of arousal means that a stimulus pattern is better able to compete with others in attracting fixation and that the subject explores it for a longer time. Stimuli bearing information that reduces subjective uncertainty, and thus relieves the condition of heightened arousal or drive (perceptual curiosity) resulting from it, can, it seems, be particularly effective when used as reinforcers to promote learning.

4. *Concentration of attention*—There is one additional variable descriptive of attention to the environment as a whole, one additional question that people using everyday language (e.g. schoolteachers) might be raising when they ask how much somebody is 'attending' (as distinct from how much he is attending to a specified object). This variable, which needs to be distinguished from the intensive phenomena discussed so far, is the degree to which attention is concentrated or diffuse. As already observed, the limited information-transmitting capacity of the nervous system allows us to give priority to what is going on in one sector of the stimulus field to the exclusion of the rest, or to keep a scattered, fallible watch over everything going on around us, or to adopt any of a continuum of intermediate distributions of attention. This variable has received little investigation so far. It is related to the problem mentioned above as one with which students of discrimination learning have concerned themselves, namely how many cues of stimulus properties become associated with a learned response. Strictly speaking, this is not quite the same question as that of how many cues or stimulus properties are determining the behaviour of the moment (which is a matter of performance rather than of learning). There have been some hints that degree of concentration may be correlated with other attentional variables, particularly arousal level. The experimental literature contains several indications that high drive (or arousal) or emotion makes for a 'narrowing of cue utilization' (25).

Wachtel (67) has recently pointed out two distinct variables to which discussions of 'breadth of attention' commonly refer. One can be identified with the range of stimuli by which behaviour is being influenced, and that is, of course, the variable that we have just been discussing. The other is how widely the external environment is scanned before selecting a stimulus to place in control of behaviour. This variable has, therefore, more to do with the exploratory behaviour that is considered below.

We must now turn to *selective* attentional phenomena, which come into play when the subject is beset by several stimuli, each of which, if acting alone, could be expected to evoke a corresponding response, but, since the responses corresponding to all of them cannot occur at once, they must compete for control over behaviour. The processes in question determine to which stimulus elements the subject will respond or, to use information-theoretic language, which element will succeed in gaining command of the organism's limited channel capacity and in sending information as far as the motorium.

1. *Selection by property (abstractions) and selection by location*—There are two ways in which this selection can be affected. First of all, selection can be made among stimulus properties belonging to the same region of the stimulus field. The subject can let his response depend on the colour, the shape, or the size of a particular object, as usually happens in consequence of discrimination training. Alternatively, the selection can be based on location.

The subject can let his response depend on stimuli coming from particular sources.

The word 'attention' is quite widely applied to both of these kinds of selection. Nevertheless, I suggest that it is advisable to distinguish sharply between them, and to use different terms to refer to them, at the present stage of research. First of all, it is by no means unlikely that quite different physiological processes are responsible for these different kinds of selection. When selection is made according to the source of stimulation, the competing inputs will enter through different sensory fibres, and one set of inputs can be made to win the contest by inhibition of the afferent processes corresponding to the other inputs. It is conceivable that something of the same sort happens when selection is made on the basis of quality. It is possible that information about say, colour, is at some stage sent off to certain neural units and information about shape to other units, so that, from that point on, processes derived from these properties are spatially separated and thus susceptible to selection by reciprocal inhibition. Several theorists have, in fact, suggested something of this sort. There is growing evidence of cells responsive to specific colours or geometrical properties at various points along visual pathways (23). Perhaps one could argue, as Konorski (37) more or less does, that there must be neurones that respond exclusively to any stimulus property or to any conjunction or disjunction of properties with which a response can become associated. Even if this is so, there may still be important differences between the physiological mechanisms of selection by property and selection by location. For instance, it is surely not unlikely that they occur at different levels. Secondly, it is still more likely that the two kinds of selection are governed by distinct, though probably overlapping, sets of independent variables.

For the first kind of selection, selection by property, there is an eminently apt term other than attention, namely *abstraction*. There is, however, no term other than *attention* that seems to fit selection by location. So I propose that the word 'attention' be reserved for this latter phenomenon, although, as is so often the case when terminological practices are critically examined, it may be already too late.

2. *Receptor-adjusting responses and attention*—Within the domain of selection according to location, there is a further distinction of some importance to be drawn. Stimuli from particular sources can be placed in control of behaviour by *receptor-adjusting responses*, sometimes known as 'orienting responses'. These include postural changes and internal physicochemical changes, which alter the direction and sensitivity of sense organs. These responses exercise selection by determining which stimuli will excite receptors, how intense the excitation from any one stimulus will be, and, since the ability to discriminate varies with intensity, how much information will be taken in from any one portion of the external environment. They also facilitate perception of certain objects and events, placing them in a favourable position to dominate behaviour, and simultaneously block perception of others.

Receptor-adjusting responses are, however, most properly regarded as adjuncts to *selective attention*. This term should, I suggest, be reserved for selection among inputs that have already entered sensory pathways, i.e. selection that is carried out independently of receptor-adjusting responses and takes place at points in the nervous system that are reached by afferent processes after they have left receptor cells.

There are, therefore, some material differences between the ways in which these two selective devices operate. Receptor-adjusting responses determine which regions of the external environment will transmit stimulation to those regions of the sensory surfaces that tend to receive priority—for example, which will have its image projected on the fovea, which will be placed directly opposite the auditory meatus, which will come into contact with the most sensitive areas of the skin, such as the finger-tips or the palms. However, although information reaching these receptor regions usually dominates motor activity, it need not do so. We can fixate one object visually but pay attention to another object that is seen peripherally. What we are calling selective attention comes in after receptor-adjusting responses have determined the mapping of environmental features on to receptors. It exercises a choice, both between and within modalities, among receptor regions and determines which region will be the one where the information controlling motor output originates.

Selective attention, in this sense, is what some contemporary psychologists are studying with several techniques that have come into use. For example, it seems to be what is involved in at least some watch-keeping experiments, where the intermittent stimulus for which the subject must be on the look-out is surrounded by other, irrelevant stimuli. It is also involved in experiments in which the subject has to respond to relevant stimuli while subjected to irrelevant 'distracting' stimuli, often from another modality.

Also of relevance are 'dual-message' experiments, in which two verbal messages are delivered simultaneously to different ears or to both ears (19). Since distinct verbal messages consist of distinct sequences of frequency combinations, and since sounds of different frequency are known to produce different distributions of excitation along the basilar membrane, the devotion of attention to one message can be classed as selection according to location of receptors. Since there appear to be no receptor-adjusting responses that can block what is heard through one ear but not what is heard through the other ear or that can pick out one pattern of sounds among several impinging on the same ear, this seems to be a clear case of selective attention in the narrow sense proposed.

Experiments using all these techniques have yielded abundant and valuable data concerning selective attention. Nevertheless, they have their limitations. Dichotic-listening experiments have generally used verbal material, and it is obvious that the analysis and interpretation of verbal stimulus patterns require processes that are not brought into play by other kinds of stimulus material. The subject is commonly also required to respond to the competing messages verbally, e.g. by repeating them or by answering them if they take the form of questions. This must introduce additional complications. To cap it all, the responses indicative of the direction of attention are frequently called for after the message has been presented, so that all the uncertainties of short-term memory play a part. So it is difficult to tell how far the conclusions drawn from experiments of this kind can be applied to other situations and how far they depend on these special features.

Investigators of watch-keeping have long recognized the difficulty of determining how far errors of omission represent deflection of attention from the stimulus, to which the subject ought to be responding, to other stimuli that are simultaneously present and how far it represents 'an interruption in the intake of all information' (Broadbent (19), page 133). In experiments where distracting stimuli are presented, it is hard to tell whether these stimuli exert their deleterious effects by taking control of behaviour away from what the subject has been concentrating on or by disrupting the central nervous system in such a way that attentiveness to the entire external environment is momentarily reduced.

A doctoral thesis by McDonnell (46) reports some experiments performed in our laboratory with a technique that has limitations of its own but overcomes some of the limitations just mentioned. It actually constitutes a modification and improvement of a technique that I used myself about 20 years ago, in Cambridge (2, 3). McDonnell's apparatus consists of a tachistoscope and, below it, a panel bearing four round buttons, arranged in the shape of a diamond, around a square button to which the subject's finger must return between trials. The inter-trial field bears a cross marking a fixation point. Examples of stimulus cards are shown in *Figure 3*. Each of them bears from one to four spots occupying corners of a diamond centred on the fixation point. When one spot appears, the subject has to press, as quickly as he can, the key in the corresponding position. When two or more spots appear, he has a free choice: he presses one, but not more than one, of the corresponding keys. The stimulus card is presented for 0·1 s. This tachistoscopic exposure, which represents the main improvement over the technique I myself used in the early experiments, served to minimize the possibility that the subject's gaze is attracted to one of the stimuli and that the choice of response is governed by fixation. According to several studies from Dodge (24) to Bartz (1), a period of 0·1 s should be too short a time for the eyes to move from the fixation point to one of the spots. If the subject obeys instructions and keeps his eyes fixed on the cross, all the stimuli will be projected an equal distance from the fovea. Even if he does not, the mean distances will be equal. Consequently, the choice of response during the free-choice trials can be presumed to depend on filtering processes that come into operation after the stimuli have been registered.

The first experiment dealt with effects of luminosity (albedo) and contrast in luminosity. Three backgrounds were used, black, medium grey, and white, and the stimuli were spots of four shades of grey, equally spaced subjectively, two of which were lighter than the medium grey background and two of which were darker. As Table 2 shows that there was a significant tendency for the subject to attend (respond) to stimuli possessing greater contrast with the background, but when contrast was equal, as when the medium-grey background was used, attention was drawn predominantly towards the lighter stimulus.

The second experiment was concerned with novelty. Half of the subjects went through a habituation phase with green spots, and the other half with

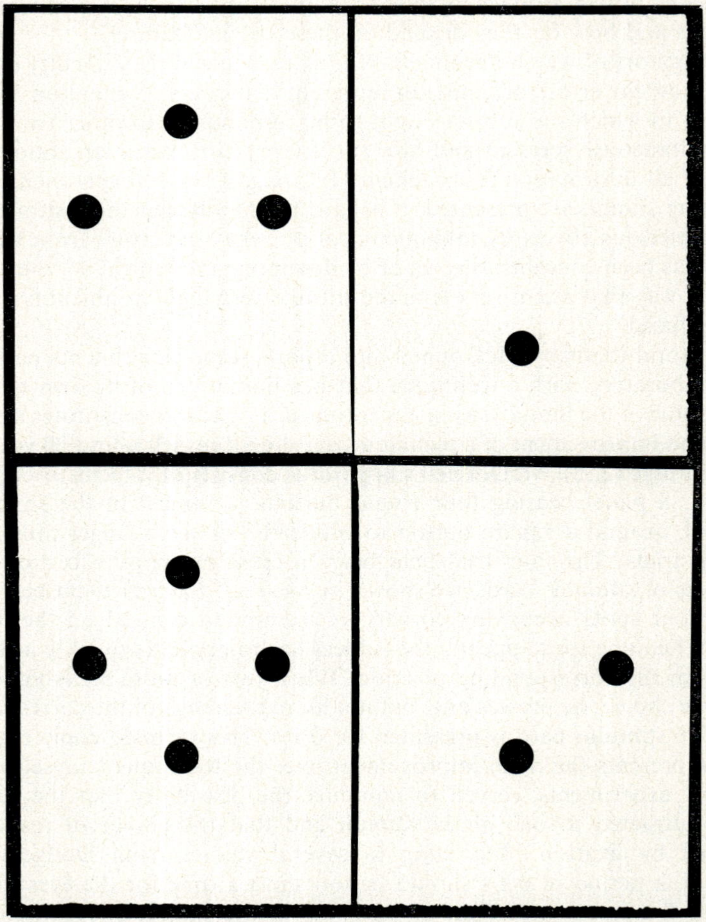

Figure 3. Sample of stimulus used by McDonnell in experiment on effects of novelty

Table 2. Experiment 1 (From McDonnell (46))

Type of trial		Mean numbers of responses							
		High-separation pairs		*Low-separation pairs*		*Quadruples*			
Munsell equivalent		3	7	4	6	3	4	6	7
Background	*No. of Ss*								
Black	29	3·1	4·5	3·2	4·0	3·4	3·3	4·4	4·9
Medium grey	28	2·6	5·0	3·0	4·2	4·0	3·4	4·3	4·3
White	26	4·7	2·8	4·2	3·3	5·4	4·4	3·1	3·1

red spots. In the second phase, there were both green and red spots, so that one colour was novel and the other familiar. Three conditions were actually used in the habituation phase. One group (M) had exactly the same motor-choice task as in the second phase. The second group (V) had a verbal equivalent to this task: they had to name the location of any one of the stimuli that appeared. The third (E) were exposed to the stimuli in the first phase without having to make any choice between them: they were told that they had to estimate how many spots were present on any trial. As can be seen from Table 3, there was a significant tendency to respond more often to a

Table 3. Experiment 2 (From McDonnell (46))

Mean numbers of responses on choice trials in experimental phase

Group	M		V		E	
Stimulus	*Novel*	*Familiar*	*Novel*	*Familiar*	*Novel*	*Familiar*
First half	10·1	6·6	6·9	8·8	9·8	7·8
Second half	9·0	8·7	9·0	8·8	9·4	8·8
Whole sequence	19·1	15·3	17·9	17·6	19·2	16·6

novel stimulus in conditions M and E. Since the effect appeared in condition E, we can conclude that the effect depended on prior exposure to a stimulus that had become familiar and not on prior performance of a particular response to that stimulus. In other words, the effect seems specific to a kind of stimulus and not to a stimulus-response association. Why the effect did not appear in condition V is not clear, and we are at present investigating this question further.

CONCLUSIONS

So while it is gratifying that problems covered by the word 'attention' are coming in for more scrutiny on the part of both neurophysiologists and psychologists, the word can be rather treacherous. To decide exactly what questions are answered by experiments that purport to deal with attention requires a great deal of circumspection—more than it usually receives. Collaboration between the two disciplines may well be of value. Hitherto, there has been a tendency for neurophysiologists to assume that the psychologists know what attention is and for psychologists in turn to assume that the neurophysiologists know!

Both neurophysiological and psychological experiments commonly expose a subject to conditions in which anybody using ordinary language would say that he must be 'paying attention' to a particular stimulus. The stimulus in question may be one that calls for some verbal or manual response, or it may frequently herald some stimulus that calls for an overt response. Alternatively, the stimulus in question may be one with important biological or 'emotional' associations. The effects of these conditions on, say, electrocortical phenomena or learning are then noted. The conditions are assumed to be representative of a large set of conditions conducive to 'attention'. However, the precise membership of this set and what exactly its members are supposed to have in common are rarely, if ever, made clear.

It is true that the experimental conditions in common use require some correspondence between the presence or the absence, or the properties, of a particular kind of stimulus and the presence or absence, or the properties, of a particular kind of response. This means that information is transmitted from stimulus to response, which is certainly a necessary condition for the conclusion that attentional processes are at work. Nevertheless, some distinctions which are not always made are called for.

Above all, we should be careful not to use a blanket term to designate several distinct phenomena until we can be sure the phenomena are actually aspects of one process or of a closely intercorrelated constellation of processes. We could, for example, conclude that all the phenomena that I have been discussing are manifestations of a unitary 'attention' if empirical investigation confirmed that the phenomena tend to appear together and to co-vary, that they are affected in much the same way by the same causal factors, and that they depend on common structures and processes in the brain. As research proceeds, these requirements might well turn out to be fulfilled, but sufficient evidence is so far lacking.

In the early stages of research, it is surely advisable to draw as many distinctions as possible. Until we have grounds for concluding otherwise, we must reckon with the possibility that logically distinct phenomena obey distinct laws. Only the protracted accumulation of experimental findings can tell us which phenomena obey the same laws and can thus be treated collectively.

ACKNOWLEDGEMENT

The preparation of this paper and research cited in it were supported by a research grant from the Department of Applied Psychology, Ontario Institute for Studies in Education, by research grant APB-73 from the National Research Council of Canada, and by research grant MH-12528 from the National Institute of Mental Health, U.S. Public Health Service.

DISCUSSION

Jasper

Adrian, would you like to open the discussion of this introductory paper?

Adrian

I am sometimes inclined to regard the word attention as having rather dirty connotations. There is the schoolmaster who tells us at school that we are not attending and tells us to carry out various impositions, and so on. And the drill sergeant who scolds the platoon to attention. In fact, I don't know whether psychologists feel that this is a limitation at all, but it has some emotional content, hasn't it? We are inclined to use, or I have been inclined to use the word without really having any clear idea of exactly what I was meaning, but feeling that probably a fair number of other people were in the same condition, and would get some sort of information from it.

Jasper

Yes, many of us are in the same position.

Bender

I would like to ask a question. Listening to this presentation, the question of wakefulness, drowsiness and alertness was raised. I'm sure you took attention into consideration because in the tests that we perform in the neurologic patient this is a very important factor. There are subjects who show unilateral defect in perception, seemingly due to inattention. In these patients stimuli in other modalities such as noises or lights may interfere with perception of somatosensory functions on the side opposite the brain lesion. Decreased alertness will also interfere with perception on the affected side. Whereas in the alert state, in a quiet environment and single stimuli a defect could not be elicited. Under what category would you bring in 'attention' here?

Berlyne

Mainly under 'arousal'. The contrast between the behaviour of an animal when it is awake and when it is asleep was one of the first things that prompted the neurophysiological studies out of which the concept of 'arousal' grew. The alternation between sleep and wakefulness, like the smaller fluctuations in arousal that occur in the course of a day, goes together with variations in attentiveness—in the amount of information taken in from the environment and processed—and these are what some of the standard tests of wakefulness are seeking. It has often been noted that sleep is characterized by raised sensory thresholds. If a threshold is high, relatively few stimuli will exceed the threshold and be responded to at all. In addition, less information will be absorbed from those stimuli patterns that are perceived—less detail will be detected.

Bender

But neurophysiologically you will find that a good deal of information, whether input or output, is blocked or somehow lost in the central nervous system. This can be demonstrated in evoked potential studies.

Berlyne

Yes, but these blocking processes will presumably have a great deal to do with the amount of information that is transmitted through the central nervous system. One of the striking facts about somebody who is asleep is that very little of what is going on around him is being reflected in what he's doing, and this phenomenon can be analysed appropriately in the language of information theory.

Mirsky

I would like to say how much I appreciated the very scholarly review that Dr. Berlyne provided. I was hoping that he would include the area of work that I am doing in attention, and fortunately he did. I appreciate his referring to Hebb's comment—or maybe it was yours Dr. Jasper—that physiologists are dealing with 50 year old psychology. The reverse may be true as well, that psychologists have never gotten much beyond the reflex; so they are often relying on 50 year old neurophysiology. Only in the last 15 years has the concept of the reticular formation reared its head, ugly or otherwise, in psychology. One of the difficulties of course is that simply to understand the reflex in all its psychological connotations is going to be very difficult.

Mulholland

In your classification of variables that influence attentiveness you made a sort of dichotomy, one part of which involves stimulus variables. I was interested in how you treat the concept of 'relevance' of the stimulus—where an on-going task may demand selective utilization of cues which depend on the way the subject sees the task; this is both idiosyncratic and transitory. Where would you put that variable?

Berlyne

The problem of what is relevant and what isn't relevant involves a lot of psychological problems that haven't been solved. If I had to guess, I'd put it in two places. A 'relevant' stimulus is often a stimulus that has associations with rewards, punishments and other events of ecological importance to the subject. This is clear when animals are interested in stimuli that are relevant to the availability of food. The same sort of thing must, in a more subtle way, play a part when somebody is trying to solve an intellectual problem and looking for the thoughts that may be relevant. Secondly, I think we must bring in what I call the 'collative' stimulus properties; it must have something to do with such things as novelty, surprisingness and expectation. A relevant event is an event that, in some respects, is similar to what is expected. An irrelevant event is an event that produces conflict, and one can argue that conflict underlies the psychological effects of many of the 'collative' variables. An irrelevant event usually invokes 'subjective uncertainty', which is likewise closely related to the notion of 'conflict'. I think that, if we really got down to it, we should find the problem of defining 'relevance' to be even worse than that of defining 'attention'. Once again there are probably several distinct concepts underlying the word.

Jasper

There are a couple of points I would like to ask you about the relation between the arousal level and the attentive process. One is related to the sleep-wake mechanism, and the other to a more selective process at a given level of wakefulness. At times I thought you were distinguishing between these two phenomena. Then later I found that you were attempting to identify them. So I wasn't quite sure what your point of view was here.

Berlyne

I was attempting to distinguish classes of phenomena to which the term 'attention' is, in practice, applied. These classes are distinct, even though one may be included in another (just as the class of 'dogs' is distinct from the class of 'animals', even though dogs are included among animals). I suggested that, at present, it is advisable to recognize a class of phenomena—fluctuations in the total amount of information from the environment reaching the motor system—that may be labelled 'attentiveness'—and another, broader class of phenomena—including fluctuations in attentiveness—that we may label 'arousal'. The way is then left open for other distinctions to be recognized later if research shows them to be important.

Jasper

Wouldn't it be possible to defend an opposite point of view? In the attentive process there must be a decrease in the amount of information that is being taken in, not an increase, whereas arousal may increase the input of information.

Berlyne

I think that this depends on the criterion you adopt. While upward or downward changes in the total input (or, better, 'throughput') of information are occurring, changes in the distribution can also be taking place. There may well be changes in what I have called 'degree of concentration', such that the organism is beginning to react to a wider and wider range of sectors of its environment or, on the contrary, its behaviour is coming under the exclusive control of what is happening in one locality and being withdrawn from the influence of what is happening elsewhere. So an increase in attentiveness, in total intake of environmental information, could go together with increasing concentration of attention, i.e. with blocking of information from particular sources. That is why we need different terms to differentiate these things. The use of a word like 'attention' as a blanket term may well prevent us from noticing the differences between certain variables and studying how they may co-vary.

Jasper

The other exception I would take to your generalization is on the opposite end of the scale of vigilance—that during sleeping states information is universally suppressed by raising thresholds. Is this actually true? There is some evidence that certain channels are occasionally left open so that slight stimuli have a strong effect on arousal from deep sleep if they correspond to a prepared set—such as the feeble cry of a baby to a sleeping mother. Here again there seems to be some selection in the suppression of sensory information—with preservation of open channels of biological significance.

Berlyne

That is true. There certainly seems to be a surprising sensitivity to selected stimuli—the mother hearing the crying child, the cat hearing its own name, etc. But this is surely not incompatible with the evidence for transmission of less information from the environment as a whole. These phenomena are exceptional, which is why they are singled out, remarked on, and studied. Less information from the environment is being reflected in motor activity. The fact that very little motor activity is going on must, of course, have a great deal to do with this. Nevertheless, there may well be other peculiarities of sleep that are relevant to selective attentional phenomena. Sleep is likely to involve important changes in distribution, and, as far as I know, these have not been investigated as much as the other changes—but they ought to be.

REFERENCES

1. Bartz, A. E. (1962) *J. exp Psychol.* **64**:318
2. Berlyne, D. E. (1950) *Q. Jl. exp. Psychol.* **2**:71
3. Berlyne, D. E. (1951) *Br. J. Psychol.* **42**:269
4. Berlyne, D. E. (1957) *J. exp. Psychol.* **53**:399
5. Berlyne, D. E. (1958) *J. exp. Psychol.* **55**:289
6. Berlyne, D. E. (1960) *Conflict, Arousal and Curiosity*. New York: McGraw-Hill
7. Berlyne, D. E. (1963) *Can. J. Psychol.* **17**:274
8. Berlyne, D. E. (1965) *Structure and Direction in Thinking*. New York: Wiley
9. Berlyne, D. E. (1967) In *Nebraska Symposium on Motivation* (Ed. by Levine). Lincoln, Neb.: Univ. Neb. Press
10. Berlyne, D. E. (in press). In *Attention*: Contempory Theory and Analysis (Ed. by Mostofsky). New York: Appleton-Century-Crofts
11. Berlyne, D. E., and Borsa, Donna M. (1967) *Percept. Psychophys.*
12. Berlyne, D. E., Borsa, D. M., Craw, M., Gelman, R. and Mandell, E. E. (1965). *J. verb. Learn. verb. Behav.* **4**:291
13. Berlyne, D. E., Craw, M. A., Salapatek, P. H. and Lewis, J. L. (1963) *J. exp. Psychol.* **66**:560
14. Berlyne, D. E. and Lawrence, G. H. (1964) *J. gen. Psychol.* **71**:21
15. Berlyne, D. E., and Lewis, J. L. (1963) *Can. J. Psychol.* **17**:398
16. Berlyne, D. E. and McDonnell, P. (1965) *Electroenceph. clin. Neurophysiol.* **18**:156
17. Berlyne, D. E. and Peckham, S. (1966) *Can. J. Psychol.* **20**:125
18. Bower, G. H. and Trabasso, T. R. (1964) In *Studies in Mathematical Psychology* (Ed. by Atkinson) Stanford: Stanford Univ. Press
19. Broadbent, D. E. (1958) *Perception and Communication*. New York: Pergamon Press
20. Day, H. (1965) Exploratory Behaviour as a Function of Individual Differences and Level of Arousal. Unpublished *Ph.D. Thesis*, Univ. of Toronto
21. Day, H. (1967) *Percept. Psychophys.* **2**:281
22. Dewey, J. (1896) *Psychol. Rev.* **3**:357
23. De Valois, R. L. (1966) In *Frontiers in Physiological Psychology*. New York: Academic Press
24. Dodge, R. (1899) *Psychol. Rev.* **6**:477
25. Easterbrook, J. A. (1959) *Psychol. Rev.* **66**:183

26. Graham, F. K. and Clifton, R. K. (1966) *Psychol. Bull.* **65**:305
27. Head, H. (1966) *Aphasia and Kindred Disorders of Speech* (2 vols). New York: Macmillan
28. Hebb, D. O. (1949) *The Organisation of Behaviour.* New York: Wiley
29. Herbart, J. F. (1835) *Umriss pädagogischer Vorlesungen.* Göttingen: Dieterich
30. Hernández-Peón, R. (1966) *In Frontiers in Physiological Psychology* (Ed. by Russell). New York: Academic Press
31. Horn, G. (1965) *In Advances in the Study of Behavior* (Ed. by Lehrman, Hinde and Shaw). New York: Academic Press
32. Hull, C. L. (1939) *Psychol. Rev.* **46**:9
33. Hull, C. L. (1943) *Principles of behavior.* New York: Appleton Century, p. 207
34. James, W. (1890) *The Principles of Psychology.* New York: Holt
35. Kamin, L. J. (1967) *Technical Report No. 5.* McMaster University
36. Kendler, H. H. and Kendler, T. S. (1962) *Psychol. Rev.* **69**:1
37. Konorski, J. (1967) *Integrative Activity of the Brain.* Chicago: Univ. of Chicago Press
38. Krechevsky, I. (1932) *Psychol. Rev.* **39**:516
39. Krechevsky, I. (1938) *Psychol. Rev.* **45**:107
40. Lacey, J. I. (1967) *In Psychological Stress; some Issues in Research* (Ed. by Appley and Turnbull). New York: Appleton-Century-Crofts
41. Lashely, K. S. (1942) *J. gen. Psychol.* **26**:241
42. Lashley, K. S. and Wade, M. (1946) *Psychol. Rev.* **53**:72
43. Lawrence, D. H. (1949) *J. exp. Psychol.* **39**:777
44. Leibnitz, G. W. (1765) *In New Essays Concerning Human Understanding* (Trans. Langley). Chicago and London: Open Court. 1916
45. Mackinstosh, N. J. (1965) *Psychol. Bull.* **64**:124
46. McDonnell, P. (1967) 'Effects of Intensity, Contrast and Novelty on Selective Attention and Free-choice Reaction Time.' *Ph.D. Thesis.* Univ. of Toronto
47. Meyer-Eppler, W. (1959) *Grundlagen und Anwendungen der Informationstheorie.* Berlin: Springer
48. Morgan, C. T. (1960) *In Psychology—A Study of a Science.* Vol. 1 (Ed. by Koch). New York: McGraw-Hill
49. Mulholland, T. B. and Runnals, S. (1964) *Electroenceph. clin. Neurophysiol.* **17**:371
50. Nicki, R. M. (1967) 'The Reinforcing Effect of Uncertainty Reduction on a Human Operant.' *Ph.D. Thesis.* Univ. of Toronto
51. Paschal, F. C. (1941) *Psychol. Rev.* **48**:383
52. Pavlov, I. P. (1927) *Conditioned Reflexes.* Oxford: Oxford Univ. Press
53. Pillsbury, W. P. (1908) *Attention.* London: Swan
54. Razran, G. H. S. (1939) *J. expl. Psychol.* **24**: 202
55. Restle, F. (1955) *Psychol. Rev.* **62**:11
56. Schultz, D. P. (1965) *Sensory Restriction: Effects on Behavior.* New York: Academic Press

57. Sechenov, I. M. (1878) (Elements of thought) *Vestnik Evropy.* **3**:39. (English translation in I. M. Sechenov, *Collected works.* Moscow and Leningrad: Medgiz, 1935)
58. Sharpless, S. and Jasper, H. H. (1956) *Brain.* **79**:655
59. Sherrington, C. S. (1906) *Integrative Action of the Nervous System.* Cambridge: Univ. Press, and New Haven: Yale Univ. Press
60. Sokolov, E. N. (1958) *Vospriiate i uslovny refleks* (*Perception and the Conditioned reflex*). Moscow: Univ. of Moscow Press
61. Sokolov, E. N. (1964) 'The orientation reflex as an informational regulator'. In E. N. Sokolov (Ed.) *Orientirovochny refleks i problemy retseptsii v norme i patologii.* (*The orientation reflex and problems of sense-reception in normality and pathology.*) Moscow: Prosveshchenie
62. Spence, K. W. (1936) *Psychol. Rev.* **43**:427
63. Terrace, H. S. (1966) In W. K. Honig (Ed.) *Operant Behaviour.* New York: Appleton-Century-Crofts
64. Titchener, E. B. (1909) *A Text-book of Psychology.* Part I. New York: Macmillan
65. Vinogradova, O. S. (1959) 'The role of the orientation reflex in the process of establishing conditioned connections in man.' In E. N. Sokolov (Ed.). *Orientirovochny refleks i voprosy vysshei nervnoi deiatel' nosti v norme i patologii.* (*The orientation reflex and questions of higher nervous activity in normality and pathology.*) Moscow: Acad. Pedag. Sci.
66. Vygotski, L. S. (1962) *Myshlenie i rech'*., Moscow: Acad. Pedag. Sci., 1956. (*Thought and Speech*), Cambridge, Mass.: MIT Press; New York: Wiley
67. Wachtel, P. L. (1967) *Psychol. Bull.* **68**:417
68. Wilson, N. H. and Wilson, W. P. (1959) *Electroenceph. clin. Neurophysiol.* **11**:85
69. Wolpe, J. (1950) *Psychol. Rev.* **57**:19
70. Wundt, E. (1874) *Grundsätze der physiologischen Psychologie*, Vol. II. Leipzig: Engelmann
71. Zaporozhets, A. V. (1958). In L. G. Voronin *et al.* (Eds.). *Orientirovochny refleks i orientirovochono-issledovatel'skala deiatel'nost'.* (*The orienting reflex and exploratory behaviour.*) Moscow: Acad. Pedag. Sci.

Can 'Attention' be Defined in Physiological Terms?

W. Grey Walter
Burden Neurological Institute, Bristol

The great difficulty in trying to unite the disciplines of psychology and physiology is that in neither is there much respect for the other's terms, units and axioms. Psychology has been defined as art without feeling or inspiration, masquerading as a science without units or axioms. One could be equally scathing about neurophysiology, at least in human applications. Physiologists tend to use elegant techniques to perform trivial experiments and psychologists seem to delight in getting involved in experiments which are technically impossible and fall back on statistical analysis about which everyone has always been very rude.

Apart from the banter and inter-faculty jealousy, psychological concepts as seen from the physiological standpoint do appear very loose and verbal—rather like the pronouncements of the Mediaeval Schoolmen, while physiological laws and techniques which are adequate in peripheral nerve work or in studies of unconscious or mutilated animals begin to fall apart when one gets above the neck. Physiologists who get involved in the study of human behaviour learn that a word in general psychological use may not apply to any real object or event; 'dragon' and 'unicorn' are common words but no one has seen either animal and the charts of psychophysiological experimentation should be labelled 'Here be dragons'. Even the central topic of this Conference may be a dragon, a mythical concept beyond the scope of physiological probes.

These probes are admittedly coarse and clumsy compared with the delicacy of subjective experience and its expression in language, but they are more refined and versatile now than they were a few years ago and we can begin to define some of the problems in objective terms as a first step toward solving them.

The aspects of attention that I feel able to comment on are those that can be related to electrical events in the human brain. We should bear in mind that all bio-electric phenomena are connected in some way with chemical change. The converse is true also, since chemical reactions involve arrangements of electrons. We cannot yet identify the precise chemical reactions associated with the particular classes of electrical change we find in the brain, but we know that the chemical repertoire is quite large enough to match the diversity of electrogenesis. The reason for stressing the electrochemical nature of brain responses in this context is that many of the effects we have noticed in the human brain have a much longer time-course than those in peripheral structures and somehow suggest particularly strongly that they are signs of locally organized chemical transactions.

Another factor to be remembered is that in speaking of 'the human brain' we are guilty of the solecism of spurious generality. There is no useful statement we can make about the brain as a whole and our main difficulty is to restrict our assertions to the particular regions we have studied in the situations we have contrived. In another part of the brain at the same time or in the same part at another time an entirely contradictory event may occur.

RESPONSES FROM INTRACEREBRAL ELECTRODES

The regions of the brain which we have been able to study in adequate detail are the non-specific or 'silent' areas of the frontal cortex. Some of these are, of course, accessible from the scalp but we have had the unusual opportunity of recording also from large numbers of electrodes implanted in the brain for diagnostic and therapeutic purposes (*Figure 1*). Such recordings have not only helped to corroborate data from scalp electrodes but have also provided information of particular value in estimating the scale of differential responsiveness of various regions.

Multiple recordings of this order present their own problems. Our patients usually have 68 electrodes, some nearly 100, and these remain in place for several months. Of these, 20 or 30 at least turn out to be in regions showing responses to sensory stimuli in several modalities which vary with the situation and the state of the subject. The result is inevitably a very large number of recordings, and the problem is to see the wood for the trees or, worse, to decide whether there is a wood at all—or just a lot of lumber. From an embarrassing stack of such material I have selected observations from two patients which fortunately illustrate what seem to be some of the more fundamental and general facts about attention from the neurophysiological standpoint. In neither case was there any reason to presume organic abnormality of the brain and the inferences from these observations amplify those from more superficial studies of entirely normal people and I consider them to be typical of the situations concerned, rather than of the mental disturbances for the treatment of which the electrode implantations were performed. For this reason the clinical histories and condition of the patients will not be described here.

The observations in the first case are concerned with the variations in responses in three situations: when a stimulus was novel and irregular, when it had been repeated monotonously many times and when it was again presented irregularly (*Figure 2*). Recordings were made simultaneously from 14 electrodes and the stimuli were clicks of moderate intensity delivered through earphones from a generator synchronized by a Digitimer with the operation of a LINC 8 computer. The computer was programmed to collect eight samples of the responses and plot the average of these on a Calcomp digital plotter. Both the analogue signals and the digital averages were stored on

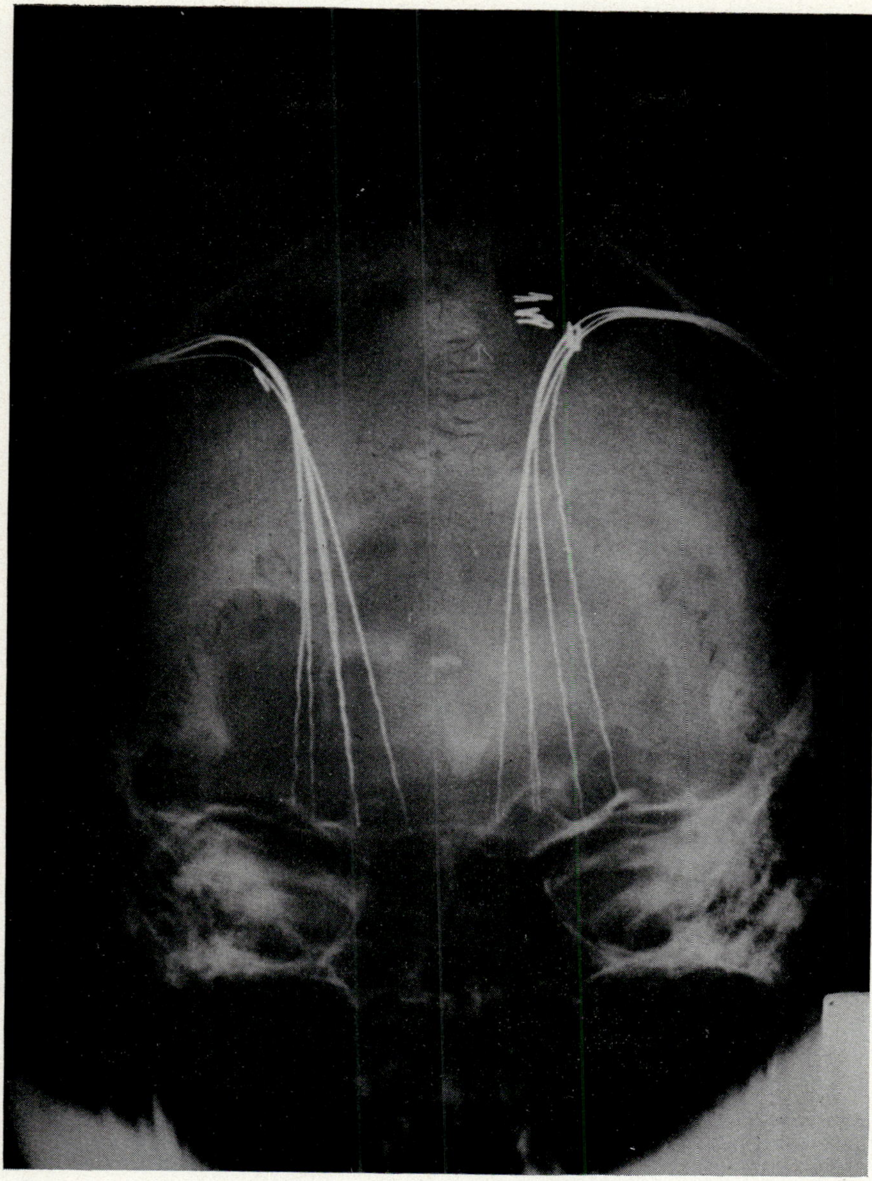

Figure 1. Electrodes implanted for diagnostic and therapeutic purposes

magnetic tape. Of the 14 channels thus analysed two have been selected for illustration since these provided the most obvious differences between adjacent areas. At the same time the responses from the scalp vertex were also averaged, and these are plotted as a thicker line for comparison with the more intimate data from the intracerebral electrodes. The reference for the scalp

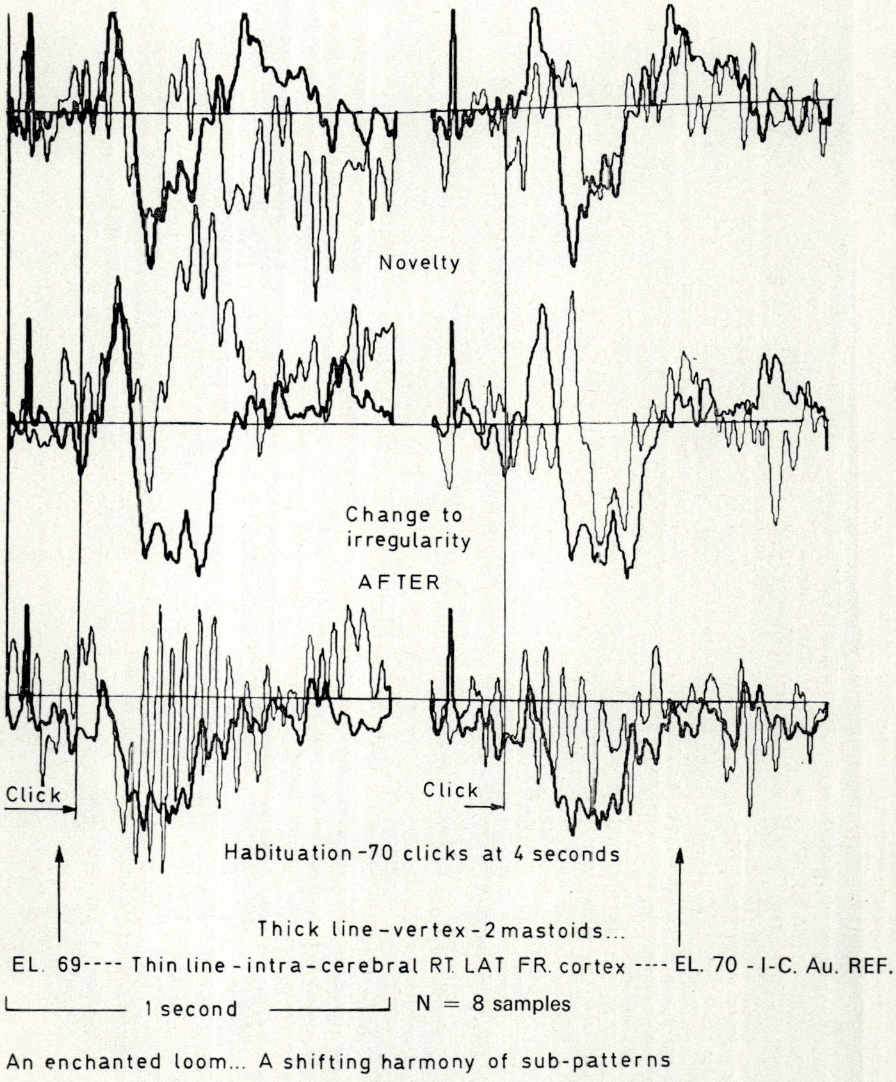

Figure 2. Comparison of scalp and cortex responses before, after and during habituation

recording was electrodes on both mastoid processes, and for the intracerebral ones the network average of all electrodes other than those found to be in responsive regions. Negativity of the probe electrodes is plotted upward. A 20 μV calibration signal is injected automatically at the start of every sample and appears at the left of the plots. The subject was given no warning or instructions before or during the experiment.

The left-hand column of thin line graphs are responses from the electrode 69 which was in the lateral frontal cortex of the right hemisphere and the right-hand column is from electrode 70 adjacent to it. Each electrode had a diameter of 150 μ and was bare for 4 mm. The contact areas of electrodes 69 and 70 were 4 mm apart. The top row of graphs are averages of the responses to eight clicks at varying intervals of about 15 s from these two electrodes with the vertex response superimposed on both. There is some resemblance between the response at electrode 69 and the vertex particularly in the positive component, but the peak of the negative component is about 20 ms later at electrode 69 and the second, slower, negative wave is about 150 ms earlier. There is considerably more residual activity in the cortical derivation, as is usual, since the scalp itself performs a spatial average in which minor rhythms and transients are attenuated.

Electrode 70 in the first situation shows an initial negative component which starts and peaks synchronously with the scalp but the duration is prolonged. However, the return from the positive component is again congruent and the second negative wave, although smaller, seems to peak at about the same time.

The responses in the second row were recorded after those in the third and are plotted in this position for comparison because the stimuli were identical with those in the first row in their irregularity , different only in that they were presented *after the habituation* illustrated in the third row. Electrode 69 shows a dramatic change. The first negative peak is only a few milliseconds earlier but the positive component is very brief and attenuated while the second negative wave arises at the same time as before but is about three times larger —and actually appears out of phase with the extended positive wave at the vertex. An even more peculiar effect is seen at electrode 70. The first cusp of the first negative wave which was synchronous with the vertex in the first set has sunk below the zero line while the third cusp has become larger than the vertex response and again gives the appearance of being out of phase with the vertex. The positive wave is unchanged.

These tracings are examples drawn from a large population but similar effects were seen in other situations and also in other regions. These two small cortical zones, only a few millimetres apart, have an entirely different 'attitude' to irregularity and distinguish between this and novelty. The question arises—how did these changes come about; what happened during the process of habituation which preceded the shift to irregularity of presentation?

The third row of responses illustrates some of the features of this mechanism. At electrode 69 and at the vertex the responses to the same clicks (after 70 monotonous presentations at 4 per second) have lost the first prominent negative wave, leaving only the slower positive component at the vertex, while a new feature appears at electrode 69. This is a rhythmic burst of waves at about 26 Hz lasting for over 250 ms. At electrode 70 the effect is simply a disappearance of the response beneath the background activity, with a faint trace of the rhythm at electrode 69.

This repertoire of response patterns, evoked by very simple changes in the stimulus sequence, illustrates the difficulty of defining even the most elementary mental states in physiological terms. The only encouraging aspect of such observations is that they indicate an orderly, though complex mechanism. After computing, plotting and scanning the responses in this case it was easy to tell by looking at a particular set of responses not only what experience the subject was having but what experiences he had recently had. The set of patterns seen in the second row of *Figure 2* cannot arise with a novel stimulus or during a monotonous series of stimuli; only when monotony gives place to irregularity. However, discrimination between these conditions is not a universal or uniform property of frontal cortex. If one takes the responses from the vertex as a spatial average of the various local patterns from an area of about 25 cm^2 of cortex, one can say that the cortex around electrode 69 has an 'average' attitude to novelty, a peculiar rhythmic mechanism which may erase responses to predictable stimuli, and another mechanism which is activated when the predictability of a stimulus is diminished by irregularity. Around electrode 70 there is a less than average concern with novelty, indifference to monotony, but a greater than average (although sluggish) response of a particular mechanism to unpredictability. Whether it is worth expressing such inferences in terms of information theory at this stage is doubtful, but it is worth recalling that other experiments have shown that the size of responses in frontal cortex are generally related to the selective information content rather than the intensity or modality of the stimuli. The difficulty here is that it is the fine detail of the responses in local regions which seem to reflect information content, rather than their gross amplitude.

THE IMPLICATIONS OF HABITUATION

This study also illustrates one of the apparent paradoxes of higher nervous activity which should provide a clue to its mechanism. The phenomenon generally called habituation is familiar and is obviously essential to adaptive integration of experience. A brain cannot learn to identify and store important events unless it can ignore and erase trivial ones. The responses to the first set of clicks in the experiment described were reasonable as indicating an expectant attitude to novelty—the clicks *might* have been important. After 70 such events every 4 s with no conditional or contingent development, however, they could safely be classified as trivial and the responses waned accordingly. When the sequence of clicks became irregular, however, the responses returned in a new pattern. The stimuli were identical but the rhythm was different—the time interval between them was unpredictable. This is such a common experience in everyday life that we tend to ignore it or take it for granted but the implications are startling. The same effect can be obtained with almost any change in the stimulus presentation; a rise or fall in intensity, a variation in timbre or duration of the clicks (or of visual or tactile stimuli) even an alteration in the irregularity itself—all can evoke a reappearance of some response feature characteristic of each region. It is difficult to avoid the conclusion that when a response to a particular stimulus presentation has habituated, a large amount of information about the situation has been stored, simply in case some slight change should appear that might indicate fresh importance. Such a change could obviously not be detected unless the original parameters were available for comparison. As long as each stimulus matches the canonical form or template established for it during the habituation process, no response is generated in these zones of frontal cortex. But if a particular stimulus does not match the template in any respect, even in its time relation to its predecessor, then an error signal is generated, with a little delay, that re-establishes the state of attention and mobilizes the mechanisms of expectancy and decision. This change again may be subject to second order habituation and so on, so that the brain can build up a stock of templates, comparison with which permits economical and effective action, with economy when no action is relevant. This is the paradox, for it would seem that in order to achieve this desirable, and indeed essential, basis for contingent adaptation the brain must store an enormous amount of intrinsically useless information about trivial events and situations, just in case one or other of them should be involved in a slightly different form in another experience. As to how slight the alteration need be to restore an habituated response, further quantitative experiments must be planned. The answer to this question is important for if we could estimate the minimum effective variation for a wide range of stimuli in all modalities we could perhaps make reasonable conjectures about the probable capacity of the 'rubbish store' of the brain. This could be a valuable figure on which to base further predictions of the relation between habituation and related effects and mental state. The rate of habituation in a particular situation varies widely between people and with modality but is fairly constant in a given person in a particular modality. Simulation of these effects with computer programmes might be instructive since the actual scale of the store necessary for such analysis may not be as large as would seem at first sight (provided that allocation of memory is sequential rather than simultaneous). There is some evidence that this may be so in the brain since an habituated response can be restored by presentation of a novel stimulus in another modality. This suggests that the system providing the first habituation procedure can be cleared and reset for application to another set of experiences. The 'rubbish store' would then only have to be large enough to cope with limited sets of trivial information at any one time instead of having to deal with all perceptible variations in any combination of stimuli over long periods. The interaction of habituation effects has not yet been investigated in detail. Difficulties would arise if several sets of monotonous and meaningless stimuli were delivered at the same time and a secondary process would be useful. This may exist in the form of the sleep which tends to supervene in such circumstances, and which is described by Pavlov as 'irradiation of inhibition'.

The relation of these effects to what, for want of a clearer phrase, may be called 'conscious attention' is uncertain. In the experiment described above, the subject was given no instruction and asked no questions, but in another set of trials he was asked to count the clicks even when they were presented monotonously. This had the expected effect of reducing the rate of habituation and also of restoring responses which had already habituated. Whatever, or wherever, the rubbish store may be, it is not simply related to the class of memory which is used for the storage of useful associations.

RESPONSES TO ASSOCIATED STIMULI

This raises the next question; how does the brain deal with signals which are intrinsically unimportant but are assigned significance because of their conditional relation to other events? Some of the details of these mechanisms are illustrated by studies in another patient (*Figures 3* and *4*). Dr. McCallum will be describing in some detail the effect which we have called the Contingent Negative Variation (CNV) and which is sometimes referred to as the Expectancy Wave (E-Wave). This is a sustained rise in surface electronegativity of many regions of frontal cortex in the period between the presentation of a conditional stimulus and the succeeding unconditional or imperative one. Because of its slow time-course the CNV cannot be recorded satisfactorily

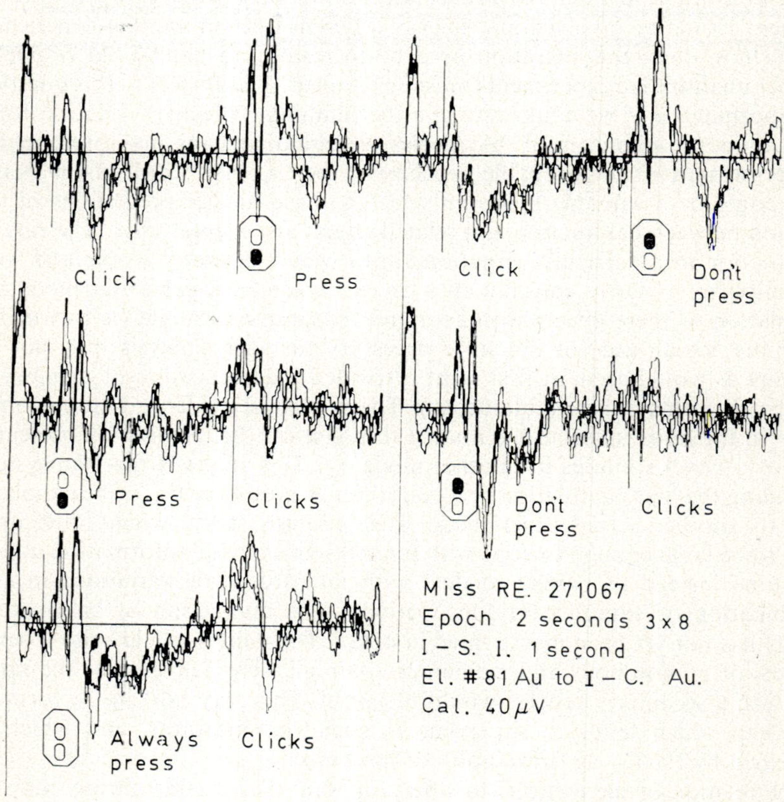

Figure 3. Responses evoked by associated stimuli

with the gold electrodes used for intracerebral implants; these have an intrinsic time constant of about 0·2 s. However, the responses evoked by associated stimuli can be recorded without appreciable distortion and examples are given in *Figures 3* and *4*.

The purpose of this experiment was to see how several cortical regions responded in a variety of situations in which the subject had to decide whether or not she was going to make an operant response. The electrodes from which the responses in *Figures 3* and *4* were recorded were both in the subdural space over superior anterior frontal cortex, electrode 81 on the right and electrode 34 on the left. The averaging epoch was two seconds with an interstimulus interval of one second. The number of samples in each average was eight and three averages are plotted superimposed, equivalent to responses from 24 trials. This last technical detail is important because the consistency and coherence of the responses, which is a remarkable feature in these situations, cannot be demonstrated in any single average.

The stimuli were clicks and brief exposures of a simple pattern on a closed circuit television monitor synchronized with the computer. The pattern con-

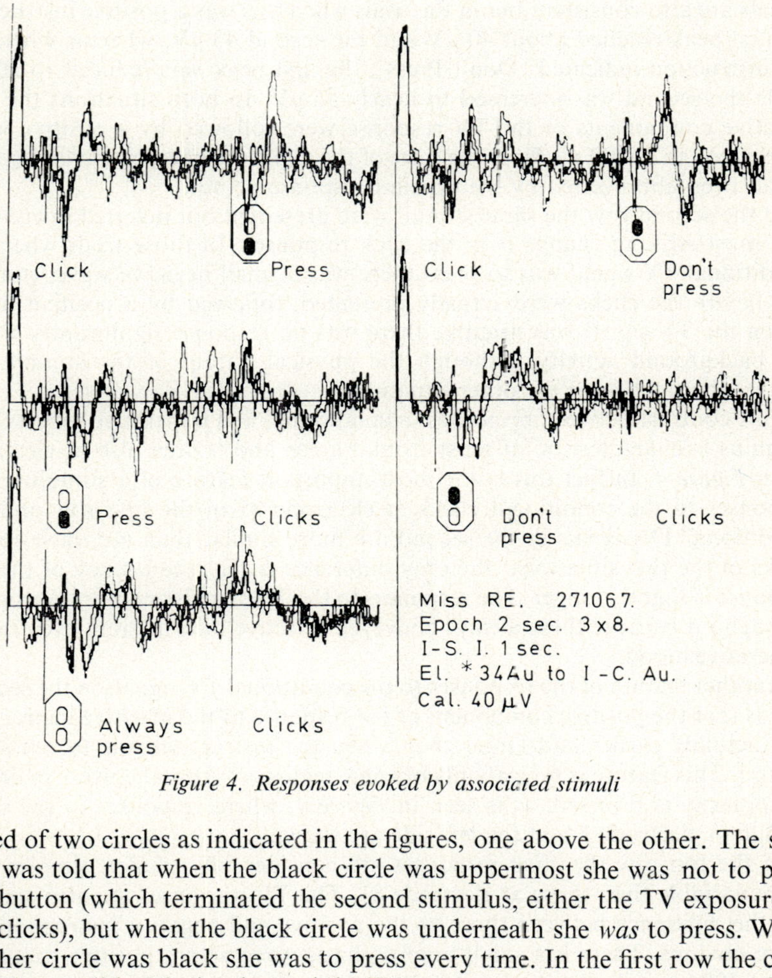

Figure 4. Responses evoked by associated stimuli

sisted of two circles as indicated in the figures, one above the other. The subject was told that when the black circle was uppermost she was not to press the button (which terminated the second stimulus, either the TV exposure or the clicks), but when the black circle was underneath she *was* to press. When neither circle was black she was to press every time. In the first row the click was the conditional stimulus and the TV exposure the imperative one, and the up or down position was varied in a random sequence over the 16 trials, the eight trials in each category being averaged separately. This was repeated three times. In these sets therefore, the subject could not know whether a particular click was a signal for action until the TV exposure appeared one second later. In the second row the TV exposure was the conditional stimulus, again with 'down-press' and 'up-don't' presented in a random sequence. In this situation the subject did not know which instruction would be received but was able to make her decision as soon as the exposure appeared. In the bottom row the TV exposure was constant and was always a positive instruction similar to any other simple conditional stimulus.

At electrode 81, on the right cortex (*Figure 3*) the responses to the conditional clicks in the top row have a very consistent form. The first negative peak is at 60 and the second at 150 ms after the click, the second superimposed on a positive component lasting about 400 ms. The responses to the imperative TV signals are also consistent but in the trials when this was a positive instruction the first peak reached about 40 μV and the second 45 μV, whereas when the TV instruction indicated 'Don't Press', the first peak was reduced to 20 μV while the second was increased to nearly 55 μV. In both situations the two negative components of the TV response were followed by a positive wave with a trough at 400 ms from the start of the exposure irrespective of whether it had been abbreviated by the operant response or not.

In the second row the same stimuli were presented but inverted in context. The most striking change is in the click responses. In those trials when the conditional TV signal was to press there was a small negative wave, starting just *before* the clicks were actually presented, followed by a positive wave. When the TV signal was negative there was no response significantly above the background activity, although the physical nature of the stimuli was exactly the same as in the other situations.

This contrast between responses to conditional and imperative aspects of a stimulus is characteristic of most frontal areas and is seen also at electrode 34 in *Figure 4*. In fact this is the most important feature of a stimulus. The responses to the conditional clicks at electrode 81 in the first row and the conditional TV signals in the second are more similar than are those to the clicks in the two situations; the only difference is that the latency of the TV response is slightly longer. The responses to the TV signals are more consistent throughout because these always conveyed selective information, even in the imperative mood.

Another feature of the responses to the conditional TV signals in the second row is that the positive component of the response to the negative instruction was actually earlier and larger than when the instruction was to press the button. This feature is typical only of some regions of frontal cortex; in others the polarity is reversed, as is seen in *Figure 4*, where responses to the same stimuli at electrode 34 on the left side are shown.

In the top row the click responses are considerably smaller, simpler and more variable than those at electrode 81. The TV responses are also smaller but the difference between those to the positive and negative instruction are more obvious. There is a small brief diphasic wave with a positive trough at about 100 ms in both situations, followed by a larger negative wave peaking at about 230 ms. In the positive instruction trials this feature was rather variable, two averages being only at 10 μV and one at 30 μV. In the negative instruction sets all three averages reach 20 μV with a slightly longer latency and this negative wave is followed by a positive deflection of the same size. These records illustrate the advantage of the average superposition method.

If a single average of 24 trials had been taken in each case the amplitude of the negative peak in the positive set would have been 17 μV and that of the negative set 20 μV—an insignificant difference.

In the second row, when the TV signal was conditional, the responses to the positive instruction are small, although consistent, and there is almost no response to the imperative clicks. In the negative instruction set, however, the small early responses are followed consistently by a slow negative wave lasting about 400 ms and there is no trace of a response to the clicks. In electrode 81 also the click responses were almost invisible in this situation; this indifference to a stimulus which is known to require no response is the only feature common to these two regions in the situations described.

In the bottom row of *Figure 4* the responses to the 'always press' signal are indistinguishable from those in the record above in which positive signals were randomly mixed with negative ones. There is no sign of the ramp function seen at electrode 81.

Before considering the inferences that can be drawn from these observations it is worth recapitulating the main features of the experimental conditions. The first question is the extent to which one can rely on the recording and computing procedures. The electrodes through which the potential differences were recorded were of gold, with a surface area of about one millimetre. With this area the time constant of the electrode surface is too short to transmit changes with an effective duration of more than 0·2 s without considerable attenuation. The slower components of the responses illustrated here were therefore larger and longer than they seem. The examples in *Figures 2–4* were selected from simultaneous records from 14 channels to illustrate specific effects; other regions exhibited their own local versions of the response patterns but none in or on frontal cortex displayed features inconsistent with the foregoing description. The dependence of response pattern on context, significance and operant implication is a general property in these regions although it may be expressed in a variety of ways. The computation and plotting of averages was performed on-line by the LINC 8 computer so that the results of each set could be seen before the next trials were begun. In the experiments with alternative TV signals, the positive and negative instructions were mixed in a random sequence so that the subject could not know which to expect. The related sets of averages were collected and plotted over one another later, in order to estimate the range of variation in each situation. This seemed surprisingly small in the original plots in spite of the high level of background intrinsic activity which is usual when recording direct from cortex. This observation is particularly important taken in conjunction with the short distance between regions which show consistent differences. Adjacent electrodes on frontal cortex in the patient of *Figures 3* and *4* were over zones with characteristic local responses which varied only with situation and attitude.

The first inference from such observations is that the non-specific cortex of the frontal lobes is an integral part of the sensory system, receiving signals in all modalities but consisting of a large number of domains each with its own attitude to the incoming signals. We are accustomed to the notion of specific parameter receptors in the peripheral organs; on and off, edge, line or movement units in the retina for example. It seems possible that at the other end of the sensory system also there is some degree of parameter specificity. The responses in *Figure 2* show adjacent regions responding in distinctive ways to novelty, monotony and irregularity. The complex patterns in *Figures 3* and *4* are examples of what one might call the syntax of brain language. Each domain displays a characteristic inflection according to whether it is dealing with a conditional, indicative or imperative statement and also whether the imperative is positive or negative.

From interpretations of this type one can build up a set of paradigms relating the diversity of brain responses to the operational requirements of attention and adaptation. In these paradigms the 'root' would be the modality of the signal or statement but the root by itself could have no meaning. The implication of this is that there is no such thing, for example, as a 'click response'. From the standpoint of the frontal lobes an indicative click or other signal is either novel, predictable or irregularly related to other similar signals or conditional or imperative, and the form the response takes is inflected accordingly.

In these fanciful but familiar terms the responses in the first row of *Figures 3* and *4* correspond to a rather involved statement: '*If* there is a click, there *will be* a TV picture and *if* the black circle is below, *press* the button'. This sentence contains a conditional followed by an indicative phrase leading to another conditional and terminated by an imperative.

Returning to the general question of whether 'attention' is a concept which can be defined in physiological terms or is a semantic dragon, we can see one possibility at least. If we start by considering what can actually be observed and measured in physiological experiments, the diversity of brain responses, although large, is not infinite. We can identify a finite set of response patterns, each related to a relative parameter of the sensory signals. Unfortunately, the specification of this set must include regional characteristics as well as voltage-time dimensions so that a single measure will rarely be adequate.

THE RELATION OF EVOKED RESPONSES TO INTRINSIC RHYTHMS

This is just one aspect of the basic sampling problem that arises in all such work. Another example is the familiar and convenient method of averaging by computer. This procedure extracts and enhances regular events otherwise obscured by random ones and is almost indispensable for the sort of experiment described here. None the less the relative attenuation of irregular components destroys information when these may be related to the phenomena of primary interest. The best known effect of sensory stimuli on brain activity is *suppression* of intrinsic rhythms—alpha blocking for example. Although more variable, and less obviously related to attention and adaptation than the features described above, this effect is sometimes of great interest and might be a valuable clue to some of the more elusive details of higher nervous function: it is almost completely effaced by averaging. How rhythmic processes can play some part in the inflection of brain responses in various states of attention is illustrated by the coherent rhythm evoked during habituation at electrode 69 shown in the bottom row of *Figure 2*. Such coherent rhythms are rather rare, particularly in relation to auditory stimuli, but coherence of an alpha component following a conditional visual signal is not so unusual and is sometimes related to development of the CNV. The important features of such rhythms is their coherence, that is constancy of phase relations with respect to stimuli. The degree of coherence is unrelated to average amplitude and local studies of posterior cortex with implanted electrodes has shown that this is usually a property of zones as limited as those characterized by a particular form of evoked response.

The occasional appearance of coherence in situations requiring special treatment of sensory signals has suggested that the rhythmic processes which can be phase-locked may have a special function in the storage and retrieval of information. The analogy with formation of a hologram by coherent light from a laser is obvious. This process is one of the most versatile, rapid and compact methods of filing and retrieving large quantities of complex information and it would not be surprising if organic living systems had evolved something similar. They seem to generate rhythmic waves in profusion and have had a rapidly increasing storage problem as the complexity of central nervous organization has risen over the eons.

This fanciful but not impossible suggestion is included here to illustrate two points. First, the danger that an easy way to obtain clean records by 'data reduction' may be a first class method of data destruction. Second, that consideration of the various things the brain does actually do may lead to a better definition of what we mean by attention and similar states than will an effort to find objective correlates of verbal assumptions.

The Contingent Negative Variation as a Cortical Sign of Attention in Man

Cheyne McCallum
Burden Neurological Institute and Department of Psychology, University of Bristol

THE CONTINGENT NEGATIVE VARIATION

At the beginning of the last paper Dr. Grey Walter painted a vivid picture of the linguistic and quantification problems of psychology, and the lure of the technical, as opposed to the scientific, problems in neurophysiology. As a psychologist, working principally with neurophysiological techniques, I find myself between the multiple heads of an electrophysiological Scylla and the semantic whirlpool of a psycholinguistic Charybdis. Navigation in such waters is hazardous, but some assistance is gained by keeping sight of the immediate goal, namely the need to establish reliable correlations between measurable events in the brain and behaviour as exhibited in a variety of relatively controlled situations and environments.

The principal measure of brain activity employed in the study to be described was of the contingent negative variation (CNV). It will be recalled that the CNV is a slow change in electrical potential recorded from the fronto-vertical regions in man. It occurs in the period between the cerebral evoked responses to pairs of stimuli *when the second stimulus requires a decision or action by the subject* (15). Normally some form of averaging technique is required for adequate recording and measurement of this slow potential change.

In the majority of the experiments, which will be described, recording of the EEG was carried out using a 16-channel Offner Type TC electroencephalograph, coupled with an Enhancetron 2-channel averager and a Barrier-grid 2-channel averager (3). In addition to the 4-channel on-line averaging thus provided, an 8-channel tape recorder was used to store information for subsequent processing. During the latter stages of the study the facilities were supplemented by a LINC 8 computer, which enabled seven channel on-line averaging to be undertaken and extensive processing and analysis of data to be accomplished within minutes of the conclusion of each experiment.

During the main experimental series the stimuli used were clicks, tones and single or repetitive flashes. In a second, smaller series a variety of visual material was presented to subjects via a closed circuit television link. Recording was from five silver-silver chloride electrodes equally spaced along the mid-

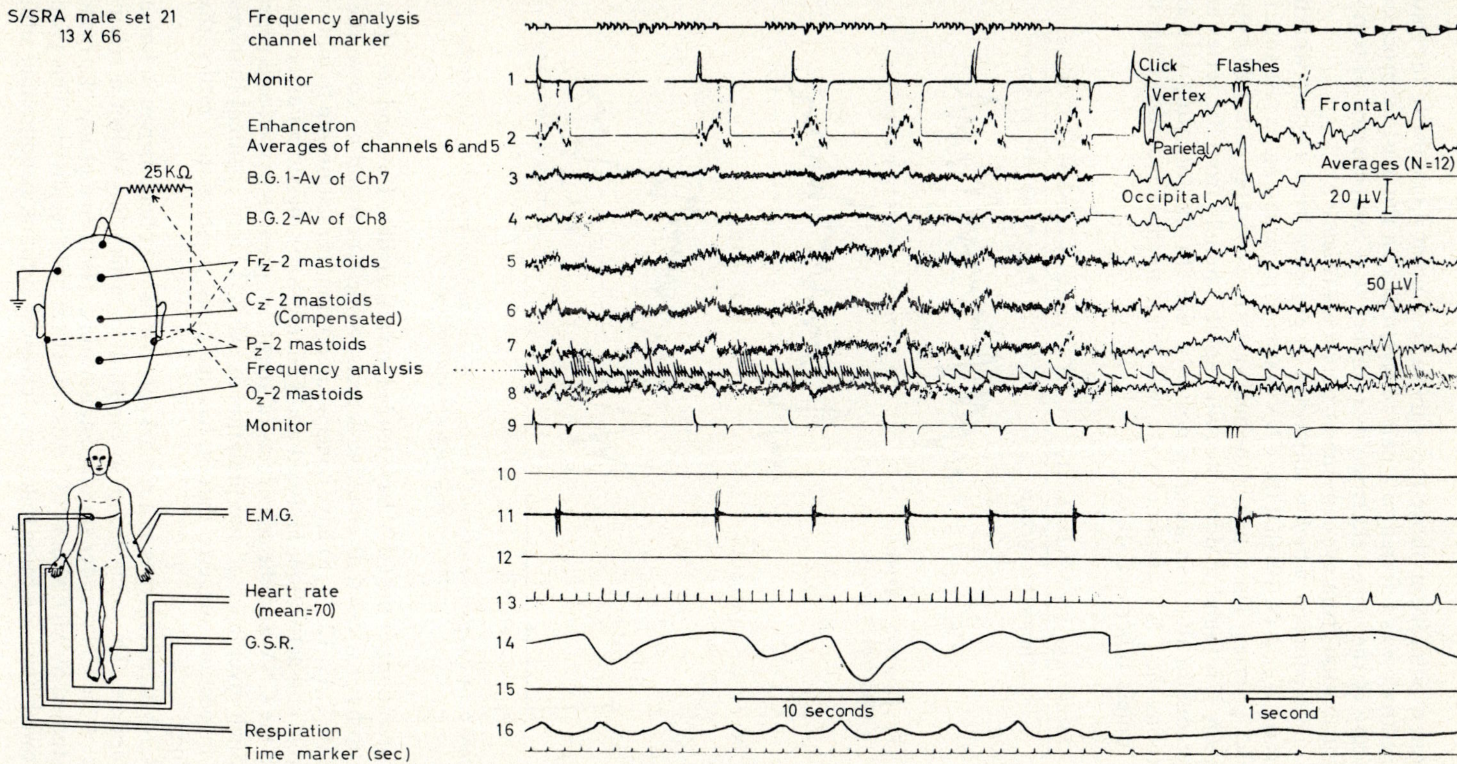

Figure 5. An extract from the record of a non-patient subject showing the standard form of recording employed in the main series of experiments. The last six trials of a series of twelve are displayed followed on the right of the figure at higher paper speed, by the averages

line from a point 4 cm above the glabella to a point 4 cm above the inion. Each electrode was referred to a pair of electrodes on the mastoid processes. In the case of the vertex electrode its reference was compensated to remove any contribution from eye-movement potentials (*Figure 5*).

It is our experience, which is supported by evidence from a group working in Houston, Texas (9, 10) that the CNV is seen at its largest at, or just anterior to, the scalp vertex. The extent of the field varies considerably with individuals, but usually it is found to diminish quite rapidly laterally from the vertex and more gradually along an anterior-posterior line. In the following text and illustrations the CNV amplitudes etc. will relate to those recorded from the vertex, unless otherwise stated.

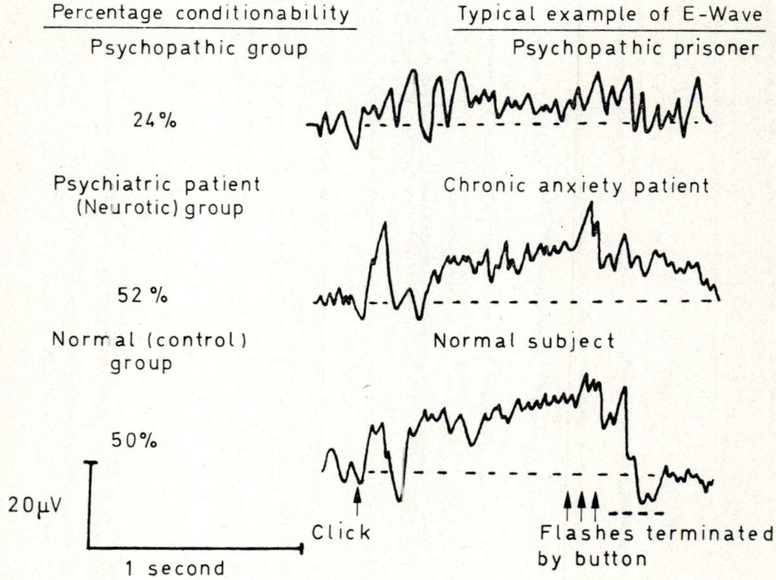

Figure 6. A comparison of the CNV (or E-Wave) and conditionability in psychopaths, anxiety patients and normal subjects

Interpretations of the CNV have tended to vary slightly. Low and the Houston group have stressed its conative aspects, while Knott and his co-workers at Iowa (6, 7, 11, 12, 14, 8) have concentrated upon its relationship to motivation, drive and activation. Hillyard and Galambos (5) suggest an association between the CNV and the constructs of attention, preparatory set and expectancy. Such divergence of interpretation as exists is more apparent than real. Closer inspection suggests that it has its basis in semantics rather than physiology.

The CNV itself emerges as a remarkably consistent phenomenon in human adult subjects. The only group so far studied, which has shown a relative absence of the CNV under standard conditions, has been a group of psychopaths, studied by McCallum (13), most of whom were recidivists. This group also showed how conditionability in a classical eyeblink conditioning situation compared with a group of anxiety neurotics and a group of non-patient controls (*see Figure 6*).

THE EFFECTS OF SIMPLE DISTRACTION

In early experiments with the CNV it was found that, in spite of its general consistency, there were occasional instances when its amplitude unaccountably fluctuated. When, from time to time, a subject, known to have a consistent CNV at about the mean amplitude for the non-patient population, namely about 20 μV, showed a marked reduction for no apparent reason, an attempt was made to find out what factors might be influencing the situation. Subjective reports strongly suggested that the reductions were related to lapses in attention or to positive distractions, either from the external environment or internally, for example from a full bladder or preoccupation with personal worries. It also was noted that certain psychiatric patients tended to be more susceptible to these fluctuations than non-patients.

A systematic study was carried out on 64 psychiatric patients and 40 non-patient controls to determine the effects of a mild degree of distraction on the CNV. The psychiatric patients were selected from categories in which the disorder is generally held to have a detrimental effect on attention. Thus 40 patients were neurotics suffering from a chronic anxiety state, many of them also exhibiting severe obsessional features. A further four chronic neurotics were classified as primarily obsessional. These displayed only limited secondary anxiety. A further 20 patients were diagnosed as schizophrenic. This group displayed a wide range of the symptoms generally classified as schizophrenic, but contained only patients in whom the psychiatric diagnosis of schizophrenia was held to be unequivocal. All patients in this study were hospital in-patients at the time of the tests. The 40 non-patient controls were drawn from as wide a range of ages, socio-economic backgrounds and educational attainments as possible.

The experiments reported formed part of a more general study of the contingent negative variation. This larger study included recording of certain autonomic measures, such as galvanic skin response (GSR), heart rate, respiration and an electromyogram (EMG) of the muscles of the forearm involved in making the motor response.

In the principal series of experiments the stimuli used were a single click (S_1) and repetitive flashes at 16 Hz (S_2). These were first presented on their own at irregular intervals. Averages were taken to establish the evoked response pattern to each. Following this they were each presented regularly once per second for 48 trials to investigate the effects of habituation. The CNV was evoked by presenting S_1, followed one second later by S_2, the subject being required to stop the flashes (S_2) as quickly as possible by pressing a button (*Figure 7*).

Once the CNV had been established, a mild distracting stimulus, in the form of irregular tones, was introduced between trials, although never between S_1 and S_2. Finally, the tones were withdrawn and the CNV re-established in its original form. The method adopted for measuring CNV amplitude was to measure the mean amplitude above base-line during the 150 ms immediately prior to S_2. The base-line was established as the mean level of activity during the one second immediately prior to S_1. In the later experiments normalizing of the base-line was carried out by the LINC 8 computer automatically as part of the averaging procedure.

The use of an auditory signal as a distracting stimulus was adopted after

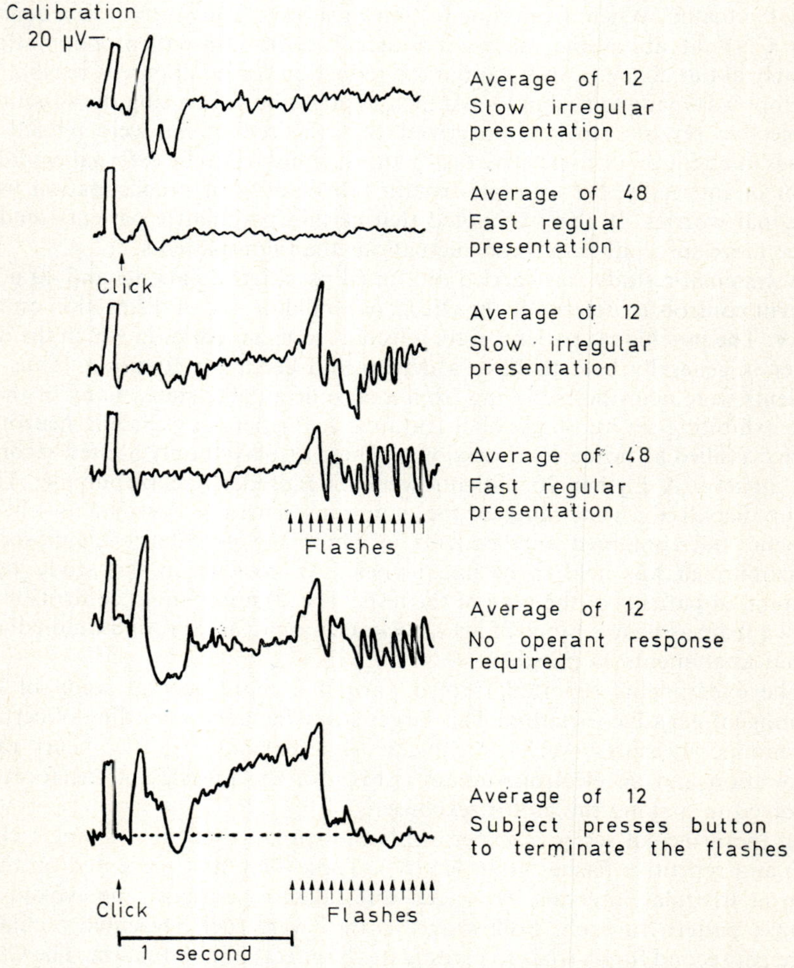

Figure 7. Averages of evoked responses showing the effects of habituation, simple contiguity and association involving the making of an operant response

a small pilot experiment in which several non-patient subjects were given both auditory and visual stimuli in between normal CNV trials. It was found that, while none of these subjects showed any dramatic reduction of the CNV in the distracting condition, the evoked response to S_1 was markedly modified when the distracting signals were in the same sensory modality as S_1. This effect was most noticeable when S_1 was a click, in which case the main negative component of the response, having a latency of about 100 ms, tended to disappear. When S_1 was a single flash, and a second, although readily discriminable, series of irregular flashes from a separate source was used as distraction, the effect on the 100 ms component of the flash-evoked response was less marked. However, there was some suppression of later response components, such that the onset of the CNV itself was delayed. (*See Figure 8.*)

These findings were interesting inasmuch as they showed that the association made between S_1 and S_2, as indicated by the CNV, was independent of the size of the evoked response to S_1. As this distinction appeared most clearly

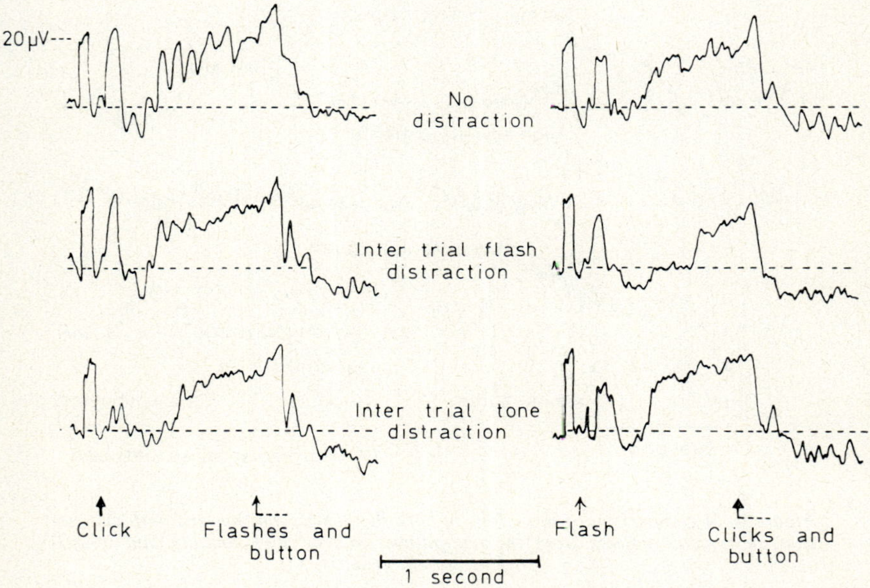

Figure 8. The effects of inter- and intra-modality distraction on the CNV of a normal subject. Such trace is the average of 12 trials

in the auditory modality it was decided to present both S_1 and the distracting stimuli in this modality during the main experimental series. In this way it was hoped to ascertain whether this same occlusion of the evoked response occurred in patients as well as non-patients.

The results from the various patient and non-patient groups taking part in the main distraction study were subjected to a Kruskal–Wallis One-way Analysis of Variance. Where significant differences in variance were encountered, a Mann–Whitney U test was applied to the results of each pair of groups within the sample. In the case of the click evoked response there was no significant difference between the means of the groups (all being within the range 18–20 µV) with regard to the amplitude of the main 100 ms negative component. The high anxiety and schizophrenic groups showed less habituation than the non-patients, but the difference was not significant. The flicker evoked response also showed no significant difference between any of the groups, although the mean amplitudes were all smaller (within the range 13–15 μV) than for the click response, and showed considerably less habituation.

The Kruskal–Wallis H test showed significant differences between the patient and normal groups at the 0·01 level.

The differences were greatest between the non-patients and the two main patient groups—i.e. the high anxiety and schizophrenic patients. The four essentially non-anxious obsessional patients showed a mean CNV amplitude similar to that of the non-patients.

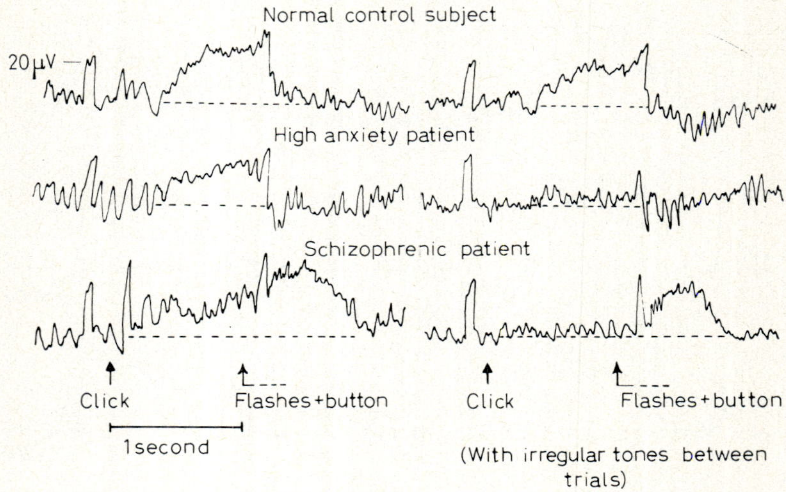

Figure 9. A comparison of the CNV, in both the basic condition and with distraction, in typical subjects from the non-patient control, high anxiety and schizophrenic patient groups

Distraction considerably accentuated the differences between the groups established in the CNV acquisition condition. Virtually all subjects showed a marked reduction of the CNV for the first few trials after the introduction of distraction. In the case of the two main patient groups the CNV tended to disappear completely. With continued presentation, the CNVs of the non-patient controls steadily recovered until by the end of 25 trials the mean amplitude was approximately 70 per cent of its original value. The two patient groups, however, showed little recovery, about half of the patients showing none at all. At the end of 25 trials the mean recovery of these groups was only just over 20 per cent (*see Figure 9*). Table 1 shows the values obtained.

With the exception of the non-anxious obsessionals, who were too small a group for statistical analysis to be appropriate, the drop in CNV amplitude was significant for all groups. For the non-patients it was significant beyond the 0·01 level, while for the high anxiety and schizophrenic groups it was highly significant—beyond the 0·0001 level. The four non-anxious obsessionals showed little effect from distraction, their records showing considerably greater similarity to those of the non-patient, than to the other patient groups.

Little difference was found between the groups with respect to the autonomic measures recorded, with the exception of heart rate. The mean resting heart rate of the high anxiety group was 80 beats/min, 10 beats faster than that of the non-patient controls and 7·5 beats faster than that of the schizophrenic patients. The four essentially non-anxious obsessional patients showed a mean heart rate of 70, i.e. the same as that of the non-patient group. A Spearman Rho test showed a significant negative correlation between the mean heart rate level for each patient and the CNV amplitude in the distracting condition. Other significant correlations were found between the CNV distraction amplitude and the factors of anxiety, phobia, obsessionalism, somatic disorder and depression as measured by the Middlesex Hospital Questionnaire (4).

Table 1. Comparison of the Mean CNV Acquisition and Distraction Amplitudes for the Patient and Normal Control Groups

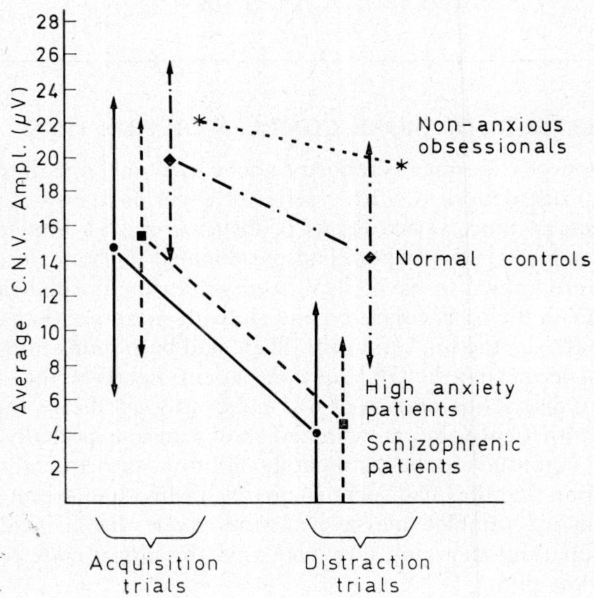

During the course of the study eleven of the high anxiety patients completed their treatment which consisted of leuco-coagulation designed to relieve the symptoms of anxiety and fear. In all cases the treatment was assessed by the psychiatrist in charge as having given some relief of symptoms and in the majority of cases recovery was considered to be nearly complete, including resocialization.

In these eleven patients a second CNV test was carried out at or slightly after discharge from hospital, the first test having been undertaken prior to treatment. The period between the two tests varied from 4 to 12 months.

It was found that on the post-treatment test the CNV showed an increase in amplitude in both the acquisition and distraction conditions over that recorded on the first test. In the case of the acquisition condition the increase just failed to reach significance statistically, although at 19·8 μV the post-treatment mean amplitude was virtually identical with that of the non-patient sample. The increase of the CNV in the distraction condition was significant ($p < 0{\cdot}03$). *See* Table 2.

Table 2. Comparison of the CNV in the Acquisition and Distraction Conditions before and after Treatment by Electrocoagulative Leucotomy

High anxiety patients (N=11)	*Mean Acquisition CNV*		*Mean Distraction CNV*	
	μV	*Significance of difference* (Mann-Whitney U)	μV	*Significance of difference* (Mann-Whitney U)
Before treatment	16·5	p 0·06	3·4	p 0·03
After treatment	19·8		9·4	

THE EFFECTS OF MORE COMPLEX DISTRACTION

The main series of experiments reported above used only one simple and fairly mild form of distraction. A further series of experiments was carried out to test the effects of more varied forms of distraction on a limited number of non-patient subjects, all of whom had participated in the main experimental series. All were known to have CNVs well within the normal range, that is, close to 20 μV in the basic condition and showing not more than a 30 per cent reduction in the distraction condition. There had been indications in the main experimental series that the CNV of non-patients could be much diminished if the form of distraction employed was sufficiently significant for the subject. Thus two non-patient subjects, who had shown an unexpectedly large reduction in CNV amplitude in the distraction condition, subsequently volunteered the information that they had spent some time trying to interpret the irregular tones used as a scrambled message in morse code. Both, incidentally, had been in occupations in which they had used the morse code regularly, one being an airline pilot.

Distraction by Television

Four different situations were involved in the second experimental series. In the first situation six of the subjects watched television programmes during the course of the test. At irregular intervals the experimenter cut off the picture for several seconds. One second prior to each cut-off a warning click signal was given to the subject. By pressing a button when the screen went blank, but not before, the subject could restore the picture. All subjects produced CNVs in this situation very similar to those recorded from them in the standard Click-Flicker situation. Amplitude was in the range of 15–25 μV. A subjective account of the interest value of the programme was obtained from each subject after the experiment. Only in one case was the interest engendered sufficiently great to engage the major part of his attention for quite long periods. This subject had an interest in horse racing and watched two horse races during the course of the experiment (*see Figure 10*). While comparatively uninteresting race preliminaries were being shown, the average amplitude of the CNV was 23 μV. Once the race started the CNV decreased in amplitude to 11–13 μV and, during the final stages and finish, there was a

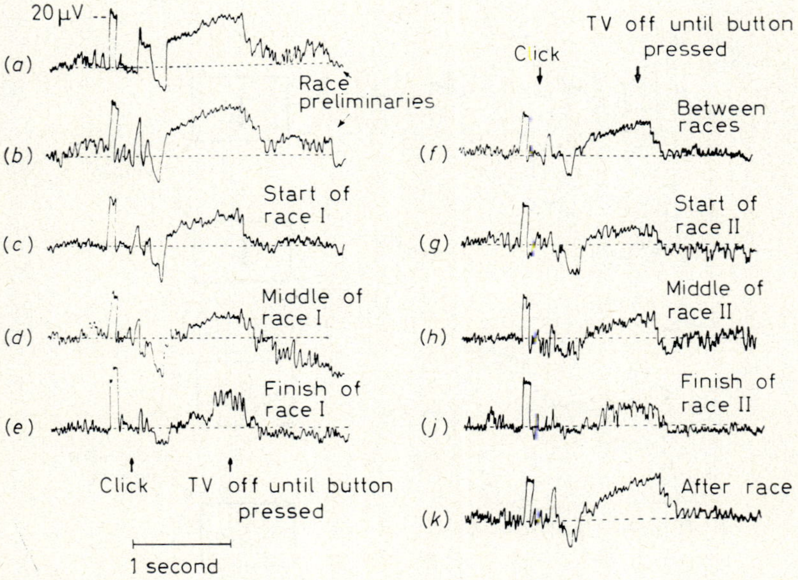

Figure 10. The effects of distraction on the CNV. S watched two horse races on television. Following a warning click the screen went blank from time to time. S was instructed to press a button to restore the picture. Each trace is the average of 12 trials

noticeable lengthening of the latency of onset of the negative variation (from 360 to 760 ms). In between races the CNV reappeared in its original form with an amplitude of 18 μV, but diminished in the same manner—to 9 μV—once more during the course of the next race. The CNV was finally re-established at 18 μV during the relatively uninteresting post-race proceedings and presentation of trophies.

Speech and Music Distraction

In the second situation two subjects were provided with distraction by the experimenter talking to them and, in one case, playing music, during an otherwise standard series of click followed by flicker trials in which the subjects pressed a buttom to terminate the flicker. In both subjects the speech distraction produced a small reduction in CNV amplitude ranging from 4 to 6 μV. In the case of the subject to whom classical music of his own choice was played during the test the CNV was reduced in this condition by 10–12 μV.

Intrinsic Distraction

The third situation, involving seven subjects, employed a one second presentation of a simple line drawing on the closed circuit television screen as S_1, followed immediately by a series of repetitive clicks which had to be terminated by the subject pressing a button. The drawings presented as S_1 varied considerably in their interest value, the purpose being to investigate any division of attention between preparation to respond to S_2 and a prolonged in-

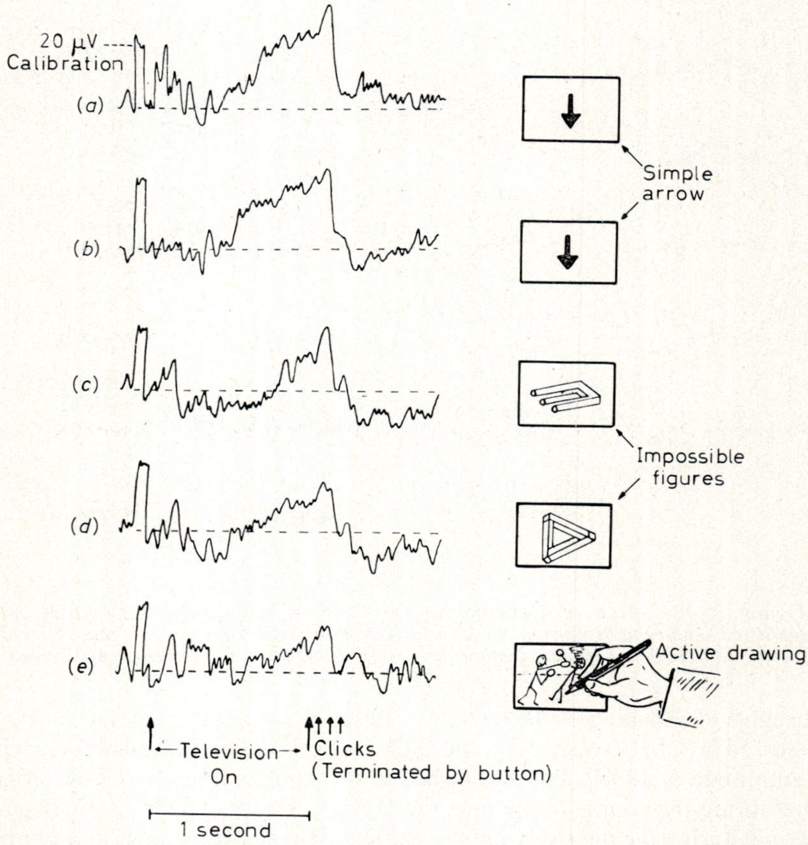

Figure 11. The CNV of a non-patient subject. S_1 was a one second exposure of a picture on closed circuit television. S_2 were repetitive clicks which subject terminated by pressing a button. The traces show the effects of varying the degree of intrinsic interest in the picture used as S_1. Each trace is the average of 12 trials

spection of the intrinsically interesting features of S_1. The results of this experiment showed a clear relationship between the content of the television picture and the amplitude of the CNV (*see Figure 11*). When the picture was a simple arrow figure, the CNVs of all subjects were in the range 15–25 μV. When this picture was replaced by the more complex, and for most subjects more interesting, drawing of an 'impossible' figure there was a reduction in CNV amplitude which ranged from 17 to 55 per cent. There was also in this condition a marked increase in the latency to onset of the negative variation. Finally, when a dynamic picture involving live drawing in front of the television camera was used as S_1, the decrease in amplitude was even more marked in all subjects.

Voluntary Concentration

In the final situation five subjects were used to investigate voluntary variation in attention. They were given a series of about 50 standard click-flicker trials, the flicker being terminated by pressing a button. Before beginning they were asked to concentrate particularly on some trials, of their own choice, with a view to obtaining the shortest possible reaction time.

No indication was required by the experimenter in advance of the intention to concentrate, but the trials concerned were to be signalled retrospectively by the subject giving two additional presses on his button. Subjects were told that the overall proportion of trials on which concentration was directed should be in the order of one in five. These trials were later averaged from tape and were compared with averages of random samples selected from the remaining 'non-concentration' trials. The main differences observed in the 'concentration' averages were (*a*) an increase in the amplitude of the non-specific evoked response to S_1, (*b*) a small increase (3–7 μV) in the amplitude of the CNV and (*c*) a more rapid rise in potential at the onset of the negative variation. (*See Figure 12.*)

THE CNV IN CHILDREN

So far all of the experiments quoted were carried out on adult subjects. If indeed attention is implicated in the CNV one might well expect to find maturation of this function reflected in the development of the CNV throughout childhood. Few studies have yet been carried out of the development of the CNV in children. Cohen and Offner (1) and Cohen, Offner and Palmer (2) describe the beginnings of the CNV at 6 years and report a progressive development in form and amplitude towards the adult pattern, which is usually well established by the age of 20 years although may be later in some subjects. These findings are consistent with those of our own laboratory in Bristol, with the exception that rudimentary CNVs are sometimes recordable in children younger than 6 years. In the context of attention it is also interesting that the CNV is found to be more diffuse and widespread in children, being proportionately quite large at the occiput until the age of 12 years when it tends to concentrate more anteriorly.

We have been able to carry out a longitudinal study of just one normal child over a period of 4 years between the ages of 7 and 11 years. Routine CNV testing was carried out each year and the opportunity was taken to apply a number of other CNV situational tests including one in which pressing a button at S_2 produced a brief exposure of a picture on a television screen. The results of this study are illustrated in *Figure 13*. At the age of 7 years the tentative beginnings of a CNV can be seen, although in a variable form. However, after a careful explanation of the way that the click could be used

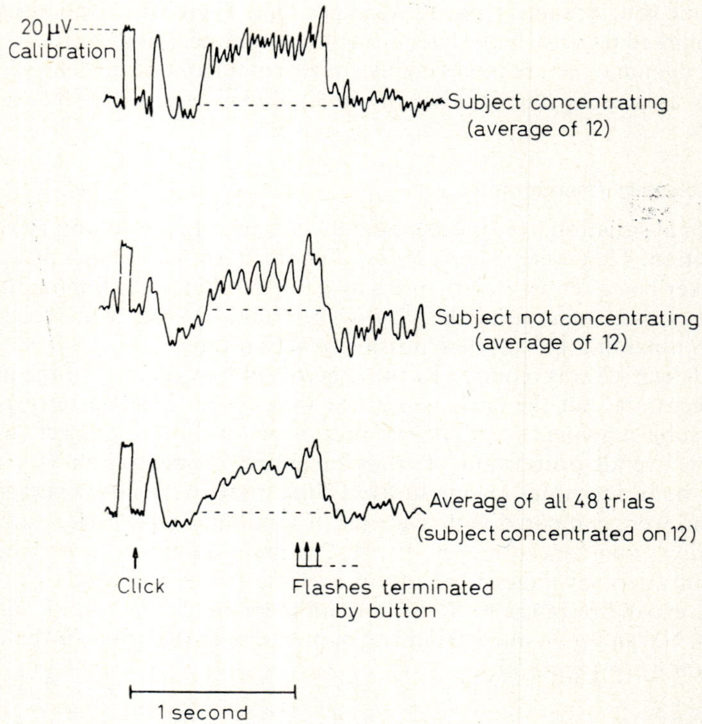

Figure 12. The averaged CNV of one non-patient subject in a series of 48 trials during which he voluntarily concentrated on some trials and not on others. The first trace is the average of 12 trials on which he concentrated. The second trace is the average of 12 trials, taken at random, on which there was no special concentration; the final trace shows the average of all 48 trials. Note the more rapid rise at onset and greater amplitude of the CNV in the 'concentration' trials

as a warning signal to prepare for the button press, a CNV more like the adult pattern appeared. At the age of 8 years the CNV appeared more consistently although the form still tended to be variable. At the age of 9 years a slightly different procedure from the usual Click–Flashes + Button was used. On this occasion the button pressing response gave a short exposure of a picture on a screen in front of the subject. The general form of the CNV which resulted was very similar to that of the previous year, although its termination and return to baseline was delayed until after the picture exposure ended. At the age of 10 years the CNV appeared well established, large, even by adult standards, at around 30 μV, and followed by a somewhat smaller secondary negative wave after the usual cut-off. At this age the tone distraction sequence, employed as part of the routine adult experiments, was used for the first time. The CNV was retained, but its amplitude reduced to about half of its former value and the secondary negative wave appeared to increase in size. At 11 years the same routine was followed. The trend towards an adult pattern of CNV had continued and the amplitude, at 20 μV, was around the normal adult mean. On this occasion there was virtually no reduction in amplitude in the distraction condition.

Two further points of interest emerged from this study. The first was the tendency of the latency to onset of the negative variation to be longer than that usually seen in adult records. This could be interpreted as being due to longer central processing time for assessing the significance of the warning click. Such an interpretation would be consistent with the regular appearance of late components (i.e. at 200 ms or longer) in the evoked response complex

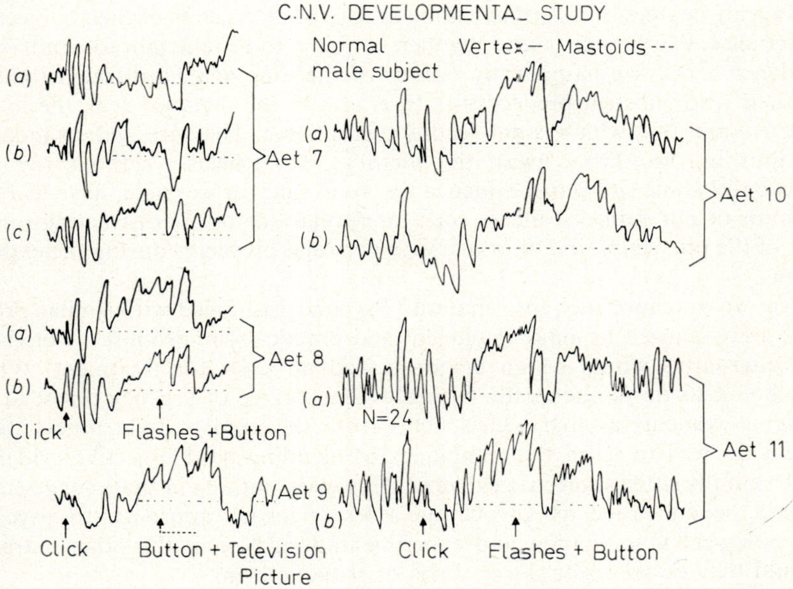

Figure 13. A longitudinal study of the CNV in one male child between the ages of 7 and 11.

Trace b. at ages 10 and 11 shows the effect of tone distraction presented between trials

to the click. It is perhaps significant that although by the age of 11 years the CNV of this particular child was similar in most respects to the adult pattern, it was recorded against a background of high amplitude intrinsic activity characteristic of the juvenile EEG. This might suggest that the CNV reflects an aspect of intellectual development, occurring in the frontal lobes, which is independent from general physical and emotional maturation, the latter being reflected more in the maturation of the intrinsic rhythms.

THE CNV AS AN INDICATOR OF ATTENTION

The evidence presented, both from patients and non-patients, strongly suggests that the CNV has a relationship to what at the psychological level we are pleased to call attention. It further suggests that changes in amplitude and form of the CNV are closely related to moment-to-moment changes in the focus of what for want of a better expression one must term 'conscious attention'. While such expressions may give rise to shudders and apprehension among many psychologists it has to be faced that concepts such as attention, distraction and concentration stretch our facilities for objective behavioural assessment to the limit. They raise problems with regard both to the quality and the quantity of the experience. At some point in our experiments we are obliged to realize that we are dealing neither with a black-box nor with an animal devoid of speech. Our subjects are human beings capable, in varying degrees, of communicating at least some aspects of their subjective experience. Psychologists tend, by their training, to be reluctant to admit such evidence. They are haunted by visions of the long and hard-fought battles of their distinguished predecessors to eradicate the metaphysical, the introspective and that which is not readily quantifiable. They are understandably anxious not to throw away the victory. Nevertheless, without the admission of some subjective evidence we could not, for example, have learned of some of our subjects' morse code interpretations of the distracting tones, nor of the preoccupation of others with personal problems during the experiments.

Are we to ignore such information? Psychiatrists, faced with similar problems have learned to make intelligent and practical compromises. For them the alternative would be no diagnosis and no basis for treatment. While psychologists do not generally have the same degree of urgency thrust upon their assessments as that which stems from the need to treat illness, they might do well to learn the techniques of blending hard objective evidence with skilfully sifted subjective evidence. However, nothing in these suggestions should be construed as advocating a less rigorous approach to psychophysiological investigation and experimentation. Nor at the other extreme should they be seen as a covert form of reductionism.

Whatever the ultimate interpretation of the psychological concomitants of the CNV, it has been demonstrated as capable of differentiating selected groups of psychiatric patients both from non-patients and from certain other kinds of patients. As yet it is a crude tool, but its diagnostic potential would seem to be well worth further investigation. One may also hope that its study will lead to an increased understanding of central mechanisms operating at a cortical level.

Slow and sustained negative shifts seen in animals and observed occasionally in the current study to follow verbal instructions, given during the course of the experiment, raise unanswered questions. Is the CNV perhaps only one aspect of larger and longer d.c. shifts which are taking place in the brain? Conclusive answers may have to await improvements in our techniques for recording d.c. both from the scalp and the depths of the brain. Meanwhile studies in the field of sleep suggest that gross changes of this kind may reflect fluctuations in arousal. Interaction between the CNV, evoked responses and more general, sustained shifts might well offer an explanation of both evoked response habituation and the reduction of CNV amplitude seen in distraction.

DISCUSSION ON GREY WALTER/McCALLUM

Jasper

These are a couple of very fascinating studies introducing some objectivity in what has otherwise been a very speculative field. It's quite remarkable the manner in which these variations change, with extraordinarily complicated psychological processes, presenting problems for the neurophysiologist which are challenging, to say the least.

Lindsley

I found Dr. McCallum's paper very interesting, and I want to comment that Dr. Kellaway's group in Houston, Texas have published an article in Neurology (18) confirming the evidence of this change—a slow build-up of the CNV in children and the difficulty, apparently, of establishing this build-up in children. They have not studied it longitudinally as you have, but in a 4 year old child they found it very persistent and resistant to change. I wonder whether you would interpret this as an indication, possibly, of the difficulty that a child may have in establishing a set towards a given experimental situation such as this, and then gradually, perhaps, learning to attend or to follow instructions and establish a set in which the CNV build-up would come more rapidly and completely. And that raises a question with regard to your psychopathic prisoners and psychiatric patients, both anxiety and schizophrenic, in which there was also a delay in the starting of the CNV and some irregularity and slowness in the build-up of the absolute rise in amplitude. Would you say that in a prisoner, because of habitual denial or negativity to commands, or demands, that this might be responsible in part, for the kinds of difficulty evident in terms of the CNV build-up? Likewise in psychiatric patients with the possible conflicting stimuli, both external and internal, that they are supposed to experience, let us say in chronic anxiety, or in schizophrenic states where there may be difficulty in breaking through a stimulus barrier into whatever mental activity is exhibited, might these psychological conditions be responsible for difficulties in establishing the CNV?

McCallum

I did not have time to develop this aspect as fully as I would have liked. If I may take your second point first; our original hypothesis with regard to the psychopathic group studied, and our interpretation of the results was slightly differently based from the one which you put forward. Eysenck and Franks have shown that psychopaths conditioned less readily than most other members of the population. It therefore occurred to us that if this were due to some constitutional deficit in their ability to form associations, one might well expect to find a corresponding decrement in the CNV. As I indicated previously, my own experiment confirmed the reduced conditionability of the psychopaths in a classical eyeblink conditioning situation and revealed a markedly smaller CNV than that seen in any other group so far studied. In the case of just one psychopath we also recorded the CNV after the administration of a stimulant drug (methedrine) and after a placebo for the methedrine had been given. With the methedrine there were small, but significant, increases in both the amplitude of the CNV and the percentage of conditioned eyeblink responses shown. This also related to a reported subjective heightened awareness of his surroundings by the subject.

As far as the psychiatric patients are concerned many are undoubtedly preoccupied to a very high degree with their symptoms even during the course of our experiments. This may well account, at least in part, for the fact that they show a slightly lower average CNV than the normal controls. If the effects of various forms of distraction are summative it may also account for why even such a small distracting factor as the irregular tones we use can result in such a large reduction in CNV amplitude.

To return to your first point; we too, like Low and Kellaway in Houston and Cohen and Offner in Chicago, have examined the CNV in a limited number of children, although only in the case I illustrated has a longitudinal study been possible. Our results have been very similar to those of the Houston and Chicago groups. We have found that it is just possible to see the rudiments of a CNV at as early as 3 or 4 years. However, at ages below 7 or 8 years results seem to be highly dependent upon the child having a very clear grasp of what is happening in the experimental situation. We have for example found that, when the mother has explained to the child in very simple terms exactly what pressing the button will do to the light, a CNV has appeared where on previous trials it was absent. Of course there are technical difficulties, inherent in the use of d.c. recording or very long time constants, for recording CNVs from children in the youngest age group, who tend to be restless.

Lindsley

Well, this would not be to say that a child aged 4, 5 or 6 years could not attend to a stimulus but perhaps the tactics and strategy one uses would be important. The child, having a very limited background of experience stored with which to trigger the kind of instructional set that one gives him, might find difficulty with the type of click-flash arrangement, whereas if you presented a different kind of stimulus more appropriate to a child's level of experience, it might be quite feasible to demonstrate a more persistent type of CNV and expectancy.

McCallum

I entirely agree. We find that for most children one of the best situations is to make their operant response switch on a closed circuit television screen, having on it a picture of themselves.

Beck

At what year level did you find the largest changes in the CNV taking place? Why I ask is that in the perceptual development of the child or in the development of his ability to learn, marked changes in performance are often noted at about the age of 5–6 years. We have also noted this with amplitude changes in the evoked response. Are there such changes in the CNV in this age group, that is, aged 5–6 years, or do these come later?

McCallum

I don't think we have enough evidence to be dogmatic. It is our impression that the development of the CNV seems to be more closely related to the mental age of the child than the chronological age. A major step forward in its development seems to occur at around the age of 7 years.

Beck

There are numerous studies indicating that the ability to read changes markedly between the ages of 5 and 6 years. In one year the child's ability to comprehend has developed greatly. For example, Sheldon White using an operant conditioning technique to study learning in children observed that a fairly basic and important change in the character of learning appears to take place after the age of 5. Before this age, the pattern of performance with children resembles that obtained when animals are used in a like procedure. After this age the pattern of performance approximates that found for human adults.

Grey Walter

May I supplement this reply? This work actually started with a survey of evoked responses in normal children, in order to see whether there were any clear anomalies in so-called autistic children. Apart from the results in the clinical group we had two groups, one of entirely normal family children, and another of institutionalized but clinically normal children in local 'family homes'. Incidentally, there were striking differences between the absolutely normal family children and the institutionalized children. But we had one normal child whom I remember very well. She was aged about 2½ years and very precocious; I would guess that her mental age was probably 4 or 5 years. She produced a CNV which was almost within the adult range for quite long periods. Then she got physically tired of pressing the button because her hands were very small, and she couldn't press the button easily so she gave the button to her mother and said, 'Now mummy, you press the button when I tell you'. So she organized the whole laboratory around her to solve her little problem. When she saw the flashes she told her mother to press the button and still produced her CNV—I suppose the mother had one too! There is a chicken and egg problem here. We would like, for example, to see what happens with the evoked response in CNV in very young children, say 1½ years or younger, but obviously in a child of below speech age you cannot tell the child to press the button. Whether they can't learn to press the button because they haven't got a CNV, or they haven't got a CNV because they can't be told to press the button is a question one can argue over until dawn. But the two do seem to be closely related and the attending of a child in relation to their optimum capability seems to be closely related to both the amplitude and the general responsiveness of details of the CNV development.

Beck

Have you studied the autistic child Dr. Grey Walter?

Grey Walter

Oh yes. One of these groups was in fact autistic children, and in general those who were autistic and not suffering from some other disorder had no CNV.

Beck

Dr. Rimland has an interesting point regarding the autistic child. Up to the age of about 6 years they are very easy to identify, but at about the age of 6 years they suddenly become a sort of heterogeneous type of child. It would be interesting to see what the CNV correlates would be at that age in the autistic child, say aged 4–7 years.

Grey Walter

Yes, this was not a longitudinal study of course, and it was a very heterogeneous group; more than half of the children had some interfering factor, either a bad family history or some evidence of perinatal trauma. Only 12 of these were autistic by the nine point standard of the Great Ormond Street Group, and of these none showed either clear evoked responses or CNV. However, they had marked signs of autonomic excitement, such as very high pulse rates. Often the stimulus would evoke very marked GSR or tachycardia, but no evoked response in the brain. In some, on the contrary, there were signs of evoked response but no autonomic changes. This suggests a sort of dissociation between the neurocortical and the hypothalamic autonomic channels. We can make nothing of this except to describe it and I can't relate it helpfully to the autistic state.

Mirsky

Have you ever recorded CNV directly from dura or cortex?

Grey Walter

Yes, I thought I made this clear, that we have recorded from electrodes implanted subdurally over the frontal cortex, and there we have seen CNVs which are exactly the same as the scalp CNV but not very much larger. This is because the effect occurs in patches on the cortex. If you have, say, a dozen electrodes over the frontal cortex subdurally, of these perhaps three or four will show a CNV of, say, 25 μV, but the rest will not, although they are, as far as one can tell, also on the cortex, and may show evoked responses. What you see on the scalp is the average of a number of these patches and the local potential is not very much larger on the brain than it is on the scalp.

Mirsky

Do you think that this would rule out all possibility of muscular and vascular factors in the CNV effect?

Grey Walter

Yes it does. I have also observed oxygen changes in relation to CNV, and following an operant response. There is a rise in oxygen availability for about 10 s in the cortex, again locally, and this follows the CNV by about 5 s. Presumably the metabolism of the local cells increases, more carbon dioxide is released and cerebral vessels dilate. We have seen no other signs of physical change in the brain which relate to the CNV, which seems to be a real bio-electric potential change, not due to other factors.

Horn

Has anyone recorded CNV in a grave curarized patient?

Grey Walter

We haven't studied it in curarized patients, but one of the groups has done this with monkeys, and it has been published. It has been shown in the frontal cortex of the monkey and someone told me that he had curarized a monkey and trained it with the same sort of effect, until the monkey became distressed by the curarization.

Horn

They could get a CNV?

Grey Walter

Yes.

Cobb

What responses was the monkey making during this?

Grey Walter

It was trying to make a response, but as it was curarized it got limper and limper and finally couldn't make responses. The one thing which perhaps we haven't made clear is that it's not necessary to make an operant response; you need only make a *decision*. The operant response is a convenient way of seeing whether the person has responded, but the motor act is not necessary to get a CNV, you only need to make a mental decision.

McCallum

I should like to make a point with respect to animal studies of the CNV. Inevitably almost all of these studies have had to be carried out using classical conditioning techniques. There is a clear indication from some human CNV studies, which we undertook about 2 years ago using an eyeblink conditioning situation, that classical conditioning presents rather a special case as far as the CNV is concerned. In essence, what we found was that the CNV appeared during the acquisition phase of the experiment but tended to disappear when the response was well established.

Jung

My remark concerns two points. First, similar results were obtained by Kornhuber, Deecke and Becker. They found their 'Beireitschaftspotential', the readiness potential, to precede voluntary action not only for arm and leg movements but also for voluntary gaze (16). On different days and in various mental conditions the potential amplitudes were rather different. When they showed to the subject—who had to look voluntarily to one or other side—a beautiful girl on one side and an uninteresting thing on the other, the potential was larger to the side of the girl. This may be emotional facilitation and potential of the readiness potential. In their latest research Kornhuber *et al.* have distinguished three components of these potentials, which correspond mostly to the CNV. The first is the readiness potential which begins about one second before the movement. The second is the motor potential which starts 60 ms before action and this may be complicated by a third component, a positive potential which they call 'premotion positivity'; this has to be distinguished from the positive shift occurring after the movement, which is probably caused by reafferent impulses from the periphery.

The second point is concerning the slow long-lasting potentials which shift the base line. All these d.c. phenomena are frightfully dependent upon CO_2 level, and I would like to know if somebody has controlled ventilation in these conditions. I think that the overall level is very dependent upon CO_2 and I suspect that some of these negative potentials of schizophrenics could be due to over-ventilation. After diminution of CO_2 in the blood there's more surface negativity, and *more* CO_2 results in surface positivity as in sleep. Different d.c. levels at the cortex during sleep and wakefulness could partly be explained by CO_2 changes. Therefore all these experiments should be controlled by recordings of respiration and CO_2 level.

Grey Walter

Yes we do in fact record respiration as part of the polygraphic procedure, and we are particularly interested in the effects of CO_2 because we are also concerned with intra-cerebral haemodynamics—variations in cerebral blood flow and oxygen level. I agree that this factor would be interesting physiologically; not as a contamination factor but as a physiological effect, because variations of CO_2 do contribute to the control of the cerebral vascularity and possibly excitability.

Berlyne

Dr. Grey Walter said that the phenomena he was talking about were found in patches. Am I right in thinking that what Dr. McCallum was talking about is relatively widespread in its distribution?

McCallum

I think that Dr. Grey Walter's point was that when one records from the cortex itself quite large differences in waveform are seen over very small areas, although if one amalgamates the responses from a number of such electrodes the average appears very similar to that seen on the scalp. All of the traces which I presented in my paper were from the scalp vertex and as such almost certainly reflect the summation of activity over a very widespread area of underlying cortex.

Berlyne

I think this is a very interesting case in which you have the problem of how far you are getting at intensive or selective phenomena. For example, if you have distracting stimuli, there must be in some sense a redistribution of attention: the subject is attending to the distractors and not to the initial stimulus that he's supposed to be attending to. But is the CNV measuring the extent in which the subject is keyed up as a whole, or is it measuring this redistribution? In other words, is it possible that if you did your analysis by patches you'd find that, while the CNV went down in some patches, it went up in certain others, corresponding to the fact that the distracting stimulus is capturing attention?

Grey Walter

Yes. We've done some simple experiments on these lines, on the occasions when we've had electrodes on the brain subdurally over a long period of time. In studies of habituation for example, some regions show prolonged negative shifts of potential when a monotonous stimulus sequence is changed to an irregular one and this may not appear in another electrode very close to it. What I think is happening during distraction, for example, is that the synchronized development of the potential difference that we call CNV is perturbed and the cortical patches no longer develop a potential difference synchronously, so that one particular area may have the same potential in the distracting situation, but it is not synchronized with the events in another area on the scalp. The algebraic sum is smaller because of the time dispersion.

Berlyne

Is there any possibility of having an increase in some places without a decrease in some others?

Grey Walter

Yes indeed. I think that in distraction this could well be the case; in the distracting situation one area might be augmented and the other decreased, but if they're not synchronized then over the area of the physiological scalp electrode you'll get a decrease.

McCallum

It must be remembered that for all of the situations illustrated an averaging technique had to be employed. The number of trials per average was usually twelve. If one could have identified twelve other clear points of decision or action during this experimental period and have averaged these one would have been likely to find a slow potential change preceding them. Indeed, we have carried out experiments in which subjects have made voluntary button presses to acquire a picture on a television screen. We found that a slow negative 'readiness' potential preceded the motor response as described by Kornhuber and Decke (17).

Cobb

If I understood rightly what Dr. Grey Walter said, that the change of potential may go in the opposite sense then there is also a contingent *positive* variation. Is that what you were saying?

Grey Walter

No, not to the same extent during the same period. There may of course be large positive waves after the CNV. But you mean that one patch may be positive rather than going negative?

Cobb

Yes.

Grey Walter

No. Not to the same extent as far as we have seen. The negativity does seem to be an important feature of this particular situation. However, in some clinical conditions the CNV does seem to be sitting in a sort of positive hollow.

Cobb

So they're all in the same sense?

Grey Walter

Yes.

Jasper

Are you speaking of the average from the scalp now?

Grey Walter

No, this is from the cortex.

Jasper

Does that mean you never get a reversal input?

Grey Walter

Not at this stage of the CNV development. Of course if we recorded from deeper in the cortex the whole thing reverses, but that is another story.

REFERENCES

1. Cohen, J. and Offner, F. (1967) *Electroenceph. clin. Neurophysiol.* **22**:190
2. Cohen, J., Offner, F. and Palmer, C. W. (1967) *Electroenceph. clin. Neurophysiol.* **23**:77
3. Cooper, R. and Warren, W. J. (1961) *J. Physiol.* **157**:38
4. Crown, S. and Crisp, A. H. (1966) *Br. J. Psychiat.* **112**:917
5. Hillyard, S. A. and Galambos, R. (1967) *Electroenceph. clin. Neurophysiol.* **22**:297
6. Irwin, D. A., Rebert, C. S., McAdam, D. W. and Knott, J. R. (1966a) *Electroenceph. clin. Neurophysiol.* **21**:412
7. Irwin, D. A., Knott, J. R., McAdam, D. W. and Rebert, C. S. (1966b) *Electroenceph. clin. Neurophysiol.* **21**:538
8. Knott, J. R. and Irwin, D. A. (1967) *Electroenceph. clin. Neurophysiol.* **22**:188

9. Low, M. D., Borda, R. P., Frost, J. D. and Kellaway, P. (1965a) *Physiologist*, **8**:223
10. Low, M. D., Borda, R. P., Frost, J. D., Kellaway, P. (1966) *Neurology Minneap.* **16**:771
11. McAdam, D. W. (1966) *Psychonom. Sci.* **6**:435
12. McAdam, D. W., Irwin, D. A., Rebert, C. S. and Knott, J. R. (1966) *Electroenceph. clin. Neurophysiol.* **21**:194
13. McCallum, W. C. 'The EEG'. *Proc. Univ. of London Psychol. Soc. National Psychol. Conference*, Lond., 1966
14. Rebert, C. S., McAdam, D. W., Knott, J. R. and Irwin, D. A. (1967) *J. comp. physiol. Psychol.* **63**:20
15. Walter, W. G., Cooper, R., Aldridge, V. J., McCallum, W. C. and Winter, A. L. (1964a) *Nature; Lond.* **203**:380

Discussion

16. Becker, W., Hoehne, O., Iwase, K., Kornhuber, H. H. and Täumer, R. (1968) *Pflügers Arch. ges. Physiol.* **300**:105
17. Kornhuber, H. H. and Deecke, L. (1965) *Pflügers Arch. ges. Physiol.* **284**:1
18. Low, M. D., Borda, R. P., Frost, J. D. and Kellaway, P. (1966) *Neurology Minneap.* **16**:771

The CNV During Tasks Requiring Choice

Cesare T. Lombroso
Department of Neurology, Children's Hospital Medical Center, Boston, Mass.

In recent years, evidence has been presented from studies in man and in the experimental animal to support the claim that some of the cortically derived d.c. fluctuations are: (*a*) synchronized with psychological and behavioural processes, (*b*) that they may indeed form an integral part of the bioelectric activity of the underlying cortex, and (*c*) that their measurement may somehow provide information on the functional state of the cortex and on the electrophysiological substrata of these higher mental processes (1). Since 1964, Grey Walter and his associates (6, 8) have published work relating to the more rapid d.c. potential changes, recorded mainly from the scalp of man, and associated with conditioning behaviour. Among their findings, they have stressed the effect called CNV or contingent negative variation, occurring in normal adults subjected to an operant type of conditioning. Briefly, when a subject is asked to perform an act in response to an imperative (unconditional) stimulus preceded by a warning (conditional) stimulus, a surface negative d.c. shift of approximately 20 μV develops soon after the conditional evoked response and envelops the succeeding response to the imperative stimulus.

In the course of investigations carried out in our laboratories on the CNV, also dubbed the E (expectancy) wave, we have, in general, confirmed the findings published by Grey Walter and co-workers regarding the constancy and stability of this effect in normal adults, its elusiveness in children and mentally disturbed patients (and, we may add, some epileptics). We have also made a few observations on a different effect which suggests further relationships between these rapid shifts of d.c. potentials and discriminative aspects of behaviour. In this communication, we shall present only the latter observations, because they raise some questions as to the nature of the CNV itself.

METHOD AND MATERIAL

Six adult subjects were used for this experiment, four being 'naive' to the test situation and two having been exposed repeatedly before to experiments dealing with the CNV. They sat in a partially screened room, separated from the instrument area by a one-way mirror-window. One second following a flash

(Grass PS-4 at intensity 2, or Iconix Light Driver, at 2 log units above threshold), the subject was stimulated with a 10 ms tone (or longer clicks) which were delivered via earphones to either his left or his right ear, according to a programme provided by a random pattern generator, an equal number of stimuli reaching each ear. A background of noise was provided by a continuously running air conditioning fan. Series of 24 or 48 flash-tone sets of stimuli were delivered in consecutive trials, the subject resting for 1–2 min in between. Baseline trials were run first with no instructions. Next, one or two trials were run along the usual paradigm for the development of the CNV. The subject had to respond to each tone regardless of the ear it reached. Next, the subject was instructed to respond only when the tone was presented to one of his ears, and not to respond when it reached the other. After three to four trials obtained in sequence the instructions were reversed, namely the subject was now asked to shift his response to the time when the tone reached the ear to which he was *not* responding before. After another three to four such trials, final runs were made where the subject was instructed not to respond to either ear. The response required either the pressing of a microswitch (always using the dominant hand) or a serial subtraction (task). Surface electrodes (Beckman) were placed over the vertex and the right mastoid process; above and below the left eye and at both canthi as well as on the root of the nose. A.C. electrodes were placed over the arm, at the neck and over the scalp. In some instances, two pairs of electrodes were placed orthogonally (0° and 90°) 7 cm lateral to the vertex, and about 2 cm posterior to the Rolandic sulcus as determined by surface measurements. These presumably were close to the sensory projection area of the hand.

Continuous monitoring of the EEG, EOG, and EMG was obtained on a Grass 16-channel Model 7. Signals signifying flashes, tones and motor response, when one was required, were also written on the strip chart. On some of the trials, the finger activated microswitch was gated into an electronic counter in order to obtain measures of reaction time. The accuracy was close to 2 ms. Trials where the subject made mistakes or those in which obvious EOG or blink artefact occurred were discarded. The evoked potentials from the d.c. electrodes, as well as the EOG, EMG and triggers were transcribed to an FM magnetic tape for later analysis. Averaging of the EP from the vertex-mastoid as well as from the orthogonal pairs over the post-Rolandic cortex was performed with a CAT-400 B or CAT-1000. The CNV developing when the tone reached the selected ear and a correct response was made, was analysed separately from the CNV developing when the tone reached the other ear and no response was made.

RESULTS AND COMMENTS

Where the CNV is obtained under the condition where the imperative stimulus (click or tone) is presented at random to either ear and the subject is asked to respond only to stimuli reaching one ear, significant and consistent differences are found between the CNV generated when the imperative stimulus reaches the ear selected for the subject's response compared to that when it reaches the ear for which no response is requested.

These differences do not appear during the first potential evoked by the warning signal, nor during the deployment of the negative d.c. shift, reflecting the obvious fact that not until the second or imperative signal is there any difference in the programme known to the subject. Thus, the differences relate only to the termination of the CNV and can be briefly summarized as follows:

(*a*) The CNV ends earlier when the imperative stimulus has reached the ear selected for the subject's response (and the subject has responded).

(*b*) The positive deflection at the end of the CNV is notably greater when the imperative stimulus has reached the ear for which no response was requested, and often reaches or falls below 'baseline'. Conversely, the positive deflection at the end of the CNV when the stimulus has reached the ear selected for response is smaller in amplitude and fails to reach the 'baseline'.

(*c*) This difference in the termination of CNV occurs equally over both hemispheres without discrete localization. It is not due to muscle or eye movement components, although careful monitoring of the latter is essential for any study of the CNV. It occurs whether the response is motor or mental, and although it coincides in time to the muscle potentials measured from the arm when motor response is requested, the latter do not contribute significantly to the difference between the CNV. Contributions to the evoked response from sensory components of the motor response were negligible, as determined by simultaneous averaging of somato-sensory EP from the hand area of the controlateral hemisphere.

(*d*) If the response parameters are interchanged (the subject now being asked to respond only to the imperative stimulus reaching the ear *opposite* to the one to whose stimulation he was responding before), the difference in configuration between the two CNV also reverses.

(*e*) This shift in the morphology of the CNV can occur immediately after the instructions are reversed. However, in some cases, it can be stepwise. Namely, it may take more than one entire trial, after reversing response parameters, to reach a clear difference. The time basis of such shifts is measurable in minutes, while behaviourally the shift in the subject's response is immediate, and correct. Whether the subject was 'naive' or not to the testing situation seemed to make no clear difference as far as the rapidity for these changes to occur is concerned.

(*f*) Those trials in which reaction time was measured, revealed no consistent correlation between RT and either amplitude, duration or configuration of the CNV. However, we did not have enough trials to subject them to statistical analysis. *Figures 1* and *2* illustrate these points.

It is conceivable that the notable and consistently observed differences in the termination of the CNV during a random programme as described above depend upon factors of 'attending' or 'suppressing', or other aspects of discriminatory behaviour causing a second surge of negativity, or conversely, inducing a positive d.c. shift. However, it seems curious that the electrical signal accompanying such high-level processes should, at least in some subjects, so lag behind the behavioural performance, as we can see in *Figure 2*.

It has been claimed that the CNV may represent the electric sign of cortical 'priming' whereby responses to associated stimuli are accelerated and synchronized (8); or that it may indicate the diffusion of dendritic depolarization into the motor regions (6). Low *et al.* (3) also consider the CNV to be the result of cerebral electrical activity, during conation. Rebert *et al.* (4) in examining the hypothesis that the CNV represents an index of cerebral changes related

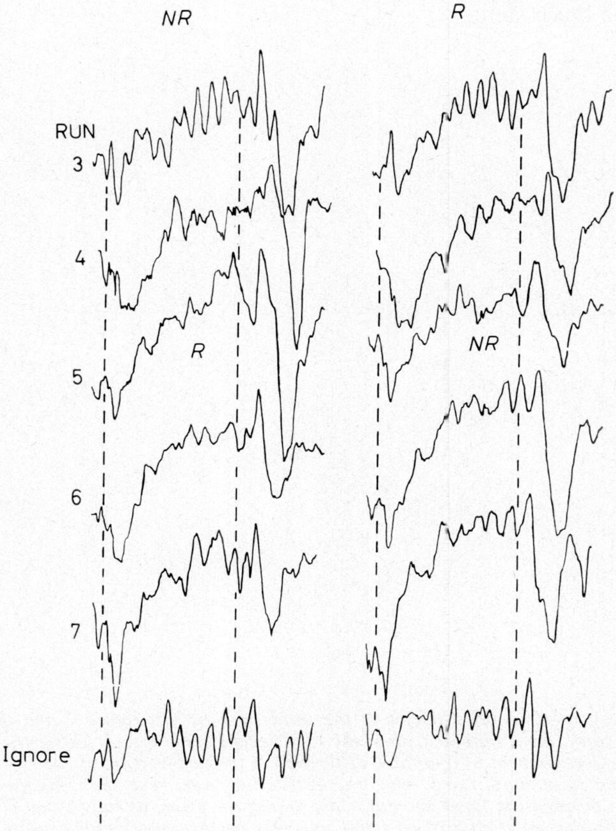

Figure 1. Broken lines indicate times of flash-tone stimuli, one second apart. The CNV's obtained when the tone reached the left or the right ear (at random) are displayed respectively in the left and right columns. During runs 3, 4 and 5, the subject was responding only when the tone reached his right ear, while during runs 6 and 7, the instructions were reversed. Note the difference in the termination of the CNV between R and NR which appears at run 4 and its rapid reversal at run 6. During the last run, the subject was told not to respond to tones reaching either right or left ears. All runs were performed consecutively and consisted of 24 flash-tone pairs, with tones distributed randomly to right or left

to the general drive state of the organism, suggest that this surface negative shift may reflect partial depolarization of apical dendrites or, alternatively, that it could reflect hyperpolarization of deeper neurone somas. The latter would introduce the concept of inhibitory neural mechanisms. Caspers (5) had previously postulated that such relatively rapid d.c. shifts are mediated mainly by neuronal elements, as opposed to the slower shifts due mainly to metabolic factors. Specifically, in situations involving processes such as arousal or orientation, the observed negative shifts would be mediated (at least preferentially) by ascending neuronal impulses from the reticular formations of the brain stem.

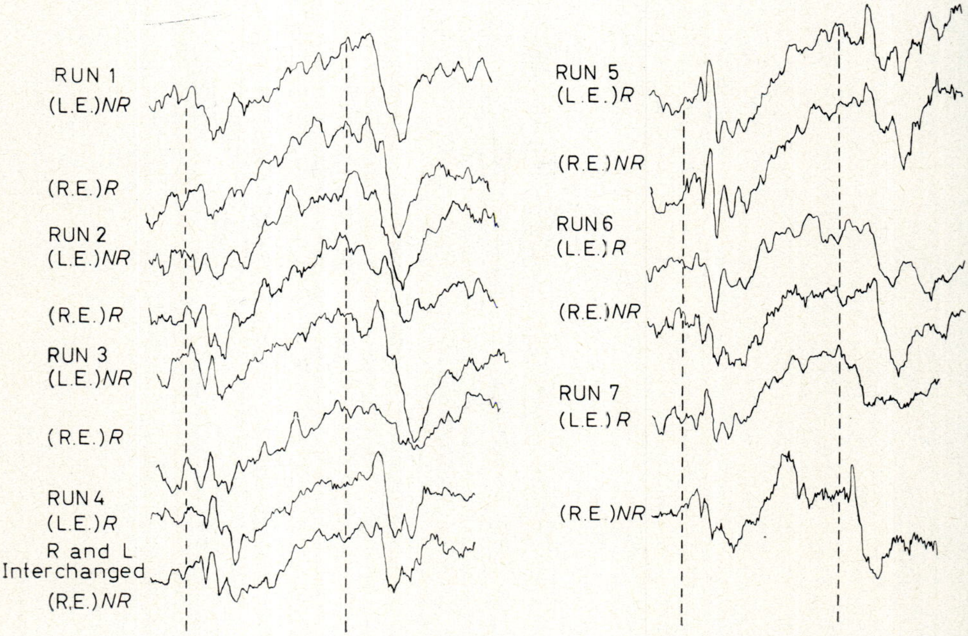

Figure 2. The experimental situation is the same here as for Figure 1 and the CNV's are displayed in pairs, those obtained when the tone reached the left ear being the first for each run. Note the development of a notable difference in the termination of the CNV as the trials progress, being maximal at run 3, when the instructions were reversed. This time, the reversal in the CNV's termination takes several trials to become clear, at runs 6 and 7. A continuous strip-chart afforded good control that the subject's performance was accurate. See text for other details

Evidence that the CNV is mainly of cerebral origin has been collected by Walter (7) and Low *et al.* (2), who have ruled out with scalp, epidural and intracerebral recording in man and monkey, some of the more evident sources of bioelectric contaminants, such as the GSR, the corneo-retinal potentials, respiratory and cardio-vascular correlates. Our observations also seem to rule out a significant contribution from sensory evoked potentials secondary to the motor response.

The origin of extracellularly recorded d.c. potentials from large populations of neurones is still uncertain, and their proposed roles as 'field potentials' in modulating or otherwise affecting the activity of other neuronal systems, are attractive but conjecturable hypotheses. Much of what has been proposed regarding these rapid d.c. shifts, seems based more upon indirect rather than direct evidence, namely their temporal relationship with aspects of integrated behaviour.

The observation of a considerable lag in time between changes in the configuration of the CNV and in the concomitant behaviour of the subject, could be interpreted as negating at least some of the attributes proposed for the CNV, and briefly summarized above. It seems possible to doubt, for instance, that the CNV represents a cortical 'priming' process within the frontal cortex for the discharge of motor neurones, when it may take literally minutes before a significant change in its configuration 'catches up', so to speak, with the change in performance. It is conceivable that the CNV represents many and different phenomena, some being 'neuronal' in origin; surely, its components relating to the cortical response evoked by the stimuli are neuronal. However, within any cortical area where populations of neurones become 'active', there are parallel metabolic processes which also involve non-neuronal elements, such as glial cells, capillary endothelial cells and the like. It is conceivable that a study of such phenomena as the graded 'direct cortical response', or measurements of changes in regional blood flow and CO_2, might reveal closer relationships with certain components of the CNV. It is known, for instance, that significant changes in the impedance of small volumes of the cortex closely parallel relatively rapid shifts of blood flow and gas exchange, and specifically that the impedance of dendritic structures seems to take place in several of those behavioural situations in which negative d.c. shifts had been measured on the cortex. This would represent a kind of intracortical GSR, whose time basis might be much less rigid than one would expect from a causally related neuronal inter-action during a stimulus-response task.

The observations reported in this study seem, therefore, to add to the complexity of the precise nature of both the psychological and physiological processes underlying the phenomenon of the CNV.

ACKNOWLEDGEMENT

This work was assisted in part by a National Institute of Health Grant (G.R.–FR–05482–04).

REFERENCES

1. Brazier, M. A. B. (1963) *Brain Function: Cortical Excitability and Steady Potentials.* Univ. of California Press
2. Low, M. D., Borda, R. P., Frost Jr., J. D. and Kellaway, P. (1966) *Neurology Minneap.* **16**:771
3. Low, M. D., Borda, R. P. and Kellaway, P. (1966) *Percept. mot. Skills,* **22**:443
4. Rebert, C. S., McAdam, D. W., Knott, J. R. and Irwin, D. H. (1967) *J. comp. physiol. Psychol.* **63**:20
5. Schulze, H. and Caspers, H. (1961) *Pflügers Arch. ges. Physiol.* **274**:55
6. Walter, W. G. (1964) *Ann. N.Y. Acad. Sci.* **112**:320
7. Walter, W. G. (1965) *J. psychosom. Res.* **91**:51
8. Walter, W. G., Cooper, R., Aldridge, V. J., McCallum, W. C. and Winter, A. L. (1964) *Nature, Lond.* **203**:380

Attention and Autonomic (Heart Rate) Regulations

Ivan Ruttkay-Nedecký

Institute of Normal and Pathological Physiology, Slovak Academy of Sciences, Bratislava, Czechoslovakia

During the past few years much attention has been paid to phenomena occurring in living subjects in connection with the arousal of their attention toward novel environmental stimuli of biological significance. We call the sum of the subject's reactions in such a situation the orienting (or orientation) reaction.

The orienting reaction is in itself a rather complex phenomenon and well deserves the interdisciplinary approach to which it is nowadays subjected. A simple comparison of the latencies of its typical components—(1) sensory (2) motor and (3) autonomic—warns us, however, not to expect to find clear-cut temporary and causal relationships among its constituent elements.

The simplest situation in which attention can be repeatedly aroused and experimentally manipulated, is a repeated exposition of physical stimuli that have, following previous agreement between experimenter and subject, acquired the quality of a signal for some simple performance. It is known that this signal meaning markedly enhances the occurrence of the autonomic component of the orienting reaction, as was demonstrated in our earlier experiments (5, 6) where the incidence of plethysmographic reactions (hand plethysmogram) satisfying Sokolov's formal criteria for an orienting reaction were compared under two conditions: first, to visual stimuli of 10 s duration, applied in regular 1 min intervals and second, to the same stimuli after they had acquired a meaning for the subject, he having been told to blink immediately after the cessation of the stimulus.

ESTIMATION OF THE HEART RATE CONTROL FUNCTION: THE RESPIRATORY HEART RATE REFLEX

In our experience—if environmental phenomena of attention arousal in its pure form are to be studied without contamination with emotional states—it is important to avoid using stimuli that are apt to forewarn a forthcoming nociceptive or emotionally meaningful stimulation, or signals for muscular work. Therefore, in our experiments, from which most of the data for the present communication are drawn, a single intake of breath served as a model of a simple performance. This natural act does not arouse anxiety, it is free of emotive associations and at the same time it elicits an easily reproduced autonomic reflex response (1, 2, 3), as observed in the hand plethysmogram and the heart rate changes: such responses may in turn be utilized to indicate the dynamic properties of the autonomic variables under study.

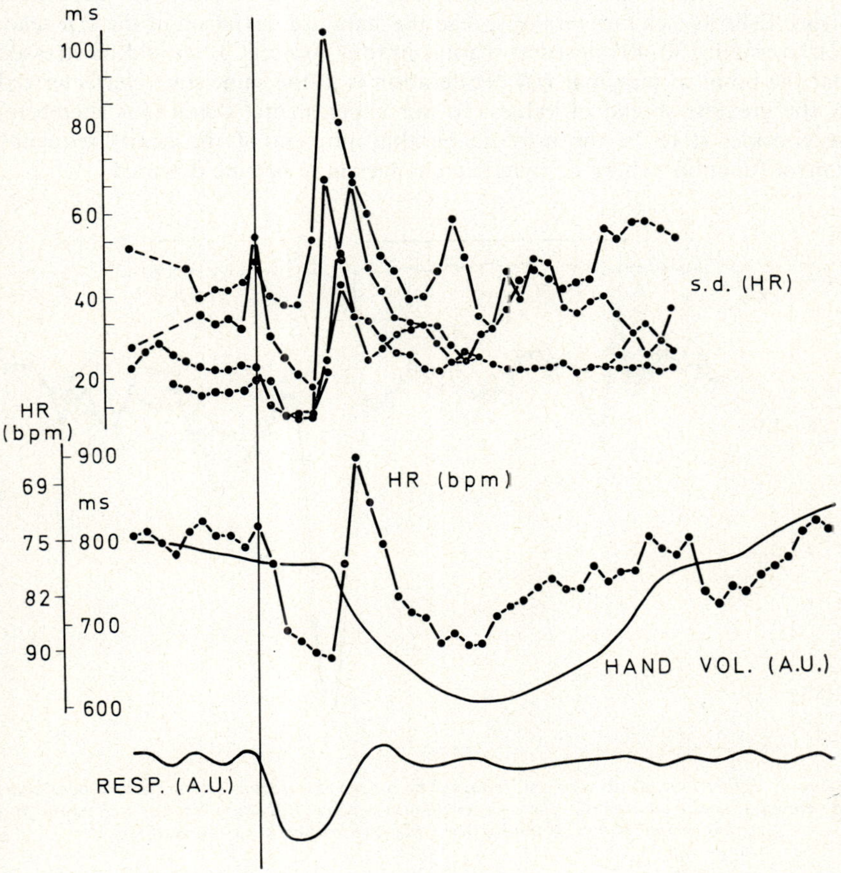

Figure 1. Beat-to-beat heart rate (HR) and hand volume (arbitrary units) changes as elicited with a single deep inspiration (Resp.)

Top: The course of standard deviations of HR changes from averaging over 30–40 trials in four subjects

Figure 1 shows the heart rate (*HR*) and hand volume changes as elicited by a single deep intake of breath in the course of a concrete trial. It should be noted that the decrement of the heart cycle duration follows the beginning of deep inspiration with practically no latency and this accelerative reflex is then invariably followed by a reactive deceleration of the *HR*. The plethysmographic reaction has a considerably longer latency, 3–4 s and outlasts the deep inspiration by a long period of time. While the heart rate responses are caused by changes in the vagal cardio-inhibitory tone, the plethysmographic reaction is an indicator of variations in the sympathetic tone, mainly in the tonic innervation of the acral subcutaneous venous bed. These two reactions are independent, nor are the blood pressure changes that also occur to some extent, clearly related to them.

Since our aim is to discuss chiefly the behaviour of the *HR*, the top of *Figure 1* shows as a measure of spread the standard deviation of the *HR* reaction averaging 30–40 consecutive trials in four subjects. It should be stressed that the point of maximal *HR* deceleration is at the same time characterized by the greatest spread of values. In our experimental conditions therefore, we consider it to be the most vulnerable moment of the heart frequency control function, where its dynamic changes may first be detected.

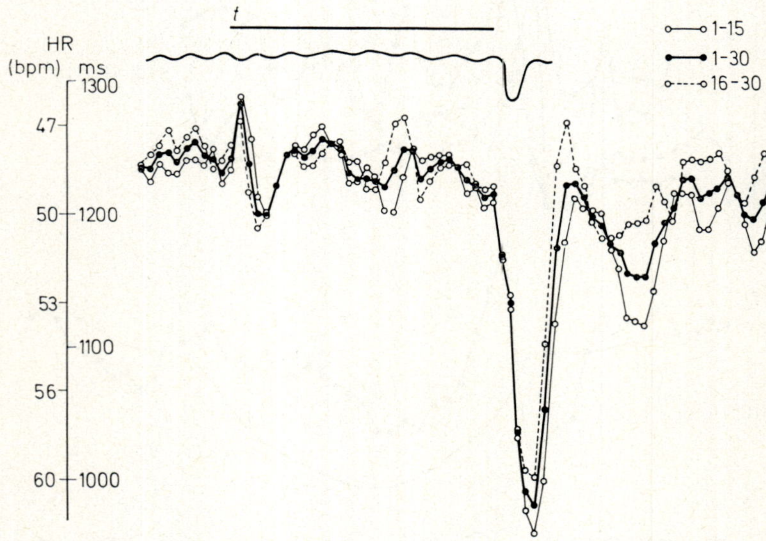

Figure 2. Effect of a 60 db *tone* (*t*), *followed by a deep inspiration on the beat-to-beat heart rate changes, averaged over the whole experimental session* (*full heavy line*) *as well as the first and second halves of the session* (*thin and broken lines respectively*)

Now tones, each of 10 s duration, are exposed at regular time intervals (1 min) to a subject reclining with eyes closed in an easy-chair in a sound-attenuated room, and the cessation of each tone presentation is agreed as the signal for taking a single deep breath, then the subject's *HR* control system is exposed to the action of two consecutively acting test stimuli: (1) the tone that arouses his attention and (2) the deep inspiration with its reproducible unconditioned *HR* responses.

During the whole course of each experimental session, a four-channel EEG (left and right bipolar *FP* and *PO* leads) was continuously recorded, together with the horizontal electro-oculogram from bitemporal electrodes (through an ECG preamplifier), the ECG, respiration (lower chest circumference changes) and the plethysomogram of the right hand.

Figure 2 shows the averaging of heart cycle duration over 30 trials in a single session of about 50 min duration. Partial averages of the first and the second halves of the session are separately expressed to demonstrate dynamic changes. The onset of the tone is followed by a biphasic *HR* response, similar to that occurring at the beginning of the deep inspiration. There are no significant differences between the averaged responses over the first and second halves of the session.

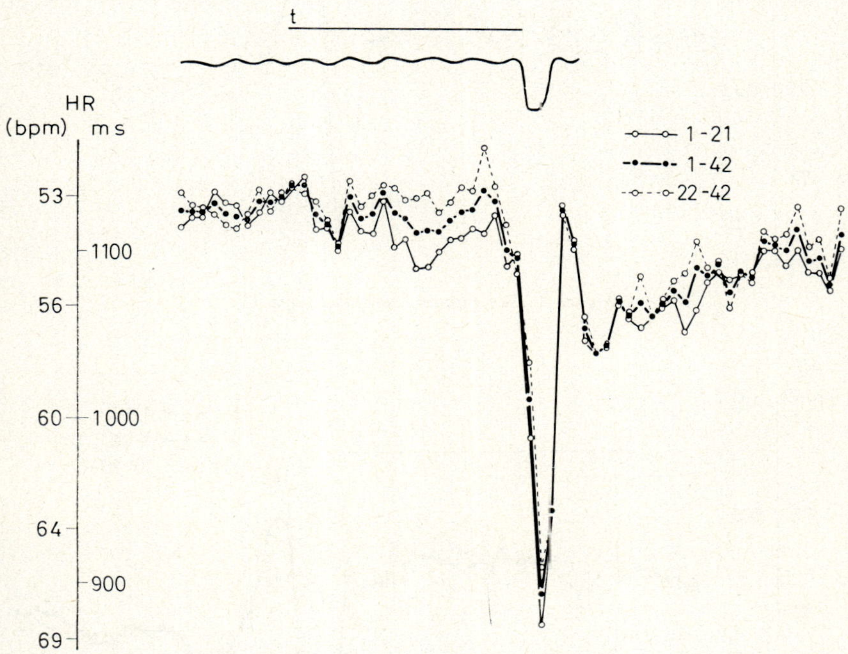

Figure 3. A similar case to Figure 2, but with the accelerating nature of the heart rate response indicated (see text page 73)

Figure 3 shows a similar case differing from the preceding one only in the purely *HR* acceleratory nature of the response.

The response on *Figure 4* is again acceleratory, but this acceleration is more marked in the second half of the session. While in the two preceding cases the amplitude of the deceleratory, deep inspiration-induced response did not change during the course of the session, in this case it is lower in its second half. The lowering of the cardio-inhibitory tone seen in the deep inspiration-elicited heart rate response goes hand-in-hand with the enhancement of the acceleratory *HR* orienting response.

The same pattern is shown in *Figure 5*, where an initial deceleratory orienting *HR* reaction changes into an acceleratory one, together with a marked lowering of the amplitude of the deep breath-induced deceleratory component of the *HR* reaction.

In order to see whether these changes of the *HR* response are not secondary

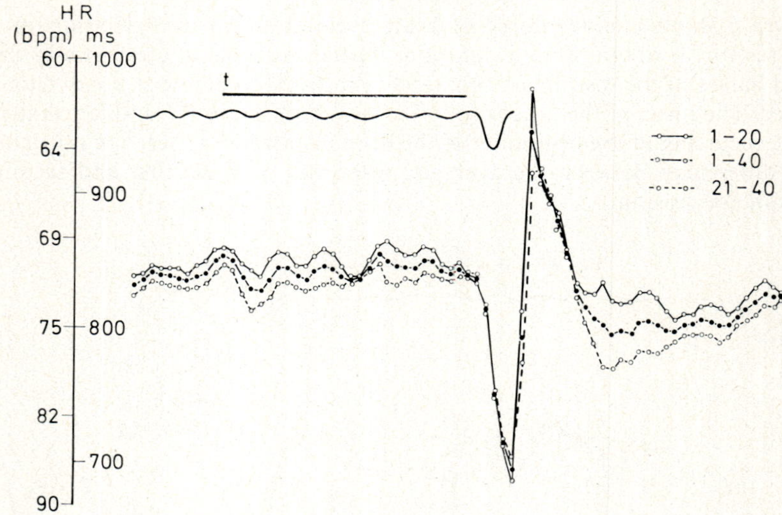

Figure 4. See Figure 2 and text page 73

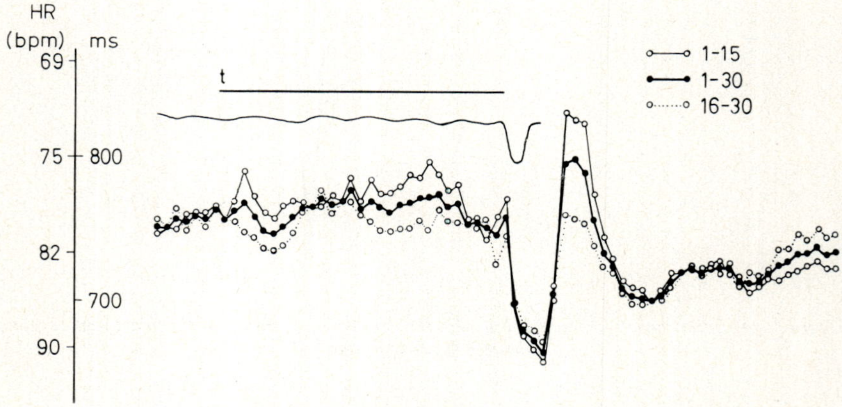

Figure 5. See Figure 2 and text page 73

to changes in respiratory dynamics, we listed the intra-individual differences of the amplitude and the angle of the first two inspirations after the beginning of the tone, the means of these variables for the subsequent respiratory movements during stimulus presentation, and the deep inspiration and the interstimulus intervals—comparing always the first half of the session with the second one. We either failed to find statistically significant differences, or those that we did find could not directly explain the heart rate behaviour. The same holds for vigilance changes as judged from the EEG and EOG, but not for the sympathetically regulated hand volume changes as shown on *Figure 6*.

Subjects with a tendency to develop more marked acceleratory responses in the second half of the experiment showed not only a decrease of cardio-inhibitory tone as tested by a deep inspiration *HR* response, but also—with one single exception—a general change of the *HR* baseline level in the sense of tachycardia (acceleration), and had higher heart rates from the beginning.

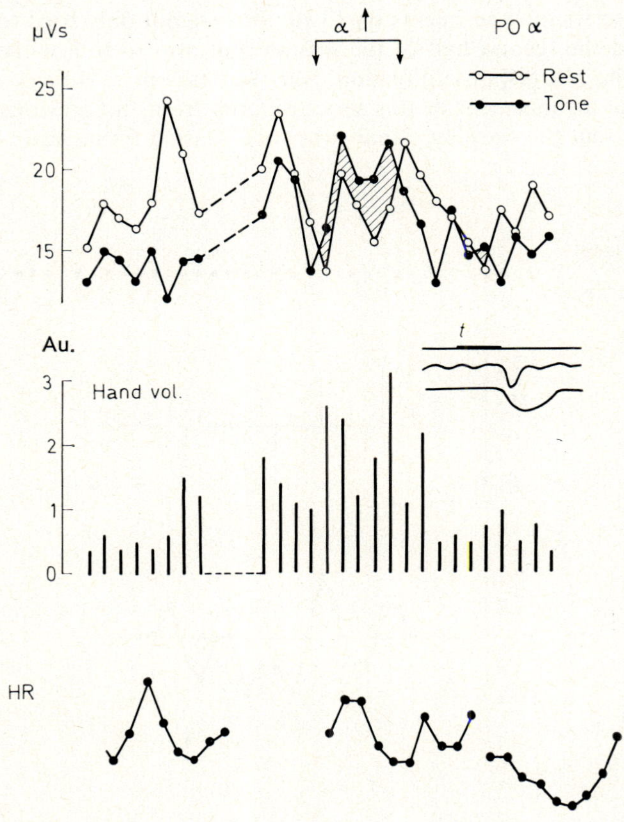

Figure 6. Top: Changes of the voltage-over-time integral of parieto-occipital EEG activity during an experimental session. Open circles: 3 s *pre-stimulus interval, full circles: first* 3 s *of the stimulus exposition* (*tone*, 60 db, *eyes closed*). *The cessation of the* 10 s *lasting tone was the sign to take a deep inspiration* (*see insert on the right*), *eliciting secondary, plethysmographically recorded hand volume changes, the areas of the downward deflection of the curve after each inspiration being indicated in the middle of the figure*

Bottom: Averages of beat-to-beat evaluated HR responses from the beginning, the middle and the end of the session respectively. Note that in the middle of the session when activation indicates sleep (*top, dashed area*), *deep inspiration elicited vasoconstrictory reactions are enhanced* (*middle*), *while the HR response maintains its deceleratory-acceleratory character*

On the strength of these four examples the need for a better understanding and assessment of autonomic variables and their dynamic behaviour should be stressed, because this may determine, at least partly, and when under the condition of non-emotional stimulation, the direction and magnitude of the heart rate response to external stimuli.

INTRA-INDIVIDUAL DYNAMICS OF THE HEART RATE REACTION

In another series of experiments on 13 subjects we not only tried to compare the first and the second half of the session, but also to follow the dynamic change of the *HR* response direction over two sessions, 2–4 days apart. The experimental arrangement of this series differed from the previous one only in the sense that the warning signal—the tone—lasted 5 s instead of 10 s and

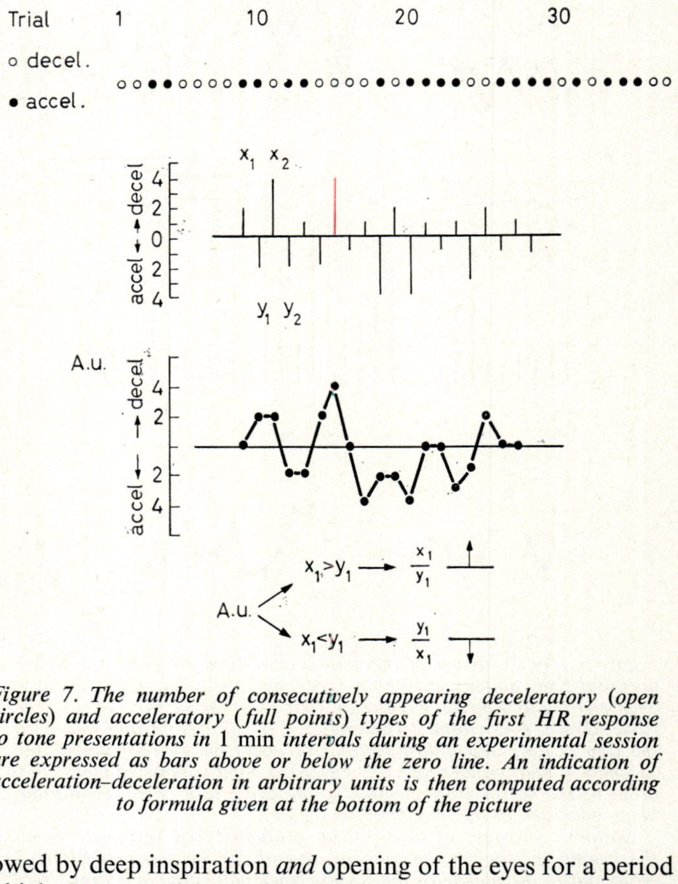

Figure 7. The number of consecutively appearing deceleratory (open circles) and acceleratory (full points) types of the first HR response to tone presentations in 1 min *intervals during an experimental session are expressed as bars above or below the zero line. An indication of acceleration–deceleration in arbitrary units is then computed according to formula given at the bottom of the picture*

was followed by deep inspiration *and* opening of the eyes for a period of 10 s, during which *ON* was elicited. The vigilance level oscillated in some subjects between the stage characterized by alpha activation upon stimulation, as well as undulating drifts of the horizontal EOG, and full wakefulness. The maximum and minimum cycle length was noted both from 5 heartbeats preceding the beginning of the tone and during the tone presentation. It was necessary to treat the behaviour of the longest and shortest cycles separately, i.e. the maximum and minimum of the variable under study—because only in 1 case could we prove by X^2 that both change in concrete trials in the same direction. The reaction to the first presentation of the tone is usually a deceleration both for the maximum and minimum in the first session (84 per cent of cases) but this figure falls to 46 per cent only in the second session where deceleration and acceleration reactions are found with about the same probability.

Figure 7 illustrates our attempt to convert the qualitative information of whether the maximum and minimum cycle lengths shortened or increased under the tone, to a sort of quantitative display of acceleratory or deceleratory prevalence in the course of the whole session.

The only relation we could detect from our polygraphic records was that subjects with low alpha activated EEG patterns responded to the first stimulus presentations with a deceleratory response to the maxima, and the subjects with abundant alpha activity, with an acceleratory response. The first response type was then followed by various undulatory patterns, and so far we have not been able to find any meaning for them or correlation with other variables.

It seems, therefore, that the primary *HR* response to arousal of attention in its purest uncomplicated form is merely a change, its extent and direction being partly determined by the actual characteristic of the heart rate control system.

This is further complicated by the fact that sensory intake in itself changes the performance of the control system. If we look, for instance, at the undulatory behaviour of the *HR* variable that remains after its stimulation by a deep inspiration, as being a certain characteristic of the dynamic state of its control function, then we can see that it is changed even by a meaningless visual sensory intake. One can show this because in this series of experiments, where optokinetic nystagmus followed the deep inspiration, we no longer saw the fine *HR* undulation that so often followed the deep breath in the preceding type of experimental session when eyes were closed throughout the experiment.

RESPIRATORY SINUS ARRHYTHMIA AND THE RESPIRATION IMAGE

The problem, however, of whether a particular dynamic state of an autonomic variable might influence sensory or higher mental processes, is intriguing and touches upon some basic psychophysiological problems. Some experiments pertaining to this question will be briefly presented (4).

If the respiratory movements are not too frequent (i.e. up to about 20/min), respiration-locked rhythmic variations of the cardio-inhibitory tone, called respiratory sinus arrhythmia, may be detected from the heart rate. Under conditions of a slight variation in the depth of respiration the amplitude of respiratory arrhythmia does not correlate closely with the amplitude of respiratory chest movement. The correlation coefficient was found to surpass the 0 value with statistical significance in only 6 out of 11 healthy subjects, and the value of r did not surpass 0·401. Further, the law of initial values operates in these conditions only as a statistical phenomenon and does not hold for each individual case. We tried, therefore, to study the nature of the residual source of this variation. The block diagram of the experimental arrangement is shown on *Figure 8*.

The depth and frequency of the subject's (*S*) respiration was displayed to him on a milliammeter connected to the mercury-in-rubber strain gauge that served for detection of the chest movements. The sensitivity of this milliammeter was changed successively and in random manner by the experimenter (*E*). The subject was asked to control the amplitude of the swings of the milliammeter pointer by regulating his own respiration, thereby forcing him to change the actual depth of respiration without his being aware of it—at least to a limited extent. We found, that subjects with an *a priori* high respiratory sinus arrhythmia made significantly more mistakes in their judgement about the actual changes in depth of their respiration, than those with slight respiratory sinus arrhythmia.

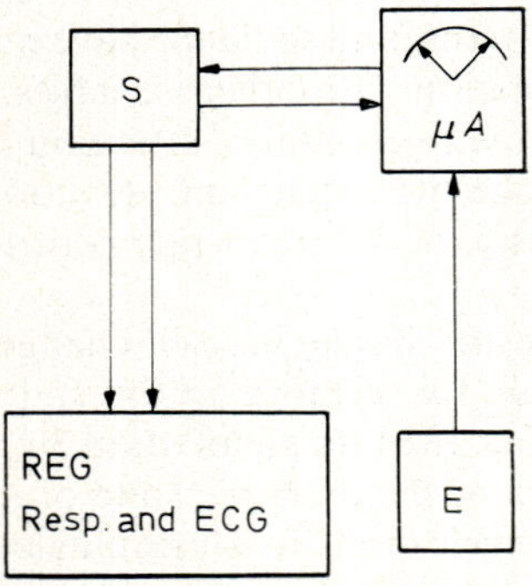

Figure 8. Block diagram of the experimental arrangement—dissociation of the subject's estimate concerning the depth of his inspiration from the actual amplitude of his respiratory chest movements. S—subject, E—experimenter

This pilot study therefore, raises the problem of possible secondary influences of the dynamic state of heart rate regulation on the mental processes involved in the judgement about respiratory changes. It may be noted that patients with neurocirculatory asthenia, where an outstanding feature of subjective symptoms is respiratory distress (properly analysed this is an inability to judge the respiratory movements as being of satisfactory depth), usually have pronounced respiratory sinus arrhythmia.

CONCLUSIONS

In conclusion, while there is evidence for reflex facilitation of sensory intake during attention arousal, via central and peripheral threshold variations, the later autonomic changes seem to be instrumental in preparing the subject for being ready for action—i.e. for the *consequences* of the sensory intake. At the same time the sharp causal relationships in our conceptual schemes of the orienting reaction, or attention arousal should be loosened and due space given to a statistical consideration of the probabilities of occurrence of their important components.

DISCUSSION ON RUTTKAY-NEDECKÝ

Koepchen

We have made similar observations on men: the mere 'imagination' of respiration, when there is firm control that no real respiratory movements occurred, induced changes in heart frequency with inspiratory acceleration and expiratory deceleration. One wonders why there is no clear correlation between the degree of the respiratory movements, and the degree of the respiratory dysrhythmia. We have experimented with the dysrhythmia, and can show that during the inspiratory phase the reflex deceleration of the heart produced by the baroreceptor input is blocked. This blocking cannot go further than a complete block of the baroreceptor reflex, and in this case no further increase of inspiratory heart frequency will take place, even if the depth of respiration is further increased. This could be one possible explanation. As to the dependence on the initial value of the heart frequency, such a dependence exists also at the heart: the slowing effect on heart rate of efferent vagal stimulation depends on the actual heart frequency (7). Therefore this phenomenon is not necessarily based only on central interactions.

Ruttkay

I agree. With normal respiration, or very slightly deeper respiration, heart rate behaviour is qualitatively different from the behaviour which occurs with much deeper inspiration, and, of course, one's not talking about amplitude but *speed* of inspiration. There is of course some limit, but its difficult to say what this limit is.

Grey Walter

I was only going to say that this reminds me of the effects described by John Lacey. I'm not sure if he's published this but he's spoken about it at a number of meetings. He has described changes in pulse rate according to whether a subject was attending to external or internal signals. There was a marked tachycardia if the subject was attending to internal stimuli and not external ones. I asked him about the respiratory effect—I assumed the tachycardia was a sort of sinus arrhythmia—and he told me that in 14 per cent of his subjects there was a correlation with respiration, but in the rest it was not so and the effect was truly independent, and he claimed that it was a *specific* change. He has quite a theory about the relation of this to the afference to the cortex from hypothalamus.

Jasper

As I recall, his conclusion was that the mechanism was more closely related to cardiac than to respiratory variables. You would not necessarily conclude, would you, that the cardiac regulation system, the central control system is linked up with attentive processes much more closely than is the respiratory system? This was one conclusion of other investigators.

Ruttkay

Some work has been published in the field of experimental psychology dealing with the fact that some people react to certain kinds of sensory input with an acceleratory reaction and to others with deceleratory reaction. Some people feel that the heart rate control system helps to explain this, and my point was that if there is any deceleration or acceleration it depends upon the heart control system functioning in a dynamic state. For instance, it can be shown in response to deep inspiration—which of course is not 'psychological input'. And of course the heart rate control system is linked closely to the respiratory control system.

Jasper

Yes, but they can be dissociated.

Ruttkay

That's true.

Jasper

Some of this is central and some peripheral?

Ruttkay

Yes, of course.

Adey

I would like to mention briefly work in our laboratory conducted by Berkhout, Walter and myself involving tests very similar to those that you have described. The autonomic and EEG responses to stressful questions were delivered initially from magnetic tape and now videotape. This allows very accurate timing from one subject to another. Thus one can look at the population response to a particular question, and also look at the response linked to the *initiation* of the question, to the completion of the question, and to the initiation of the response. Briefly, sensitive questions evoked these very sharp changes in heart rate that often began with very critical words in the questions. These questions had sexual overtones so that it was possible to load individual words in the question. By contrast the finger plethysmogram gave equally consistent responses that were linked to the subject's initiation of the verbal response. One interesting point was, that if a sensitive question was followed by three or four neutral questions, the critical questions tended to have a conditioning effect on the subsequent neutral questions, so that these also would evoke a series of declining responses.

Ruttkay

I think there is a qualitative difference if we use neutral stimuli of low intensity, but if we use stimuli with emotional load then, there is heart rate deceleration, acceleration and that's another matter. According to our experience too, the plethysmogram behaves quite differently to the heart rate.

Berlyne

I just wanted to mention, in connection with the Lacey effect, some experiments reported to the APA by Obrist. These dealt with conditioning in the dog, and he had some evidence that heart rate changes were mediated through motor activity. This has, of course, come up in the past when people have tried to bring heart-rate change, or GSR under voluntary control. There has always been the question of whether some kind of muscular activity might be producing the negative changes. Of course, respiration is one possibility but it might not be the *only* possibility. Dr. Ruttkay, is there a possibility that your results could be accounted for in this way?

Ruttkay

I think this can be excluded because the subjects rested, and if there were any motor activity it would have been detected by our apparatus which is very sensitive. As regards the dogs, there is some early physiological work showing that when they are conditioned to tones, sometimes they begin with a heart rate deceleration to the tone but this can change to an acceleratory reaction. Therefore, this question of linking the effect to movement and muscle activity is by no means a simple one.

Beck

As I recall, it was Wenger and Bagchi who toured India with an EEG and EKG machine to study the Fakir and see how he controlled his heart rate, respiration, etc. It has always puzzled me how the true Fakir has such control over his autonomic nervous system. Knowledge of how he does this would certainly be useful in the treatment of psychosomatic disorders. I have heard of situations in which he can control peristalsis, being able to swallow a small napkin and pass it through his entire digestive tract in a matter of minutes. During such performances, if I recall correctly, the resting EEG is an alpha rhythm indicating that such feats are accomplished with what appears to be little 'mental effort'. This, of course, confounds the problem even more. How does he exert such influence on his autonomic nervous system with his brain and central nervous system in a relaxed state?

Jasper

Well, I remember him reporting that some of them were Fakirs and others fakers.

Beck

Yes, and when they were fakers and not Fakirs they exhibited low voltage fast EEG during their performance.

Jasper

I think we should now move on to the paper by Dr. Koepchen.

REFERENCES

1. Bolton, B., Carmichael, E. A., Stürup, G. (1936) *J. Physiol. Lond.* **86**:83
2. Matthes, K. (1951) *Kreislaufuntersuchungen am Menschen mit fortlaufenden registrierenden Methoden*, Thieme Verlag.
3. Ruttkay-Nedecký, I. (1963) *Acta neuroveg.* **24**:414
4. Ruttkay-Nedecký, I. (1964) *Psychologické štúdie VI.* Bratislava: Publ. House of the Slovak Academy of Sciences

5. Ruttkay-Nedecký, I. (1967) *Mechanisms of Orienting Reaction in Man* (Ed. by Ruttkay-Nedecký, Cigánek, Zikmund and Kellerová). Bratislava: Publ. House of the Slovak Academy of Sciences
6. Ruttkay-Nedecký, I., Kellerová, E. and Fichtel K. (1966) *Activitas nerv. sup.* **8**:2

Discussion

7. Koepchen, H. P., Lux, H. D. and Wagner, P. H. (1961) *Pflügers Arch. ges. Physiol.* **273**:413

Vegetative-somatic Relationships in Single Neurone Activity in the Lower Brain Stem

H. P. Koepchen
Physiologisches Institut der Universität, München, Germany

Many different functions are ascribed to the lower brain stem structures outside of the specific nuclei. Thus the reticular formation of the lower brain stem is part of the ascending reticular activating system, from which EEG-desynchronization can be evoked by electrical stimulation (21, 56). Other effects on the somatic nervous system which can be elicited by stimulation of these regions are facilitation or inhibition of muscular tone and reflexes or movements (23, 25, 53, 66, 80) and influences on impulse transmission in sensory relay nuclei (26, 27, 29, 73). On the other hand, based on the classic and more recent ablation and stimulation experiments, the bulbar cardiovascular (3, 18, 49, 51, 58, 65, 76, 86) and respiratory centres (14, 15, 20, 33, 50, 52, 54, 60, 82, 85) have been localized in the same regions. Many authors have published cross-sections with more distinctly localized stimulation points for pressor, depressor, inspiratory or expiratory effects in the lower brain stem.

We have tried to combine the localization given by several authors, completing a similar comparison made by Bach (6) and have projected the findings of the various investigators upon one another in the particular levels of the brain stem. The result was, that practically no area from the level of the striae acousticae down to a few millimetres caudal to the obex was free from effective stimulation points both for cardiovascular and for respiratory effective areas. Moreover, it has been shown that several visceral and somatic effects can be produced by stimulation of the same 'point' in the brain stem reticular formation (5, 6). Therefore, the question arises: are structures with different functions closely intermingled or do the same nervous elements belong to different functional 'centres'; in other words, are some somatic and vegetative centres at least partly identical?

Because destruction and stimulation cannot be limited with certainty to regions small enough to decide this question, the answer can only be found by investigation at the cellular level. The recording of the activity of single neurones, first applied in the brain stem by Moruzzi and his school (8, 55, 71), enables the observation of the spontaneous discharge pattern of the neurone and the study of the afferent influences converging on it. By this method it has been shown, that in the brain stem reticular formation somatic afferents from many sensory inputs and also from other regions of the brain converge on single cells (4, 8, 29, 71).

BULBAR RESPIRATORY AND CARDIOVASCULAR CENTRES

Recently, the classical bulbar respiratory and cardiovascular centres have been reinvestigated with the microrecording method. Many investigators have published details of 'respiratory neurones' (1, 9, 10, 11, 16, 17, 34, 57, 69, 70) and others of 'cardiovascular neurones' (28, 42, 43, 63, 64, 68, 79). Unfortunately, the classification of these neurones is only possible on the basis of their discharge patterns and afferent connections. It is impossible however, to detect peripheral effects if only the one single cell recorded is stimulated, and therefore the efferent connections remain uncertain. On the other hand, if stimulation is extended to a greater number of neighbouring elements in such closely intermingled systems, no certain conclusions can be drawn. This means that the usual classification of brain stem neurones as 'respiratory' or 'cardiovascular' implying that their function is the control

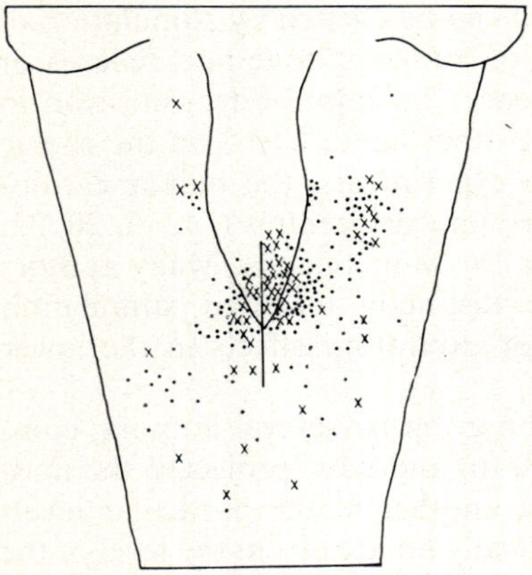

Figure 1. Scheme of the medulla oblongata of dog after removing the cerebellum. The points and crosses show the sites of insertion of microelectrodes. Crosses: Location of stabs where neurones with correlations to the cardiovascular system were found. Points: Location of stabs, where no such correlations were found

of the respiratory or cardiovascular system, is based on indirect evidence. Therefore the criteria for these classifications should be considered critically as we can illustrate by referring to our own experiments.

These were performed with P. Langhorst, H. Seller, J. Polster and D. Richter at Göttingen and Munich. The lower brain stem of an anaesthetized dog was exposed and free riding glass-platinum microelectrodes were introduced to record extracellularly the activity of single or—in special cases—of small groups of neurones. (For technical details cf. (43).) All in all the activity of 439 neurones was registered, and *Figure 1* shows the projection of the recording points on the surface of the medulla. The localization of the electrode tip was controlled histologically after marking by small electrolytic lesions. Neural activity was registered over long periods together with the more slowly changing vegetative variables on a Schwarzer direct-writing apparatus with the aid of a special device developed in our laboratory (41).

RESPIRATORY NEURONES

First let us consider the so-called 'respiratory neurones'. The commonly used criterion for this classification is discharge in rhythmic bursts or modulation of their discharge frequency in the rhythm of respiration. Other criteria are reaction to afferent vagal stimulation or, better, inflation of the lungs. There are many different types of respiratory neurones and these have been described by several authors. Some examples from recordings in the dog may illustrate such varieties.

The proper respiratory neurones in a narrow sense show distinct and fairly abruptly beginning and ending discharge bursts, synchronous with only one phase of the respiratory cycle. Until now two constant localizations have been described, where groups of such cells with the same kind of discharge pattern are concentrated, one rostro-lateral to the obex with inspiratory neurones and another more caudal and lateral near the nucleus ambiguous with expiratory neurones (9).

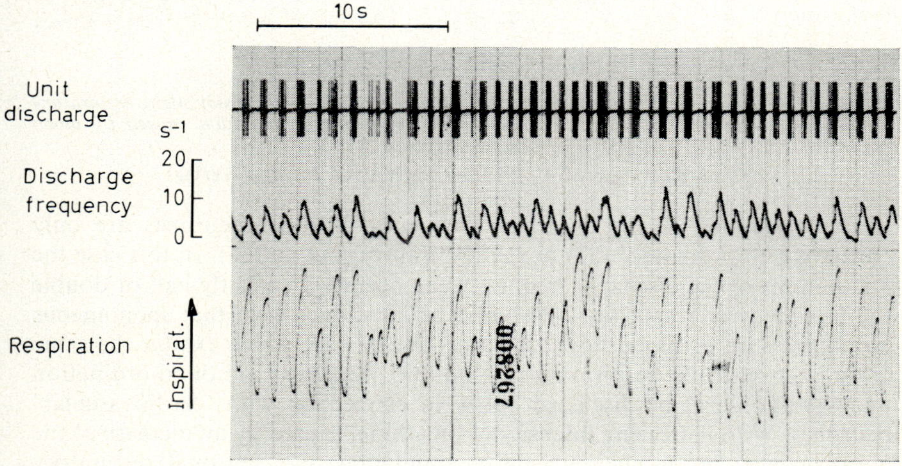

Figure 2. Record of summed activity of several neurones in the inspiratory nucleus near the obex

Figure 2 shows a record from the inspiratory nucleus (11), in which the activity of a few neighbouring neurones is derived with a special technique which allows us to sum the activity of several single units independently of spike amplitude and the distance of the recorded neurones from the electrode tip (41). The comparison of the summed discharge frequency with the various depths of breathing, shows that the intensity of the summed central activity at this site runs strongly parallel with peripheral respiratory movements.

In addition to such neurones with their strong and constant co-ordination with respiratory movements, many other respiratory neurones can be found in the lower brain stem, and these show either correlation to one respiratory phase or other modes of co-ordination with the respiratory rhythm (7, 38, 61). Here all varieties from distinct respiratory groupings of discharges to slight respiratory frequency modulations can be observed.

The degree of frequency variation in the respiratory rhythm can be altered in the same neurone. Such a change from slight respiratory variations of discharge frequency to distinct respiratory bursts is shown in *Figure 3*. The activity of this cell was inhibited by blood pressure increase. With progressive inhibition the respiratory grouping of the spikes becomes more and more marked.

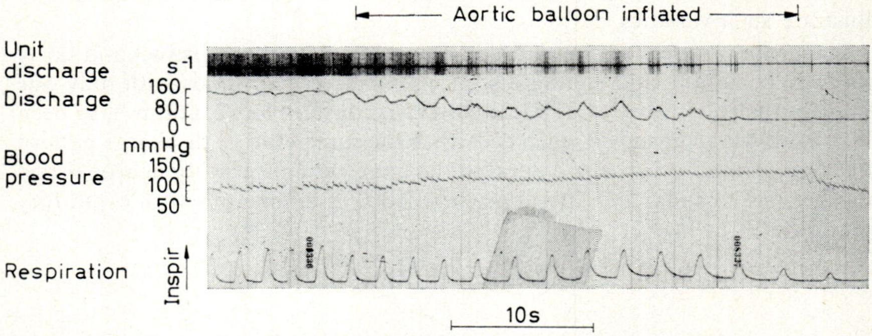

Figure 3. Transition in the discharge pattern of a reticular neurone from slight respiratory frequency modulations to distinct respiratory grouping during inhibition caused by blood pressure increase

(From Koepchen *et al.*, 1969, by courtesy of Springer Verlag)

The variations of the discharge frequency of other neurones are only 'relatively co-ordinated' (32) to the respiratory rhythm (38). In this case the fluctuations of the discharge frequency can occur with exactly half or double the frequency of respiration. It should be mentioned here that spontaneous blood pressure or blood flow variations are known, which exhibit the same co-ordination to the respiratory rhythm (39). Another kind of co-ordination to respiration will be discussed below in connection with 'cardiovascular' neurones. With increasing intensity of breathing, caused by an increase of the alveolar pco_2 or by chemoreceptor stimulation, the neuronal respiratory rhythm 'spreads' in the brain stem and the activity of neurones becomes 'respiratory', whereas previously they had shown other discharge patterns (*see Figure 8*).

CARDIOVASCULAR NEURONES

Much less information exists in the literature about the 'cardiovascular' neurones. The criteria commonly used for their classification as 'cardiovascular' are: variation of the discharge frequency parallel with either spontaneous or experimentally induced blood pressure variations, and, in particular, responses to baroreceptor activation or inactivation. Our own experiments originally started with a search for such cells in the brain stem of the dog. Indeed it was possible to detect many neurones which could be classified as cardiovascular using these criteria. Some characteristic examples of such activities will be shown in the following.

The most striking correlation of neuronal activity to cardiovascular events is the discharge in bursts synchronous with the cardiac rhythm. *Figure 4* shows the summed activity of several cells discharging synchronously with the cardiac rhythm, and recorded with the same technique as in the experiment of *Figure 2*. The frequency record closely follows the variable arterial pressure pulses. In accordance with the results of other authors in cats (28, 68, 79), this close co-ordination to the cardiac rhythm can be found only in a very limited region of the nucleus *tractus solitarii* near the obex. Many other neurones in the brain stem reticular formation, influenced by blood pressure changes via the baroreceptors, show no correlation or only some slight and inconstant correlation to cardiac rhythm (43). Considering the fact that the sympathetic outflow shows a cardiac grouping as has been already found in the first records of sympathetic activity by Adrian *et al.* in 1932 (2), one could suggest that the activity with cardiac grouping descends from the nucleus

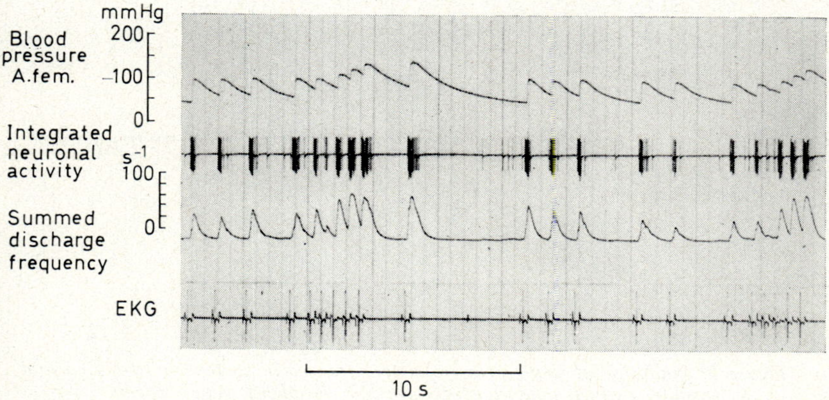

Figure 4. Summed activity of several neurones in the nucleus tractus solitarii with strong co-ordination to the pulse pressure

tractus solitarii directly down to the spinal sympathetic cells and that the cardiac rhythm in the baroreceptor discharge is lost at the reticular cells because of the many other afferents converging on them.

Beside these neurones in the nucleus of the solitary tract receiving directly the activity of baroreceptor fibres, many other neurones exist in the brain stem reticular formation with no cardiac rhythmic activity, but with other correlations to the cardiovascular system. Some of them increase their activity synchronously with spontaneous blood pressure variations. Other neurones are activated or inhibited by experimentally induced blood pressure elevation, in most of our experiments caused by an inflatable balloon in the descending aorta. 70 per cent of these blood pressure dependent neurones (b.p.d.n.) decrease their discharge frequency if the blood pressure is raised, while in only 30 per cent is the frequency increased. In the majority, whose activity could be recorded before, during and after reversible cooling of the vagi the blood pressure effects were abolished or diminished during the cold blocking period. Thus these influences seem to be mediated by afferent pathways from the cardiovascular system, most probably by the baroreceptor afferents.

More direct evidence for the influence of the baroreceptors on reticular neurones is the response to distension of the carotid sinus wall. *Figure 5* shows an inhibition of single unit activity in the brain stem, caused by bilateral distension of the carotid sinus. It is obvious that the peripheral reflex effects (cardio-deceleration and blood pressure decrease) begin later than the central inhibition. All these latter neurones could be classified as 'cardiovascular' following the usual criteria (63, 68).

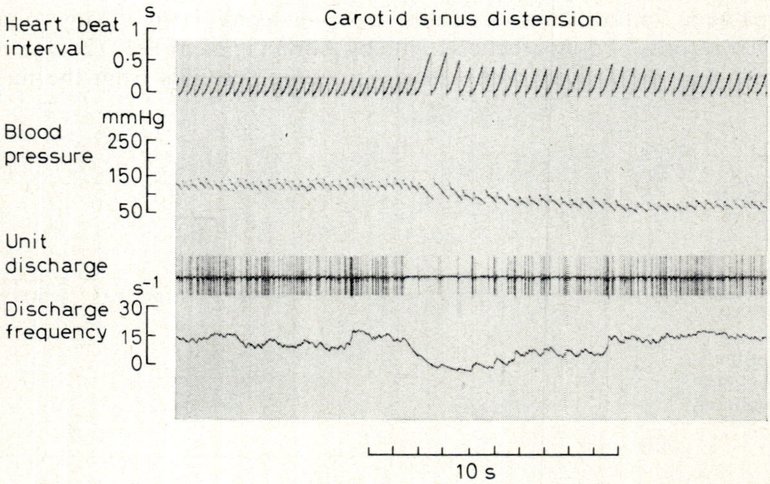

Figure 5. Inhibition of single neurone activity caused by distension of the carotid sinus. Central inhibition precedes the peripheral reflex effects

At this state one could conclude that in addition to the brain stem neurones with many afferent somatic inputs, probably belonging to the ascending reticular activating system (4, 8, 55, 71), elements of the respiratory and cardiovascular centres can also be found in the same regions of the brain stem reticular formation.

DISCRIMINATION BETWEEN RESPIRATORY AND CARDIO-VASCULAR NEURONES

Let us now consider once more the criteria for the discrimination between the different kinds of neurones, first between the cardiovascular and respiratory types. The criterion for respiratory neurones was the grouping of modulation of discharge frequency synchronous with the rhythm of respiration. It should be emphasized, however, that many events synchronous with respiration are known in other systems, and many records of sympathetic activity have shown not only cardiac but also respiratory groupings of discharge (2, 19, 22, 24, 35, 36, 72, 83). That these are correlated to cardiovascular innervation can be shown by sensitive records of the blood flow in isolated perfused vascular areas, connected with the rest of the body only by their nerves (46). In this preparation the vasomotor oscillations of the resistance to flow synchronous with respiration are well marked.

Not only in the efferent cardiovascular innervation but also in the reflex transmission of the cardiovascular baroreceptor reflexes can respiratory variations be shown (48, 67, 77). The cardio-decelerator effect of single shock stimulation of the sinus nerves depends on the respiratory phase. The stimuli are most effective during the expiratory phase, and the effect declines towards the beginning of inspiration. The duration of the inhibition of efferent sympathetic activity caused by sinus nerve stimulation likewise depends on the respiratory phase. The reflex pause in the sympathetic activity is shortest during inspiration and significantly longer during expiration. Moreover, the reflex transmitting structures are located in the same regions of the brain stem, where the various kinds of respiratory neurones can also be found.

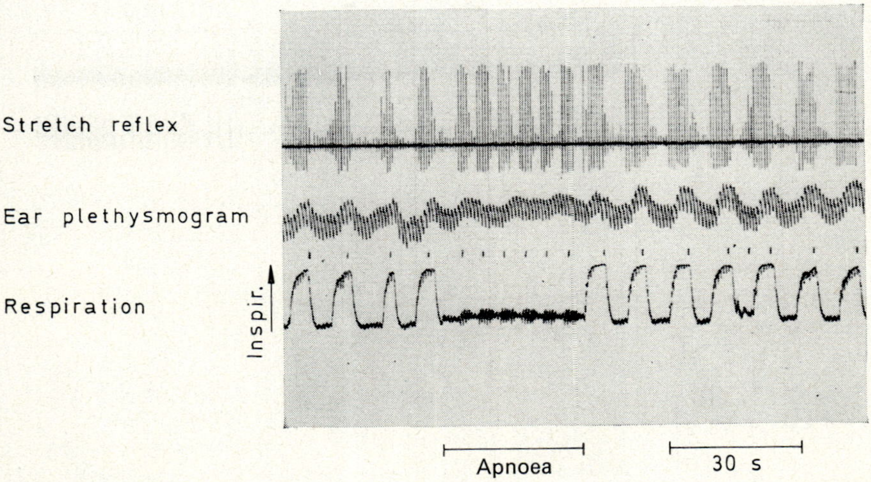

Figure 6. Oscillating amplitude of the knee jerk reflex in resting man. The uppermost record represents the muscular mass action potential of m. quadriceps derived with superficial electrodes during rhythmic stroke on the patellar tendon. Frequency of the reflex release: 92/min. *Beginning and end of the record: voluntarily slowed respiration* 6/min. *Note persistence and acceleration of reflex oscillations during voluntary apnoea to the frequency of previous spontaneous respiration* (12/min)

We also know that variations of motor innervation in the respiratory rhythm occur. We found respiratory variations in the amplitude of the knee jerk reflex in the resting man, which also followed voluntary variations of the respiratory frequency (*see Figure 6*). These oscillations persisted with the original frequency of spontaneous breathing during voluntary apnoea (74). The central nervous neurones exerting such respiratory rhythmic influences on motor innervation must also develop respiratory modulations of activity. Therefore, respiratory rhythmicity cannot be taken as absolute evidence that the function of that neurone is the control of the respiratory system.

On the other hand, the neurones classified as cardiovascular, because of their correlations to blood pressure changes or reactions to baroreceptor afferents, very often show respiratory modulations of the discharge frequency or reactions to lung inflation. Therefore, it is important also to record the respiration in investigations of cardiovascular neurones and vice versa, although this is often neglected in investigations on one specific 'centre'. The discharge frequency before the artificial blood pressure increase varies slightly with respiration, and these respiratory variations are particularly marked in the record of *Figure 3* during the course of inhibition induced by the blood pressure rise. It is known that an increase of blood pressure also can depress the respiration via the baroreceptors (31). In *Figure 7* the course of the inhibition runs closely parallel with depression of respiration and not with the experimentally induced blood pressure increase. Respiratory variations of the discharge frequency are present again.

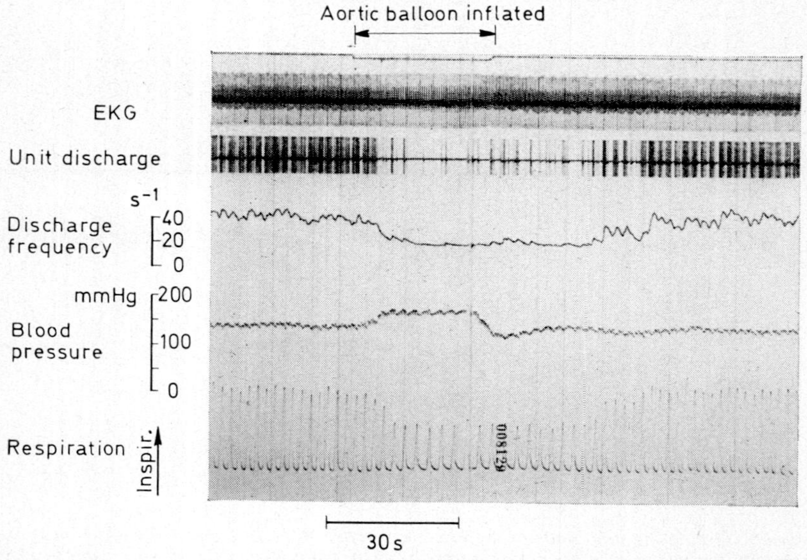

Figure 7. Decrease of discharge frequency of a single neurone induced by blood pressure increase with simultaneous depression of respiration. Note the parallelism between the central inhibition and the depression of respiration

In all such cases which can be observed frequently in the lower brain stem it is questionable if it is justifiable to classify the units as 'cardiovascular' or as 'respiratory'. Furthermore, quantitative criteria for the discrimination between respiratory and cardiovascular neurones (64) seem to be more or less arbitrary, because in our experience the relative amount of reaction to the stimuli can vary depending on the experimental conditions.

INTERACTION OF VEGETATIVE AND SOMATIC AFFERENTS

Having stated the overlapping of the properties of respiratory and cardiovascular neurones our next question was on the relationship between the 'cardiovascular' neurones and brain stem neurones with known multiple

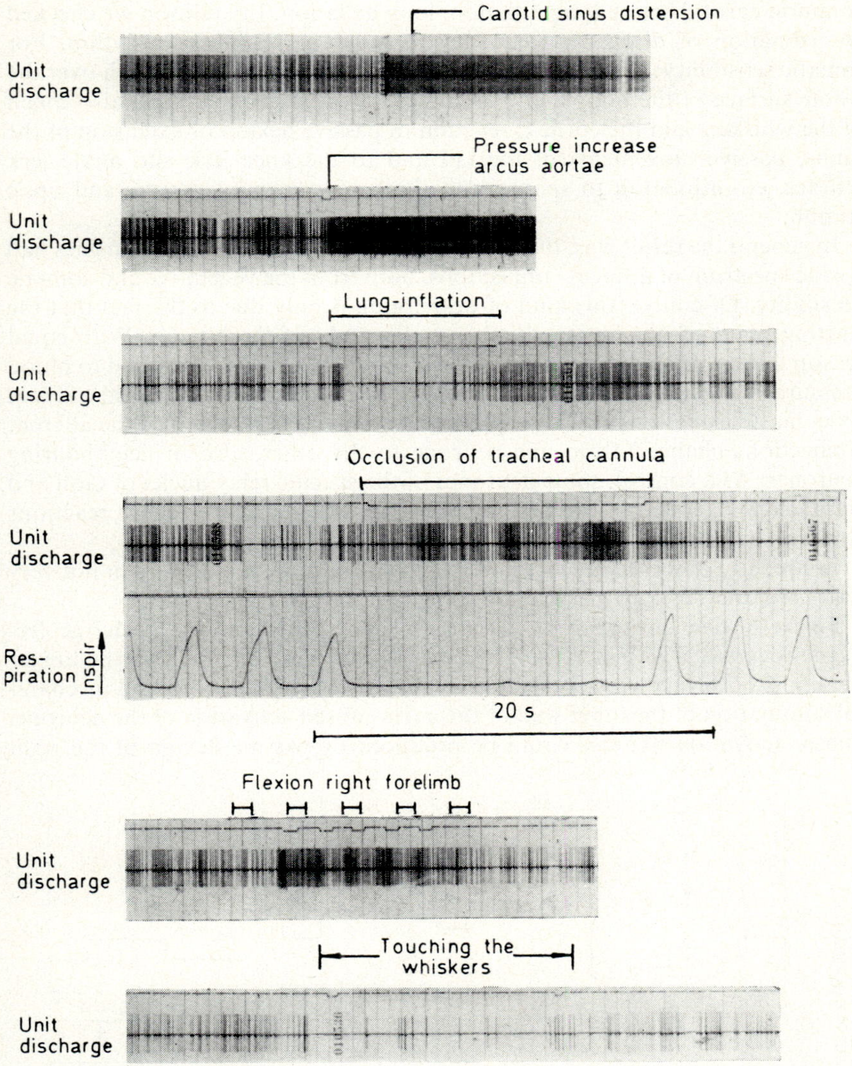

Figure 8. Convergence of various vegetative and somatic afferents on one single neurone in the brain stem reticular formation. In the fourth record from above during occlusion of tracheal cannula three abortive prohibited inspirations can be seen in the respiration record. They are accompanied by discharge bursts in the central activity

(From P. Langhorst, 1968)

somatic afferent connections. For this purpose we tried to record for long periods the activity of one single neurone and to test the reactions to as many kinds of vegetative and superficial and deep somatic stimuli as possible. The following tests were performed: For vegetative sensibility: reaction to blood pressure increase or decrease, venous infusion, stretching of the carotid sinus, chemoreceptor stimulation by injection of small amounts of NaCN into the common carotid artery lung inflation, lung deflation. In addition we checked co-ordination of discharge frequency to pulse or respiratory rhythm. For somatic sensibility: systematic exploration of reactions to local touch over the whole surface of the body and extremities, in the trigeminal area, also touch of the whiskers and the cornea; reaction to passive flexion or extension of the limbs, passive movements of the tail and to the knee jerk and ankle jerk reflexes, co-ordination to spontaneous limb movements, acoustic and optic stimuli.

In general the result was, that many blood pressure dependent neurones had a wide spectrum of afferent connections, both from the vegetative and somatic sensibility. Of course, this kind of description is only due to the fact that the starting point of our investigation was the study of the b.p.d.n. With equal reason one could state that some of the respiratory neurones respond to blood pressure variations or to limb movements, or that many of the multi-afferent reticular neurones respond to vegetative stimuli. The spectrum of the afferent connections changed from one neurone to the other, also in neighbouring neurones. As a control, some neurones of the specific relay nuclei of Goll and Burdach were tested in the same way but in these no convergence reactions could be detected.

Frequently observed reactions were activation by skin receptors in not very restricted areas and by passive movements of one or more limbs.

Figure 8 gives an example of one of these neurones. Its discharge frequency decreased with increasing arterial pressure and this effect was no longer present if both vagi were cooled. Stretching of the carotid sinus and mechanical stimulation of the inner wall of the aorta caused activation of the neurone. Phasic activation likewise could be produced by passive flexion of the right

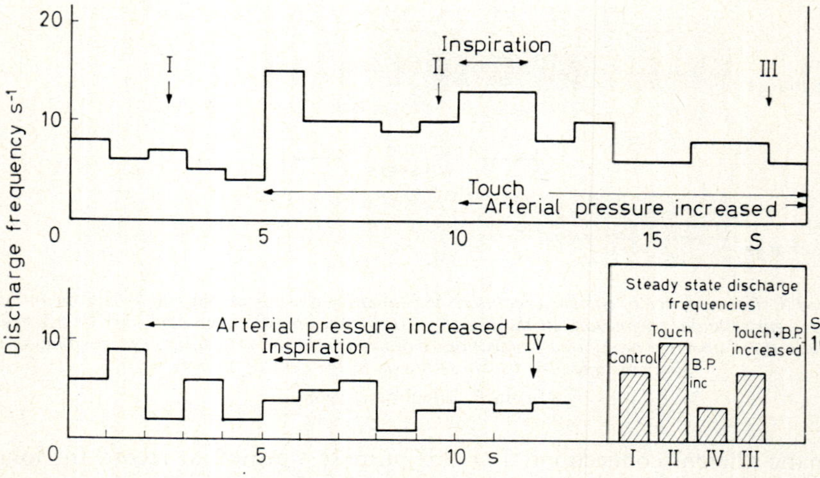

Figure 9. Course of discharge frequency of single unit activity in the lower brain stem during various combinations of excitatory and inhibitory afferent stimulation. Inset: steady state frequencies at the time indicated by the arrows. Note the mutual cancellation of the excitatory effects of touch and the inhibitory effect of blood pressure increase

(From Kuepchen *et al.*, 1969, by courtesy of Springer Verlag)

forelimb and passive movements of the tail. If the tracheal cannula was closed in the expiratory position, forced inspiratory effects caused thereby were accompanied by discharge bursts of the unit. On the other hand, if the lung was inflated, the nervous activity was inhibited. Also tactile stimulation of whiskers or cornea decreased the discharge frequency of the neurone.

The activating and inactivating influences can be added to or subtracted from the spontaneous activity. Acting at the same time they can cancel out each other as is shown in *Figure 9*. The discharge frequency during each second is plotted during application of some conflicting stimuli. Touching of the skin of the back increases the discharge frequency, artificial elevation

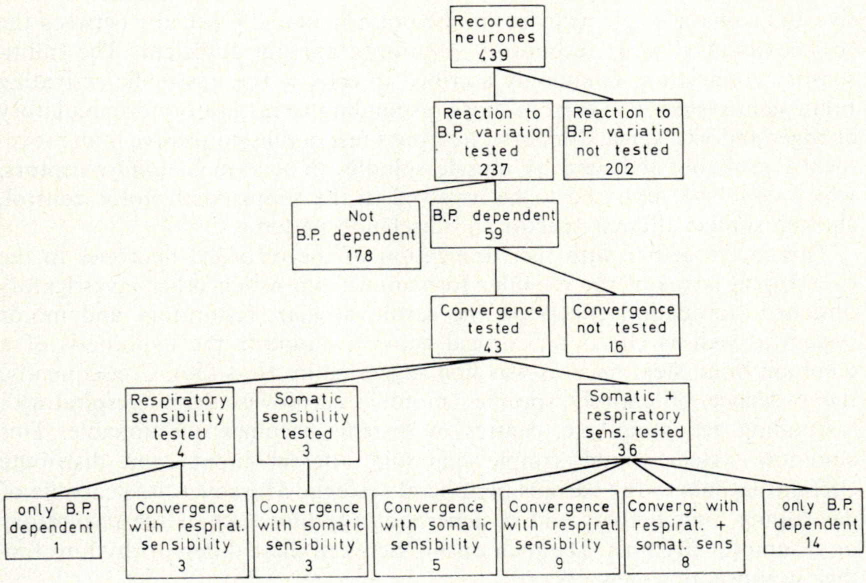

Figure 10. Numerical distribution of the reactions to blood pressure variations and of convergence reactions from vegetative and somatic sensibility in 439 recorded neurones in brain stem of dog
(From P. Langhorst, 1968)

of the arterial pressure causes inhibition. In addition each inspiration is accompanied by an increase of the activity. The comparison of the steady state frequencies (*see inset in Figure 9*) shows that the activating effect of touch is counteracted by the inactivating effect of blood pressure increase. Thus the resulting discharge frequency is not significantly different from the control value.

All the described neurones with multiple vegetative and somatic afferents were localized in the reticular formation outside of the specific nuclei.

Of course, it was impossible to test all afferent connections mentioned above in each individual cell. But considering a large number of microrecordings it is possible to obtain a survey of the numerical distribution of the convergence reactions. This is shown in *Figure 10*; 59 out of 237 neurones reacted to blood pressure variations with decrease or increase of the discharge frequency, i.e.

in the reticular formation of the lower brain stem roughly every third neurone can be influenced by blood pressure changes. From the 43 blood pressure dependent cells whose other afferent connections could be tested, about two-thirds reacted also to somatic or respiratory afferents. Probably this number is still too low, because it is likely that not all possible afferents had been tested in the time available for recording.

CONCLUSIONS ABOUT FUNCTIONAL INTERRELATIONS

Finally some conclusions as to the functional significance of the described findings can be drawn: the usual criteria for discrimination between vegetative and somatic single neurones in the brain stem and especially between the 'cardiovascular' and 'respiratory' neurones are not sufficient. The multisensory connections commonly ascribed to cells of the unspecific activating brain stem system also exist in those responding to respiratory or circulatory changes and vice versa. Similarly, neurones responding to passive limb movements, probably activated by muscle spindles or tendon or joint-receptors, which could be supposed to be involved in the supraspinal motor control, showed similar afferent spectra and discharge patterns.

This fact together with the localization of the recorded neurones in the overlapping areas of the reticular formation from which other investigators obtained stimulation effects on the cardiovascular, respiratory and motor system as well as effects on cortical activity, supports the hypothesis of a common brain stem mass as was first suggested by Hess (30). Consequently, the existence of entirely specific 'motor' 'cardiovascular', 'respiratory', 'ascending activating' etc. centres or systems becomes questionable. This common system would sample manifold afferent inputs and distribute efferent outputs to the various peripheral systems. Moreover, it is capable of developing tonic activity and of generating rhythms such as the respiratory or vasomotor rhythms. The relationship between these different rhythms further supports this view (39, 40).

The question remains open, as to how the various kinds of efferent innervation—e.g. a distinct respiratory rhythm on the one hand, and a somewhat more continous tone for cardiovascular and motor control on the other hand—can be derived from such a common system. Therefore, the whole system cannot be entirely unspecific, and the problem of the degree of specificity remains to be solved by further studies.

Concerning the special problem of the cardiovascular neurones, the relative number responding to blood pressure changes seems to be too high to represent exclusively a bulbar cardiovascular centre. In this connection some other known functions of the baroreceptors should be remembered. Distension of the carotid sinus produces a common inactivation in behaviour (37), cortical activity (13) and motor innervation (59, 75, 81, 84) by way of the baroreceptors. It is very likely that at least some of the described b.p.d.n. represent the central neurophysiological substrate for these effects. Thus the activating influences from many receptors, collected in the ascending reticular activating system are counteracted by the impulse flow from baroreceptors, generated by the ever-present source of energy from the beating heart (39).

The close functional connection between afferent and efferent cardiovascular innervation and the common brain stem system may have yet another functional consequence which is a bit speculative: each general activation of the system leads to an increase of blood pressure, which in turn, by way of the baroreceptors, is fed back into the reticular formation (12, 37). The stabilizing effect produced by this feedback acts slowly because of the sluggishness of the smooth muscle contraction of the blood vessels, which do not respond to rapid changes of vasomotor innervation (62, 78). In this way the cardiovascular system could exert a moderating effect on the central nervous system, counteracting longer lasting general changes of central nervous activity.

DISCUSSION ON KOEPCHEN

Jasper

Dr. Koepchen has pointed out a fundamental problem in our understanding of all these so-called non-specific cells, which have various specific properties but also have many non-specific ones. You started your paper by stating that with current techniques it is impossible to know the motor or efferent functions of these units. Convergence on motor cells from many sensory inputs is the rule—wherever we find them we see a great deal of sensory convergence. One wonders how much the complexity of multisensory input is related to whether the units are on the afferent or the efferent side of the system?

Horn

I wondered if you were able to discriminate different populations of neurones on the basis of differences in the latencies of response to flash stimuli, specific respiratory stimuli, and the cardiovascular stimuli?

Koepchen

There is a difficulty in the precise measurement of latencies between the adequate stimulus in the vegetative periphery and the answer of a neurone in the brain stem: the time between the onset of the stimulus, e.g. distension of the carotid sinus or the lungs, and the first appearance of afferent impulses is not very well defined, not always constant and is long in comparison with the following neuronal transmission. On the other hand, if we stimulate the afferent nerves (e.g. sinus nerve, vagus nerve) they contain mixed fibres from several peripheral systems which we simply want to discriminate. But you are right that it's an important problem which must be solved.

Grey Walter

I wonder what role you think CO_2 plays in mediating all these effects, as a sort of inorganic neurotransmitter?

Koepchen

Yes, the ground level of activity of many of these cells is affected by CO_2. In a number of experiments we have held CO_2 constant by artificial respiration. In other cases we have seen influences from variations of the CO_2-level but this refers to slower changes of central activity. The faster reactions described here related to lung distension, blood pressure changes, etc., are too fast to be caused by the sometimes concomitant changes of the pCO_2.

Jasper

I think this raises a good question—how quick, how phasic can a CO_2 response be?

Grey Walter

Well, a matter of a few seconds.

Jasper

It is possible to have a pretty transient CO_2 response, if your measuring instrument is rapid enough to detect it. The trouble is that often in measuring oxygen or CO_2, the delay in the measuring instrument is greater than the actual effect.

Hernández-Peón

I would like to add a few comments to the model proposed by Dr. Koepchen about feedback which would restrain the activity of the so-called reticular activating neurones through afferent discharges from the so-called stretch-bar receptors. In the past few years physiological evidence supports the idea that the reticular neurones within the brain stem are not all, let's say, activating or arousing neurones, but there are a group of neurones which are involved in the active production of sleep; we call them hypnogenic reticular levels. These neurones are more concentrated in the lower brain stem, particularly in the bulbopontine cordal region than in the rostral mesencephalic region. These neurones receive the afferent discharge from baroreceptors, and so forth. In addition, it has been shown that stimulation of the depressor nerve produced loss of consciousness or slow EEG. So this feedback circuit will have to take into account inhibitory linkage, synaptic linkage between bulbopontine hypnogenic neurones and the encephalic neurones.

Koepchen

This is an important problem in the interpretation of the findings: 75 per cent of the blood pressure dependent neurones decrease their discharge frequency if blood pressure is raised. In contrast the neurones whose activation by baroreceptor afferents causes hypnogenic effects or decrease of muscular tone (75) should be activated by blood pressure increase as do only 25 per cent of our b.p.d.n.

Jasper

I suppose its quite easy to set up a different model and assume that the neurones you are dealing with are tonic-activating neurones and what we think is inhibition is actually deactivation. That would make your 75 per cent do the same job in a sense, wouldn't it?

Horn

Have you found this correlative discharge between the cardiovascular events and respiratory events when recording from neurones in other parts of the brain? One of the difficulties of recording from single neurones, especially in the cortex, is the problem of damping down cardiovascular and respiratory pulsations. Certainly I have seen units in the cortex where there is a clear correlation between either respiratory or cardiovascular pulsations.

Koepchen

We have looked for pulsating activity in the brain stem and we have found it only in this very restricted region at this solitary tract, and nowhere outside this region. To avoid pulsatile or respiratory mechanical artefacts we use a special device (87): the electrodes are introduced by a microdrive and if active sites are found, the electrode is disconnected from its holder by electrically melting a drop of solid paraffin. Then the microelectrode stands free in the brain tissue and pulsates together with it and not against it.

Jasper

The question raised is whether there is any system here that's non-specific or whether it's an illusion due to the multiple interconnection between the specific units, or whether the cells are specialized to be non-specific!

REFERENCES

1. Achard, O. and Bucher, V. M. (1954) *Helv. physiol. pharmac. Acta*, **12**:265
2. Adrian, E. D., Bronk, D. W. and Phillips, G. (1932) *J. Physiol.* **74**:115
3. Alexander, R. S. (1946) *J. Neurophysiol.* **9**:205
4. Amassian, V. E. and Devito, R. V. (1954) *J. Neurophysiol.* **17**:575
5. Amoroso, E. C., Bell, F. R. and Rosenberg, H. (1954) *J. Physiol. Lond.* **126**:86
6. Bach, L. M. N. (1952) *Am. J. Physiol.* **171**:417
7. Baumgarten, R. v. (1956) *Pflügers Arch. ges. Physiol.* **262**:573
8. Baumgarten, R. v. and Mollica, A. (1954) *Pflügers Arch. ges. Physiol.* **259**:79
9. Baumgarten, R. v. and Schaefer, K. P. (1957) *Pflügers Arch. ges. Physiol.* **264**:217
10. Baumgarten, R. v. and Kanzow, E. (1958) *Archs ital. Biol.* **95**:361
11. Baumgarten, R. v., Balthasar, K. and Koepchen, H. P. (1960) *Pflügers Archs. ges. Physiol.* **270**:504
12. Baust, W. and Heinemann, H. (1967) *Expl. Brain Res.* **3**:12
13. Bonvallet, M., Dell, P. and Hiebel, G. (1954) *Electroenceph. clin. Neurophysiol.* **6**:119
14. Breckenridge, C. G. and Hoff, H. E. (1960) *Am. J. Physiol.* **160**:385
15. Brookhart, J. M. (1940) *Am. J. Physiol.* **129**:709
16. Cohen, M. J. and Wang, S. C. (1959) *J. Neurophysiol.* **22**:33
17. Dirken, M. N. J. and Woldring, S. (1951) *J. Neurophysiol.* **14**:211
18. Dittmar, C. (1873) *Math. phys. Cl.* **25**:449
19. Fischer, T., Raule, W. and Seraphim, R. (1955) *Pflügers Arch. ges. Physiol.* **262**:72
20. Flourens, P. (1851) *C.r. hebd. Séanc. Acad. Sci., Paris*, **33**:437
21. French, J. W. (1960) *Handbook of Physiol. Sect. 1 Neurophysiol. II*, p. 1281, Washington: Am. Physiol. Soc.

22. Gernandt, B. E., Liljestrand, G. and Zottermann, J. (1946) *Acta physiol. scand.* **11**:230
23. Gernandt, B. E. and Thulin, C. H. (1955) *J. Neurophysiol.* **18**:113
24. Govaerts, J. (1936) *Archs int. Med. exp.* **11**:629
25. Granit, R. and Kaada, B. R. (1953) *Acta physiol. scand.* **27**:130
26. Hagbarth, K. E. and Kerr, P. J. B. (1954) *J. Neurophysiol.* **17**:295
27. Hagbarth, K. E. and Fex, J. (1959) *J. Neurophysiol.* **22**:321
28. Hellner, K. and Baumgarten, R. v. (1961) *Pflügers Arch. ges. Physiol.* **273**:223
29. Hernándes-Peón, R. and Hagbarth, K. E. (1955) *J. Neurophysiol.* **18**:44
30. Hess, W. R. (1938) *Das Zwischenhirn und die Regulation von Kreislauf und Atmung.* Leipzig: Thieme
31. Heymans, C. and Neil, E. (1958) *Reflexogenic Areas of the Cardiovascular System.* London: Churchill
32. Holst, E. v. (1939) *Ergebn. Physiol.* **42**:228
33. Hukuhara, T., Sumi, T. and Okada, H. (1952) *Jap. J. Physiol.* **3**:138
34. Hukuhara, J., Nakayama, S. and Okada, H. (1954) *Jap. J. Physiol.* **4**:145
35. Iggo, A. and Voigt, M. (1960) *J. Physiol., Lond.* **150**:114
36. Joels, N. and Samueloff, M. (1956) *J. Physiol., Lond.* **133**:360
37. Koch, E. (1932) *Klin. Wschr.* **11**:225
38. Koepchen, H. P. (1961) *Excerpta Med. Int. Congr. Ser.* **37**:9
39. Koepchen, H. P. (1962) *Die Blutdruckrhythmik.* Darmstadt: Steinkopff
40. Koepchen, H. P. (1962) *Excerpta Med. Int. Congr. Ser.* **47**:44
41. Koepchen, H. P. and Heinich, L. (1964) *Pflügers Arch. ges. Physiol.* **280**:92
42. Koepchen, H. P., Langhorst, P., Seller, H. and Polster, J. (1961) *Pflügers Arch. ges. Physiol.* **274**:53
43. Koepchen, H. P., Langhorst, P., Seller, H. Polster, J. and Wagner, P. H. (1967) *Pflügers Arch. ges. Physiol.* **294**:40
44. Koepchen, H. P., Langhorst, P., Seller, H. and Polster, J. (in press) *Pflügers Arch. ges. Physiol.*
45. Koepchen, H. P., Polster, J. and Seller, H. (in press) *Pflügers Arch. ges. Physiol.*
46. Koepchen, H. P., Seller, H. and Langhorst, P. (1946) *Pflügers Arch. ges. Physiol.* **281**:51
47. Koepchen, H. P., Seller, H. and Langhorst, P. (in press) *Pflügers Arch. ges. Physiol.*
48. Koepchen, H. P., Wagner, P. H. and Lux, H. D. (1961) *Pflügers Arch. ges. Physiol.* **273**:443
49. Kovalev, G. V. and Bondarev, M. G. (1963) *Fedn. Proc.* **22**:T447
49a. Langhorst, P. (1968) *Dissertation.* 'Untersuchungen neuronaler Hirnstammaktivität zur Frage der Spezifität oder Unspezifität der Kreislaufzentrum.' Göttingen
50. LeGallois, C. J. J. (1812) *Expériences sur le principe de la vie, notamment sur celui des mouvements du coeur, et sur le siège de ce principe.* Paris: D'Haistel
51. Lindgren, P. (1961) *Circulat. Res.* **9**:250
52. Lumsden, T. (1923) *J. Physiol. Lond.* **57**:153 and 354

53. Magoun, H. W. and Rhines, R. (1946) *J. Neurophysiol.* **9**:165
54. Marckwald, M. (1887) *Z. Biol.* **23**:149
55. Mollica, A., Moruzzi, G. and Naquet, R. (1953) *Electroenceph. clin. Neurophysiol.* **5**:571
56. Moruzzi, G. and Magoun, H. W. (1949) *Electroenceph. clin. Neurophysiol.* **1**:455
57. Nelson, J. R. (1959) *J. Neurophysiol.* **22**:590
58. Owsjannikow, P. (1871) *Math. phys. Cl.* **23**:135
59. Pinotti, O. and Granata, L. (1955) *Archs. Sci. biol.* (*Bologna*), **39**:59
60. Pitts, R. F., Magoun, H. W. and Ranson, S. W. (1939) *Am. J. Physiol.* **126**:673
61. Polster, J. and Koepchen, H. P. (1962) *Excerpta Med. Int.* **48**:169
62. Polster, J., Seller, H., Langhorst, P. and Koepchen, H. P. (1967) *Pflügers Arch. ges. Physiol.* **296**:95
63. Preobrazhenskii, N. N. (1966) *Fedn. Proc.* **25**:T18
64. Przybyla, A. and Wang, S. C. (1967) *J. Neurophysiol.* **30**:645
65. Ranson, S. W. and Billingsley, P. R. (1916) *Am. J. Physiol.* **41**:85
66. Rhines, R. and Magoun, H. W. (1946) *J. Neurophysiol.* **9**:219
67. Richter, D., Seller, H. and Koepchen, H. P. (1967) *Pflügers Arch. ges. Physiol.*, **297**:R82
68. Salmoiraghi, G. C. (1962) *J. Neurophysiol.* **25**:132
69. Salmoiraghi, G. C. and Baumgarten, R. v. (1961) *J. Neurophysiol.* **24**:203
70. Salmoiraghi, G. C. and Burns, B. D. (1960) *J. Neurophysiol.* **23**:2
71. Scheibel, M. E., Scheibel, A. B., Mollica, A. and Moruzzi, G. (1935) *J. Neurophysiol.* **18**:310
72. Scherrer, H. (1962) *Acta neuroveg.* **23**:499
73. Scherrer, H. and Hernández-Peón, R. (1955) *Fedn. Proc.* **14**:132
74. Schmidt-Vanderheyden, W. and Koepchen, H. P. (1966) *Pflügers Arch. ges. Physiol.* **291**:R7
75. Schulte, F. J., Henatsch, H. D. and Busch, G. (1959) *Pflügers Arch. ges. Physiol.* **269**:248
76. Scott, J. M. D. and Roberts, F. (1924) *J. Physiol., Lond.* **58**:168
77. Seller, H., Koepchen, H. P. and Langhorst, P. (1965) *Pflügers Arch. ges. Physiol.* **283**:R58
78. Seller, H., Langhorst, P., Polster, J. and Koepchen, H. P. (1967) *Pflügers Arch. ges. Physiol.* **296**:110
79. Smith, R. S. and Pearce, J. W. (1961) *Can. J. Biochem.* **39**:933
80. Sprague, J. M. and Chambers, W. W. (1952) *Am. J. Physiol.* **176**:52
81. Spychala, V. (1933) *Verh. Dt. Ges. kreis/Forsch.* **6**:140
82. Stella, G. (1937) *Archs. int. Pharmacodyn.* **57**:349
83. Tang, P. C., Maire, F. W. and Amassian, V. E. (1957) *Am. J. Physiol.* **191**:218
84. Tournade, A. and Malmejac, J. (1929) *C.r. Soc. Biol. Paris,* **100**:708
85. Vassella, F. (1961) *Helv. Physiol. Pharmac. Acta,* **19**:166
86. Wang, S. C. and Ranson, S. W. (1939) *J. comp. Neurol.* **71**:437

Discussion

87. Koepchen, H. P. and Langhorst, P. (1967) *Pflügers Arch. ges. Physiol.* **294**:65

The Concept of Attention and the Electroencephalographic Alpha Rhythm

Thomas Mulholland
Perception Laboratory, Veterans Administration Hospital, Bedford, Mass.

Currently the concept of attention is widely utilized as an explanatory concept in electroencephalography. Attention is 'arrested' and 'interrupted'; there is 'wandering attention' and 'flagging attention'; subjects 'alert-to-attention'. Attention is 'directed', 'focused', and 'attracted'. The fact that many of these descriptions are metaphors for ocular and oculomotor functions is obvious. Many contemporary examples of the utilization of an attention concept could be presented in connection with studies of the EEG which do not differ significantly from the early interpretations of Berger (4, 5, 6), Adrian and Matthews (2) and Durup and Fessard (11). There is no need to review them here and a representative example may be cited from Lindsley's comprehensive review (19). In his schema, alert attentiveness, with selective attention, concentration and anticipation, is associated with a partially synchronized, mainly fast, low-amplitude EEG. Relaxed wakefulness, where attention wanders, not forced, is associated with optimal occipital alpha rhythm. Recently, attention has become part of the official definition of the alpha rhythm put forward by the International Federation for Electroencephalography and Clinical Neurophysiology (39)—'alpha rhythm: rhythm, usually with frequency 8–13 Hz in adults, most prominent in the posterior areas, present most markedly when eyes are closed and attenuated during attention, especially visual'.

Despite wide acceptance, evidence has accumulated over the years which is inconsistent with the hypothesis that occipital alpha is always associated with reduced visual attention. For instance, occipital alpha can occur with eyes open while viewing a visual pattern (11, 41); while viewing a moving stimulus (30). Increased occurrence of alpha has been reported while responding to visual signals (23); during prolonged viewing of the visual field (13, 43); during mental arithmetical calculation (22, 40, 42); while concentrating and experiencing visual imagery (33).

Attention, oculomotor and ocular functions and occipital alpha: both Adrian and Matthews (2), Durup and Fessard (11) and Adrian (1) suggested that ocular and oculomotor functions were involved in alpha occurrence and non-occurrence. Adrian (1) noted that the alpha does not occur during attention and fixation, and appeared when attention and fixation were lost. Durup and Fessard's description of the changes in alpha occurrence concomitant with changes in the fixation, convergence and accommodation illustrate this point. The results reported previously by Mulholland and Evans (27) for a subject who controlled his alpha rhythm by changing accommodation, fixation, and convergence and Oswald's (34) note that alpha occurs during intense auditory attention when accompanied by a loss of ocular fixation and accommodation are further evidence that oculomotor functions are involved in the alpha-activation cycle.

Mulholland and Evans (26, 27) and Dewan (9, 10), observed that the occipital alpha rhythm is facilitated when the eyes are elevated as much as possible, a manoeuvre which is uncomfortable and which requires continuing effort. Though not all subjects exhibit this effect, it is reliable in those who do.

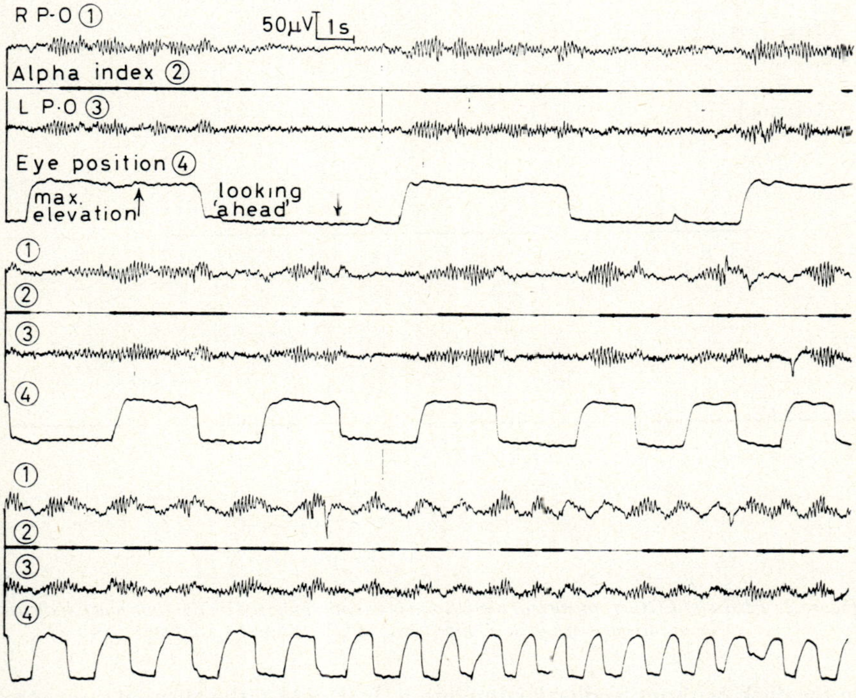

Figure 1. EEG and EOG recordings. Tracings #1 and #3 are P-O EEG. Tracing #4 is from super and suborbital electrodes on left eye. Tracing #2 is an auditory stimulus marker superimposed on a 1 s *time mark. The auditory stimulus occurred in a controlled relationship with the EEG* (*see text*). *The three sets of tracings are excerpts from a continuous recording on T.M.*

Moreover, as Mulholland and Evans have shown, this facilitation can occur whether the subject has his eyes open or closed, whether receiving visual stimulation or not, and when inspecting and reporting on a visual after-image. The facilitation of the alpha rhythm by elevation of the eyes in some subjects has been confirmed by Kris (15, 16), Lippold and Novotny (20) and Fenwick and Walker (this Conference).

In *Figure 1* an example of this effect is shown. The subject (T.M.) was in complete darkness with eyes open. Eye position was recorded using extra ocular electrodes (14). When alpha had been recorded in the right P-O leads for at least 0·5 s, a radio programme was heard over earphones providing high information auditory feedback. As the eyes were raised and lowered with increasing and decreasing rapidity, the close following of the EEG, despite auditory feedback is seen. When eye position is changed faster than 0·8 Hz the record exhibits almost continuous alpha yet rapid alternation of the eyes between elevation and 'ahead' positions requires effort and is distinctly uncomfortable. These results cannot be explained in terms of an attention hypothesis since a high level of subjective attention is required to perform this task. When eye positions were alternating most rapidly, the most alpha occurred. In *Figure 2*, similar results for a naïve subject are shown. Eyes were open

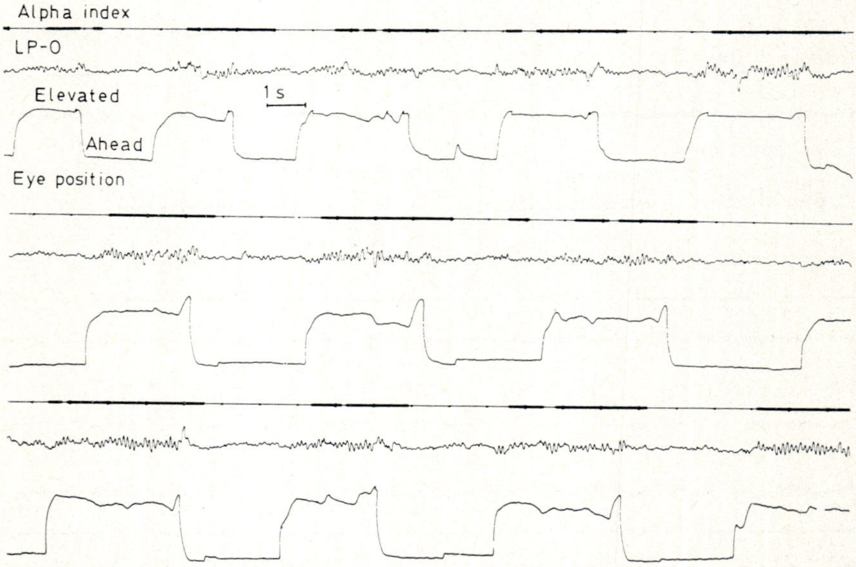

Figure 2. EEG and EOG from normal man. Solid bars superimposed on the time mark indicate the occurrence of alpha. Calibration: 20 μV/cm *and* 1 s/2·5 cm

in the dark, without feedback stimulation. In *Figure 3*, the effect of eye elevation on the EEG is shown. The visual stimulus was a very bright flickering light which was presented when alpha occurred. Despite the visual stimulus, the occurrence of alpha followed the eye position.

In another experiment on myself, the hypothesis that the changes in the occurrence of EEG alpha followed changes in eye position was tested. The correlation between the time the eye was in one position and the duration of the EEG alpha or activation (low voltage fast activity or alpha < 10 μV) which was initiated at the same time or after the eye position changed, was evaluated. The analysis showed that the following of the eye elevation by EEG alpha occurred for the following conditions: eyes open in the dark; eyes closed in the dark; eyes open in the dark and auditory feedback associated with alpha; eyes closed and visual feedback associated with alpha. Following of eye position by the EEG alpha was seen both when eye position was changed in response to an auditory click signal as well as when the shifts were made voluntarily without a signal. Both with eyes ahead and open in the dark, or eyes closed the record was often nearly continuous alpha. Alpha never occurred less during eye elevation than during eyes ahead. In my experience, drowsiness and repetition of the oculomotor exercise reduces the effect, since alpha rhythm occurs increasingly with eyes ahead.

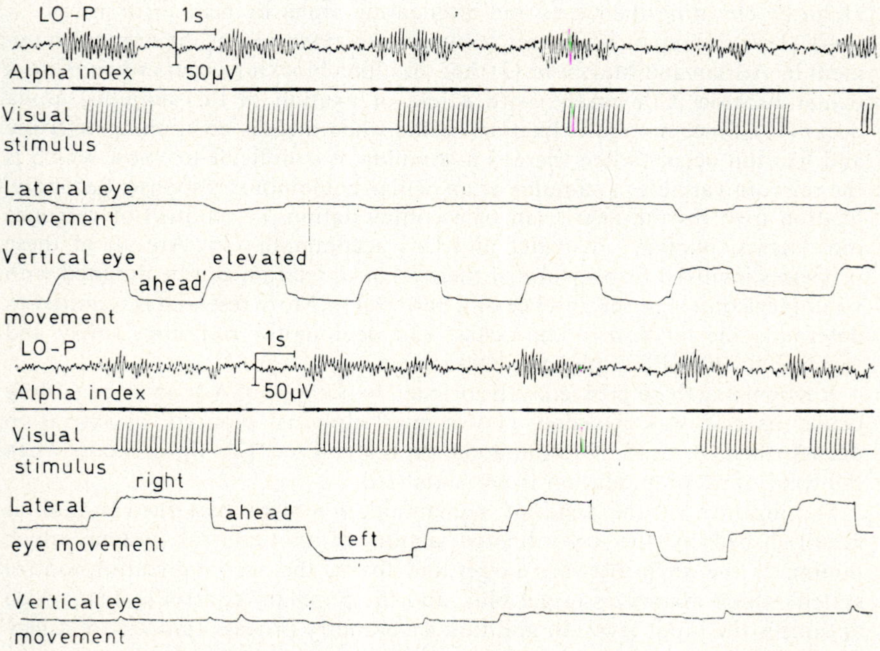

Figure 3. EEG and EOG recordings from T.M. Visual stimulation was by Grass Stimulator, maximum intensity 4 ft *from closed eyes. Solid bars superimposed on the time marker show the occurrence of alpha as determined by the feedback loop. The visual stimulus occurred in a controlled relationship to the EEG alpha rhythm. The eye positions shown were extreme deviations. Subject T.M.*

Lippold and Novotny (20) have reported that lateral movements of the eyes are related to alpha amplitude modulation; we have noted that extreme left or right or downward deviations can facilitate alpha, *see* bottom half, *Figure 3*; (Fenwick and Walker, this Conference). Lombroso and Duffy (21) at a Children's Hospital in Boston have found that a burst of synchronous activity having a frequency of alpha occurs about 200 ms after the onset of a fast eye movement. This synchronous activity is time-locked to the eye movement. The subject was sweeping his open eyes L to R, R to L while looking at a white surface. However, if the eye movement is made in the dark there is no evidence for time-locking of the synchronous activity to the eye movement. As the number of eye movements increase in an experimental session, this synchronous activity does not occur in such a definite relation to the eye movement. These are important results and indicate that research on this topic is likely to be fruitful.

In a demonstration by Dewan, subjects were trained to send Morse code by producing long and short alpha bursts which were identified by a computer. By auditory feedback of the EEG and the message sent, the subjects could appraise their own performance. In addition they had to remember the message to be sent, what had been sent, etc. Clearly this task demanded attention. Yet subjects were able to send code messages in this way, producing alpha by elevating their eyes and attenuating alpha by near fixation (10).

All these discrepancies led me to suggest an extension of the original statement by Adrian and Matthews (2) that the alpha blocking occurs when there is ocular fixation. Conversely, with a loss of fixation of the stimulus, alpha occurs (1). Since no ocular fixation occurs when there is no stimulus to fixate and fixation occurs when there is a stimulus, it is difficult to know which is the relevant variable, a stimulus or an ocular-oculomotor response. Secondly, fixation involves the near triad of accommodation, i.e. control of vergence movements, pupillary diameter and lens accommodation. Are all of these processes involved or only one of them? Third, fixation can be initiated with voluntary control in the absence of a pattern (1). More research is required to determine the relation of the ocular and oculomotor responses singly and together to the EEG alpha-activation cycle.

Recognizing these problems, the original hypothesis of Adrian (1) regarding ocular fixation was extended: (1) Occipital alpha is 'blocked' or activation occurs when the triad of accommodation is stabilized. (2) Alpha occurs when the triad of accommodation is not stabilized.

Stabilization of the triad of accommodation with patterned stimuli is accomplished by the co-ordinated action of the control system which minimizes the angle between target and fovea, the accommodation control system which minimizes target blur, and the pupillary control system which optimizes the input level. In addition a voluntary process can initiate a new fixation (38).

In the dark or in a homogeneous stimulus field or Ganzfeldt, the triad of accommodation can be momentarily stabilized by the voluntary process. However, this system may not require feedback from proprioceptors. A command function could be given which results in the adjustment of eye position, lens accommodation and pupillary diameter on the basis of a pre-computed expected triad of accommodation which would briefly stabilize the eye position (38). Such a system would be unstable over a longer time without target feedback and alpha would occur more often. Finally, it is hypothesized that the coupling between alpha occurrence and stabilization of the near triad of accommodation includes a stochastic component. These hypotheses are consistent with observations made early in the history of EEG and can be used to interpret the greater number of experimental results which are not consistent with the attention hypothesis.

In all studies of humans where visual attention has been cited as the cause of alpha attenuation (activation) it is likely that ocular fixation and accommodation to the stimulus occurred. Since fixation and accommodation involves a feedback control system requiring a target, then with eyes closed or open in the dark, or while viewing stimuli which have minimal contour or which fill the visual field, the near triad of accommodation is less stable, and more alpha occurs. Those subjects who can lose fixation and achieve a blurred, indistinct visual image can voluntarily produce more or less occipital alpha (10, 11, 27). When the subject viewed moving stimuli, trying to perceive illusions (30) he may have been voluntarily changing the stability of the triad of accommodation while trying to change his perceptions, producing more or less alpha.

When transitory occipital alpha attenuation occurs after auditory or tactile stimulation, it may be caused by a transitory ocular fixation and accommodation, i.e. a stabilization of the triad of accommodation which is a component of the orienting response (37). The rapid habituation of the EEG activation response to auditory stimulation shows that in studies of this problem visual stimuli are usually homogeneous and of low intensity with obscure contours, in order to avoid distraction or uncontrolled response to visual stimuli. Under these conditions a stable near triad of accommodation is less likely and the brief EEG activation responses and rapid habituation reflect the fact that ocular fixation following auditory stimulation is transitory, i.e., the triad of accommodation is unstable, and alpha occurs more often. A similar explanation can be given for alpha 'blocking' during mental imagery, mental calculation and visualization. Simultaneous ocular and oculomotor responses need to be recorded before the relevance of 'attention' in these cases is assumed.

In some subjects, when the eyes are elevated or deviated to extreme lateral positions there is a loss of fixation, and perhaps a relaxation of accommodation, i.e. the triad of accommodation may become unstable which thus produces more alpha. Voluntary control of the EEG which has been reported (10, 12, 27, 35) may have been accomplished by voluntary control of ocular and oculomotor functions.

Synchronous activity having the frequency of the alpha, associated with fast eye movement (20, 21) may be related to the suppression of vision just before, during and after a voluntary saccadic movement (18). With target information suppressed, the fixation and accommodation control system is deprived of an error signal, the triad of accommodation becomes unstable, and alpha occurs. When the eye comes to rest with fixation, alpha is less. As the experiment continues, the alpha begins to occur when the eye is not moving, which may mean that the triad of accommodation is becoming unstable, i.e. there is a loss of fixation and accommodation, even when the eye is not moving.

The occurrence of alpha rhythm in cases of binocular enucleation poses problems for both the visual attention and the ocular-oculomotor hypotheses. Adrian and Matthews (2) reported that a case of binocular enucleation showed no alpha rhythm. Later, Callahan and Redlick (8) reported cases of binocular enucleation from which occipital alpha could be recorded. Recently we recorded alpha from a recovered psychotic person having a traumatic self-produced binocular enucleation with no associated neural or brain disease. Recordings were made with glass eyes in place and removed. In addition he was tested with auditory stimulation. Occipital alpha was recorded under all conditions, increasing as the experiment continued. Brief alpha blocking in response to auditory stimulation occurred. At present we have no useful interpretation of these results. The fact that some sighted subjects do not have alpha recordable from scalp electrodes is also a problem for both hypotheses.

In sum, the hypothesis that alpha occurs when there is a loss of fixation and accommodation and alpha is blocked, i.e. the EEG exhibits low amplitude fast activity when fixation and accommodation occurs, seems consistent with more of the experimental results on the occipital alpha rhythm than is the hypothesis that the occurrence of alpha depends on visual attention. Both of these hypotheses were given by Adrian (1) but the evidence accumulated since then indicates to me that the attention hypothesis cannot be correct. Considering sighted persons who have recordable alpha, if we accept as valid the fixation-accommodation hypothesis, then a temporal series of occipital alpha and activation durations would be associated with a temporal series of ocular

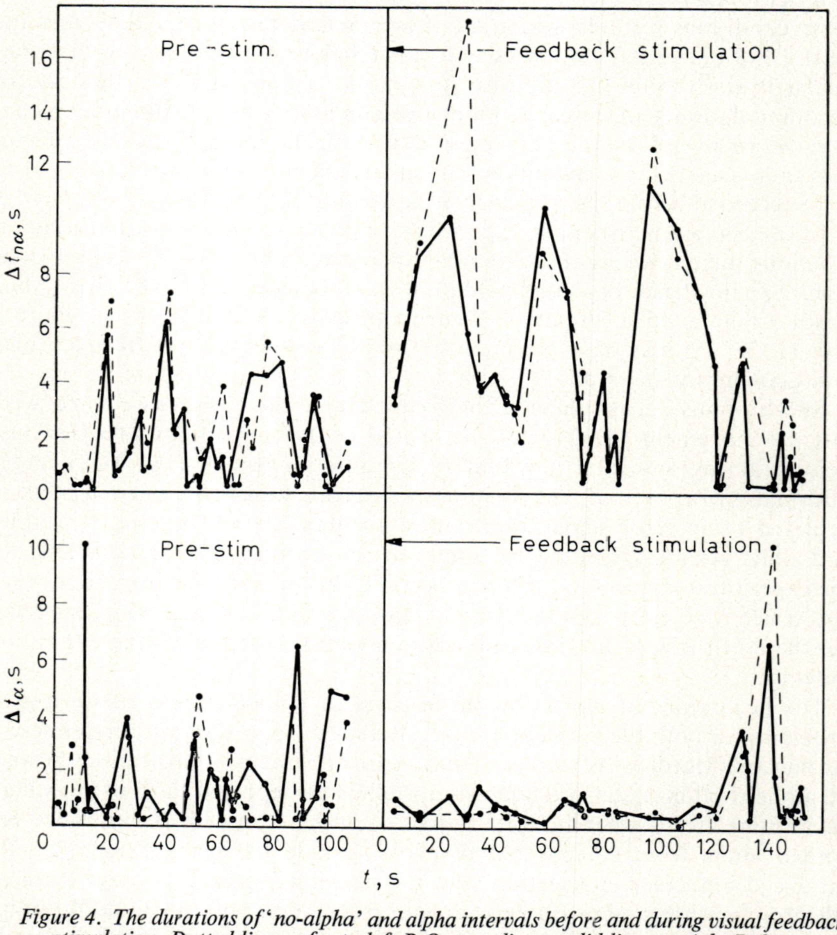

Figure 4. The durations of 'no-alpha' and alpha intervals before and during visual feedback stimulation. Dotted lines refer to left P-O recordings; solid lines to right P-O

fixation and accommodation responses alternating with loss of fixation and accommodation. The next section of this paper presents an interpretation of the alpha-activation cycle as a reflection of the control system which regulates the triad of accommodation.

DYNAMIC RESPONSE OF THE ALPHA-ACTIVATION CYCLE

The shift from occipital alpha to alpha attenuation or activation is a recurring phenomenon, even though stimulation remains constant, as Bagchi reported (3). He termed the alternation between alpha and activation (or less alpha) 'intermittent adaptation'.

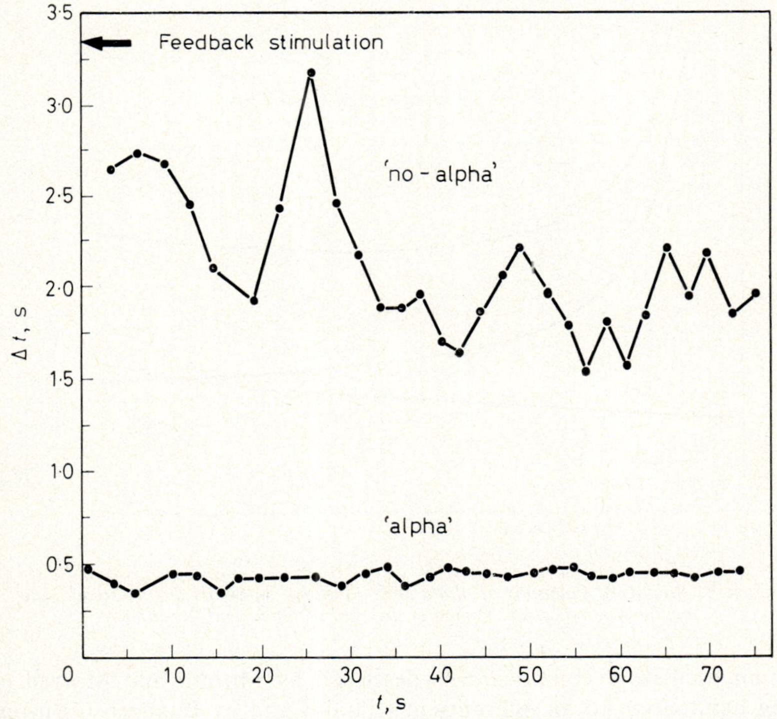

Figure 5. Mean EEG alpha and 'no-alpha' intervals from normal adults during feedback stimulation. Each point is the average of four trials on each of 24 subjects

A number of researchers have noted the non-random fluctuations in the successive durations of EEG activation and of alpha bursts in response to stimuli both during steady stimulation (25), intermittent (24) and feedback (7, 25) stimulation. In many cases an evident regular fluctuation in the period of the alpha-activation cycle is seen from pen recordings both before and after stimulation.

The period of the alpha-activation cycle can be decomposed into the Δt

alpha and Δt 'no-alpha', and though perhaps not reflecting physiologically independent states, it is useful to describe them separately. They may change conversely or inversely to each other; they may generate different time series. In general, a step change in stimulation is followed by an initial increase in the period of the alpha-activation cycle followed by a gradual decrease over time to a limit. The Δt alpha is initially short, and gradually increases, the Δt no-alpha is initially long then rapidly decreases to a limit, i.e. the EEG response is habituated. In *Figure 4* a cyclic variation of the period of the alpha-activation cycle occurs during feedback stimulation. In *Figure 5*, even average data

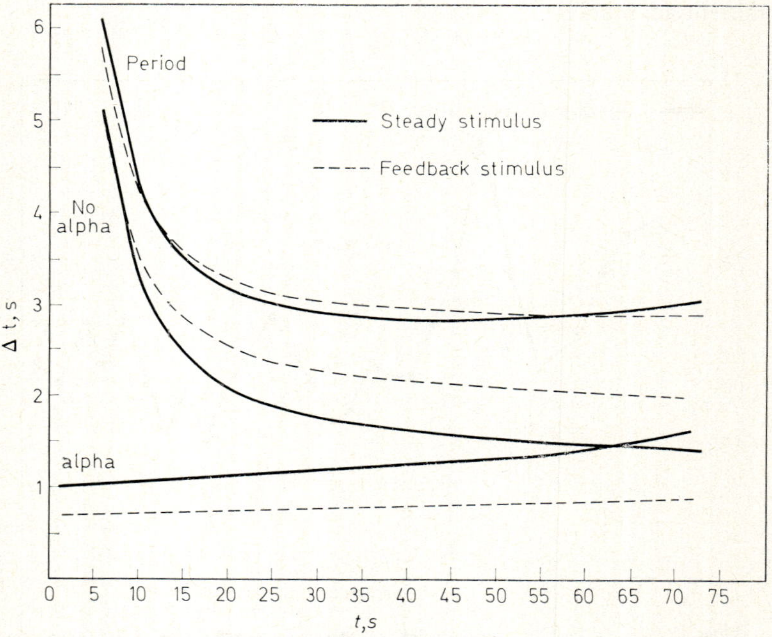

Figure 6. Smoothed functions of alpha and 'no-alpha' intervals during steady and feedback stimulation. Period is the sum of alpha and 'no-alpha'

show an oscillatory component as described by Morrell and Morrell (24) during habituation to an intermittent stimulus and by Bundzen (7) using a feedback stimulus.

Continuous visual stimulation following a step increase and intermittent stimulation, which is controlled by the occurrence of EEG alpha by means of an external feedback path, yield similar EEG changes though the time required for the EEG response to decrease to an asymptote, the period of the alpha-activation cycle at asymptote, etc. can be altered by varying the parameters of the feedback stimulation (*see Figure 6*). The original data presented by Mulholland and Runnals (31) has been fitted with a continuous function, $\Delta t = f(N)$, N being the arbitrarily continuous function of number of successive cycles of alpha and activation. These smooth functions were replotted here as $\Delta t = f\,(t)$ for Δt alpha and Δt activation. The similarity between the effects of a steady, continuing stimulus and of intermittent, feedback stimuli suggested that the external feedback path permitted a modification of feedback, rather than introducing feedback *per se*. This increased feedback could result in an increased stability of the system, whether the change of the stimulus were an increment or decrement. In *Figure 7*, the response of the average alpha-activation cycle to a change from no feedback to feedback is shown. The decreased variability under feedback is evident, even in these average data from a single subject (28).

When the feedback stimulus is an increment or decrement of stimulation from a level which is not zero, changing the sign of the feedback stimulus makes little difference (31). However, if the stimulus changes between zero and some large value, changing the sign of the feedback stimulus may change the stability of the system.

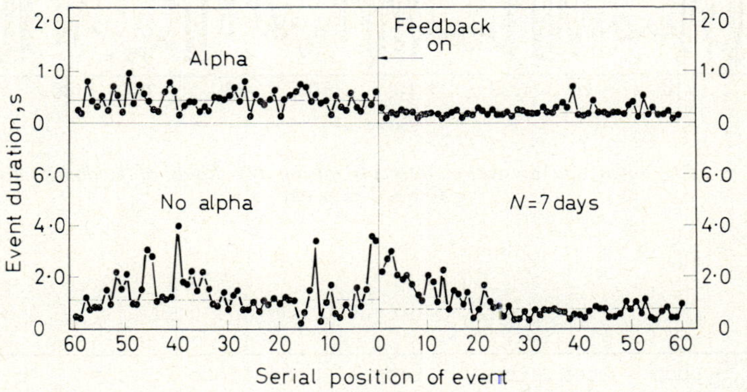

Figure 7. Stabilization of the EEG alpha-activation cycle by feedback

The following examples are taken from a study of children. The stimulus was a silhouette of human figures which had been presented many times previously. The apparatus was so arranged that the presentation of the stimulus occurred either during alpha (Loop 1) or, when activation occurred (Loop 2). The shift from one condition to the other was accomplished by a hand switch. As shown in *Figures 8* and *9*, Loops 1 and 2 were alternated, permitting a comparison of the EEGs under each condition. For these young children, there was an immediate and definite change in the EEG with a change in feedback. The alternation between alpha and activation was unstable for Loop 2 and more stable for Loop 1. The durations of activation were less variable from Loop 1 to Loop 2 compared to the duration of alpha.

A SCHEMA OF THE ALPHA-ACTIVATION SERIES

To describe the EEG cycle of alternation between alpha and activation states, a hypothetical control process is suggested which is continuously oscillating between upper and lower limits. Between these is a threshold. Above threshold, alpha is attenuated or disappears, below it, alpha occurs. To account for the

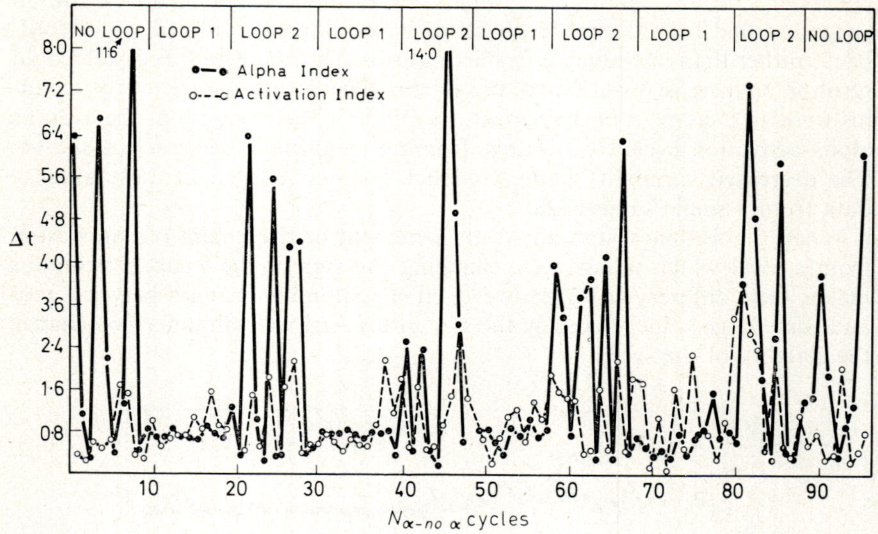

Figure 8. EEG alpha and activation intervals during two kinds of feedback stimulation. Child, 7 year *old*

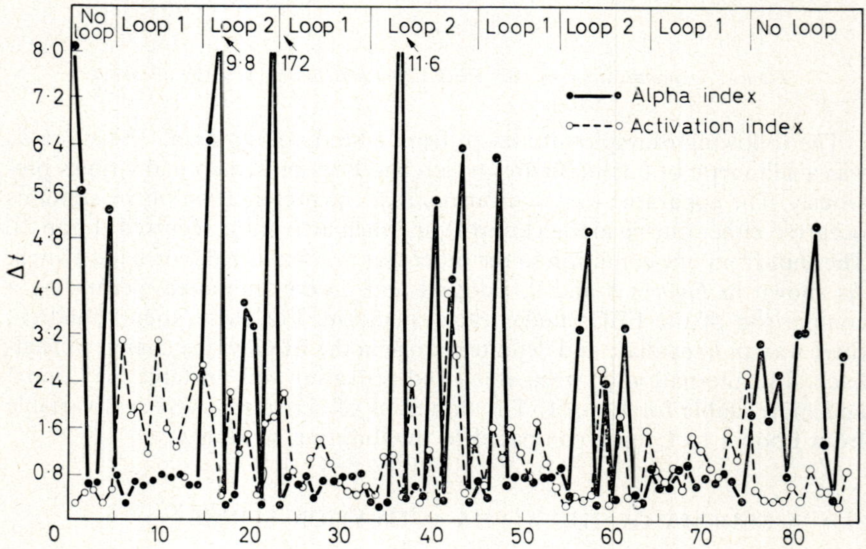

Figure 9. EEG alpha and activation intervals during two kinds of feedback stimulation. 11 year *old boy*

observed periodic variation of the alpha and activation durations, the upper and lower limits have a periodic component added to the exponential function. In *Figure 10* the recovery of the hypothetical system from a step increase in stimulation is shown.

The variations in the alpha-activation cycle which have been observed in experiments are included in this schema: the alternation between alpha and activation during steady stimulation; the initial increase in the period of the alpha-activation cycle with a step change in stimulation followed by an oscillatory decrease in the period of the alpha-activation cycle as stimulation continues; the asymptote period which is approached as stimulation is continued;

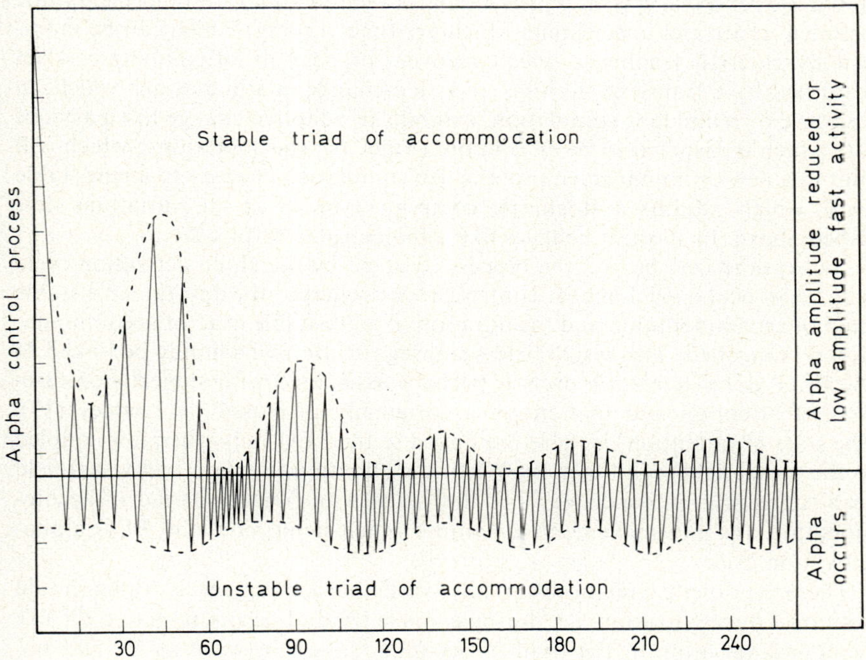

Figure 10. Hypothetical schema describing the temporal changes of Δt alpha and Δt no-alpha during steady visual stimulation

the different rate of change for Δt alpha compared to Δt activation as stimulation continues; the convergence of Δt alpha and Δt no-alpha with repeated stimulation.

It is clear that a transient stimulus or intermittent stimuli will have effects reflecting the dynamics of the system as well as the parameters of the stimulus. If an external feedback path connects the EEG alpha-activation cycle to the stimulus, then the system behaviour will reflect the complex resultant of the external feedback parameters interacting with the dynamics of the system. It is clear that Δt alpha or Δt activation are neither meaningful nor useful measures unless the time series information is preserved.

The concept of latency of EEG activation also requires a new interpretation here. If the alpha-activation system oscillates then the delay between stimulus presented during alpha and the following activation will be a function of the state of the system as well as the stimuli. For instance, if the system is near threshold for activation, a stimulus may be associated with a brief latency of activation. From *Figure 10*, it is obvious that in studies of the habituation of EEG activation, the contingency between the EEG and the stimulus should be controlled, to reduce the contribution of a variable contingency to the variability of the measured EEG response. An external feedback path between EEG and stimulus provides a convenient method for doing this (29).

This line of speculation may be summarized as follows: The alpha-activation cycle reflects a system which oscillates relative to a threshold magnitude between upper and lower limits which are time-varying; which can be modified by external feedback; which recovers to near its original state when disturbed by a transient stimulus or a step-change in stimulation; which, in response to redundant stimulation, exhibits an adaptive change to an asymptote which is assumed to be an optimum value for the conditions; which will shift to a new optimum (asymptote) when stimulation changes to a new stable level; which exhibits a stochastic component in all of the variations described above. In short, it behaves like a biological control system.

As hypothesized before, the process reflected by the alpha-activation cycle may be an ocular-oculomotor control process whereby the duration of a stable triad of accommodation and the duration of an unstable triad of accommodation is controlled. This visual field scanning function which may be linked to the EEG alpha-activation cycle is perhaps associated with subjective awareness of attention and inattention as originally proposed by Lashley (17). The state of 'attention' is therefore linked to the EEG only when it is coupled to the visual field scan function. When the attention function and visual field scan function are dissociated, it is the visual scan function (triad of accommodation and eye movement functions) which determines the EEG alpha-activation cycle.

These hypothetical relationships have verifiable consequences. Alpha should not occur during fixation, i.e. during a stable triad of accommodation (2, 11). Temporal variation in the triad of accommodation between stable and unstable states should be correlated with temporal variation of the occipital EEG between activation and alpha states. The familiar habituation of the EEG activation response (36) should have its counterpart in the time series of successive fixations and accommodation responses. Habituation of the oculomotor component of the orienting response should closely parallel the habituation of the EEG component. Aberration of the process controlling the triad of accommodation should be reflected in aberration in the alpha–activation cycle.

Visual perception functions which vary with changes in the fixation accommodation process should vary with the alpha-activation cycle. In general, alpha occurrence should be associated with a *decrease of visual accuracy, precision, and efficiency* relative to activation. This would *not* be true for non-visual stimuli.

Variations in visual perception which have been linked to attention may be linked to the oscillations in the limits of the alpha-activation control process rather than being linked to alpha-activation *per se* (*see Figure 10*). This would be consistent with the idea that fluctuations in attention are linked to the non-random oscillation in successive EEG activation intervals which have been described by Morrell and Morrell (24) and confirmed by Mulholland (25) and Bundzen (7).

The observations reported earlier which have been accumulating in the EEG literature over the years, as well as the more recent ones described here, pose serious problems for the attention hypothesis. However, they permit no conclusion concerning the origin of the alpha rhythm and might be considered independently of that question. In my view, this is neither a tenable nor useful position. The experimental determination of those brain processes causing the alpha rhythm will permit us both to understand the failure of the attention hypothesis in the interpretation of the occipital alpha rhythm and to explain its relationship to the ocular and oculomotor control systems.

DISCUSSION ON MULHOLLAND

Jasper

Well, the organizers of this meeting, Dr. Evans and Dr. Mulholland have set off a bomb that's upsetting a lot of ideas about the relationship between alpha rhythm and attention. Now, are we to consider it as merely reflecting oculomotor functions, or do you feel that oculomotor functions reflect attention?

Mulholland

If you were able to account for attention phenomena in terms of what the subject is doing, how he's selectively scrutinizing, where he's directing his eyes, and how well stabilized the eye control system is, that would be a major step forward. One reason would be that you wouldn't really need the attention hypothesis with its ambiguities and difficulties. What I'm not saying is that 'attention' is not a useful concept, but I *am* saying that it is probably not as useful as one that can be tested. I would say that to the degree that attention is coupled with oculomotor control, to that degree it is liable to be linked with alpha.

Jasper

Didn't Adrian show a shifting in alpha with a shifting in attention from auditory to visual stimuli in one of his experiments?

Mulholland

Durup and Fessard did that, and noted a lot more alpha with auditory attention compared to visual attention.

Jasper

Well, his interpretation was that it was only visual attention that was related to the occipital alpha, that you can't speak of 'the alpha', that there are many, many alphas, and to identify it with 'one attention' would be a mistake.

This was visual attention and when he shifted attention just as intensely on another sensory modality the occipital alpha returned.

Dewan

In the experiment where you put your visual attention on an after-image that you are not actively fixating—just paying attention to—that should block the alpha rhythm if Lord Adrian was right in saying that it was visual attention.

Jasper

Yes.

Dewan

But what happened to it? Does 'eyes up' make the alpha rhythm stay in spite of visual attention?

Mulholland

Yes. But I wonder, if when a subject is paying attention to auditory stimuli, that he may not relax his accommodation as it becomes background and not central to his interest. In fact he may even close his eyes somewhat which would be an additional problem. So I think that, until we see what happens to the oculomotor system during attention-getting stimuli, it's an open question.

Horn

If I remember the Adrian experiment correctly he wore +10 dioptre lenses because of the great difficulty he had of 'not attending' to a visual stimulus when his eyes were open. The purpose of the lenses was to blur the field. He kept his eyes open when attending to the visual environment and also when attending to the ticking of a watch. So it is entirely conceivable that he did lose fixation.

Jasper

Well, this wasn't inconsistent with his conception of this being a *visual* attending phenomenon, and not an attentive phenomenon generally speaking. You see you're introducing a conception that we are talking about attention as a special reaction of the occipital alpha rhythm to all types of stimuli, which wasn't his interpretation.

Mulholland

That wasn't what I was intending to convey. I was specifically directing myself to the problem that when one is scrutinizing visual stimuli then the alpha is blocked, but some subjects with eyes elevated can receive visual stimuli, unpatterned or with a visual after-image, or even with an intermittent light such as is produced by a strobe, and not block the alpha rhythm.

Grey Walter

When you are dealing with a complex problem or system which is open to a number of interpretations, people are nearly always right in what they assert and wrong in what they deny. In other words I think that all that you, Dr. Mulholland and Dr. Evans, say is true; the effect that you describe exists. But I think it's also true that alpha is blockable in some people by what one might call attention or concentration irrespective of the position of the eyes. For example, in the classical demonstration everyone does in their laboratory, getting a subject to close his eyes and then do sums, you get an alpha block when their eyes are closed and they are thinking.

Mulholland

When the eyes are closed, who is to say that there isn't any relaxation of accommodation, or shift in eye position?

Grey Walter

But they won't be accommodating with their eyes closed; the alpha blocks when they are doing a sum as though they were consciously fixating. You can't fixate on anything very much with your eyes closed.

Mulholland

Right. But you see until the measurements are made you don't really know, because you don't know what the eyes are doing during a mental calculation.

Grey Walter

All I'm saying is that what you assert is true and I think that other facts are also true relating to attention. This is really a general comment on our difficulty in dealing with a complex process of which the physiological nature is obscure. Another point is that I think a great deal of confusion is caused in the literature by talking about 'the alpha rhythm'. Herbert Jasper said this just now, and it cannot be said too often or too clearly or too explicitly; even the most elementary analysis in terms of space or frequency or time domain will show that alpha rhythms are plural and complex in every sense. They occupy a large area of visual and associative cortex in those people with whom they occur, and involve usually three or four or more separate components. We've managed to check this with intracerebral electrodes in the occipital-parietal cortex; in regions which are practically adjacent and almost congruent one finds a variety of alpha rhythms, some of which are blocked by opening and closing eyes, some are not, some are driven by flicker, some are not, some respond in some way to mental activity, some do not. What one sees on the scalp is a spatial average of a large number of components, and whether you see an alpha rhythm of a particular type or not depends upon which component happens to be the most highly synchronized process over the largest superficial area; there are complex rhythms in everybody. If Matthews had first put electrodes on myself or on one of his laboratory assistants whom I knew at Cambridge, they would have discarded the Berger rhythm; I am one of those people who don't have alpha rhythms at all, which leads me to my next point. Adrian had an alpha of a very pure frequency, so that the records they published long ago are particularly neat and clear ones. But this is not always true, and it is quite hard to find a person in an ordinary laboratory with a good demonstration alpha rhythm which shows the classical responses. There are many people in our surveys—about 15 per cent of the population—who have no alpha rhythms at all, at any time. I personally regard the alpha rhythm as a mildly pathological, or atavistic condition, as I don't have any myself. And I don't think my eyes are constantly fixated; whether open, shut, thinking, drowsing, and so on, I have never produced a single alpha wave. There are quite a few people like this. So the phenomenon may not exist in some brains. We thought at first this might be an anatomical accident. But again in patients with intracerebral electrodes implanted for long periods we have been able to record from the occipital cortex in people who have no scalp alpha rhythm, and sure enough, even with electrodes in the brain, they seem to have no alpha rhythm there either. One has to bear in mind that this is a phenomenon unlike the pulsation of the heart or the lungs, which happens in everybody; it's as though one were dealing with variations in pulse rate when some people don't have a pulse at all.

Mulholland

I think there's a lot more flesh on the concept of alpha, as indicated in the glossary of terms put forth by the International EEG Federation. The definition includes a particular location and frequency. In these terms, alpha should be recordable from occipital placements. Were you not to find it there questions would be raised in that context alone. Now you raise further questions, not only regarding deviations from expected phenomena, but about the concept of the alpha rhythm *per se*. I'm not willing to get into how valid this concept of alpha rhythm is from a neurophysiological point of view, but rather the more limited question which arises by taking these definitions at their face value and taking the accepted generalizations about these definitions at their face value, and finding exceptions to the alleged relationship of attention to the phenomenon of alpha attenuation. That's the particular question I was posing. Of course, if 'alpha' goes, then the whole issue must go.

Jasper

What do you say of attention in that respect? You mean visual attention don't you?

Mulholland

Yes.

Jasper

Shouldn't you always say visual attention, for that's quite a different statement.

Mulholland

Yes, that's just the point because some subjects can turn alpha on by elevating their eyes yet alpha is not blocked by visual stimulation when their eyes are up. That's the incongruity for the attention concept. We could raise the question about validity of the concept of alpha, whether there really is such a thing, whether one should qualify it by calling it occipital alpha, but that's a different question.

Grey Walter

All the rhythms I've been discussing are occipital. In an area say 2 or 3 cm square, if you analyse carefully you'll find several different components, all behaving differently. Some of them do and some do not change with the opening of the eyes, some of them are affected by flicker, and some not, but all are occipital.

Mulholland

Then what is meant by a definition that includes the statement that this rhythm is one that is blocked or attenuated by visual attention? With these qualifications the definition is even more difficult.

Grey Walter

I absolutely agree. I was never happy about a definition of *the* alpha rhythm for I don't think it is a single process at all. This is one of the definitions that I quarrelled with on the Federation Committee, but it is useful for recordists. They can identify it as opposed to a wave and spike, for example.

Jasper

What would you consider to be a scientific definition for technicians then? Recently we had a paper given in Montreal by a man from Oslo by the name of Andersen. He was re-examining the relation between the thalamus and cortex in the control of barbiturate spindles. With micro-electrodes he examined the precise relationships between points in the thalamus and micro-areas in the cortex. With this kind of analysis the degree of multiplicity of rhythms is perfectly astounding. He was finally able to find a small group of cells—I think he estimated not more than 200—in the thalamus which were synchronous and really phasically related to a column of cells in the cortex. This required a great deal of searching and if he moved 2 mm away from this column of cortex he found cortical cells which were not related to the spindle rhythm that he had found in the thalamus. In other words, the fine structure of this rhythmic process was perfectly remarkable in these micro studies, so that, to speak of *the* alpha rhythm becomes contrary to facts.

Grey Walter

May I just add a warning apropos of what you say. Norbert Wiener was a very great man, whom we all respect deeply, but in two separate publications he talked quite glibly about 'the brain wave' in referring to alpha rhythm and this has found its way into a good deal of technical and information theory literature. I think this is a very dangerous concept, and Norbert Wiener was not particularly well advised, or well informed, about 'the brain wave'. But he did give the impression to a great many engineers, particularly mathematicians, that there was a thing which could be called 'the brain wave', as though this had a characteristic frequency, like the vibrations of an atom that the Director of this Laboratory described as being a basis of time measurement. Nothing could be more misleading, and it's unfortunate that it's so widely disseminated in non-physiological disciplines.

Mulholland

I think you can add that it is even more misleading when associated with an undefined concept, such as attention.

Lindsley

These interesting phenomena you have described, involving generation of alpha sequences upon upward gaze, seem to me to be related to work of Per Andersen and his group, which I understand will be published soon in book form (44). Their work has been contrasted with that of Morison and Dempsey and also Jasper with regard to the role of midline thalamic nuclei in the control of electrocortical rhythms. They claim to be able to remove the middle 3 mm of the thalamus without disturbing the rhythmicity of the cortex, or of rhythms recorded more laterally in the thalamus. Andersen and Andersson maintain that any of the nuclei examined, including specific relay nuclei, will generate something comparable to a 10 Hz alpha rhythm. They explained this in terms of 'recurrent inhibitory neurones' linked to a series of thalamic neurones which project to the cortex. They use some of your concepts, Dr. Creutzfeldt. They find that if an afferent volley or even a single shock is applied to almost any thalamic nucleus it will produce a series of these rhythms, very much like a spindle burst. They have recorded this by microelectrodes in individual neurones of the thalamic nuclei. Hyperpolarizing intervals in between these unit discharges tend to synchronize their firing upon post-inhibitory rebound, producing a series of rhythmic pulses which trigger or generate waves on the cortex.

I want to ask a question about these rhythms which are generated when the eyes are suddenly turned upward. This is a position of strain for the eyes, and one that is difficult to hold very long. Could it not be possible that by the turning upward of the eyes there is a discharge of impulses, let us say into basoventral nuclei, just as Andersen and Andersson were doing by nerve stimulation, and perhaps thus producing a burst of alpha rhythm? Would not such a concept be reasonable? It seems to me that there is also another factor. What one does with one's eyes may be related to the pattern of speaking. In other words, when some people are asked a 'thought' question many times they look up at the ceiling before they give an answer. It may be that there has developed a kind of conditioned association between the process that one uses in establishing an attentive set in order to answer a question, and the kind of eye movements made. Just as some people gesture with their hands in connection with speech, and if you tie their hands they may be very much slower thinkers. There was a recent unpublished study in which human subjects were curarized in order to determine whether thinking and perception were possible in the absence of motor action and therefore proprioceptive input. Some people have been concerned about a motor theory of consciousness or a motor theory of thinking, and in this case, though they didn't have complete curarization, apparently, there was no interference with thinking or perception.

Beck

This was done in Salt Lake City. I recall discussions about the experiment and the procedure. Believe me, the subject was well curarized. He was given

500 units of *d*-tuborcurarine chloride, this is about two and one-half times that required for complete paralysis of respiration. This study may be found in *Anesthesiology*, 1947, Smith, Brown, Toman and Goodman. Scott Smith was the man curarized. Not only was he completely conscious, he also was sufficiently aware to be annoyed, because, so I've heard, his co-workers agreed to aspirate him every few minutes. They became so involved in the experiment they did not hold to their schedule. Dr. Smith, after the experiment, gave a detailed and accurate account of the whole procedure. The EEG during curarization was low voltage fast. The authors concluded that complete paralysis had no cerebral effects at all.

Mulholland

If one followed this up and tried to associate EEG phenomena with things such as eye movement patterns, of movement patterns in general, as you suggest, and use these, which can be measured and defined with some agreement, then we find we don't need the attention concept. This is what I'm getting at.

Bender

I have something to say about the oculomotor system which may have some bearing on what Dr. Mulholland has said. We found in our laboratory that the oculomotor system provoked by vestibular input, either thermal, vestibular or optokinetic stimulus, does reflect in some of the so-called sensory structures within the CNS, namely the lateral geniculate body and the occipital lobes. Of course, when the open eye moves, objects within the environment act as visual stimuli input and that gives rise to nystagmus. The quick movement or the jerk of the eye is reflected as a beat. The same reponse is obtained during nystagmus while in the dark when there is no visual input. The deflections are still present but the difference is that the lateral geniculate retains a very high amplitude, whereas in the occipital lobe there is still a beat but the amplitude is lower; curare abolishes eye movements but the 'beat' or the nystagmus-like action is still evident in the electrical recordings from the lateral geniculate body. We also find that the oculomotor impulse for eye muscle action first originates in the pontine reticular formation; the impulse reaches the lateral geniculate lobe about 30–40 ms later. Therefore, if the reticular formation has anything to do with attention or alerting, the impulses are transmitted through the oculomotor system and to the occipital lobes. Now as to the upward movements of the eye, we found, years ago, that unilateral stimulation in the pontine reticular formation (just rostral to the olive and slightly dorsal to it) produced upward movements of the globes and pupillary constriction. With pupillary constriction there was always a change in accommodation, and at the least a change in the lens. Of course, it's not really relaxation but there is change in the lens system when the eyeballs move up. One must note that upward movement of the eyes is not the same act of closing the eyes, or testing the eyes in the dark. Therefore, I would think that there is something in what Dr. Mulholland proposes, that the alpha rhythm has something to do with oculomotor activity.

Jasper

Can we clarify what you mean by the beat?

Bender

In the movements of nystagmus, a rhythmical to and fro motion of the eyes with a resultant jerk or beat. In the electro-oculogram the rhythmic spike-like deflections appear as a rhythmic beat.

Jasper

Are you speaking of an electrical response?

Bender

Yes. It is also the electrical response in the lateral geniculate body which corresponds to the beat of induced nystagmus. You can induce nystagmus with stimulation in the lateral canal. Plant the electrode in the lateral canal, let's say on the right ear, and this will produce eye deviation and also nystagmus.

Jasper

The beat is an electrical response in the nervous system? This is a beat in your electrical recording—in response to your nystagmoid movements?

Bender

Let's assume we observe ocular nystagmus. Corresponding to the nystagmus you will see electrical activity, that is in the lateral geniculate and the occipital lobe. There will be a beat, synchronous with eye movement. We can abolish eye movement with Flaxedil but there will still be activity in the lateral geniculate body.

Jasper

So a beat is electrical activity in the brain, and an electrical rhythm that's in time with the nystagmus, and it occurs even when the movements are not present.

Dewan

I'd like to pose a question in regard to this after-image experiment. How would you explain the presence of alpha if a person has his eyes up and he's paying attention to and looking at an after-image? How would you explain that the alpha rhythm is abolished by visual attention?

Evans

Could I say something that might help to complicate matters? My interest in alpha really started about 2 or 3 years ago when I first started working with perfectly stabilized retinal images. These I achieved by flashing a bright light behind a pattern and bleaching a portion of the retina, in the form of the pattern, so that the pattern itself is to all intents and purposes locked on the retina (45). Now it immediately occurred to me that this technique could be used to see what would happen to the alpha component of the EEG if one looked at a visual pattern with one's eyes closed. It was the first way of tackling this that I had ever come across, and I realized that it would make a simple alpha no-alpha experiment. So Dr. Grant Smith (who's now at McMaster University) and I set up this experiment rather crudely and, to cut a long story short, what we found was that during the periods when the patterns faded (stabilized images fade by the way, in case anybody doesn't know that; there are periods when they're seen and periods when they're not seen) there was an increase in alpha, and a decrease when the patterns came back. Actually this was a rather crude experiment and we simply measured the probability of alpha being present and didn't note the specific pattern of its waxing and waning. It wasn't a particularly marvellous experiment, though we did publish it in *Nature* at the time, and it's probably worth repeating thoroughly (46). Fender in California did, I think, do a rather similar experiment a year or so later, but he used contact lens to stabilize which isn't quite the same thing. Now the difficulty, of course, is that this doesn't seem to fit too well with the experiments that Mulholland and I did, and which I suppose kicked off most of this discussion, because there we used an after-image as our visual stimulus with eyes closed and turned up, and of course we found that we got *more* alpha with the eyes up, whether there was a pattern there or not.

Mulholland

There's a problem here too, which I had neglected to mention. When we did the after-image experiment that Chris has described, we found, in general, much less alpha than you would get if you had just a homogeneous field, so that while the subject was looking at the pattern the overall amount of alpha was decreased. But the differential between eyes ahead and eyes up was still maintained, so it's not an either-or thing. But it's quite clear that some serious complications have developed here for the attention hypothesis.

Jasper

Well there's no doubt that if you don't raise your eyes up you get blocking and unblocking of the alpha rhythm with a sequence of after-images. This was one of the very early observations we made with Cruikshank. So to throw out that observation because it's less obvious when you roll your eyes up—well the logic is not quite clear to me.

Mulholland

When you are looking at an after-image normally, when you are not elevating your eyes, they are freer to move and looking at an after-image may produce all of the over-learned responses, which have occurred in association with fixating an image. It would be interesting to evaluate the successive fixations as these after-images come and go. Perhaps when the eyes are up, they are less able to fixate. Also in these experiments you can't say that the eyes just stay up there. There are momentary lapses when they come down and this should be measured and evaluated too. All of this is related to the hypothesis that fixation, and the stabilization of the triad of accommodation, determines whether alpha is going to be there or not.

Evans

Could I say, incidentally, that I wasn't suggesting that the two things were totally incompatible, or that one would necessarily have to throw out one in order to have the other. But the finding that if you look at a patterned after-image (as opposed to a diffuse one) and discover that when the pattern vanishes alpha starts to come back, seems to me to tie the whole thing clearly with something to do with *looking at patterns*. And that was the point of the Evans and Smith Paper, which I referred to earlier. But we're still stuck with the Mulholland–Evans observation which seems to show something quite different.

Jasper

That's right. Yes, that's unfortunately true.

Fenwick

There are just three quick points I'd like to make. The first is that we've been looking at people who showed the Mulholland–Evans effect. We had to search through a large number of people before we found those who showed it. The effect is uncommon, although it quite clearly occurs. It seems to me to be very dangerous to generalize, or to set up an hypothesis which is general, on an effect which only occurs in a few people. With regard to the fixation triad, in my clinical experience, one would certainly expect to find an alteration in those people who show a third nerve palsy; they would show changes in accommodation. You don't, in these people, get any change in the alpha rhythm. So again, although the ideas about accommodation and the alpha rhythm may be true in certain cases, it doesn't look as if it's general. We have examined one or two people by putting atropine in their eyes and blocking their accommodation, and in some people this does decrease the alpha rhythm, in others it doesn't. So I think we are dealing with a multifactoral problem.

Grey Walter

Could I introduce an anecdote at this stage in the discussion? I had a colleague some years ago who had the unusual but useful capacity for demonstration purposes of producing a really juicy alpha rhythm when he was reading aloud from a book. He could read aloud from a textbook and produce an almost continuous sine-wave alpha rhythm. When he did this, however, he didn't remember or recall any of the contents of what he had been reading, though he read perfectly accurately. When we told him to read and *remember* so as to produce an abstract, then his alpha blocked. So although he was reading accurately from the point of view of visual fixation and his visual acuity was good, yet he was not extracting semantic content and his alpha did not block, but when he did extract the semantic content his alpha *did* block in the normal way. He later went mad, by the way, but I'm not sure whether this had anything to do with it.

Mirsky

A question for Dr. Mulholland. In this eyeless patient you studied, leaving aside the question of visual attention, have you, or has anybody, studied the effect of other kinds of stimuli and other parameters on blocking?

Mulholland

Yes, we did apply odd tones, noises, and we'd get a transitory attenuation of the alpha. So this picture has a double incongruity. I was going to ask Dr.

Bender if it's possible with cases like this, where the patient has plucked out his own eyes, that there might not be adhesions to the tissue behind the eye and that there might be some load on the remnants of the extraocular muscles, enough load to permit a rudimentary kind of fixation process to occur in muscle?

Bender

I can't answer that question, but we know that if we eviscerate the two eyeballs in monkeys, that oculomotor activity is still very prominent in the lateral geniculate body and occipital lobes. This occurs even though the eye muscles are not there to contract. I mean no matter how one induces the nystagmus in the vestibular system, calorically or by rotation, these still activate the oculomotor system as if the monkey had eye muscles.

Jasper

You mean you get vestibular activating responses?

Bender

Yes, even without nystagmus or corresponding eye muscle actions there is activity in the lateral geniculate body.

Mulholland

There's sometimes enough left though—this is another anecdote—a friend of mine has a monocular enucleation and is able to give a movement to his glass eye, and he has told me that other people he knew could do this because there was enough muscle there.

Grey Walter

There are one or two things about the control of eye movements that are interesting; it's one of the few muscle systems in the body that is not included in a servo loop from itself. Fixation of the eyes on an object involves an elaborate prediction process in the cortex, as opposed to a simple feedback system such as we use when we put one of our limbs in a particular position, or touch our nose and so on. This might be relevant to what you're talking about: that the eye movements are a special case of motor control in which a very large mass of cortex seems to be necessary in order to predict and co-ordinate the movements exactly; to lay both eyes on the same target and so forth. This must involve an intricate system in the neocortex, including prediction of eye movements in which the anticipated positions of the eyes have to be registered in the brain, after which the muscles take up the appropriate tension. It's like an incremental plotter as opposed to a servo plotter, in which there's no knowledge of the origin or zero, and full information has to be provided beforehand, and this must be absolutely accurate without a servo feedback from the muscles.

Mulholland

Would it be possible that all this computation of the anticipated eye position, or the anticipated triad of accommodation is involved in breaking up the alpha rhythm? One sees an emphasis on sensory processes to explain attenuation or blocking of alpha, yet oculomotor computations may also play their role in desynchronization.

Grey Walter

I think the impression one might get from what you said is that a rather trivial function, eye movements in other words, are related closely to alpha activity. But this is not trivial from the standpoint of the neocortex—it involves a very elaborate process of position prediction and control. This could very well involve *some* of the processes involved in *some* people in *some* alpha rhythms.

Mulholland

I think what puzzles me too, is the idea—I may have misinterpreted this—that the attenuation of alpha, or the so-called desynchronization is the opposite of the alpha process and this may not be quite true. If the desynchronizing activity does not go on, then the stage is set for some other process that is not necessarily on the same continuum, a process which causes alpha. I think that the notion that desynchronization and alpha are opposite, fitted too neatly with the contrast of attention with inattention and added another dimension of ambiguity to the whole concept. It doesn't necessarily follow at all that the opposite of attention would be related to the occurrence of alpha rhythm.

Bender

I want to make one more comment about these eye movements in the dark and recordings in the occipital lobe. Eye movements evoked by vestibular stimulation and optokinetic after-nystagmus but also spontaneous eye movements were easily detected and were recorded electrically in the eye while in the dark. Of interest is that the action potentials in the occipital lobe seemed larger when the monkey became drowsy.

Berlyne

Dr. Jasper, I think it was said that learned associations might play a part. People sometimes elevate their eyes when they're asked embarrassing questions. Don't many people also elevate the eyes when going to sleep? In some people, might the eyes not begin to go up when they're dozing off?

Bender

As they're actually going to sleep? Yes.

Evans

I believe Dr. Kennedy has an observation to make, but could I just say something quickly about Dr. Berlyne's point. It relates in fact to one of the points we made in our first paper—which we called the 'Improper alpha' paper—and which we put in as a bit of a joke, but perhaps not too much of a joke. That is that when one closes one's eyes normally, one's eyes do tend to roll up somewhat, and this could be a rather fantastic way of looking at the production of alpha in the eyes closed situation.

Kennedy

I don't see the necessary connection between your argument and the bilateral enucleation case. Surely you are proposing some hypothetical reduction mechanism or control mechanism that is influenced by fixation or the triad. So it wouldn't be particularly unusual to encounter a case of bilateral enucleation that exhibited a reasonable normal alpha wave.

Mulholland

Well it exhibited blocking too, and then we have Adrian's case where there was no alpha, so this didn't fit.

Kennedy

This matter of individual differences is the number one concern here.

Jasper

The differential attention that we've observed with Dr. Penfield with electrodes on the cortex, is quite clearly a case of being able to block local areas of the cortex, depending on what's attended to. In that case attending to a movement would block activity in the sensorimotor region, without affecting the parietal occipital alpha. If the person was able to attend specifically to the movement, it didn't affect the alpha rhythm. So the generalization of this concept is quite a mistaken idea. The second thing is that it's been suggested that you might consider cases with what we call paradoxical blocking reactions. I think everyone who's used EEG has run across patients who show alpha when they open their eyes instead of blocking. Some of these are cases of narcolepsy and this is a characteristic of a narcoleptic patient.

Mulholland

I think we're still undecided on the question of what is alpha, and yet we ask the question of what's the difference between alpha that has an amplitude of practically zero and desynchronization. Some like to talk about attenuation, but actually there are a great many varieties of alpha reduction here.

Jasper

Anybody else like to complicate the question?

Lindsley

One more point related to alpha and the CNV. Low and associates described situations where the CNV was not produced under conditions of upward gaze, whereas lateral or downward gaze did produce it. Now, I think it was Herbert Jasper, who, in 1936 (47) called attention to the fact that negative base-line shifts were associated with alpha activity whereas positive shifts were accompanied by blocking or absence of alpha activity—Herbert, you may remember?

Jasper

I haven't forgotten.

Lindsley

In any case, the period of the CNV would normally be the period of alpha blocking if one were attending to a stimulus. I think Kellaway and Low were testing for CNV in the absence of eye movements and they found it.

Jasper

Very interesting. I remember that they were trying to control the effect of eye movement artefact on the CNV. Now I believe Dr. Evans has a movie of what he calls the 'Improper Alpha Effect'.

REFERENCES

1. Adrian, E. D. (1943) *Ophthamological Soc., U.K.* **63**:194
2. Adrian, E. D. and Matthews, B. H. C. (1934) *Brain*, **57**:355
3. Bagchi, B. K. (1937) *J. Psychol.*, **3**:463
4. Berger, H. (1929) *Arch. Psychiat.* **87**:527
5. Berger, H. (1930) *J. Psychol. Neurol.* **40**:160
6. Berger, H. (1931) *Arch. Psychiat.* **94**:16
7. Bundzen, P. V. (1966) *Fedn. Proc. Translation Suppl.* July–August, **25**:No. 4, Part II, T551–T554
8. Callahan, A. and Redlick, F. C. (1946) *Am. J. Ophthal.* **29**:1522
9. Dewan, E. (1966) *Proc. Symp. Biomedical Engineering*, Marquette University, 1:849
10. Dewan, E. (1967) *Nature, Lond.* **214**:975
11. Durup, G. and Fessard, A. (1935) *Annee Psychol.* **36**:1
12. Kamiya, J. (1965) 'Conditioned discrimination of the EEG alpha rhythm'. Cited in: *Current Research in Sleep and Dreams. Public Health Service Publication No. 1389*:57
13. Kreitman, N. and Shaw, J. C. (1965) *Electroenceph. clin. Neurophysiol.* **18**:147
14. Kris, E. C. (1960) In *Medical Physics*, Vol. 3, Year Book Publishers, Chicago, 692
15. Kris, E. C. (1966) NASA Report CR2372
16. Kris, E. C. (1968) *Electroenceph. clin. Neurophysiol.* **24**:189
17. Lashley, K. S. (1954) (Ed. by Adrian, Bremer and Jasper) *Brain Mechanisms and Consciousness*, Blackwell: Oxford, 422
18. Latour, P. L. (1962) *Vision Res.* **2**:261
19. Lindsley, D. B. (1960) In: (Ed. by Field, J., Magoun, H. W. and Hall, V. E.) *Handbook of Physiology, Section I: Neurophysiology III*, Am. Physiol. Society: Washington, 1553
20. Lippold, O. C. J. and Novotny, G. (1966) *Laboratory Report, Dept. of Physiology*, Univ. College: London
21. Lombroso, C. and Duffy, F. (1967) Personal communication
22. Lorens, S. A. and Darrow, C. W. (1962) *Electroenceph. clin. Neurophysiol.* **14**:739
23. Morrell, L. (1966) *Electroenceph. clin. Neurophysiol.* **21**:552
24. Morrell, L. and Morrell, F. (1962) *Electroenceph. clin. Neurophysiol.* **14**:724

25. Mulholland, T. (1964) *Electroenceph. clin. Neurophysiol.* **16**:394
26. Mulholland, T. B. and Evans, C. R. (1965) *Nature, Lond.* **207**:36
27. Mulholland, T. B. and Evans, C. R. (1966) *Nature, Lond.* **211**:1278
28. Mulholland, T. and Runnals, S. (1962a) *J. Psychol.* **54**:69
29. Mulholland, T. and Runnals, S. (1962b) *Electroenceph. clin. Neurophysiol.* **14**:847
30. Mulholland, T. and Runnals, S. (1962c) *J. Psychol.* **54**:317
31. Mulholland, T. and Runnals, S. (1964a) *Electroenceph. clin. Neurophysiol.* **17**:371
32. Mulholland, T. and Runnals, S. (1964b) *Cortex*, **1**:225
33. Mundy-Castle, A. C. (1957) *Electroenceph. clin. Neurophysiol.* **9**:643
34. Oswald, I. (1959) *Electroenceph. clin. Neurophysiol.* **11**:601
35. Runnals, S. and Mulholland, T. (1964) 'Selected demonstrations of voluntary regulation of cortical activation'. *Perception Laboratory Report*, Veterans Administration Hospital, Bedford, Massachusetts
36. Sharpless, S. and Jasper, H. H. (1956) *Brain.* **79**:655
37. Sokolov, Y. N. (1963) *Perception and the Conditioned Reflex*, Oxford: Pergamon. Trans. from Vospriyatiye Uslovnyi Refleks, Moscow: Univ. Press, 1958
38. Stark, L., Kupfer, C. and Young, L. R. (1965) *NASA CR-238*, Washington
39. Storm van Leeuwen, W., *et al.* (1966) *Electroenceph. clin. Neurophysiol.* **20**:306
40. Toman, J. E. P. (1943) *Fedn. Proc.* **2**:49
41. Walsh, E. G. (1953) *J. Physiol.* **120**:155
42. Werre, P. F. (1957) *Relations between EEG and Psychological Data in Normal Adults.* Leiden: Leiden Univ. Press
43. Williams, A. C. (1940) *J. exp. Psychol.* **26**:413

Discussion

44. Andersen, P. and Andersson, S. A. (1968) *Physiological Basis of the Alpha Rhythm.* New York: Appleton, Century, Crofts
45. Bennet-Clark, H. C. and Evans, C. R. (1963) *Nature, Lond.* **199**:1215
46. Evans, C. R. and Smith, G. K. (1964) *Nature, Lond.* **204**:303
47. Jasper, H. H. (1936) *Cold Spring Harbor Symposia on Quant. Biol.* **4**:320

The Effect of Eye Position on the Alpha Rhythm

P. B. C. Fenwick* and S. Walker
Department of Applied Electrophysiology, National Hospital, Queen Square, London, W.C.1

Recently, Mulholland and Evans (1, 2) reported a marked increase in the alpha rhythm of a small number of subjects when they moved their eyes in an upward direction. The aim of the present experiment was to investigate their results and to explore this effect in rather more detail.

METHOD

Fourteen medical students were chosen as subjects for the experiment. They sat in front of a plain white screen, behind which were placed five neon pilot lamps, each of which could be illuminated separately, though only one lamp was illuminated at any one time. The background illumination of the room was reduced so that the lights on the screen were clearly visible. The subject was instructed to fixate on the lamp which was illuminated. Each lamp was on for 30 s, and the sequence of fixations was scheduled by a randomized latin square. This was so arranged that the subject gazed for 30 s at the centre light, 30 s at a peripheral light, and again for 30 s at the centre. Each movement away from the centre was repeated four times. When fixating on either of the lights in the vertical plane, the direction of gaze subtended an angle of 41 degrees, and when in the horizontal, 57 degrees. Seven channels of information were recorded, using a standard Ediswan EEG machine. Channel 1 recorded vertical eye movements, channel 2 horizontal eye movements, channel 3 a signal marker which indicated when the central lamp was illuminated. Channel 4 a bipolar right parietal channel (C_4–P_4) and channel 5 was a bipolar right occipital channel (T_6–O_2). Channel 6 was a unipolar right parietal channel and channel 7 a unipolar right occipital (C_4 and T_6) channel, both of which were with reference to the opposite ear. Channels 6 and 7 were analysed by a B.N.I. Faraday analyser, set to analyse all frequencies in the alpha band together with those at 7 and 14 Hz. The spike heights produced by the analyser in the alpha band were summed and this value was taken as a measure of the total alpha activity.

* Present address: The Maudsley Hospital, London.

RESULTS

Figure 1 shows one type of result obtained. The activity at each point plotted was the total activity in the alpha band summed over three 10 s epochs, with the eyes held in one position. The changes in alpha activity which were seen with changes in eye position were small and in no case amounted to more than 10 per cent of the total 'eyes-shut' alpha amplitude. In this subject, the away

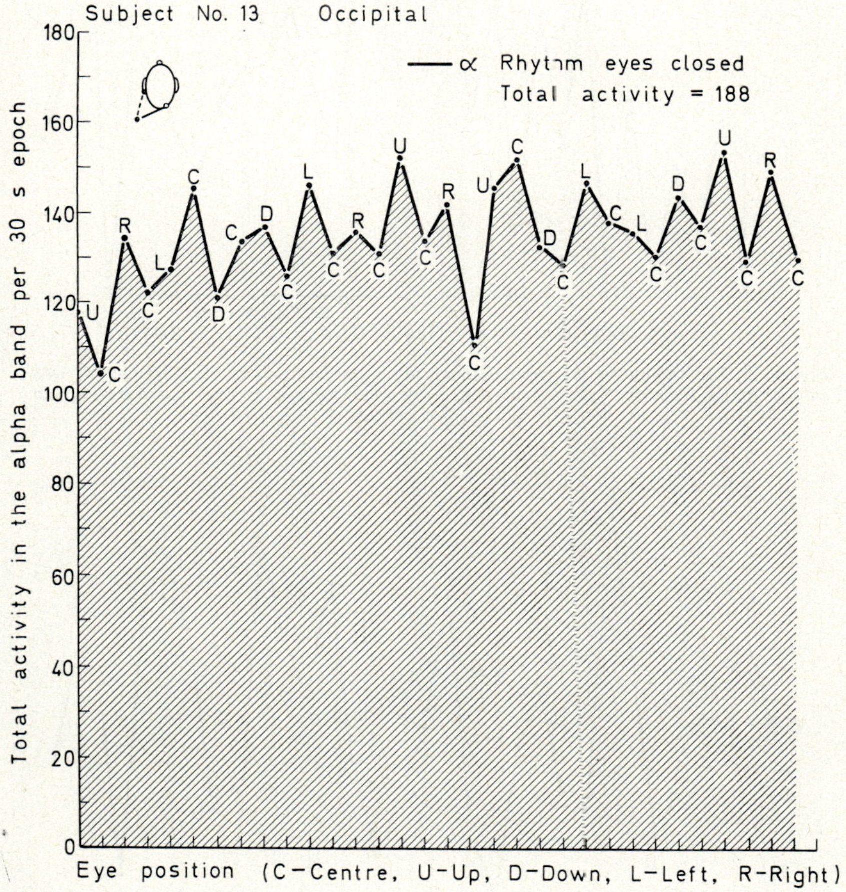

Figure 1. The change in the total activity in the alpha band in 30 s *epochs with small changes in eye position. In this subject the alpha activity increased on average when the eyes were deviated away from centre*

from centre epochs were larger than the centre epochs. This result tended to confirm the work of the previous authors.

Figure 2 is plotted in the same way as *Figure 1*. In this figure the zero level has been removed to demonstrate the change in alpha rhythm more clearly. The centre epochs were higher than the away-from-centre epochs. This result negated the hypothesis being tested and is opposite to the results shown in

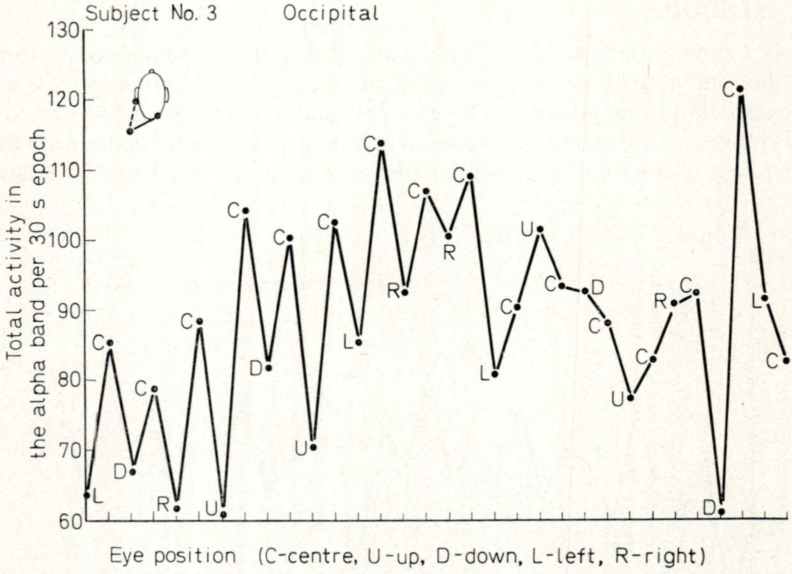

Figure 2. The change in the total activity in the alpha band in 30 s *epochs. In this subject the alpha activity decreased on average when the eyes were deviated away from the centre*

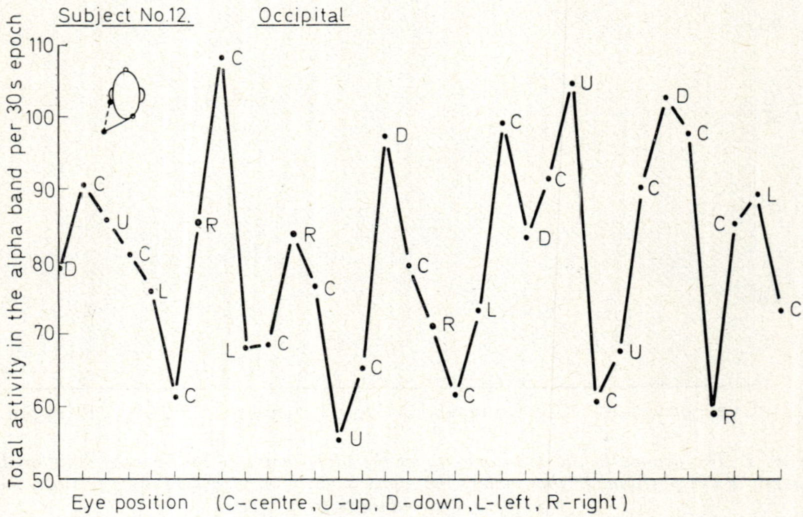

Figure 3. A change in the total activity in the alpha band in 30 s *epochs. In this subject the alpha changes appear to be random and showed no constant relationship to eye position*

Figure 1. The changes remained small, never exceeding 10 per cent of the total amplitude of the eyes closed alpha activity.

Figure 3 shows the third class of results. In this class, large intercurrent changes in alpha activity, due to uncontrolled variables, swamped the small changes that were produced by alteration in eye position.

In view of the three different classes of results that had been obtained, it was decided to test statistically whether or not the hypothesis that the alpha rhythm increased on looking away from the periphery could be confirmed for the group. This was done by scoring the results binomially. If the alpha rhythm increased on looking away from the centre, *and* decreased on looking back, a 1 was scored. This was called the double condition. If on the other hand, the alpha rhythm increased on looking up, *or* decreased on looking back to the centre, a 1 was scored for each response. This was called the single condition. The distribution of correct responses for the group were analysed by the method of Kincade and Lancaster. The levels of significance achieved for different eye positions is seen in Table 1. This table shows

Table 1. Probability that the Alpha Rhythm Increases on Looking Away from Centre. 14 Subjects. 4 Trials in each Patient. Method: Kincade–Lancaster (30 seconds)

	Double	*Single*
Centre up and down 41°		
Up	0·028	0·2322
Down	0·8781	0·0498
Centre left and right 57°		
Left	0·9561	0·2936
Right	0·003 sig.	0·00001 sig.

that the level of significance reached by the hypothesis that the alpha rhythm increased when looking away from the centre reaches the 0·01 level of significance only when looking to the right. It is difficult to see why this result should be so, and it is felt that it is probably caused by an artefact of method.

In summary this first experiment showed:

1. The alpha rhythm does not alter in a consistent way with changes in eye position in the group of subjects tested, except on looking to the right.
2. Certain subjects do show an increase in alpha rhythm on looking away from the centre.
3. The amplitude increase is never greater than 10 per cent of the amplitude of the eyes closed alpha rhythm.
4. The amplitude change in alpha activity in the away-from-centre eye position tends to be the same in all directions in any one subject.

Evans (personal communication) mentioned that in his experimental situation extreme deviation of the eyes was used; sufficiently extreme on occasions to cause discomfort to the subject. The experimental situation was then altered so that this time the subject, instead of looking directly at the illuminated peripheral neon light, made an extreme eye movement in that direction. Six

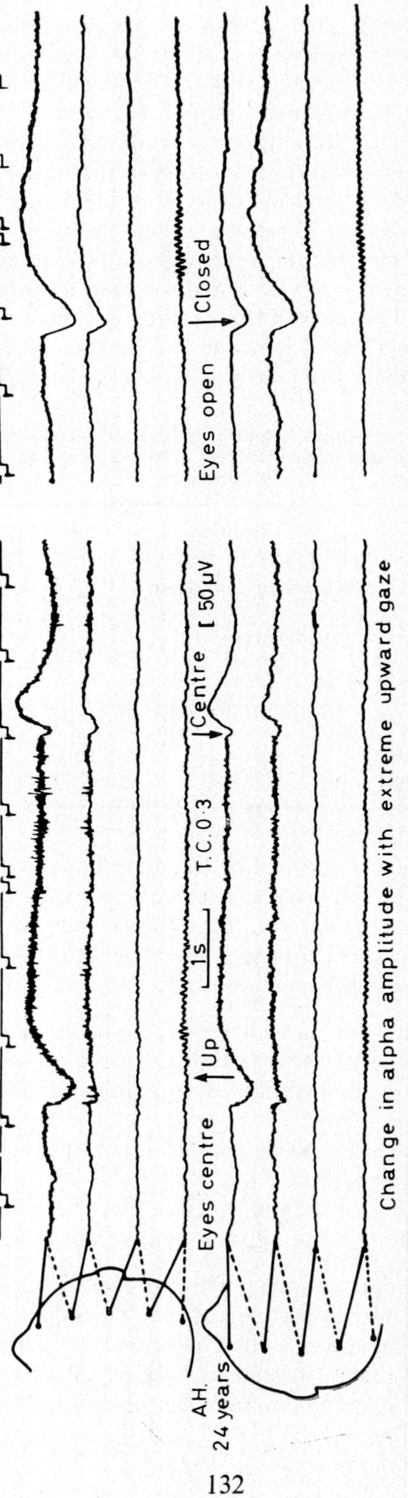

Figure 4. Changes in alpha activity are clearly seen with extreme upward deviation of the eyes. The right-hand side of the figure shows the evoked alpha activity with eye closure and its similarity to the alpha activity evoked by upward gaze

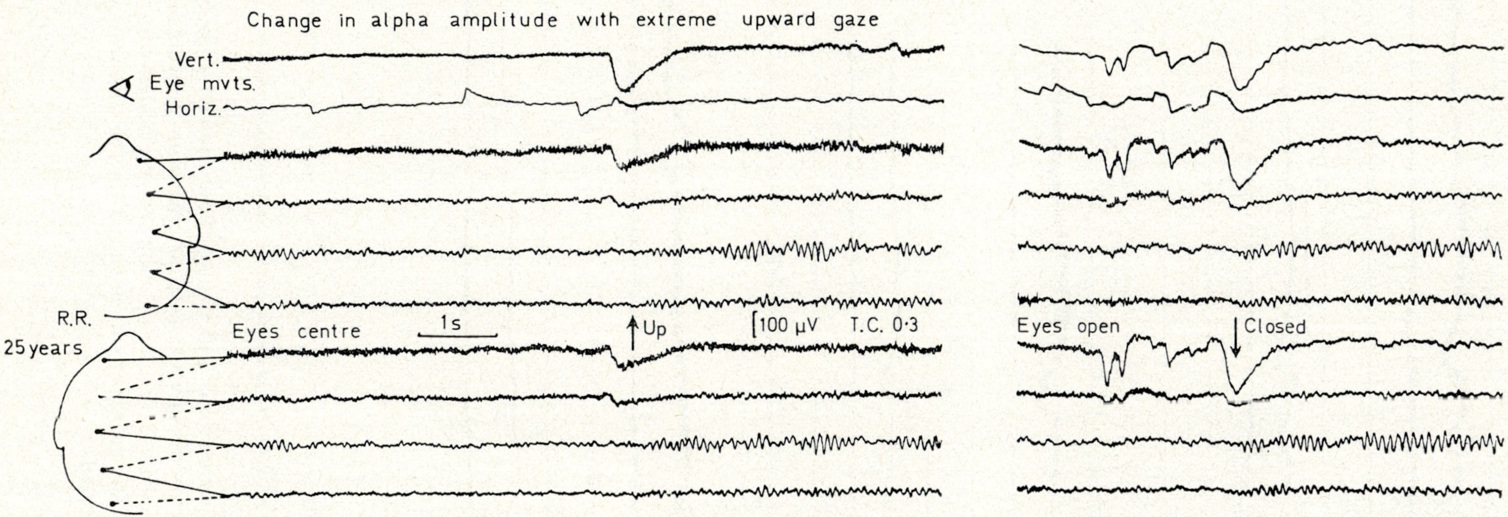

Figure 5. Changes in alpha activity are clearly seen with extreme upward deviation of the eyes. The right-hand side of the figure shows the evoked alpha activity with eye closure and its similarity to the alpha activity evoked by upward gaze

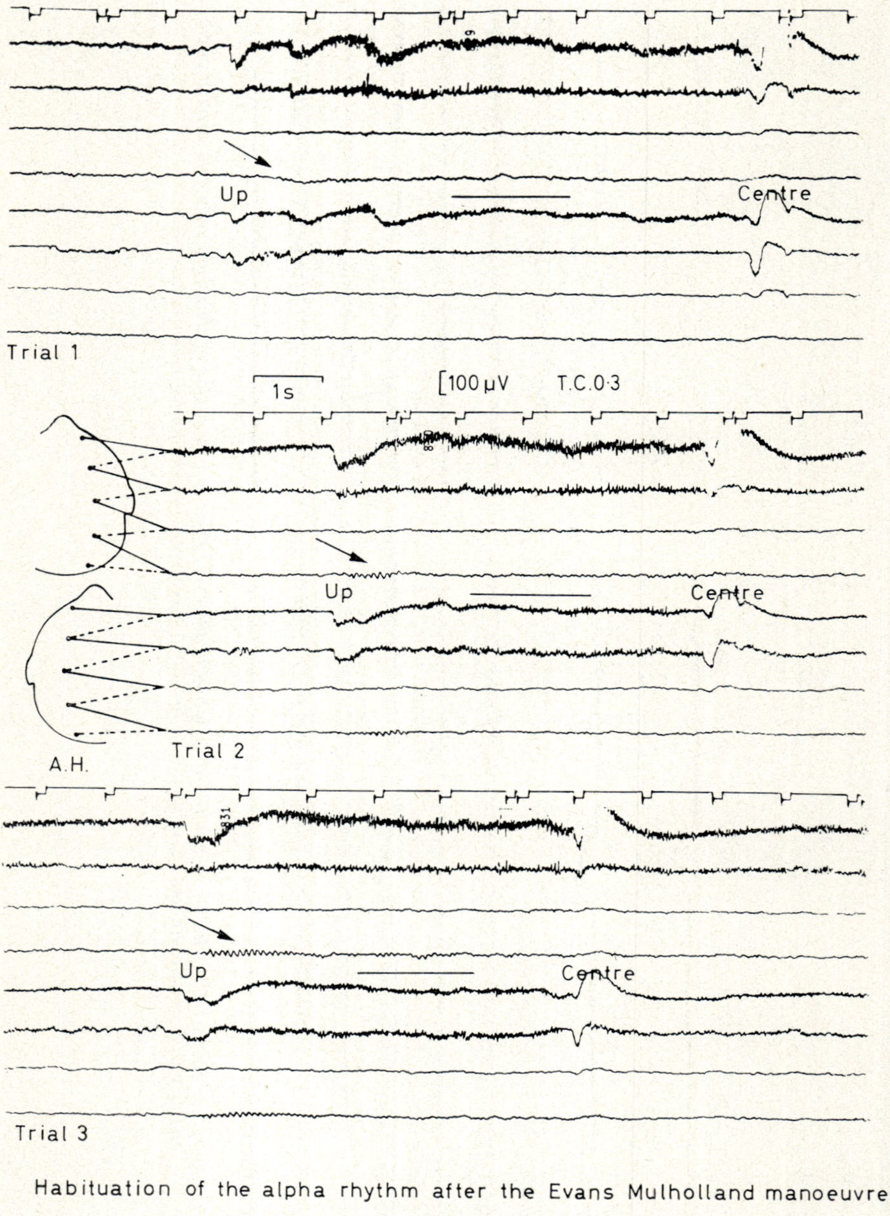

Figure 6. A further property of the 'induced alpha rhythm'

additional subjects were tested and two of these showed a large increase in alpha activity with extreme upward gaze, whereas four subjects failed to show a significant change. Later, three more subjects were added who were known to show this effect. The group of subjects used for the following experiments was thus highly selected.

Figure 4 shows the induced alpha rhythm in the raw record. The induced alpha rhythm was of the same order of magnitude and had the same distribution as the eyes closed alpha rhythm. *Figure 5* shows this effect in a different subject, when again the induced alpha rhythm remained similar to the normal 'eyes closed' alpha rhythm. The exact nature of this effect is difficult to explain. Mulholland and Evans have shown that it is independent of visual stimulation by asking the subject to look at a retinal after-image and by shining a bright light into the subject's eyes. Despite these manoeuvres the induced alpha activity was unaltered. We have been able to demonstrate the effect in total darkness and while reading cards. This has confirmed the independence of this effect in relation to visual stimulation. In people with scanty alpha rhythms it is possible to show that the effect can also be induced when the eyes are shut.

Figure 6 shows a further property of the induced alpha rhythm. Its amplitude progressively increases through the first three extreme eye movements and then reaches a plateau. If the movements are discontinued for 5 min, on recommencing them the induced activity behaves again as previously described.

The amplitude of the induced alpha rhythm is not the same along the major visual axes. This was investigated by means of a perimeter. *Figure 7*

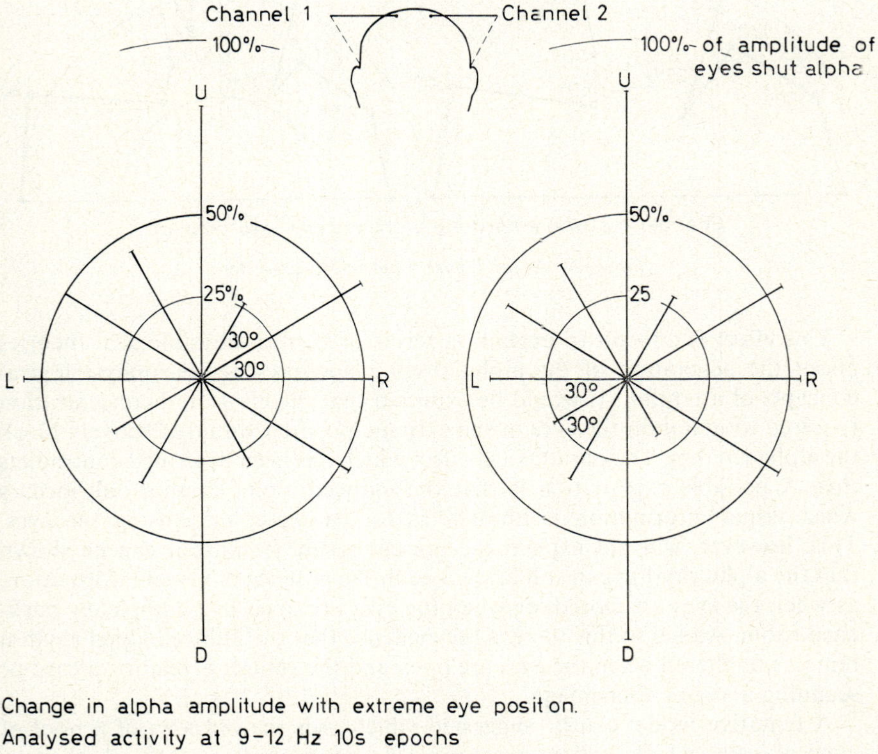

Figure 7. Increase in alpha amplitude

shows the increase in alpha amplitude which is obtained on looking away from the centre to the extreme of movement in steps of 30 degrees around the visual field. The length of line in this figure represents the increase in alpha activity which occurs when the eyes deviate from the central position. The circles show the 25, 50 and 100 per cent level of the total 'eyes shut' alpha amplitude. The figure indicates that deviation in the up-down axis evokes a much greater response than in the left-right axis.

Figure 8 is plotted in the same manner as *Figure 7*, but this time in 5 degree steps. It can be seen that the maximum induced alpha activity falls off sharply about 20 degrees on either side of the vertical visual axis.

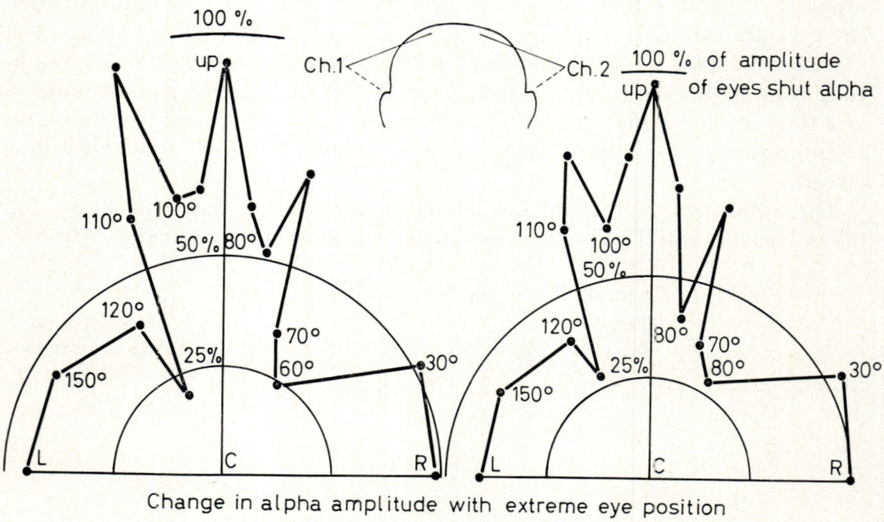

Figure 8. As for Figure 7 but in 5 degree steps

This effect is difficult to explain in terms of current physiological theories about the generation of the alpha rhythm and present neurophysiological concepts of attention. It would be expected that the effort and concentration required to maintain the eyes in the extreme position would effectively block the alpha rhythm. The paradoxical effect which has been described contradicts this. A possible explanation is that the induced alpha rhythm only occurs when visual information is limited, as for example, on closing the eyes. This, however, will not explain the present result, because it can be shown that the alpha rhythm can still be evoked in the absence of visual information, as when the eyes are closed, or when the eyes are open in a completely darkened room. Added to this there is the evidence that the induced alpha rhythm remains unaltered when the eyes are open and the subject is reading a card or scanning a visual after-image.

A tentative model can be suggested which uses the old idea of a pool of cortical cells which, when synchronized, produce the alpha rhythm as measured on the scalp. This pool can however be modified by several factors, some of which are central. Using this model, the concept of attention would be a centrally acting influence. Eye position has now to be added to this list, and the suggestion is that in extreme eye positions in some subjects, the central pool of neurones becomes free and synchronizes to produce an alpha rhythm. This could occur for two reasons. Firstly, because the act of maintaining the eyes in the extreme position causes a shift of attention from the visual field, or secondly, because the act itself in some way causes a modification of the central neurone pool. This phenomenon underlines our lack of understanding of the mechanisms responsible for the generation of the alpha rhythm and suggests as Mulholland and Evans have done that there is considerable danger in using the alpha rhythm indiscriminately as a measure of the psychological state of an experimental subject.

DISCUSSION ON FENWICK

Grey Walter

This is a very helpful commentary on the Mulholland–Evans effect.

Jasper

Yes I think that it's a very nice study of this odd phenomenon. The suggestion at the end that this is a burst of afferent volleys going into control a central system in certain subjects with extreme movement of the eyes is, I think, very reasonable. It is saying much better than what I was trying to say last night —that we had an opposing volley going in and blocking the usual blocking reaction. The first time such things were observed, Fessard was trying to photograph with a moving picture camera the blocking reaction of the alpha the first time that he saw it. He soon discovered that if he started to photograph the subject before he opened his eyes to block the alpha rhythm, he blocked before he opened his eyes! This was the original experiment in which after a number of photographs, the click of the camera blocked the alpha—accidentally discovering conditioning the alpha by trying to photograph it. In the alpha conditioning experiments we carried on with Shagass' work, we first conditioned the alpha blocking reaction to a tone and then omitted the tone and asked the subject to *imagine* that he heard the tone. This in a sense is like Adrian's original instruction to the subject to imagine he was looking at something in a dark room. Well this worked with nearly all subjects. After conditioning the reaction to a tone by tone-light combinations (this required 20 or 25 trials) most subjects could then voluntarily block the alpha by just *imagining* the tone, with the eyes closed and in a completely darkened room. We did not analyse what the closed eyes were doing, but we thought that this was a voluntary control of the alpha, conditioned by a previous association, with a sound which originally did not block it. I think that Dr. Fenwick might have put his finger on the probable relationship of the reaction.

Grey Walter

Fenwick has shown the importance of practice, and I wonder what this means. It's easy to say 'practice' but could this be a type of conditioning?

Fenwick

The increase in evoked alpha rhythm over the first few trials is one effect which we found to be constant in all subjects tested. The evoked alpha activity has always been of lower amplitude in the first trial than in the third. One can look at this in terms of practice if one wants, e.g. each time the eyes are elevated the alpha increases up to its maximum level. We think of this in terms of habituation, drawing a parallel between this effect and the way the alpha rhythm habituates to a tone. If the eyes are put in an alpha provoking condition, during the first trial little alpha is provoked, the second trial provokes more and so on until the full evoked alpha amplitude is reached. This again suggests a central rather than a peripheral mechanism.

Grey Walter

You suggest that this is a double effect; that the rolling up of the eyes needs effort and attention for the first few trials, but once you've got the knack of it you no longer pay much attention to rolling your eyes up?

Fenwick

Yes, I think this is probably correct.

Bender

Dr. Fenwick, what do you mean by a peripheral effect? I know that Dr. Jasper understands it, but I'm not very clear. Do you mean that closing the eyes and looking upward initiates something from the periphery into the central mechanism? If so, how does it do it?

Fenwick

What I mean by a peripheral effect is that by affecting structures on the periphery you alter the setting of a central mechanism.

Bender

How?

Fenwick

Well the answer to this is that I don't know.

Bender

Why can't the act of looking upward or closing the eyes be a reflex of itself?

Fenwick

It may indeed, and I think we must differentiate this. Whether it is the *intention* to look up which produces the alpha activity or whether it is the actual change in eye position itself. No experiments have yet been done to show which is which.

Cobb

I think we sometimes ignore what is involved in eye closure. It isn't just pulling a blind over the eyes, it's a profound psychological effect which is induced. We relinquish maybe 50 per cent of our contact with the outside world in doing this, and when you roll up your eyes in this disagreeable manner equally you give up effective vision, I think, and to that extent it has a good deal in common with eye closure.

Bender

Does this occur in darkness *without* rolling up the eyes?

Fenwick

No. One of the subjects who was put in a darkened room didn't get very much alpha, but if he performed this manoeuvre of elevating his eyes in a darkened room then the alpha was evoked, showing that it was the manoeuvre and not the darkness which increased the alpha activity.

Lindsley

Well there are one or two paradoxes here. It's very difficult to look up with your eyes closed because you have to inhibit the lids. If I look up, my eyes are wide open, if I look down its very natural for my lids to come down along with the eyes. Now suppose you give a series of repetitive flashes, you'll get a large electroretinogram, and good visual cortex responses to the first flash. But you can find a certain frequency (15–20 Hz) and certain intensity of light (moderately bright) which will block out completely the responses to the next several flashes, and only gradually will these responses return and stabilize at a normal level. Subjectively, one cannot see these flashes; you see only the first flash and then gradually you see the flicker. So there's a period of suppression which we think, from some studies we've done recently, is due to lateral inhibition within the eye and there is only a gradual recovery from that. I think this needs to be studied very carefully with respect to the contrasting features of the eyes-up, lids-wide-open, and eyes-up-with-lids-closed. Now if you try to do that your lids will flicker as you look up. And I noticed this in the movie; it was almost as if there was a little seizure phenomenon.

Evans

I'd like to say two things if I may. First of all, the fact that the effect takes place in darkness does narrow down the possible peripheral components of this. The second thing is that one notices that even when people have fairly abundant alpha in the dark, the raising of the eyes in these subjects who *do* exhibit the effect produces a kind of surge of the alpha—almost overlaying the basic alpha—which is rather odd. Now it seemed to me when we first come across this phenomenon, that whatever the nature of this particular effect, it really does mean that it is important if one is going to perform experiments in which alpha is brought in and tied to some variable or other, to keep a constant monitor of eye position. That's for the future. For the past I think there can be no doubt that a reappraisal of quite a number of experiments in which eye position was not monitored will have to be made.

Lilly

The enhancement of the alpha by the raised eyes may be caused by more effective closure of the eyes. The pupils are raised below the bony orbit. There may be feedback, including a neuro-muscular type programme such that even in the dark, this enhances alpha in some subjects. If one is in total darkness and opens the eyes, there is a definite change in perceptual computations as can be seen by trying the experiment on one's self. If now one raises one's eyes up under the bony orbit, one can see a quick change in the perceived space in spite of lack of external reality stimulation. There is a change in the perceived space from the black 'eigengrau' which one sees when the eyes are turned up, to the eyes open and looking straight ahead, when one 'sees' a three-dimensional blackness in front of one's self, i.e. there is a projection of assumed physical space in front of one's self. Experiments should be done seeing if this perceived space blocks the alpha.

Grey Walter

But Dr. Evans mentioned that even in total darkness, raising the eyes will augment the alpha rhythm, and rolling up the eyes can't *increase* the darkness.

Byford

If we could forget the rather restricting definition of alpha for a moment, alpha is available in a number of sites and under a number of conditions; corneoretinal potential recording during fixation may produce alpha rhythms, but this has nothing whatever to do with eye movements. There is, however, a physical movement at alpha frequencies but it's rather small, much less than 1 degree. Although, as Dr. Evans has suggested, it is desirable that eye movements should be recorded, it would be difficult to produce measurements having the required accuracy.

Lindsley

Is it not true that there are trigger rhythms up as high as 60 per second?

Byford

Yes, and some are even smaller still, of the order of seconds of arc.

Kennedy

Back in the old days that Dr. Berlyne referred to, when there were a good many psychologists studying the phenomenon of attention, there was a paper published (4) describing experiments conducted on a man with a hole in his head. He was a trephined subject. The investigator in this case thought he had demonstrated a very high correlation between some physical changes in the brain and the various manoeuvres that have been described here, and others such as mental multiplication and so on, that have often been observed to block the alpha rhythm. What he was saying was that there was a change in the volume of the brain related to these kind of activities. I think this might be an interesting lead to pursue. He found the same kind of habituation as has been reported with the alpha, he found the same kind of unpredictability of the amount of this effect. I've suggested (3) that this might be a kind of variation in the brain that would be interesting to look at in relation to eye movement 'control' of alpha.

Grey Walter

Lord Adrian, your original experiments have been quoted quite extensively. Would you like to close the discussion on this?

Adrian

It has always seemed to me to be a puzzle that the alpha rhythm should be so closely tied up with the eyes. Ocular fixation seems to stop it, but do we know what happens to one's eyes when one goes into complete darkness and gives up the attempt to see what is there? In my own case the alpha rhythm would come back as soon as I ceased 'looking'. I should like to know what changes, if any, take place in the eyes and the eye muscle when one gives up the attempt to see.

Adey

We've had quite a lot of experience of monkeys put into dark boxes immediately after birth and kept in the dark for 4 or 5 years, observed with infrared TV as well as having implanted EOG electrodes. Neither in acute nor chronic animals are the eyes ever stationary. The occurrence of what one might call nystagmoid movement appears gradually in monkeys that are raised from birth in darkness, but is never characterized in these animals by the full development of the miner's type of nystagmus. However, there are recurrent oscillatory movements of the globe and at the same time there is a great deal of blinking. Both blinking and movement of the globe often have a place in ritualistic behaviour which characterizes the chronically light-deprived animals. They will slap their faces and thereafter there is a pattern of enhanced or exaggerated eye movement.

Cobb

With respect to you, Lord Adrian, I think one does know what one's eyes do in the dark. Pat Merton in the Physiology Department at Cambridge has shown that one has very good knowledge, within a degree or two, of one's direction of gaze in the dark. That is to say, one can point in the direction of one's gaze and this is accurate within a degree or two.

Adrian

Yes, no doubt one can do that when one is gazing, but I was thinking of what the eyes would be doing if one was not consciously directing one's gaze anywhere, or conscious of its direction.

Cobb

Well I would suggest, sir, that one would know if one were doing something definite with one's eyes.

Adrian

If I was aware of that I am fairly sure that my alpha rhythm would be absent.

REFERENCES

1. Mulholland, T. B. and Evans, C. R. (1965) *Nature, Lond.* **207**:36
2. Mulholland, T. B. and Evans, C. R. (1966) *Nature, Lond.* **211**:1278

Discussion

3. Kennedy, J. L. (1959) *Psychol. Rev.* **66**:347
4. Shepard, J. F. (1906) *Am. J. Psychol.* **17**:522

Cerebral Potentials evoked by Pattern Reversal and their Suppression in Visual Rivalry*

W. A. Cobb,† H. B. Morton† and G. Ettlinger‡
†National Hospital for Nervous Diseases, Queen Square, London
‡Institute of Psychiatry, Denmark Hill, London

Since the introduction of techniques to improve the signal to noise ratio (4), the number of publications on human visual evoked potentials has been considerable; usually the stimulus has been a brief flash or series of flashes (flicker) on an unstructured field, but recently the interest of patterned fields has become apparent (2, 6, 11). Following a suggestion by Gross *et al.* (5) we have demonstrated the effectiveness of 'pattern reversal', without change in the total luminous flux, in evoking visual cortical responses and have applied the method to the investigation of binocular rivalry.

METHOD

As in our previous experiments on binocular rivalry (3), the target area was limited by a hole 3·75° (2·5 cm at 37 cm distant) in diameter, or less, in a black surround, usually viewed by one eye. The target was a thin transparent 'Cellophane' sheet with a regular pattern of black horizontal bars printed on it, lying immediately behind the hole and moved vertically up and down by a linear transducer. It was transilluminated by a tungsten filament lamp behind a diffusing screen of white paper; the luminance of the clear parts of the target was 200 ft-Lamberts or less. For rivalry experiments the second eye viewed a congruent target, similar in all respects except that its parallel black bars were vertical; it could be stationary or moved from side to side by a second transducer.

The electroencephalogram was recorded from an electrode on the midline 3 cm above the inion, with reference to an electrode at the vertex or on one ear lobe; in the former case an ear-vertex derivation was used to check the absence of response from the vertex lead. After amplification the responses were averaged by a Biomac 500 averaging computer, usually 128 sweeps of 500 ms duration. The averaged responses were stored in digital form on

* This paper is reprinted from *Nature*, **216,** Dec. 1967 by permission of the Editor.

punch tape and written out by an *X-Y* plotter. The target used most often had fifty-eight alternating black and white bars of equal width and was 2·5 cm in diameter; the transducer was adjusted to move the target by one bar width, a distance of 0·43 mm or 4 min of visual angle, so that the black–white pattern was wholly reversed; the rise or fall time was 3 ms (10–90 per cent amplitude).

RESULTS

In these conditions good evoked responses were obtained in all the twelve subjects examined for frequencies in the range from 2 to 20 Hz (*Figures 1* and *2*). It should be noted that the frequency of movement was 1–10 Hz; upward and downward displacements of the transducer gave similar responses so that the response frequency was twice the transducer frequency. Exactly similar responses were obtained when the bars were placed vertically and moved horizontally (*Figure 1*, channel 3). The maximum amplitude of response was about 12 μV peak to peak, obtained at 12 Hz.

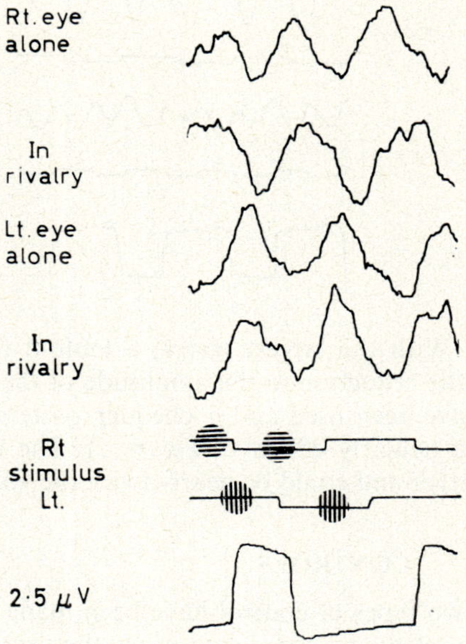

Figure 1. Pattern reversal in visual rivalry. Each trace is the average of 128 *sweeps and lasts* 250 ms*; response frequency,* 12 Hz. *The targets,* 6 mm *in diameter* (56′), *have alternating black and white bars* 0·43 mm *wide; for the right eye they are horizontal and move vertically by one bar width; for the left eye they are vertical and move laterally. The two traces above the calibration signal, if viewed at* 37 cm, *show the actual size of, and change in, the targets and the temporal sequence of their movements, which are arranged so that the responses from stimulating the eyes separately are* 180 *degrees out of phase* (*channel* 1 = *right eye, channel* 3 = *left eye*). *When the eyes are stimulated together rivalry occurs and the subject sorts the responses by pressing either of two switches, according to whether the right eye* (*channel* 2) *or left eye* (*channel* 4) *is dominant. It will be seen that the responses during rivalry are as large as or larger than during single-eyed viewing, which suggests that the non-dominant eye has not contributed to the trace*

Three sizes of target (25, 12 and 6 mm in diameter) have been used and the same trends are clear in all subjects although the effect of target size has so far only been studied systematically in one. When the visual angle of the target was increased from 56′ to 1° 52′ the response amplitude rose by 55–75 per cent, depending on frequency; a further increase to 3° 45′, however, caused only a 16 per cent increase in amplitude, or less, except at the highest frequency (20 Hz) at which the responses with the smaller targets had been ill defined. In this and other subjects, looking at the edge of the target caused marked decrease in response amplitude; it fell almost to zero when the fixation point was a few degrees to either side. Thus the response, like those to flash stimulation (10), must be largely the result of foveal stimulation and probably requires good resolution of the pattern; this is further suggested by the sharp decrease in amplitude when the target is thrown out of focus by a suitable lens.

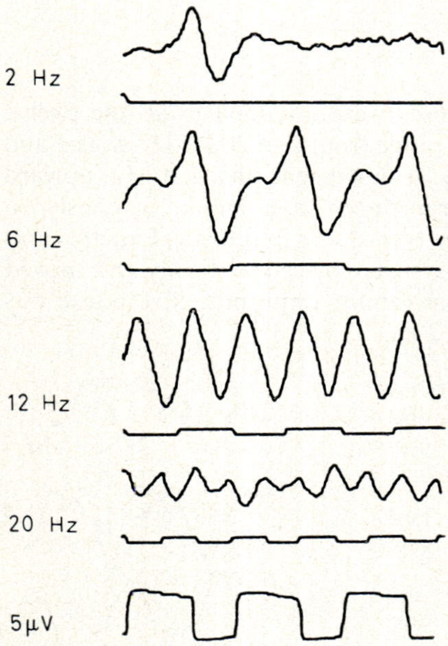

Figure 2. Responses in one subject to stimulation of the right eye by vertical movement of a chequer pattern through one square, or 13·5′. *Field diameter*, 12 mm *or* 1° 52′. *Each trace is the average of* 128 *responses and lasts* 500 ms. *Channel* 1 *is the response to downward movement of the pattern, indicated by the trace immediately below it; the remaining channels show the similarity of responses to up or down movements at various frequencies. Note that the transducer frequency is half the response frequency, the latter being given in the figure*

With the largest target, a tenfold reduction of the luminous flux caused little reduction in the amplitude of the response. Some other target materials have been used and a chequer pattern (one black–white pair = 27′) seemed particularly effective (*Figure 2*). The response to two-eyed stimulation was larger and could be nearly twice the voltage of that to a unilateral stimulus.

CONTROLS

Two types of control have been made. The theoretically more satisfactory is to set the transducer to move the target by a black–white pair width so that there is no apparent movement. In these conditions, the responses were greatly reduced although a small component at the stimulus frequency might remain, possibly caused by target overshoot, resulting in some visual stimulation. An alternative control was to cover the target with a sheet of plain white paper; when the movement of the target was through one bar width, no response was recorded at the stimulus frequency.

VISUAL RIVALRY

The term binocular rivalry implies that when different images fall on congruent areas of the two retinae the subject usually perceives first one image, then the other; one eye is said to be dominant while the other is said to be suppressed. It has been reported (1, 9) that flicker to the dominant eye causes larger occipital evoked responses than when it is applied to the suppressed eye. In our experiments (3) we used the same target material as that described except that the patterns of horizontal or vertical bars were stationary and transilluminated by gas-discharge flashes of 100 per cent square wave modulation of a fluorescent source. We found no consistent differences in amplitude between responses to stimulation of the dominant or suppressed eyes, confirming the negative results of Kaufman *et al.* (8).

Using pattern reversal as the stimulus, we have so far investigated seven subjects, two of whom found it impossible to sort dominance and suppression reliably. In one type of experiment only one of the targets moved and the subject indicated by two switches which eye was dominant, causing the responses to be stored in the corresponding channel of the Biomac (3). Thus the amplitudes of the two stored signals represented the mean amplitudes of the responses to stimulation of the one eye when it was dominant or when it was suppressed. In twenty-four experiments of this kind on five subjects (2, 2, 5, 10 and 5) a clear reduction in amplitude, easily seen without measurement, was present in twenty-two cases; one subject gave two doubtful results out of ten; in no case was the opposite effect seen. No such consistent effects were found with conventional flicker.

In a second type of experiment, stimuli were presented 180 degrees out of phase to the two eyes (*Figure 1*); the subject sorted the responses as described. If suppression were complete and if this sorting were perfect, it would be expected that the two traces would be similar to those obtained for stimulation of either eye alone, whereas without suppression or without sorting each channel would contain an equal mixture of the out-of-phase responses and both traces would approximate to straight lines. This was the result when using conventional flicker, whereas with pattern reversal the four subjects investigated (twenty experiments: 1, 8, 4 and 7) gave clearly positive results apart from three out of seven in one subject which were equivocal. Two of the subjects gave out of phase responses as large as or larger than those obtained when the eyes were stimulated separately (*Figure 1*). Sorting cannot be perfect because of indecision and reaction time delays, so this is a surprising result which, if confirmed, raises interesting possibilities regarding the interaction of these responses at the cortical level.

If, as seems likely, the amplitude of the visual evoked response is dependent, in the first instance, on the number of approximately synchronized post-synaptic potentials in a given area, the work of Hubel and Wiesel (7) should have given an indication that a large uniformly flashed or flickering field was unlikely to be the optimum stimulus; possibly it will be found that the greatest number of clearly resolvable contrast borders gives the highest probability of activating the largest number of cortical cells.

DISCUSSION

Spehlman (11) has shown that the addition of pattern to a previously uniform flashing field can modify the form of the evoked potentials, but the present work suggests that reversal of the pattern, although without any change in the overall luminance of the field, is a particularly effective stimulus. This observation goes some way to explain the occurrence of lambda waves, as potentials evoked by scanning patterned fields, which have always seemed disproportionately large in comparison with the potentials evoked by bright flashes in the same subjects.

ACKNOWLEDGEMENT

The apparatus necessary for this work was in part provided by the National Hospital for Nervous Diseases and in part by the Medical Research Council.

DISCUSSION ON COBB

Mulholland

There is evidence that the suppression of vision which occurs with rapid eye movements has different consequences for the evoked potential than the kind of suppression one gets with the stabilized image, or with binocular rivalry. Suppression *per se*, it would seem, needs to be examined in the context of what the eye itself is doing.

Cobb

First, I do not think that actual movement is a factor in this experiment. We have not yet set up our own optical system, but the original report by Vaughan and Valenstein was of a purely optical method of switching the pattern. Generally speaking, we did not use a fixation point, but we have used a dot on a piece of perspex in the middle of the target to fixate on. We have not attempted to record eye movements because they are too small for simple methods to be successful—only 4 min of angle if you accept that the eye moves with the target. It can be done, but I do not think the ocular potential would be of any use for the purpose.

Mulholland

There seems to be a contrast between kinds of suppression relative to the evoked potential. For example, between visual suppression during saccades and that experienced with the stabilized image. There might be a difference in the stability of the eye control system. Dr. Stark has suggested that the eye is open-loop during a saccade.

Goff

At a recent meeting of the American EEG Society it was pointed out that work is being done in the early diagnosis of amblyopia in young children, and it was found that when simply presenting flashes to two eyes, there was no difference in the visual evoked potential, but if one used pattern vision one could frequently detect an amblyopic condition at a very early age.

Lombroso

It's quite true that in a study of amblyopic children, Dr. Frank H. Duffy and I found it quite difficult to recognize abnormal VEP when stimulating the normal and the amblyopic eye with lights of just different intensity. This, even when analysing the evoked responses for RMS power. If we used pattern stimuli of matched intensity, however, several interesting differences between amblyopic and normal eyes became apparent.

Adrian

Does movement do it too?

Lombroso

These were stationary stimuli. In respect of what Dr. Cobb was saying about the experiments of Gross and collaborators, I think he is right to rule out retinal displacement. But even by doing this, I don't think you totally eliminate the possibility of retinal blur.

Jung

The only thing that disturbed me was the difference between one-bar and two-bar displacement. It's clear that intensity must have some influence.

Cobb

In the control there is no apparent movement, because each black bar has moved to the position of an adjacent one.

Jung

But you don't see movement?

Cobb

No.

REFERENCES

1. Balen, A. T. M. van (1964) *Docum. Ophthal.* **18**:440
2. Bickford, R. G. and Klass, D. W. (1962) *Trans. Am. Neurol. Ass.* **87**:176
3. Cobb, W. A., Ettlinger, G. and Morton, H. B. (1967) *Electroenceph. clin. Neurophysiol.*, suppl. **26**:100
4. Dawson, G. D. (1954) *Electroenceph. clin. Neurophysiol.* **6**:65
5. Gross, E. G., Vaughan, jnr., H. G. and Valenstein, E. (1967) *Electroenceph. clin. Neurophysiol.* **22**:204
6. Halliday, A. M. (1967) *Electroenceph. clin. Neurophysiol.*, suppl. **25**:178
7. Hubel, D. H. and Wiesel, T. N. (1962) *J. Physiol.* **160**:106
8. Kaufman, L., Pitblado, C. and Miller, jnr., J. F. (1965) *Sperry Rand Research Centre*, RR 65, 101
9. Lansing, R. W. (1964) *Science*, **146**:1325
10. Rietveld, W. J., Tordoir, W. E. M. and Duyff, J. W. (1965) *Acta physiol. pharmacol. neerl.* **13**:330
11. Spehlman, R. (1965) *Electroenceph. clin. Neurophysiol.* **19**:560

Changes of the Basic Rhythms of the EEG during the Performance of Mental and Visuomotor Tasks

O. Creutzfeldt, G. Grünewald,* O. Simonova and H. Schmitz
The Department of Neurophysiology, Max-Planck-Institute of Psychiatry, München 23, Germany

The blocking of the human alpha rhythm during eye opening or during mental activity in its broad sense is such a 'classical' finding (2, 3, 5) that it might be considered a waste of time to invest more research into this phenomenon. The observations on humans have been correlated with experimental results in animals and from this a general hypothesis about the relations between brain activity, wakefulness and mental activity has developed. Repeatedly, however, some papers have raised doubt about the general validity of the statement of decreased alpha-activity during mental activity (9, 11, 13, 17). During this meeting, we were reminded again of the now classical experiments of Adrian on the relation between alpha-activity and visual and oculomotor activities. Furthermore, the discussion after Mulholland's paper pointed again to the polyfactorial causality of alpha rhythms. It was clear from this that our knowledge of the functional significance and the physiological determinants of the alpha rhythm is still incomplete.

Owing to this situation, a reinvestigation of this aspect of human neurophysiology seemed to be indicated using some of the newly available methods of telemetric recording and automatic analysis of the EEG. In the following paragraphs I shall report on the effect of different mental and visuo-motor tasks involving different degrees of 'attention' or 'concentration' on the EEG. The first part of the lecture on the relationship between postsynaptic activity of visual cortex neurones and the visual evoked potentials will not be included in this report, as its presentation would involve too many methodological details. The findings will be published elsewhere (8).

METHODS

The experiments were done on 16 healthy adult subjects (mostly students) at the age of 21–24 years.

* Present address: Abteilung für Klinische Neurophysiologie, 78 Freiburg/Br.

Recording

The EEG was recorded by telemetry, the records were stored on tape. Bipolar records were taken on both sides from the temporo-occipital and praecentro-central electrodes. (T5–01, T6–02, F3–C3, F4–C4.) At the same time, the EMG of one gastrocnemius and the EKG were recorded. During the experiment, paper records were taken in order to discover artefacts and for protocol.

Experimental Programme

A set of 8 tasks was prepared which was designed so that each task involved a different degree of 'attention' or 'concentration'. They involved different forms of visuo-motor performance and mental activity. The tasks were presented in a balanced fashion (8×8—Latin square), and a written instruction was given before each task. At the end of the experiment the subjects scaled the tasks according to the attention or concentration needed, using the method of comparison of pairs (Thurston). *Figure 1* shows the different tasks and the result of the scaling. The tasks were as follows: (1) Drawing a simple wavy line. (2) Continuous writing of the word 'momom'. (3) Writing the word 'Dampfschiffahrtsgesellschaft' with ocular control. (4) Writing the same word without visual control but with the eyes open. (5) Maze-drawing. (6) Mental arithmetic tasks (simple addition and subtraction). (7) Mental arithmetic (multiplication of two 2-digit numbers). (8) Maze-drawing while seeing the maze in a mirror.

The experimental procedure was as follows: the subject sat comfortably at a table, the experimenter was seated beside him. The recording was done in a different room. At the beginning of the experiment, the subject was instructed to perform the tasks according to the instructions as well as he could. Then the written instruction for the first task was given, which was followed by a resting record (designated as R_1 in the following figures). During the resting periods, the subject had to keep his eyes open, but was not allowed to look around. The task started when the subject was given a sheet of paper prepared for the corresponding task. The task was continued for 120 s and was stopped no matter how far the subject had got (record T in the figures). Then another 90 s-period of rest with eyes open was recorded (R_2), and this was followed by reading the information for the subsequent task. Thus, the following recording periods resulted: 60 s information reading—90 s R_1—120 s Task—90 s R_2—60 s information reading, etc. During the information reading the EEG was not always recorded. The total experiment lasted about 45 min.

Evaluation of the Records

The records were evaluated automatically with a Schwarzer-Analyser, a Toennies-Analyser, a CAT 1000 and an IBM 1130. The Schwarzer-Analyser was used to transform single waves into impulses. The method works on the principle of zero-crossing of a filtered record. The record is filtered electronically. If the filtered potential crosses a threshold value (10 μV in our experiments) in one direction, an impulse is triggered, which opens and closes a

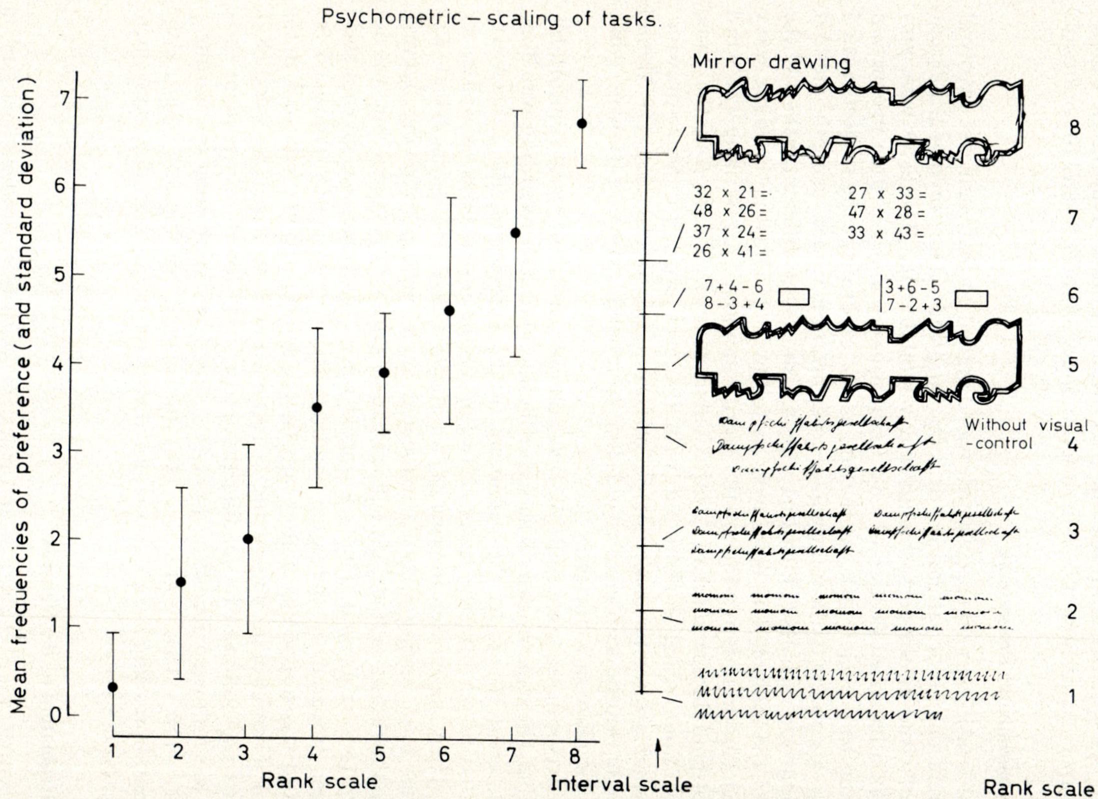

Figure 1. Set of tasks (right) and the result of their psychometric scaling (left). The tasks are put in the order of the rank scale which resulted from the scaling of 16 individuals. The dots in the graph give the mean value of the scaling with its variance (σ, vertical lines). During the experiments, the different tasks were presented randomly (see text)

gate. When the gate opens, a capacitor is charged continuously until the next cycle (impulse) starts. The amplitude of the capacitor discharge is a measure of the time elapsed between opening and closing. The output is such that the impulse is high after a short cycle (high frequency) and low after a long cycle (low frequency). The gating impulses can thus be used in two ways: (1) They were recorded on paper, lines of different amplitudes corresponding to the different frequencies drawn and the number of pulses representing the different frequencies counted by hand. (2) The impulses were used only as trigger impulses to produce an interval histogram with the CAT-computer, thus disregarding their amplitude. The second method of evaluation is similar to that described by Lettich and Margerison (10), but the minimal intervals were shorter (2 ms).

The histograms were punched on tapes, which could be put into the IBM 1130 computer and the number of waves belonging to certain intervals (or frequencies) could thus be counted.

With this method it was possible to count the number of waves of a defined frequency during a given period. The amplitudes could not be analysed. In order to get a precise measure of amplitude, slow records (0·5 mm/s) were used which condensed the whole record into a thick line whose width gave a good measure of the average amplitude.

The filter band used in these experiments was from 6 to 16 Hz. This band was chosen, after careful analysis by eye and the use of different filters had shown that waves within this frequency range were never superimposed on each other in healthy waking subjects, and in the areas from which the records were taken. In other words: only one wave within this frequency range was present at the time. With this method the well known high stability of the individual mean alpha-frequencies was confirmed.

(Since this paper was presented, a Fourier-transformation has been performed on the same records and the results of this will be described elsewhere.)

RESULTS

In contrast to the usual finding of complete blockage of alpha during short opening of the eyes, longer EEG-records during rest with open eyes—possible with telemetry—showed a considerable amount of alpha-waves. The alpha-blockage was 'complete' only during the first seconds after opening of the eyes, but after a while alpha returned. This is in agreement not only with Adrian's observations but also with those of Walter *et al.* (16) and Oswald (14). With the eyes closed, the alpha rhythm could be enhanced by simple repetitive motor tasks such as walking up and down on the spot. The μ-rhythm in the central records was also present during rest or walking with closed or open eyes—on the other hand, the amount of alpha waves was always lower with open than with closed eyes even if the subject was at complete rest (*Figure 2*). It can be seen, that the differences of alpha activity we are dealing with in the following, are at a much lower level than those seen between eyes open and closed.

During task performance the EEG may show drastic changes of pattern in some subjects. Nevertheless, group statistics showed only small changes in the average number of alpha waves in the occipital records during rest and task (*Figure 3a*). Only in the central records does the group show a strong depression of the 8·5–12·5 Hz waves. This depression was lowest during mental arithmetic and most pronounced during task 8 (mirror-drawing), a difficult motor performance. This difference was small, but could be statistically validated (7). Waves between 12·5 and 16 Hz became more frequent in the

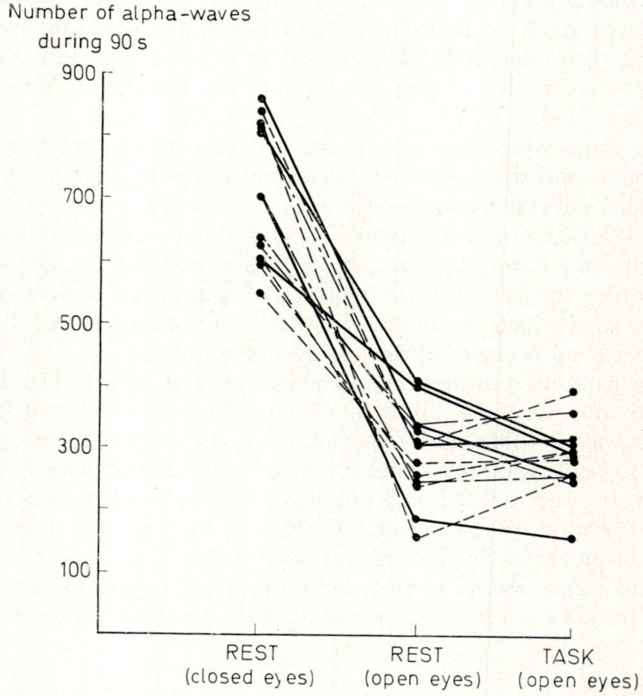

Figure 2. Occipital alpha-rate of 13 *individuals during rest with eyes closed (left), during rest with eyes open (middle) and during task performance (right). Each line connects the values of one individual; continuous and interrupted lines were used for better discrimination of the different individuals. All waves between* 8 *and* 13/s *were counted. Rest, closed eyes: Number of alpha-waves during a randomly selected period of* 90 s. *Rest, eyes open: Mean alpha-rate/*90 s *from all* R_1 *during an experiment. Task, open eyes: Mean alpha-rate/*90 s *from all T during an experiment*

occipital but did not change significantly in the central leads (*Figure 3b*). It may be noticed, that the occipital beta enhancement was least pronounced in task 4, i.e. writing without visual control, but maximal in task 8. This difference was also statistically significant though small. Waves between 6·0–8·5 Hz did not change their number significantly in the occipital or the central leads.

Inspection of the individual results confirmed that the μ-rhythm was reduced in all subjects during the tasks, but that the occipital alpha was

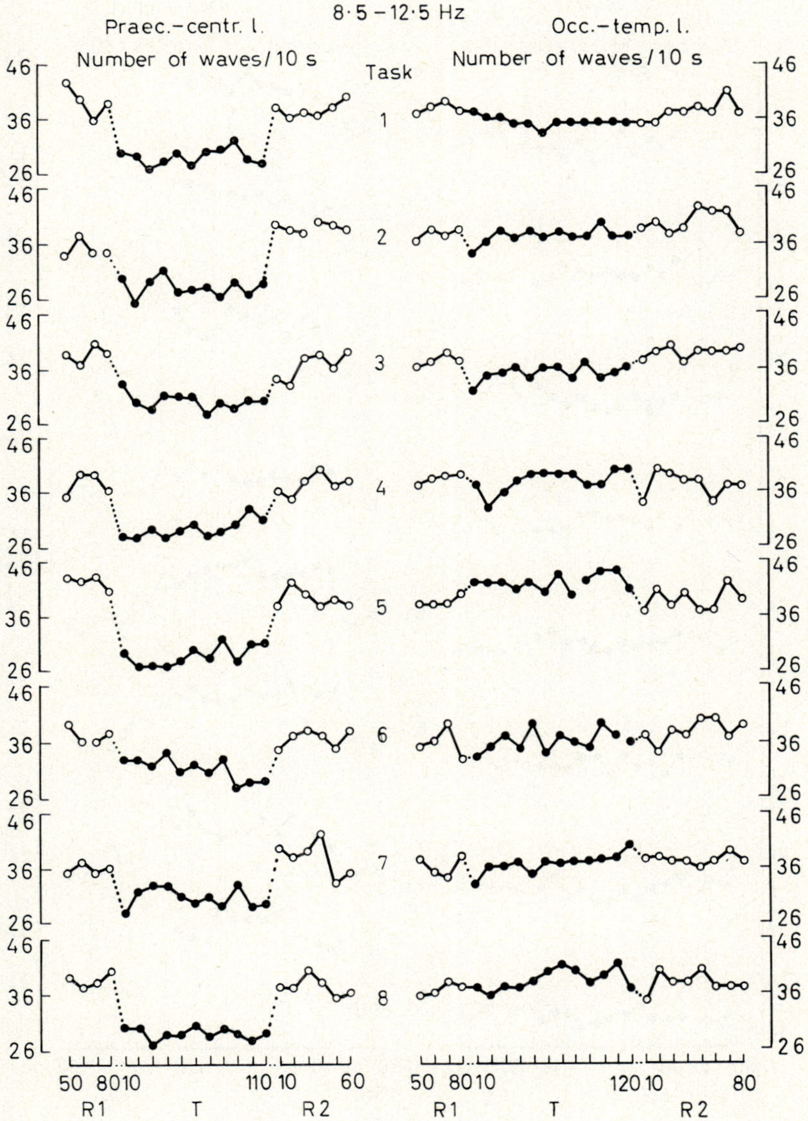

(*a*)

Figure 3. Number of waves with a frequency of 8·5–12·5/s (*a*) *and* 12·5–16·5/s (*b*) *during rest and performance of different tasks* (*number of tasks according to Figure 1*). *Mean values for sequential* 10 s-*periods from* 16 *individuals. Open circles: Values during* R_1 (*left*) *and* R_2 (*right*), *closed circles: Values during task performance. Left columns in a and b: Praecentro-central records, right columns: Occipito-temporal posterior records. Abscissa: Number of waves per* 10 s. *Note increase of occipital alpha and decrease of central* μ *during task performance. No consistent change of occipital alpha in the total group*

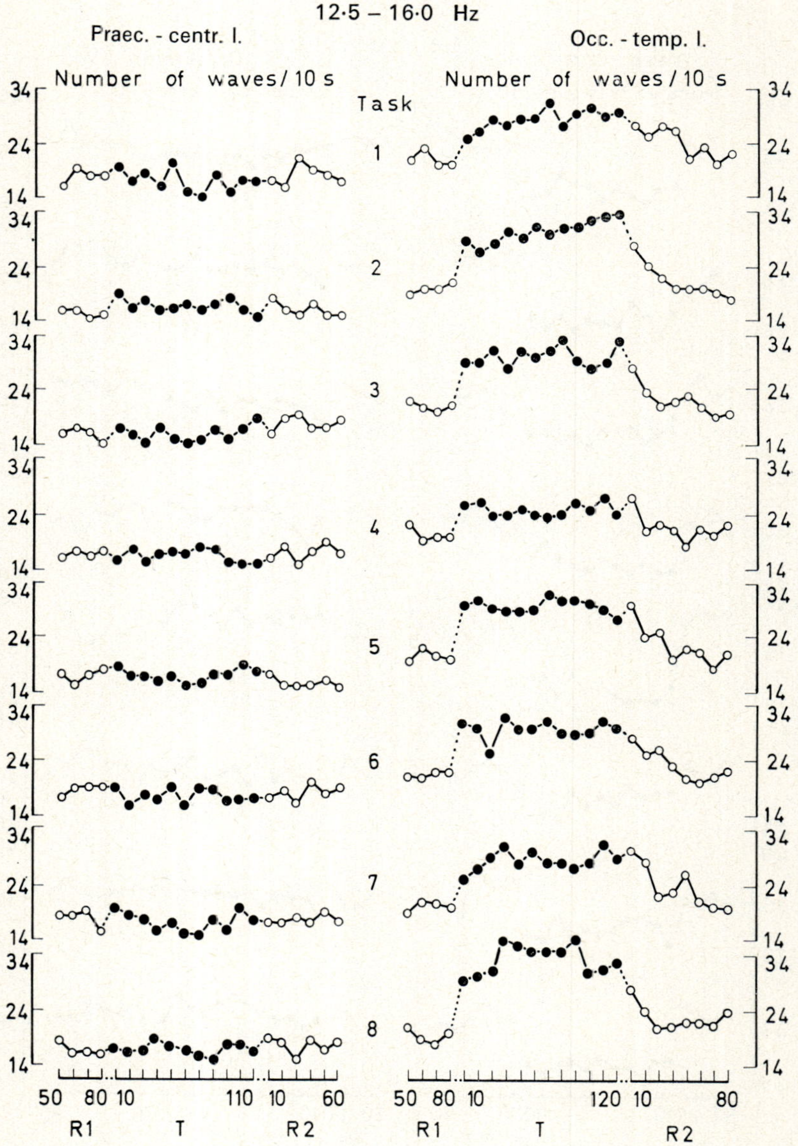

Figure 3 (*b*)

diminished in some subjects and enhanced in others. *Figure 4* shows an example of occipital alpha-reduction and beta-enhancement during task 7. It can be seen that this subject showed pronounced alpha activity of the 'spindling type' during rest (eyes open). During the task performance not only the number of alpha waves but also their amplitude decreased considerably. The

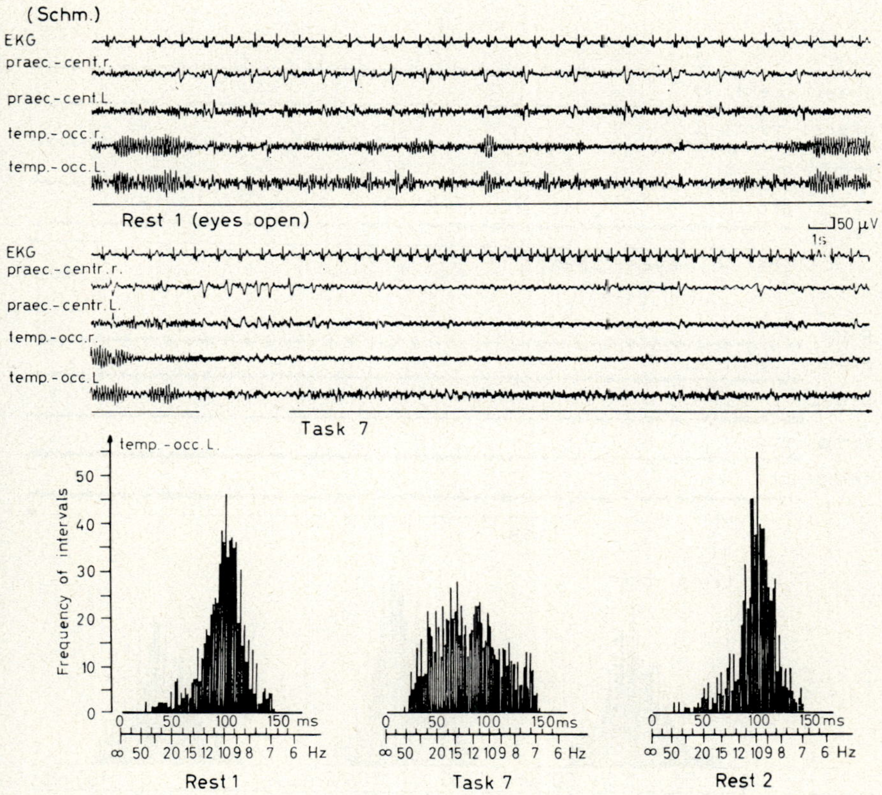

Figure 4. Original record and interval histogram of alpha-decrease during task performance. The original record was taken during rest (R_1) *with eyes open and during the beginning of the task. Alpha- and μ-reduction during performance of task* 7 *(see Figure 1), slight increase of heart rate. Between* R_1 *and the beginning of the task the subject is given the sheet of paper with the arithmetical tasks (orienting reaction with artefacts of eye movements in the praecentral records). Below: Interval histogram of the zero-crossing of one occipital record (see methods) during* R_1 *(left),* T *(middle) and* R_2 *(right). Abscissa: Intervals in* ms *(top) and corresponding frequency,* Hz *(bottom); ordinate: Number of intervals per* 90 s. *Note decrease but not complete disappearance of intervals in the alpha-range and increase of beta-intervals during* T

increase in heart rate at the beginning of the task may be noted. *Figure 5* shows an example of occipital alpha-enhancement during task performance. In contrast, the central μ-rhythm is strongly diminished. A slight increase of the heart rate can also be noted during the task performance.

The evaluation of the single experiments of all subjects revealed, that in some subjects alpha-enhancement was seen in occipital areas during all tasks, in some an alpha-reduction during all tasks and in some different alpha-reactions during different tasks (*Figure 6*). For the following analysis subjects were therefore divided into three groups. (1) Alpha-diminishers were those who showed alpha-reduction in 8, or in 7 out of 8 tasks (closed circles in *Figures 6–9*; (2) alpha-enhancers were those who showed alpha-enhancement

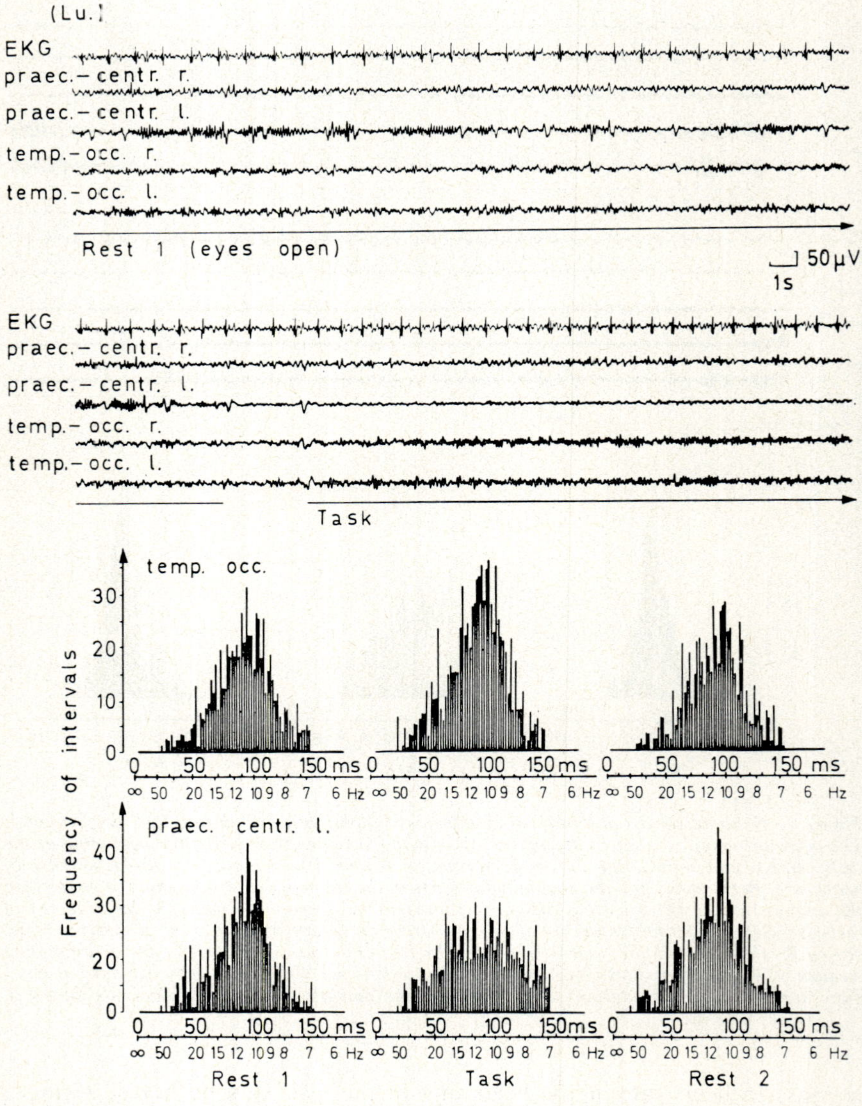

Figure 5. Original record and interval histogram from the occipital alpha (above) as well as the central µ (bottom) of an individual with alpha-increase and µ-decrease during task performance. Histograms as in Figure 4

in 7 or 8 out of the tasks (open circles in *Figures 6–9*); and (3) the 'indifferent group (crosses in *Figures 6–9*). Of 17 people investigated, 5 alpha-diminishers, 7 alpha-enhancers and 5 'indifferent' subjects were found.

In *Figure 7*, the alpha rate (i.e. number of alpha waves in 90 s) during R_1 was plotted against that during the tasks. *Figure 7B* shows the mean alpha rate of the different subjects during all resting states preceding the tasks (R_1,

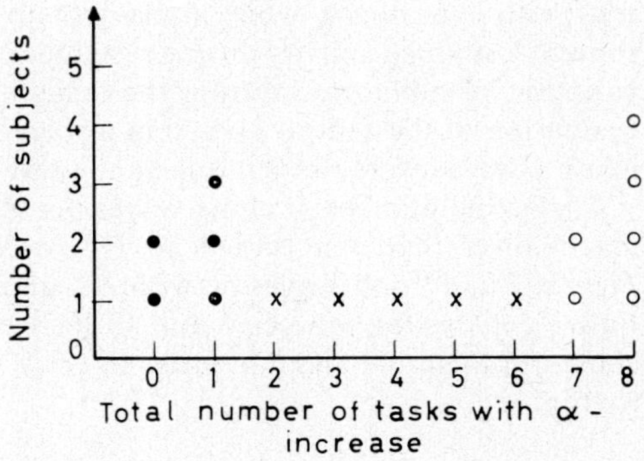

Figure 6. Number of individuals showing an increase of alpha-activity during tasks. Ordinate: Number of individuals. Abscissa: Number of tasks during which an increase of alpha-activity was found. Read for example: 4 *individuals (out of* 16) *showed an alpha-increase during* 8 *out of* 8 *tasks,* 1 *individual showed an alpha-increase during* 4 *out of* 8 *tasks etc.*

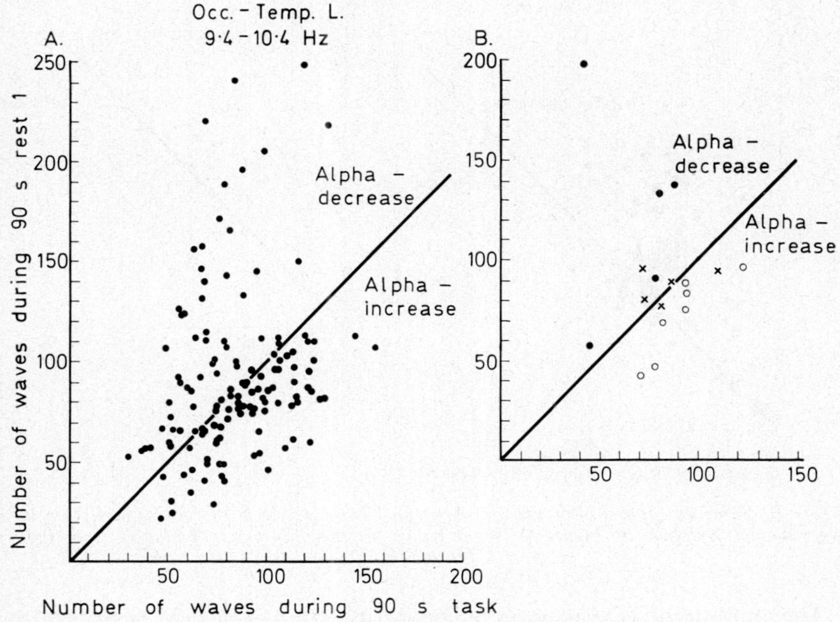

Figure 7. Relation of occipital alpha during R_1 *(rest before the task, eyes open, ordinate) and during the task (abscissa)*

A. Plot of all tasks of all individuals. The R_1*-values are those of the resting period preceding each task. Number of waves from* 9·4–10·4/s *per* 90 s. *Points lying on the* 45 *degree-line correspond to identical values during* R_1 *and* T, *those above the line to decrease and those below the line to increase of alpha*

B. Plot of the mean values for each individual during all R_1 *and* T. *Dots: Individuals who showed alpha-decrease in* 7 *or* 8 *out of* 8 *tasks; open circles: Individuals showing alpha-increase in* 7 *or* 8 *out of* 8 *tasks; crosses: Individuals showing alpha-increase in* 3–6 *out of* 8 *tasks (see Figure* 6)

ordinate) and during all tasks (T, abscissa); *Figure 7A* shows the relationship for each individual task. The diagram confirms the impression which we had already gained from looking at the individual records, namely that alpha-diminution was most pronounced in persons with a high alpha rate during rest (with eyes open), while alpha-activation was seen in such persons who showed less alpha activity during rest. One can also see that the inter-individual variation of alpha rate during the tasks is less than during rest. The same evaluation of the central μ-rhythm always showed a blocking no matter how many μ-waves were present during the rest (*Figure 8*).

The actual number of alpha waves per 90 s in *Figures 1* and *8* is apparently much lower than that shown in *Figure 2*. This is due to the fact that in *Figures 7* and *8* only waves between 9·4 and 10·4 Hz were considered, while in *Figure 2* all waves between 8 and 13 Hz were counted. The narrow band was chosen in *Figures 7* and *8* in order to get a representative sample of the alpha rhythm.

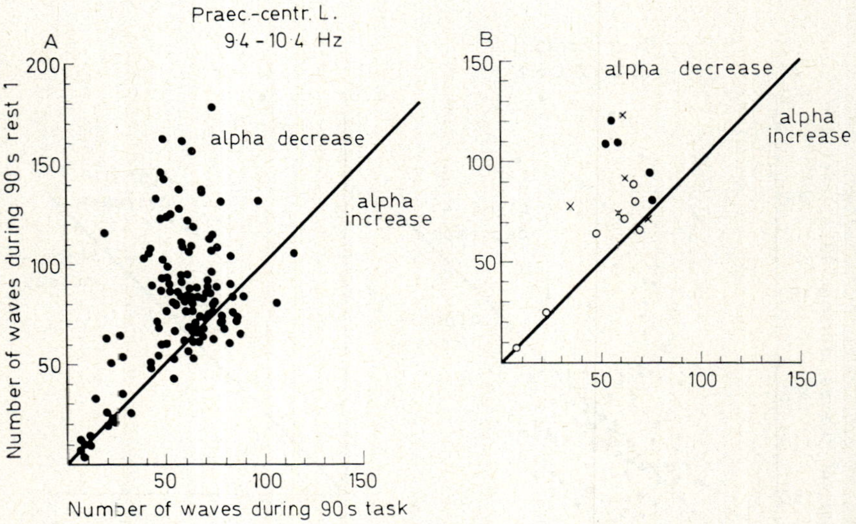

Figure 8. Same as Figure 7 for central μ-rhythm. Note decrease of central μ rhythm in all experiments. Symbols of Figure 8b (mean values for the different individuals) as in Figures 6 and 7b

The amplitudes were assessed by measuring samples of slow records during R and T. The results of these measurements are shown in *Figure 9*. It can be seen that the average amplitude of the EEG as determined by this method shows a slight increase in subjects with alpha-enhancement and a slight decrease in those with an alpha diminution. Even though evaluations by eye of records with normal speed (1·5 or 3 cm/s) showed that the measured amplitude change was mainly on account of alpha amplitude, a statistical evaluation would necessitate amplitude histograms for the different frequencies.

Finally, the mean alpha frequency was calculated from the interval histograms and plotted against the alpha-reaction. No relation between alpha frequency and alpha reaction was seen in the central and the occipital areas. Furthermore, the mean alpha frequency did not change during task performance.

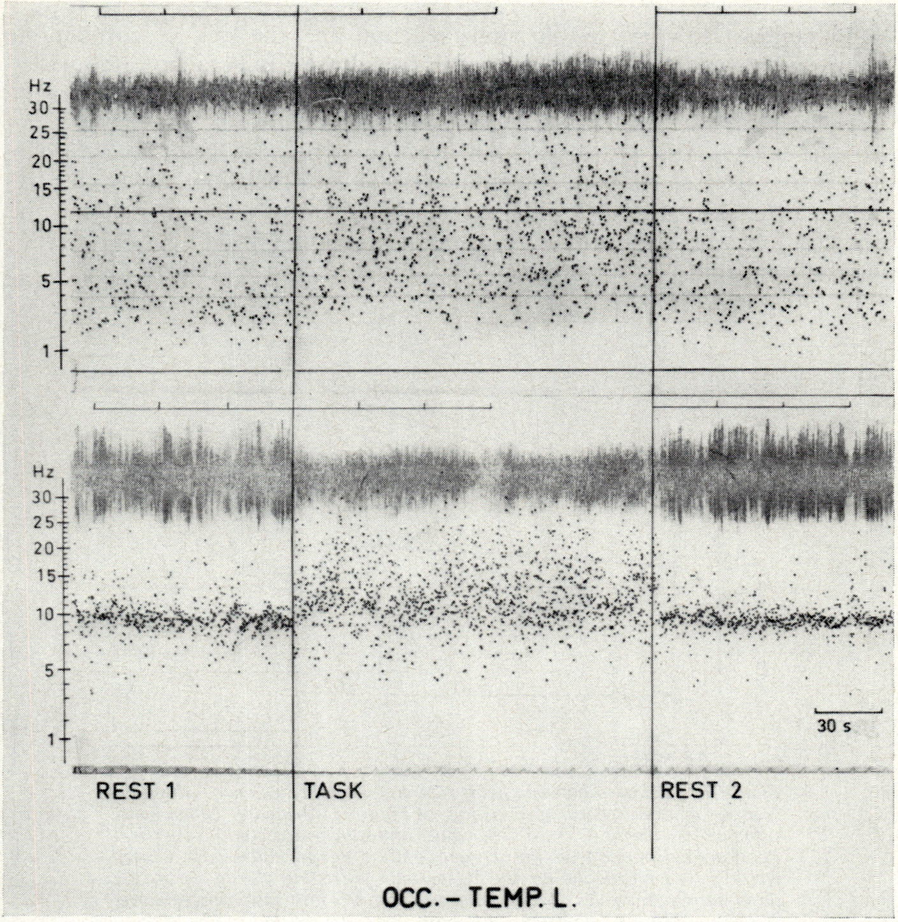

A

Figure 9. Amplitude changes of occipital alpha during task performance in two individuals. A: Recording with Toennies-analyser. Slow recording time (see time mark 30 s). *The record on the top of each curve is the occipital EEG. The points mark the zero-crossings, their height corresponds to the time elapsed since the last point (ordinate in frequency per second). Upper record: alpha-increase during the task (more dots in the alpha-range), increase of amplitude. Lower record: alpha-decrease during the task, fewer dots in the alpha-range, more in the beta-range. B: Plots of the mean EEG-amplitudes during rest* (R_1 *and* R_2 *ordinate) and task* (T, *abscissa). The amplitudes are given in relative values* (mm) *because of different calibrations for the different individuals. They are mean values from measurements taken at different times before, during and after each task as indicated above the original records in A. Symbols as in Figure 7B*

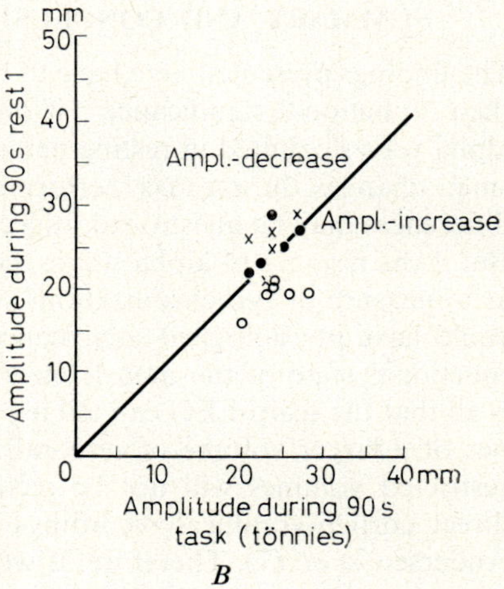

B

In contrast to the variable alpha reaction and the lack of correlation between the scaling of the task (*Figure 1*) and the EEG-reaction, the heart rate increased during all task performances and a high correlation was found between heart rate and the scaling of the tasks (*Figure 10*). The correlation co-efficient was 0·8. No correlation was found between heart rate during rest before the task (R_1) and the number of the task following this R_1-period (correlation co-efficient 0·26). Since the subject had read the instructions on the task following the recorded R_1-period, expectation could not have played a significant role.

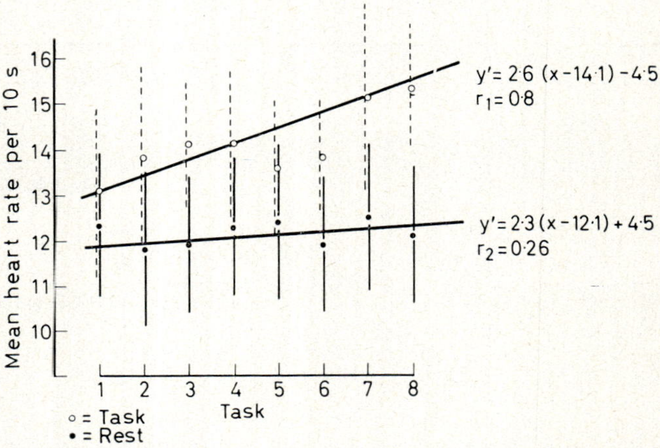

Figure 10. Relation between heart rate and task performance. Abscissa: Number of task according to scaling of Figure 1. Ordinate: Mean heart rate per 10 s *during* 1·5 min. *The values and standard deviations (vertical lines) are calculated from the averages of* 12 *different individuals. Closed circles: Heart rate during the* R_1*-periods preceding the tasks on the abscissa. Open circles: Mean values during* 1·5 min *task performance. The correlation coefficients for both lines are written at the right*

SUMMARY AND CONCLUSIONS

The findings presented here have to be interpreted with a general notion about their 'functional significance': if one considers the low absolute number of alpha waves counted in resting persons with the eyes open and the absolutely small changes during task performance, the physiological meaning of such differences may be questioned, especially if single alpha waves are concerned. But if the number of alpha waves, counted in a certain time, were taken only as a measure of 'synchronization' of cortical activity, even small differences could have physiological significance, as they would be an expression of the functional state of the area from which they were recorded. This, anyway, is all that the scalp-EEG can tell us, as it measures only gross averaged activities of a larger volume of cerebral tissue, while discrete activity changes of restricted volumes will not be recognized. This limitation is even true for direct cortico-graphical recordings and was pointed out again recently by Andersen *et al.* (1). Therefore, if we see more or fewer alpha waves or even larger or smaller ones, this does only mean that a relatively larger or smaller volume is 'synchronized' such that a visible change can be recognized in the EEG record by the human or electronic analyser.

The results presented here show that the 'blocking reaction' of the occipital alpha rhythm cannot be correlated in a simple manner to the task performance which involved different degrees of mental effort and different visuo-motor performances. It rather turned out that the blocking reaction of the occipital alpha rhythm was less dependent on the type of task than on individual differences: in some individuals (about one-third of our group) an augmentation of the occipital alpha was seen during the tasks, while only one-third showed a consistent blocking reaction, and about one-third showed inconsistent reactions. These different reactions confirm earlier observations (9, 12, 17). Our data suggest furthermore, that the alpha-reduction was most pronounced in subjects with much alpha during rest and eyes open, than in persons with little alpha, which rather tended to augment their occipital alpha rate. This might suggest a hypothesis, that too much 'synchronization' would not be compatible with the task performance while a certain degree of 'synchronization' may even be needed. This also becomes evident if one realizes that a complete 'blocking' of alpha is never observed, even when under stress situations, as the analysis of Walter, D. O., *et al.* (15) shows. Therefore, the terms 'alpha-blocking' as well as 'alpha-activation' should be used as little as possible and might be replaced by the terms 'alpha-reduction' and 'alpha-enhancement'.

No clear correlation between occipital alpha rate and the different tasks could be found in our experiments, although they could be scaled satisfactorily by subjects according to the degree of 'psychical effort' needed for their performance, and in contrast to the heart rate changes, which showed a high correlation to the subjective scaling. If we assume that this 'effort' is at least a rough measure of 'attention' or 'concentration', we may conclude that no simple correlation exists between the amount of attention or concentration and the degree of the alpha reaction within the range of our experimental situation. This is in agreement with the observations of Kreitman and Shaw (9). On the other hand, it may be noted, that occipital alpha was always reduced at the time the subject read the instructions. This may be related to the visual-oculomotor activity as well as to a different form of 'concentration' involved in reading.

The reaction of the central rhythmical activity (μ-rhythm) was different in our experiments in so far as a significant reduction was observed during all tasks and in all persons. Even though motor activities were not always excluded in the different experiments, the amount of motor activity varied considerably and was not related to the 'difficulty' of the tasks. It thus appears, that the μ-rhythm seems to react more sensitively to 'problem solving activities' of the brain than does the occipital alpha. This does not exclude a higher responsiveness to more localized activities in the somato-sensory and motor system (5, 6), but needs a critical interpretation of such findings as to their physiological relevance.

The occipital beta-increase needs some comment. The small but significant differences during task 4 (smaller increase) and task 8 (larger increase) may suggest some special, local factor. Because of the small differences in actual number of beta waves, however, we hesitate to put too much physiological significance on this finding, particularly as some admixture of muscle potentials could not always be excluded.

In conclusion, it can be said that our experiments support the notion that the occipital alpha rhythm seems to have a multifactorial relation to the activity states of the brain as a whole. Individual differences must be taken into account as well as certain dominating factors (e.g. eye closure or opening, oculomotor activity, orientation, etc.). The findings furthermore support the independence of the central and occipital rhythmical activity of the alpha band, also shown experimentally in animals recently (1, 4). The 'arousal' hypothesis of the alpha reactions therefore needs some reconsideration.

ACKNOWLEDGEMENT

The investigations were carried out with financial support of the Deutsche Forschungsgemeinschaft, which gave a Research Fellowship to one of us (G. Grünewald) and financed the IBM 1130 (Grant Cr30).

DISCUSSION ON CREUTZFELDT

Grey Walter

Dr. Creutzfeldt's paper is open to discussion. I am particularly interested in his analysis of alpha variations in various situations. May I first ask if you think that the baseline crossover, period-analysis type of detection is really a suitable way of providing data for your computations?

Creutzfeldt

I think it is a very good way of looking at one aspect of alpha activity or any sort of rhythmic activity. None of the methods, neither Fourier analysis nor power-spectrum analysis, tells us anything about the *number* of waves appearing during a certain period. The question we are asking with this method is: how much time does the brain spend in producing a given type of rhythmic activity? Of course, the method of interval analysis alone does not give a complete picture of brain activity, and one has to add at least an amplitude histogram.

Grey Walter

If you choose to ignore or dispose of one parameter of information such as amplitude this is perfectly permissible—in fact, we must all do this because we obviously can't deal with all the information all the time. But the baseline cross-over period analysis not only ignores amplitude but also admits certain types of error which you did mention, but then dismissed.

Stark

This has been studied quite thoroughly, and is what one would call a one-bit power-spectrum, and it approximates to an ordinary power-spectrum very well for most relatively well-behaved functions like the EEG.

Grey Walter

Is the EEG a well-behaved function?

Stark

Well for these purposes I would say it is. I mean it's not a discontinuous function and has a restricted bandwidth. I mean well-behaved mathematically, by the way.

Grey Walter

Yes, I know what you mean, but I should have thought it rather ill-behaved.

Stark

Well, I shouldn't.

Adey

I'd like to make a plea for retention of amplitude information from a number of points of view. I think in the first place the spectral density function—the preparation of the spectral analysis by one or other of the Fourier methods, does give a very important handle that is missing here—namely the question of the alpha density in the resting condition as opposed to that in the performing subject. Our own data shows very clearly that the relative density in the alpha band is dependent upon the difficulty of the task performance by man in relation to the alpha density at rest or across a series of situations. In this regard the question of stationarity of the EEG is something in which the sensitivity of the spectral tool to stationarity as such, invites its use in assessing whether or not the subject has in fact changed state between, say, one 5 s epoch and another. It would seem to be very difficult to justify the use of a simple period analysis when so much more can be learned, and in fact the descriptors of the alpha band are that much better if you use a true spectral analysis. I think you're throwing away far too much in this way.

Stark

Well you may not be, because one can see in an intuitive way how even amplitude information enters in the one-bit power-spectrum. Supposing you have a 15 Hz wave which is very high amplitude, then all its zero-crossings will appear in the one-bit power-spectrum, whereas if it's a very small amplitude wave, sometimes it will be riding high on some other lower frequency and won't appear. However, at times it will contribute a zero-crossing. So I don't believe one can say the one-bit power-spectrum throws away amplitude information, although it certainly reduces information content. One can compute it in terms of bits with the help of Shannon's Information Theorem. This zero-crossing assessment is a reasonable thing: forget about multi-level thresholds, just take the one-bit power-spectrum and obtain a reasonable well-behaved approximation to the entire process.

Creutzfeldt

May I add to that, that we also have analysed the EEGs by visual inspection. If a subject had many alpha waves they were also large, but if they had little alpha, then the waves were small.

Grey Walter

Oh, I wouldn't agree with that. Just looking at a lot of EEG records goes against that.

Creutzfeldt

Well, in our experiments a high number of alpha waves was correlated with high alpha amplitudes.

Grey Walter

I should say that there might even be an *inverse* correlation if anything; people who have very large alpha rhythms usually have it in bursts, while those with small alpha have it continuously.

Byford

For a signal of wide and largely unspecified bandwidth, the interpretation of zero-crossing points is made more difficult when filters are used; largely because the time relationship between the several components of a complex signal are changed. Low frequency interference and baseline sway pose additional problems, whilst the use of the first or second differential only aggravates an already undesirable noise situation.

Grey Walter

Making use of the first second derivatives.

Byford

But it's a very difficult technique because you have two processes, each of which is proportional to the frequency.

Mirsky

I think that the proof of the pudding is in the eating; if you have a method for reducing the data of the EEG in order to compare with behaviour, the important question is, what is the correlation with the behavioural measure? A related question concerns whether or not you can improve this correlation by using more refined EEG methods, and this is essentially what we've been talking about. I'd like to ask also whether or not Dr. Creutzfeldt considered relating variation in performance of the subjects to the EEG measure. It seems to me, that what he did was to vary the difficulty of the task, at least as determined by the subject in terms of how much attention he thought it was necessary to use while performing. I'm sure that there must have been some variability here, with some subjects perhaps finding the mirror drawing extremely difficult, and others being able to cope with it somehow. Does taking into account this variable improve in some instances the understanding of the variability in the alpha activity?

Creutzfeldt

Our aim was to devise a test battery with tasks of different difficulty and in doing so we completely disregarded other aspects which might also have been important. The scaling of the tasks differed somewhat from person to person as you may have seen from the standard variations but generally you can say that tasks 1 to 4 were all considered easier than tasks 5 to 8. Comparing these two groups we did not see any significant differences. By careful statistical analysis, however, slight differences could be found between tasks which involved different 'systems' as, for example, mainly visual-motor, sensori-motor or purely mental tasks (calculation). The most striking fact was that the reaction of one individual was almost identical during all the tasks.

Mirsky

In a way you've anticipated my presentation. If you use a task that permits the study of such lapses of attention it provides a measure of omission errors such as Dr. Berlyne referred to in his review. Then you might have a means of relating behaviour to moment-to-moment changes in EEG, by baseline crossings or some other methods. This would permit you to study the variability more fully. I think with most of the methods you used it was not possible to know if a subject was blocking or failing, or making an error at a particular time. You do not have a serial record of performance which you could relate directly to the EEG.

Adrian

Might I ask whether you tried to get them to solve their problem in a given time or not? I always got the impression that it was *urgency* of the problem rather than the *difficulty* of it that was one of the important factors.

Creutzfeldt

We tried to avoid this factor by letting the subjects do one task for some time (2 minutes) without urging them to do as much as possible. But they were all motivated to do their task as well as possible.

Dewan

I am not clear whether you used the one-bit spectrum or zero-crossing or whether you use a threshold above zero to make your measurements.

Creutzfeldt

We used the zero-crossing with a set threshold of about 10 μV.

Dewan

So you raise it up to filter out noise, as it were?

Creutzfeldt

Yes.

Grey Walter

I have a query about the zero too, and this seems to me to be the basic question.

Creutzfeldt

Again, as long as you know what you are doing and as long as you are aware of the defects of your method, you may use it. The threshold was the same throughout the experiment. The absolute number of waves counted may have been slightly wrong because of the threshold above the zero, but relative changes came out clearly.

Dewan

Does this introduce artefacts? I mean in the problem of filtering out noise.

Stark

I don't think it's a big problem. He's using bipolar electrodes and he's just looking at a bandpass of 6–16 Hz. I don't think the mathematical details of that are really relevant at the level at which the brain is being approached.

Dewan

You don't think this is introducing some sort of artefact that could give you the kind of result he's reporting?

Stark

No.

Adey

Then if I read you correctly Dr. Stark, you're inferring that the rest of us are wasting our time in doing a more detailed spectral analysis?

Stark

I don't know that anybody's doing a more detailed spectral analysis.

Adey

Well, my group is certainly doing more than a one-bit spectral analysis.

Stark

Well, I don't know what the redundancy is in computing these statistical functions. One always has a lot of redundancy anyway, and you might be doing it by taking a longer record, by calculating more amplitude values, or by taking more frequency samples, and so on. It's just more work for the computer.

Adey

I think that the detail of the answer is relevant qualitatively to the answer that you are getting. There are certain manipulations of the data in the realm of population characteristics that become exceedingly difficult to justify on the basis of a one-bit spectral analysis; and that it is necessary to have the analysis in more detail than that in order to evaluate the population characteristics.

Stark

But certainly not for the data he showed earlier.

Adey

Why do you say that?

Stark

Well, just showing this narrow bandwidth at the alpha frequency. It just had the gross features of the shape of that bandwidth—it was showing large changes, confirming his ideas. Nobody was looking at fine details in that curve so why do a more precise analysis?

Adey

Well, my contention is that there are certain other interpretations that can be put on data gathered in these circumstances if the analysis is more detailed and the question of task-related levels of physiological change are those which are beyond that which one would see with a one-bit spectrum.

Creutzfeldt

This may be. Our findings just relate to one aspect of the EEG, and that is its periodic frequency contents. You may discover other aspects if you use aperiodic analysis methods, etc.

Dewan

But, from what Dr. Stark says, you put this through a bandpass filter before you do the analysis?

Stark

Yes, and it differentiates before you put it into the bandpass filter.

Dewan

Why does it have so much noise?

Stark

It doesn't have much noise. Except the EEG, which is noise, isn't it?

Lindsley

We are all agreed that many of these activities will block the alpha rhythm, but I am sure that in these very dominant alpha rhythm people we've all seen cases where alpha will be practically 100 per cent and show no signs of blocking. When you ask such a subject to image a breakfast scene and report afterwards what he had imaged, or get him to do associations you find no blocking of the alpha waves. New processes can go on without apparently any disturbances in the alpha rhythm. I have found this in a normal subject, a professor, if such can be called normal, so this raises an interesting question. Certainly more often than not one finds blocking under such conditions in most people, but the fact that the mental processes involved in solving problems can go on in some persons without blocking also raises the question of how essential it is that alpha reflects this change.

Grey Walter

I am afraid I must close this discussion in order to be fair to later speakers. I am sorry, for I should like to have heard some more from the participants. Just to close, Dr. Creutzfeldt's observations seem to have produced a number of hypotheses which can be tested by other methods. There seems to be one advantage of this type of data reduction, that he's produced a hypothesis and suggestions that the rest of us, using other methods, can proceed to study and check, and see whether in fact the results do relate to the electrogenesis of the brain or to some other source.

REFERENCES

1. Andersen, P., Andersson, S. A. and Lomo, T. (1967) *J. Physiol.* **192**:283
2. Berger, H. (1929) *1. Mitt.-Arch. Psychiat. NervKrankh.* **87**:527
3. Berger, H. (1931) *3. Mitt.-Arch. Psychiat. NervKrankh.* **94**:16
4. Buser, P. (1967) Personal communication
5. Cobb, W. A. (1963) *Electroencephalography*, 2nd edn. London: MacDonald (Ed. by Hill and Parr), p. 232
6. Gastaut, H. (1952) *Revue neurol.* **87**:176
7. Grünewald, G., Simonova, O., Schmitz, H. and Creutzfeldt, O. (1968) *Arch. Psychiat. Nervkrankh.* **212**:46–69
8. Ito, S., Rosina, A. Creutzfeldt, O. and Probst, W. (1969) *J. Neurophysiol.* **32**:127
9. Kreitman, N. and Shaw, J. C. (1965) *Electroenceph. clin. Neurophysiol.* **18**:147
10. Lettich, E. and Margerison, J. H. (1961) *Electroenceph. clin. Neurophysiol.* **13**:606
11. Mulholland, T. B. and Runnals, S. (1962) *J. Psychol.* **54**:317
12. Mundy-Castle, A. C. (1951) *Electroenceph. clin. Neurophysiol.* **3**:477
13. Mundy-Castle, A. C. (1957) *Electroenceph. clin. Neurophysiol.* **9**:643
14. Oswald, I. (1962) *Sleeping and Waking*, New York: American Elsevier
15. Walter, D. O., Kado, R. T., Rhodes, J. M. and Adey, W. R. (1967) *Aerospace Med.* **38**:371
16. Walter, W. G., Cooper, R., Crow, H. J., McCallum, W. C., Warren, W. J., Alderidge, V. L., Storm van Leeuwen, W. and Kamp (1967) *Electroenceph. clin. Neurophysiol.* **23**:197
17. Williams, A. C. (1940) *J. exp. Psychol.* **26**:413

Evoked Potential Correlates of Perceptual Organization in Man

William R. Goff
Neuropsychology Laboratory, Veterans Hospital, West Haven, Conn.

Utilization of the average evoked potential technique to study the effects of attention on sensory processes is based on the assumption that these potentials reflect neural activity whose psychological result is sensory perception. The term sensory perception, as I use it, includes psychological processes subsequent to and more complex than simple detection of peripheral receptor activation. It includes, for example, recognition of the quantity and quality of the stimulus; its association with other concurrently impinging stimuli and with previous stimuli, that is, past experience. We further expect, or at least hope if we are working in this area, that alterations in the attention, or in other words, in the perceptual receptivity of the subject will be reflected by co-varying changes in one or more components of the sensory evoked response. In short, then, the independent variables of evoked response-attention experiments are operations designed to organize perceptual receptivity in a given way with respect to specified stimuli; dependent variables are alterations in responses evoked by these stimuli. Attention, then, is an intervening variable, a name given to the hypothetical subjective effect of the operations.

There is general agreement in the literature that it is the so-called late components of the averaged cerebral evoked response—those occurring later than 80–100 ms—that are most labile under experimental operations designed to alter attention. This is not surprising, because numerous reports, some of them from our laboratory, show that it is these late components which are maximally affected by any procedure which alters the 'state' or, as I prefer to call it, the perceptual organization of the subject. At the same time, there is considerable disagreement as to what effect alterations in attention have on these late components. The majority of reports say that response amplitudes increase with increased attention, but a significant number say that there is no change or a decrement in amplitude. Finally, there are reports of both effects in the same experiment.

Research in our laboratory has been directed to a general analysis of the neuropsychology of cerebral evoked responses and I have been asked to evaluate the concept of attention in evoked response studies. I will attempt this by presenting results from our own and other work and indicating my opinion of the implications of these results for methodology and concepts in attention experiments. Our work has been primarily in the somatosensory system, but I believe that the main conclusions can be generalized to other senses.

METHODOLOGICAL PROBLEMS

First, let's look at the evoked response components we will be talking about. In *Figure 1* we compare somatic, visual, and auditory late responses evoked

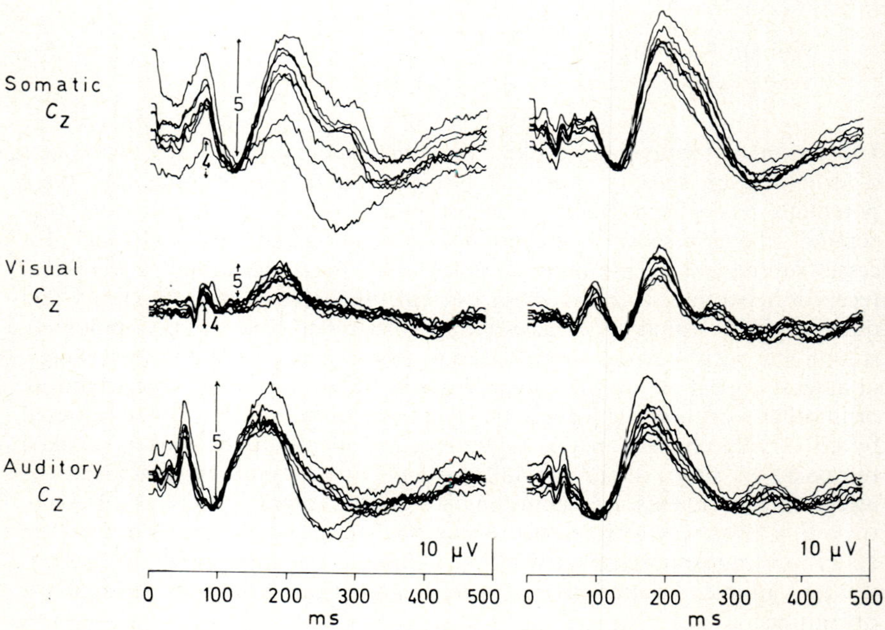

Figure 1. Somatic, visual, and auditory evoked responses recorded from the vertex in two subjects. In this and subsequent figures, positivity at the active electrode produces an upward deflection

under identical experimental conditions in two subjects. The active electrode was at the vertex—C_z in the international 10–20 system. The reference was the earlobe contralateral to the stimulus. In preliminary tests, conducted by recording from this location referred to other non-scalp locations, this reference was found to be inactive with respect to evoked potentials in all three modalities at stimulus intensities greater than those normally used. Each trace is the average of 64 responses, the time base is 500 ms, and positivity at the vertex is upwards. The somatic stimulus is a shock to right median nerve at the wrist; the visual stimulus is a flash presented in Maxwellian view to the right eye; the auditory stimulus is a click presented by earphone to the right ear. Auditory, visual, and somatic stimuli were equated subjectively. The averages were obtained over several sessions in which the three types of stimuli were presented randomly at a rate of 1 per 4 seconds. Stimulus presentation and response averaging were done by an on-line LINC computer. The traces are the X-Y plots from the computer.

In a 1962 study of somatic evoked response distributions (11), we separated the later aspects of the somatic evoked response into two components. These are labelled components 4 and 5 in the figure. As is now well known, visual and auditory response components are also recorded in the latency range and with the same polarities as somatic components 4 and 5. Possible homologues to somatic components are correspondingly labelled. An auditory component 4 is not consistently distinguishable in all subjects. The component seen more likely corresponds to somatic component 3, as I will show in a moment.

Dr. Yoichi Matsumiya in our laboratory is completing an intermodality comparison of waveform and distribution in the three sensory systems, in 12 subjects, under identical experimental conditions. Eight averages of 64 responses were obtained for each of the three modalities at 19 standard 10–20 system locations. Additionally, post-auricular, trapezius muscle, and outer canthus placements were used to assess possible contributions of retinal, oculomotor, and myogenic evoked activity.

To compare amplitude distributions of response components, iso-amplitude contour maps were drawn on the basis of peak-to-peak measurements made as shown by arrows in *Figure 1*. These are presented for the late components in *Figure 2*. The location of maximum amplitude is shown by the filled circles. Amplitudes at other locations were normalized to this maximum taken as 100 per cent and the 75 per cent iso-amplitude contour was estimated as shown. It is clear that despite a similarity in latency and polarity between somatic component 4 and its visual counterpart, their distributions are quite different. The somatic component has a central-frontal projection; the visual projects mainly to the occipital region. The two maps at the bottom of the figure show the focus and distribution of component 5. This component shows identical foci at the vertex in all three modalities. The distributions of the somatic and auditory potentials are highly similar. The visual potential encompasses the same area as the somatic and auditory but is more widespread particularly towards the occipital region.

These long latency evoked potentials are variously called secondary, diffuse, non-specific, or most often, vertex potentials. It is some aspect of these potentials which is most frequently measured and compared both within and between modalities in attention experiments. The vertex potential is classically described as having a maximum amplitude focus in the vertex region, as being diffusely projected, and as being modality non-specific, that is, evoked with similar characteristics by auditory, visual, and somatic stimuli. Our results indicate that only the diphasic potential that we label component 5 may properly be called the vertex potential with its implied modality non-specificity. The earlier diphasic sequence, which we label 4, is not consistently distinguishable to auditory stimulation, is not focused at the vertex, and has modality specific distributions.

Since some laboratories have been interested in the effects of attention

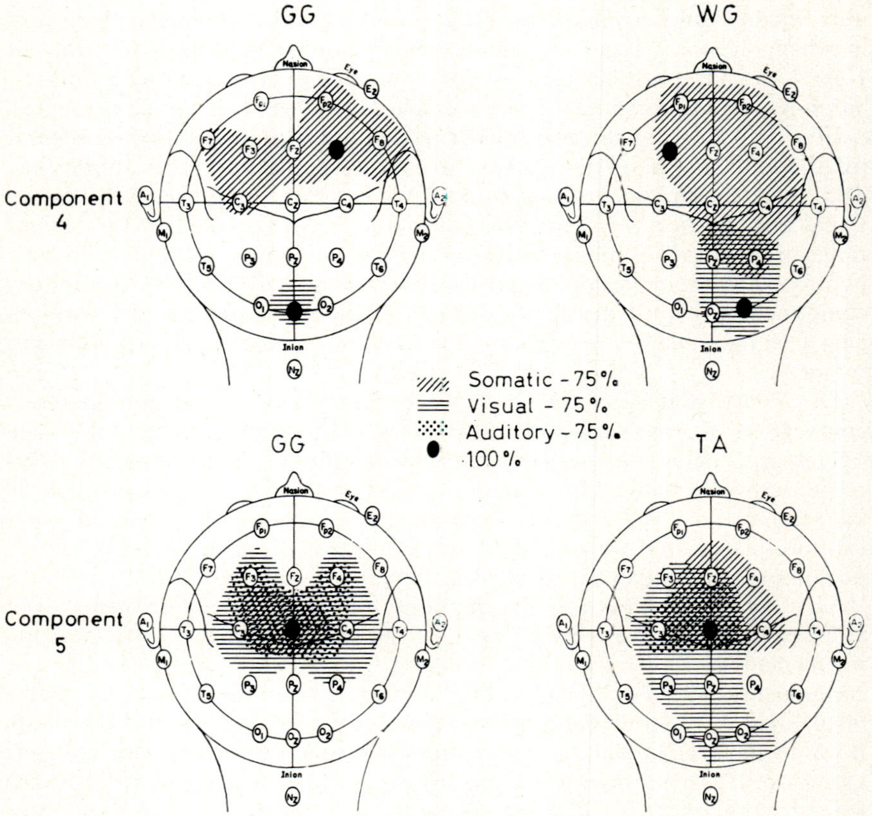

Figure 2. Comparison of amplitude distributions for long latency components of somatic, visual, and auditory evoked responses

on early responses, it is instructive to compare them to later components. In *Figure 3*, the first 100 ms of responses for the three modalities are shown at locations where they are maximum. Somatic components are labelled according to our 1962 system and possible homologues are labelled correspondingly in the auditory system. Very similar appearing components are recorded to somatic and auditory stimuli although with some latency differences. For reasons as yet unknown to us the visual system early response waveform does not resemble the other two senses.

Figure 4 compares amplitude distribution for several of the early components shown in the previous figure. The specific components mapped and how they were measured base-to-peak are indicated in the evoked response reproduced below the schematic of each head. Distributions from somatic and auditory component 3 and somatic 2 are from the same subject; the visual distribution is from a different subject. We do not record prominent early responses in all modalities in all subjects. Distributions displayed were selected as representative of the subject group studied. Auditory component 2 is not plotted as we are having trouble separating it from extracerebral potentials apparently originating in the eye region. The figure shows that early components generally have a modality specific focus and relatively restricted distribution.

Subject to further analysis of auditory early components, these data indicate that all components of cerebral evoked responses, with the possible exception of vertex potentials, have differential characteristics of waveform, latency, and distribution both within and between modalities. Which, if any,

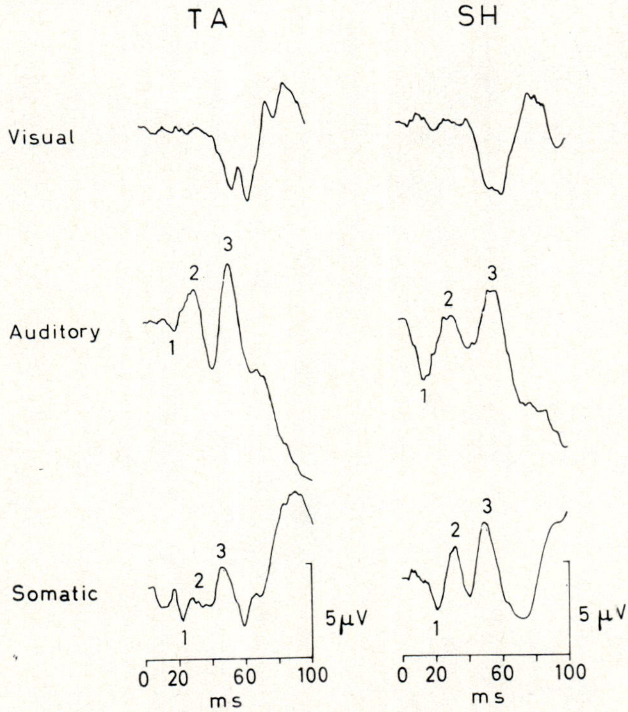

Figure 3. Comparison of early components of visual, auditory, and somatic evoked responses

of these components are neurally or functionally homologous across modalities requires further research. Of the late components, the vertex potential is most likely homologous and thus is the most legitimate potential to compare among the three sensory systems. Before discussing the implications of these results for attention experiments, I would like to review two other studies.

In 1962, Dr. Truett Allison in our laboratory did a study of recovery functions for somatic evoked responses (2). Our other work had suggested to us that the total somatic evoked response was a composite of cortical activity mediated by multiple neural systems. In animals, complex evoked response components attributable to different areas of the brain had been shown to display different recovery functions. The recovery technique consists of presenting a conditioning stimulus followed at various time intervals by a

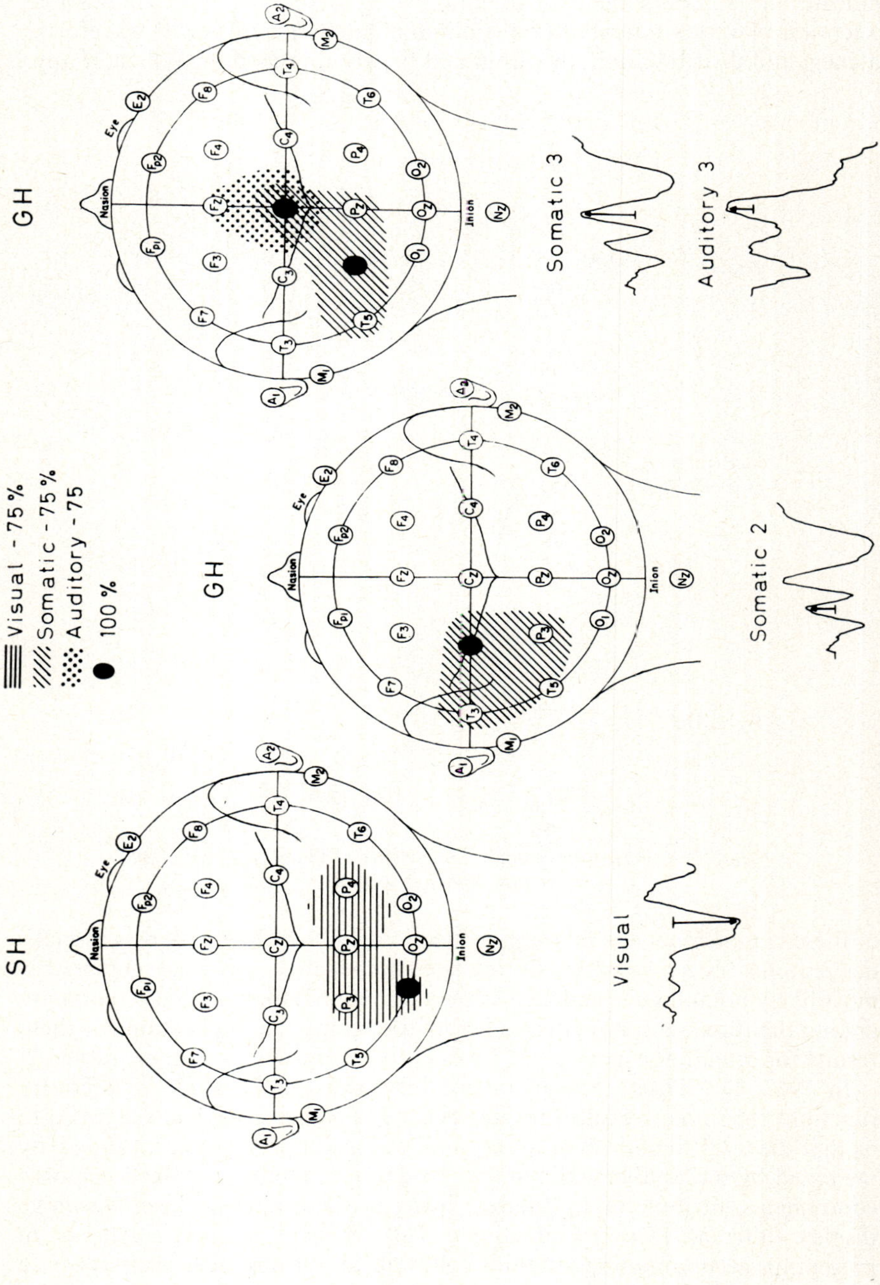

Figure 4. Amplitude distributions for several evoked response components shown in Figure 3

test stimulus. The test response is altered by the occurrence of the conditioning response. Relative test response amplitude plotted against interstimulus interval yields the recovery function. Applying the recovery technique to man, Allison found that different components did have different recovery functions. Time to full recovery—that is return to control amplitudes—varied directly with the latency of the component ranging from about 300 ms for component 1 to about 4 s for the latest component, the vertex potential—a better than tenfold difference. Davis *et al.* (5) have done a similar study for the auditory system. They report depression of test response amplitude at intervals of at least 6 s and perhaps as long as 10 s. They

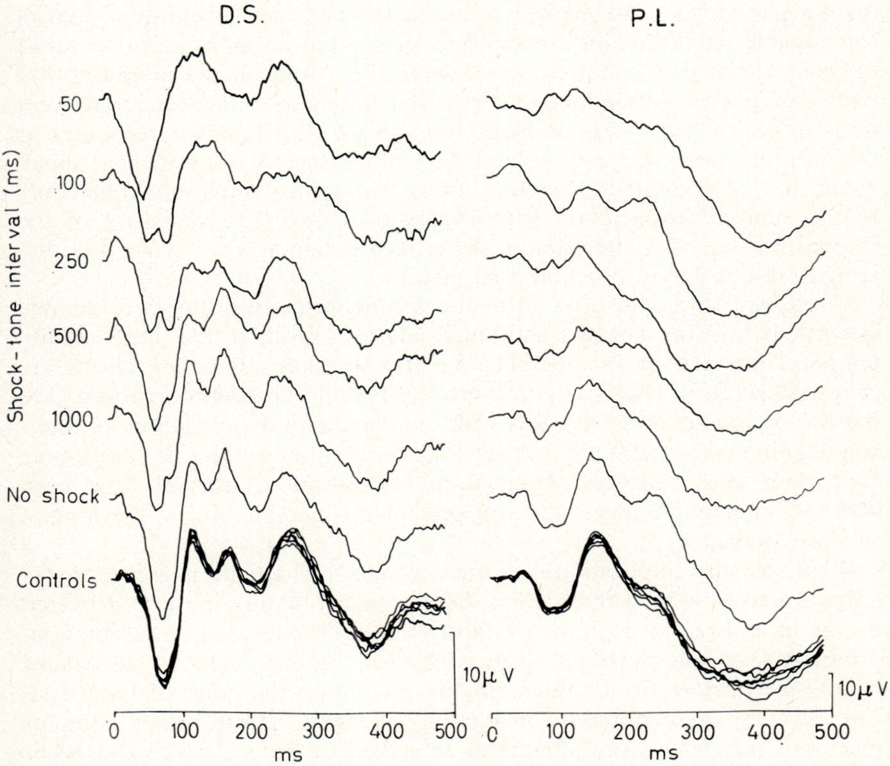

Figure 5. Effect upon a tone-pip evoked response of a shock preceding it by various interstimulus intervals. Two subjects are shown

also make the point that the relation of response amplitude to interval between stimuli depends not only on the single interval between one stimulus and the stimulus immediately preceding it, but also on the average number of stimuli per minute during the entire period of summation.

The recovery work shows that the neural systems giving rise to different components have different temporal capacities. This in turn suggests that there will be differential temporal interactions between repetitive stimuli.

Interactions should be greatest in the same sense modality but may also be significant between modalities. *Figure 5*, from an experiment Dr. G. D. Goff and I have in progress, is an example of temporal interactions between shock and tone pips. Results for two subjects are shown. A median nerve shock preceded a 10 ms tone pip by the time interval indicated. On alternate stimulations the tone was not presented and the evoked response to shock alone, after appropriate delay, was subtracted from the shock-tone composite. Thus the tone-evoked response shown here is uncontaminated by superimposition on the shock response. The single traces are computed grand averages of 384 responses recorded in blocks of 64 on each of 6 different days. Shock-tone intervals were varied randomly but the duty cycle was always maintained at 5 s. The evoked response labelled 'no shock' is a control for possible alterations in the subject's perceptual organization as a result of being shocked. Conditions were identical to those of the preceding five responses except that randomly interspersed among the shock-tone trials were trials in which shock was omitted. For the six superimposed responses at the bottom, summated over several days, the subject was aware that shock would not be presented. Therefore, these two control situations differ only in the subject's expectation with regard to shock. The similarity of the responses shows that alteration in the evoked potential when shock was presented are not due to psychological factors.

Comparing the controls to the shock-tone responses, the considerable alterations in form, latency, and amplitude as a result of the intermodality temporal interactions is apparent. We also see that interaction effects are subject dependent. Examining the vertex potential, for example, subject D.S. has a U-shaped depression effect with the maximum depression at an interstimulus interval of 250 ms; subject P.L. shows a monotonically decreasing depression as a function of increasing interstimulus interval. For both subjects, interaction effects are still present at intervals of 1 s, the longest we have studied so far.

What are the implications of these results for attention experiments? I would like to suggest three. First, because many stimuli must be delivered to obtain an average, legitimate concern about changes due to habituation, drowsiness, or other changes during the resulting long sessions has caused many experimenters to use interstimulus intervals in the range of 1 s or less. The recovery and stimulus interaction evidence indicates that stimulus repetition rate may cause alterations in evoked responses which can occur both within and between modalities and which are differential as a function of the latency of a given component—the later a component the more it is apt to be altered. I suggest that these purely neurophysiological effects must be assessed in each experiment before one is secure in attributing alterations in the evoked response to effects of psychological variables such as attention.

Second, in reviewing attention-evoked response literature, I found that many reports considered only changes in the overall amplitude of the total response—that is, the total response under attention conditions was compared by superimposition or other technique to the total response under inattention conditions and gross changes in response amplitudes were noted. In other cases, a single component was measured, frequently the peak-to-peak difference in the largest fluctuation. This usually, but not always, corresponds to the vertex potential. I have presented some evidence illustrating that the multiple components of the total or compound evoked potential have different distributional and temporal characteristics and therefore probably reflect activity in multiple neural systems. I find it comfortable to assume, as do others, that these different neural systems have different functions in the perceptual organization of the brain. I suggest, therefore, that some of the inconsistency of results from different laboratories as to whether attending to a stimulus increases, decreases, or does not change evoked response amplitude, stems from the fact that incomplete measurements are made of the total response, or that if specific components are measured, there is sometimes inadequate specification in terms of latency and polarity of which components. As a result, in different reports, different components and therefore different neural systems are being compared. Relevant to this point is a recent paper by Wilkinson and Morlock (21). They measured six components of the total auditory evoked response to clicks approximately 40 db above threshold under three conditions: first, with the subject relaxed and passive to the stimulus; second, a condition they called low incentive in which the subject pressed a button as quickly as possible after the stimulus but received no feedback or differential reward for performance; and third, a condition of high incentive in which there was feedback and differential reward for reaction times below 250 ms. *Figure 6*, adapted from their paper with the kind permission of Dr. Wilkinson, summarizes their results.

The bar graphs represent base-to-peak amplitude measurements of the components of the response reproduced. The response was recorded from the vertex (C_z) against a mastoid reference. Component latencies are given for each bar graph. Comparing the passive condition to both response or attention conditions, it is apparent that which component is chosen for measurement affects the result. Responding produced a statistically significant increase in the amplitude of P_1, N_1 and P_3; no change in P_2; and a significant reduction in the amplitude of N_2 compared to the passive situation. There is also a significant increase in certain components in the high incentive situation with its implied greater attention compared to the low incentive situation. I ask you to remember this high-low attention difference as I will return to it in another context.

A third point is that the differential characteristics of response components magnifies the problem of electrode placement. The evidence I have summarized indicates that even within the same sense modality we are recording the weighted algebraic sum of activity in different neural systems differentially affected by attention variables. When two modalities are used, different stimuli give rise to components with different latencies and distributions. In reports from different laboratories, even when the 'active' electrode location is similar, variations in location of reference electrodes hinder and often prevent comparisons of results.

I do not suggest that these factors necessarily invalidate the results of individual experiments. The typical experiment controls for them to some degree by comparing attention to inattention while holding other factors constant. However, the fact that the results from different experiments are in conflict indicates that further procedural refinement is necessary. In this context, I would like to suggest, first, the desirability of evaluating the total evoked response by measuring changes in all components; second, precise specification in terms of latency and polarity of what components are measured and how, giving due consideration to the possible effects of electrode placement especially in cross modality comparisons; and third, evaluation of temporal interactions which may be causing us to seek effects of psychological variables on response components already altered indeterminately by such interactions.

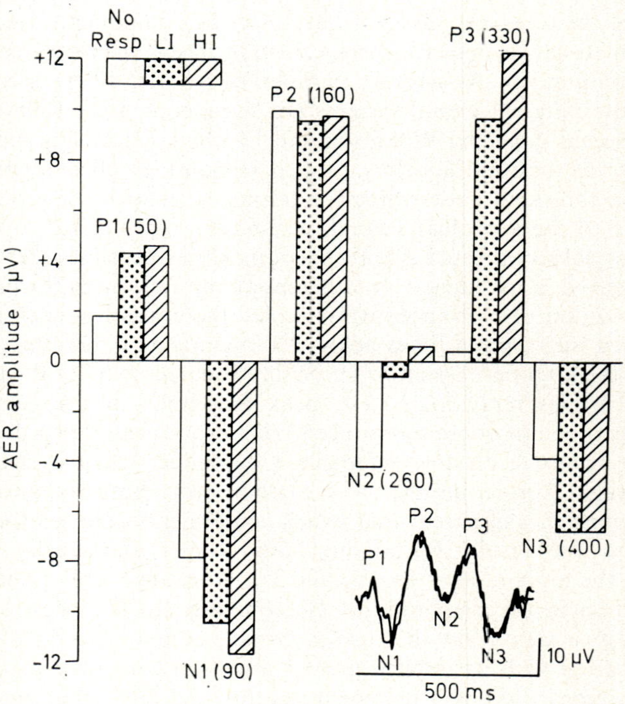

Figure 6. Amplitude of auditory evoked potential components under conditions of high incentive responding (HI), *low incentive responding* (LI), *and no response*

(From Wilkinson and Morlock, 1967, by courtesy of the Editor, *E.E.G. Journal*)

Overshadowing all these procedural problems, however, is a problem that I think is the most serious confronting evoked response-attention experiments. That is the lack of adequate independent determination of the efficacy of operations designed to alter attention. Some experiments have relied on the intuitive judgment that when a subject is required to count stimuli, or detect a change in some stimulus parameter, he is attending to that stimulus in contrast to remaining passive or attending to other stimuli. In other words, two or more stimuli are presented to the same or different modalities and differences in attention are defined in terms of the condition that one stimulus is made meaningful and the other is not. In some experiments, no measure of performance is taken; the efficacy of independent variable manipulation is defined in terms of the dependent variable effect, that is, if changes in evoked responses are seen, they must be due to changes in attention. Therefore the operations used to alter attention must be effective. Expressed in these terms, the circularity of this reasoning is obvious. Other experimenters have tried to assess the effect of their manipulations of attention by making careful measurements of performance. The question then becomes: do the changes in performance really correlate with changes in attention, which are in turn really correlated with evoked response changes? An experiment of Satterfield (18) is relevant in this context. In 47 subjects, he presented alternating click and shock stimuli at interstimulus intervals of 500 ms and 1 s. The subject was required to detect and report verbally any change in the intensity of one type of stimulus. He was said to be attending to the type of stimulus requiring this response and not attending to the other. Measurements were made of the diphasic vertex potential evoked by shock and click in the attending and non-attending situation. Performance on the detection task was used as a behavioural measure of attention. The typical result for both types of stimuli was an increase in the amplitude of response for the attended stimulus. However, Satterfield found that seven of the 47 subjects showed amplitude decrements for one or both vertex potentials although they showed no difference from other subjects in their performance on the detection task. One must infer, then, either that the vertex potential as measured does not vary with attention in a consistent manner across subjects, or that performance was not an adequate measure of the efficacy of operations designed to alter attention.

I acknowledge my own inability to be constructive on this point. I do not know how we can assure that our psychophysical determinations really measure attentive variables. My purpose is to bring the problem up for discussion. I think it reflects a lack of clear definition of what attention is. Perhaps, if this conference can clarify the concept, we will then see how to measure it.

NEUROPSYCHOLOGICAL CONCEPTS

Leaving methodology, let us turn to concepts in attention-evoked response research. To date, the research has been predominantly concerned with establishing and quantifying correlations between changes in attention and changes in evoked responses. Such research lays the foundation for conceptually more complex experiments and current disagreements show that the foundation is by no means complete. However, we must eventually progress to understanding these changes in terms of the neurophysiological mechanisms and psychological significance of evoked responses as they relate to perceptual organization in general. Therefore, I would like to discuss the concept of attention in the perspective of present knowledge in this area as I understand it. In my discussion, I will lump all components occurring after 100 ms as late components although emphasis will be on the diphasic vertex potential since it is the most often studied. Parenthetically, I remind you that analysis of the neurophysiology of human late components is hampered by the lack of established homologues in lower animals. We are thus denied the investigative techniques which have been so informative with regard to shorter latency, primary evoked potentials.

As I have said, attention is only one of numerous psychological variables which alter long latency evoked potentials. Systematic relations have been reported between these components and stimulus uncertainty, 'meaningfulness' of the stimulus, difficulty of discrimination, reaction time, as a function of conditioning, and even intelligence. Most important for our interests here, these potentials reflect neural activity which is intimately associated with levels of consciousness. Illustrative evidence for this point comes from an early attempt in our laboratory to test a theory about the neural basis of the somatic evoked response. This theory said that the specific projecting, faster recovering, short latency evoked response components resulted from cortical activation by impulses travelling in the so-called lemniscal afferent system, while the diffusely projecting, slower recovering components resulted from activity in the non-specific extra-lemniscal system. This division of afferent sensory systems was first suggested by Morison and Dempsey at the thalamic level and its characteristics in the thalamus and brain stem have been investigated extensively by several of our colleagues here today. It was concluded that in addition to the classical somatosensory pathways there exists a secondary route to the cortex activated by collaterals from the lemniscal system. This secondary system relays in part through the reticular formation and is apparently multisynaptic since relative to the lemniscal system, its evoked potentials have longer latencies and recovery times. It was also found that the reticular portion of the system is very sensitive to barbiturate anaesthesia. The longer latency and recovery were compatible with the characteristics of human late components. The sensitivity to barbiturates suggested a way of testing the theory further and also of correlating evoked potential changes with the altered perceptual capacity of the subject produced by anaesthesia. We obtained a group of patient volunteers undergoing elective non-neurological surgery and recorded from them as they were gradually induced into pentathol anaesthesia (1, 3). The results upheld the theory. Human late responses are exquisitively sensitive to barbiturate; early components are very resistant.

One thing we noted were the long latency components resulting, we thought, from extralemniscal activation. Control records were taken prior to administration of anaesthesia and then averages were taken between intravenous administration of graded doses of thiopental. At doses as low as 0·34–0·36 mg/kg, late components were suppressed. At this level the patient is still semiconscious and communicating to some extent. With greater doses, the patient loses consciousness and the late responses are obliterated.

Components 1 and 2 were either unaffected or potentiated by doses of anaesthetic up to where the anaesthesiologist pronounced the patient ready for surgery and the EEG was showing intermittent burst suppression.

This experiment, in conjunction with other data from animal research seemed to tell us two things: first, on the neural side, the total somatic evoked response resulted from activation of two types of pathways. Early components were mediated by the lemniscal system. Late components appeared extralemniscal—probably relaying through the reticular formation. Second, on the psychological side, the extralemniscal components appeared to be associated with consciousness, or more strictly, their suppression was associated with loss of consciousness.

Thinking in these terms, our next experiment was obvious. Another condition which alters the perceptual capacity of the subject and creates a behavioural state not unlike drug-induced loss of consciousness is natural sleep. We studied changes in the somatic evoked response during the stages of sleep in 11 subjects for a total of 27 nights (10).

Figure 7 illustrates the changes occurring during sleep in the long latency components to median nerve shock. The sleep stages were categorized from electrocorticographic, myographic, and eye movement records according to

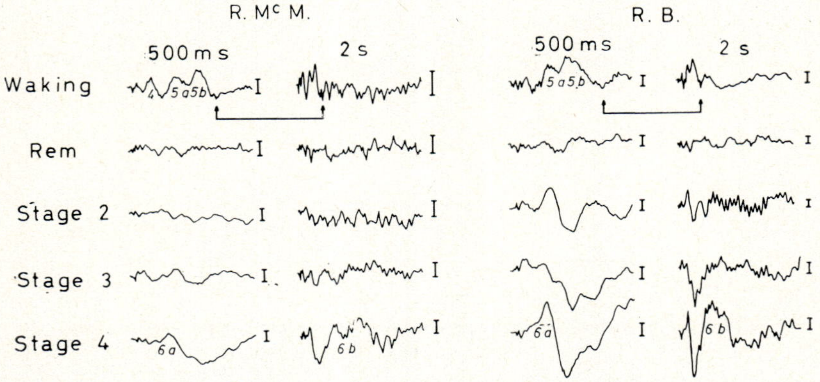

Figure 7. Changes in long latency somatic evoked response components in sleep
(From Goff *et al.*, 1966,
by courtesy of the Editor, *E.E.G. Journal*)

the criteria of Dement and Kleitman (6). Responses are shown for two subjects; the same response is presented on two time bases. The components marked 5a and b constitute the vertex potential point. The vertex potential as seen in the waking state is absent in all stages of sleep and is sharply suppressed in drowsiness. The REM state is characterized by a suppression of late activity, but beginning in stage 2 and becoming most prominent in stage 4, large potentials appear with much longer latencies than any waking state component. In stage 4, the latency of the latest, most prominent positive component varies between 600 and 800 ms in different subjects. Its maximum is at the vertex and it is diffusely distributed over the scalp. Similar findings have been reported in the auditory and visual systems. This slow wave sleep vertex potential corresponds to the K-complex originally described by Davis *et al.* (4). Somatic response components occurring earlier than 100 ms are shown for three subjects in *Figure 8*. Components 1 and 2 are only moderately altered by sleep in contrast to the marked changes we have just seen for the late components.

The theory that early versus late components reflect activity in dual neural systems conveying sensory information to the brain seemed supported by the dichotomous effects of drug induced and natural sleep. The results also indicated that the late components, especially the vertex potential, are intimately related to those neural mechanisms which determine the level of consciousness, which in turn determines perceptual receptivity.

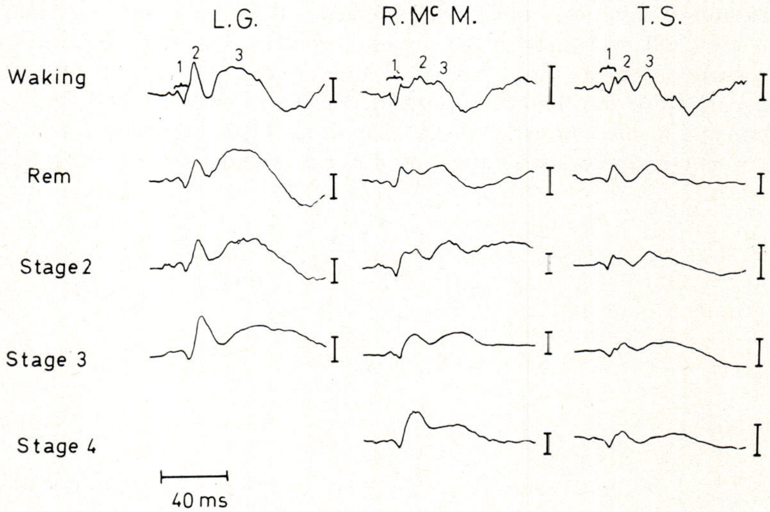

Figure 8. Changes in early somatic evoked response components in sleep
(From Goff *et al.* 1966, by courtesy of the Editor, *E.E.G. Journal*

Thinking about the concept of attention in this context, I was reminded of Dr. Lindsley's (16) suggestion of several years ago that it is possible to view the stages of attention and consciousness upon the same continuum with stages of wakefulness and sleep. He related these stages to changes in patterns of spontaneous activity. Extrapolating from his idea, I would like to suggest that attention can be regarded as one end of a continuum along which perceptual receptivity and long latency evoked potentials co-vary. The other end of the continuum, which we might call zero receptivity, is the condition of deep barbiturate anaesthesia in which primary input—the early components—is present, but late components are absent, and intense, even injurious stimulation will not behaviourally arouse the subject. Next we ascend to the depressed perceptual organization of sleep in which late evoked activity is suppressed or markedly altered, but not absent, and subjects can be aroused or can perform certain types of behavioural responses (22). Then we progress to the waking state where the late components and especially vertex potentials are a prominent part of evoked responses even in the passive, relaxed subject. Make the evoking stimulus meaningful so as to direct attention to it, and the general consensus is that the amplitude of vertex potential evoked by the attentive stimulus is enhanced. Finally even within attention, the experiment of Wilkinson and Morlock to which I referred earlier indicates that high incentive with its implied greater attention increases some components of the evoked response compared to the theoretically less attentive situation. Of course, not all evoked response components vary systematically with consciousness and attention and analysis of the functional significance of these deviations is important.

The intimate relationship between late potentials and the neural mechanisms of consciousness is compatible with the dual sensory projection theory which says that these potentials are mediated through brain stem reticular and non-specific thalamic structures. The relation of this so-called reticular activating system to levels of consciousness and arousal is well known. The human evoked potential evidence I have summarized indicates that the specific lemniscal system continues to function when the secondary, extra lemniscal system is depressed by drugs or sleep. If the two systems are truly independent, then conversely, the integrity of the lemniscal projection system rostral to the brain stem should not be critical to the function of the extra lemniscal system. Animal research on recruiting responses suggests that this is the case. Stimulation of the thalamic portion of the reticular system still produces cortical responses after destruction of specific sensory thalamic relay nuclei (12). Although the pathways to the cortex of this reticular activating system are not certain, it is known that its cortical projection is bilateral and diffuse (16). Therefore if the two systems are truly independent in man, we should be able to record late components from at least some cortical areas in the absence of early components resulting from disruption of lemniscal pathways rostral to the brain stem. It has been difficult to test this in man because of the resistance of early components to experimental manipulation. Recently, however, evidence has begun to accrue from neuropathology and neurosurgery experiments. The weight of this evidence does not support the independent dual projection theory. Instead it suggests that the integrity of the specific, primary projection, possibly all the way to the cortex is necessary for the activation of late components.

Liberson (14, 15), recording median nerve shock evoked potentials in patients with aphasias resulting from middle cerebral artery thrombosis, found that in all cases when early and late responses were absent or markedly depressed on the affected side, all responses were similarly absent on the non-affected side. In some patients in whom somatic evoked responses were suppressed in the parietal lobe of the affected hemisphere, visual and auditory stimuli could induce normal late evoked potentials in the same area. Liberson pointed out that his evidence was not compatible with the theory that late components reflect a non-specific bilateral projection. He suggested that the bilateral representation seen in normals is most probably due to transmission to the ipsilateral hemisphere from contralateral cortex through the corpus callosum, or the commissural structures in general rather than the callosum specifically. Liberson (14) also cites evidence reported verbally by Ferriss showing that in two patients with congenital agenesis of the corpus callosum, the secondary evoked potentials were quasi absent on the ipsilateral side.

A study by Domino *et al.* (7) presents indirect evidence against the independent dual projection theory. Recording from the scalp and epidural surface during placement of cryogenic thalamic lesions to relieve dyskinesias, they found that massive lesions in ventralis posterolateralis encroaching on ventralis posteromedialis and the centre median nucleus produced sensory loss and reduced all components of the evoked response. They suggested that the ventral posterior nuclei play a critical role in the genesis of late somatic evoked response components and therefore they might all be considered to be lemniscally mediated. It is important to note, however, that Domino *et al.* used an analysis time of 125 ms which includes only components we labelled 1, 2 and 3 and the beginning of 4. Thus this study suggests, but does not directly show, the lemniscal mediation of later components 4 and the vertex potential.

Collaborating with Dr. Peter Williamson of Yale's Department of Neurology, we have in progress a study of neuropathology patients. Primary emphasis is on patients with unilateral parietal lobe sensory deficits. But we are also including any unilateral central nervous system disease with clinically detectable sensory deficit. By choosing unilateral abnormalities and recording responses evoked by peripheral stimulation of both sides, each patient serves as his own control. Responses are also being compared to the extensive normative data we have available. To date, 14 patients have been selected and their somatic evoked potentials to randomized left and right median nerve shock have been recorded from a bilaterally symmetrical array of eight electrodes. Interest here focuses on nine patients presenting clinically detectable hemisensory deficit of varying degree. Of these, seven had right-sided or non-dominant hemispheric lesions. The clinical location of these lesions varied from the thalamic-internal capsule level to the relatively pure parietal lobe disease. The eighth patient had a superficial lesion on the dominant hemisphere. The ninth was judged to have an hysterical hemianaesthesia. We expected on the basis of the dual projection theory to find in at least some of these patients a disruption of the primary, lemniscal projection to parietal area with sparing of the secondary diffuse projection system. In other words, we expected alteration or suppression of early evoked response components, probably components 1 and 2, but relatively intact late components especially outside the affected parietal lobe area. In preliminary results to date, these expectations have not been realized. With the exception of the hysterical patient, we consistently found diminished or absent evoked response early components when stimulating the side with sensory deficit and all late components were similarly attenuated or absent bilaterally at all electrode locations. Apparently normal early components were present at appropriate electrode locations when stimulating the non-affected side and late components were similarly present and normal appearing at all locations.

Figure 9 illustrates these findings. Patient G.P.D. is the one judged to have an hysterical hemianaesthesia. Comparing early response components in parietal areas contralateral to the affected and non-affected sides, we see normal appearing responses on both sides. We also see that late responses recorded from the vertex were almost identical regardless of which side was stimulated. Patient J.P.M. has a superficial parietal lobe lesion in the dominant hemisphere and presented mild sensory impairment such as decreased two-point discrimination, position sense, and stereognosis. Early components contralateral to the affected side were present but considerably altered compared to the control side and this alteration is mimicked by the late components. Note that the normal and altered late responses are both recorded from the same parietal location on the affected side of the brain. This shows that the alteration in the affected side response is not due to a generalized suppression of neural function in this area. Patient F.H.D. presented a dense sensory loss and moderate homogeneous motor impairment on the non-dominant side. Dr. Williamson suspected a small vascular lesion in the posterior limb of the

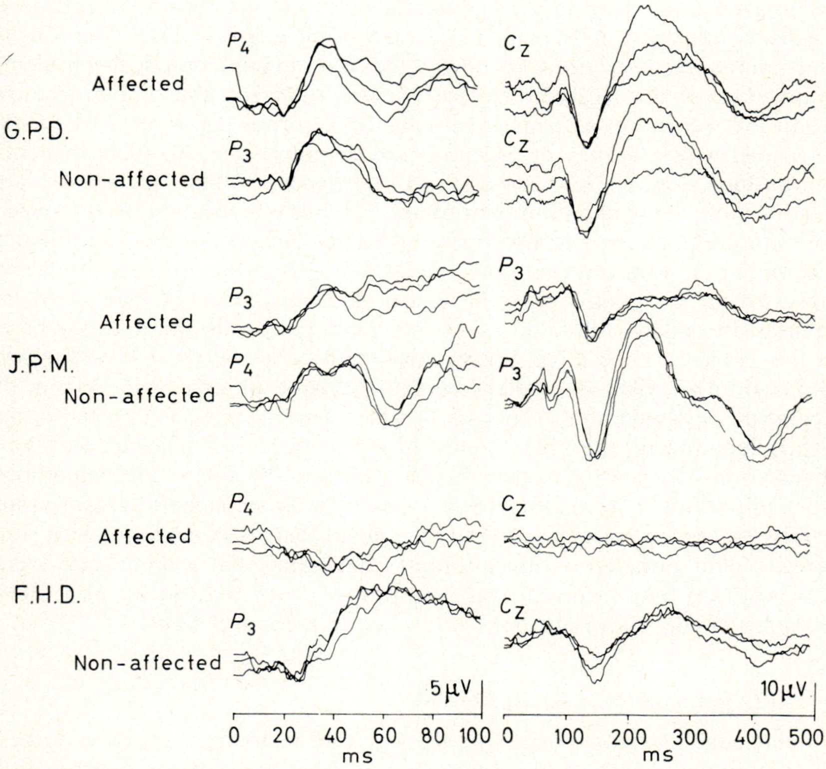

Figure 9. Somatic evoked potentials from patients presenting different types of neurological symptoms

internal capsule with possible involvement of the lateral aspect of the thalamus. There was a complete absence of normal early components to stimulation of the affected side mimicked by a complete absence of late components. Relatively normal early and late components were evoked on the non-affected side. In summary, although our patient population is as yet small, there have been no exceptions to the finding of a one-to-one relation between the presence, alteration, or absence of primary lemniscal components, and the presence, alteration, or absence of late components.

Thus recent evidence suggests that long latency evoked potentials do not reflect cortical activity mediated via a non-specific functionally independent extralemniscal system. Instead, the activation of these responses is apparently dependent at some level above the brain stem upon the integrity of the same pathways which mediate primary evoked responses, i.e. the specific lemniscal system. Our evidence suggests that the critical pathways are either in the thalamocortical projections to specific sensory cortical areas or in the cortical areas themselves.

It has been suggested that the corpus callosum specifically, or commissural structures generally, are involved in the mediation or generation of these potentials and that their functional significance is the interhemispheric transfer of sensory information. In man, a direct test of the relation of late components to commissural structures is obstructed by obvious practical problems. Opportunities exist in cases of callosal agenesis or in recent surgical procedures producing split-brain patients. Agenesis of the corpus callosum is frequently associated with a variety of psychological disorders (19, 20) such as mental retardation, apathy, slow associations, deficiency in bilateral transfer and sensory motor co-ordination, emotional instability, and inability to concentrate attention. Since it is also frequently associated with other neurological abnormalities, however, caution is required in attributing any causal relationships. So far as I know only the report of Ferriss which I mentioned has attempted to relate callosal agenesis to evoked potentials and that was based on two subjects. Gazzaniga, Sperry, and their associates (8, 9) working with split brain man conclude that the corpus callosum plays a major role in the mediation of behavioural responses in which sensory input is directed to one hemisphere and primary motor control lies in the other. Their comparative observations suggest that the corpus callosum becomes correspondingly more important in mediating some types of interhemispheric sensorimotor activity as one ascends the evolutionary scale. This is compatible with comparative data gathered in our laboratory indicating that waking state vertex potentials are found only in man and possibly primates. So far as I know, evoked potentials have not been studied in split brain patients.

CONCLUSION AND SUMMARY

If I may conclude with a brief speculation, it seems to me that any hypothesis about the neural basis of late components must account for their intimate relation to levels of arousal and consciousness. If late components are attributable to commissural activity subserving interhemispheric communication, then late response characteristics suggest that this communication involves complex integrative functions of perceptual awareness and memory. Late components are altered or absent under conditions in which perceptual awareness is altered or absent—such as sleep or anaesthesia. A recent paper on the effects of barbiturates on learning and memory (17) found that subjects administered Thiopental showed a subsequent loss of memory for events while under sedation. In various recognition and recall tests, subsequent memory loss was correlated with concentration of Thiopental in venous blood at the time the material was learned. I have shown that late evoked response components are depressed by very low doses of Thiopental.

In a recent symposium, Dr. Jasper (13) raised the following question: 'Is there an anatomically distinct neuronal system in some central location, with widespread functional connections with all parts of the brain, involved in the selection of the ever-changing momentary patterns of neuronal activity of perceptual awareness, as opposed to those systems constantly engaged in the unconscious processing of information and in the execution of automatic movements?' The characteristics of late evoked responses suggest that they might reflect activity in such a neuronal system.

In summary, I have reviewed the 'state of the art' with regard to the neuropsychology of evoked responses relevant to perceptual organization in general, and attention in particular. I have made some suggestions for further refinement of methodology in the interests of more consistent results from different laboratories. I have discussed the problem of devising independent measures of the effectiveness of operations designed to alter attention. I have suggested that attention, at least in its relation to evoked potentials, be viewed in the perspective of one end of a continuum along which the neurophysiological correlates of consciousness and the psychological correlates of perceptual organization co-vary. Finally, I have reviewed evidence suggesting that late evoked response components, most frequently measured in attention experiments, may not reflect activity mediated by an independent extralemniscal projection system as previously thought, but appear to be dependent on the integrity of the lemniscal system at the thalamocortical or cortical level. They may result from, or be generated by, activity in commissural structures and function in the interhemispheric transfer of sensory information, or they may subserve still more complex processes of perceptual organization and memory. I think it highly desirable that conceptualization of future attention-evoked response experiments consider possible neural bases for and functional significance of the electrical signals being measured. Determining that certain evoked response components change in relation to what we define as alterations in attention is only the first step toward the goal of specifying the neural mechanisms which permit the biological adaptation of selecting and attending to relevant stimuli while suppressing other sensory input.

ACKNOWLEDGEMENT

Research supported by USPHS Grant MH-05286, NSF Grant GB-3919, and by the Veterans Administration.

DISCUSSION ON GOFF

Jasper

I'd like to say how much I admire this fine piece of work. It's a beautiful analysis of a difficult problem. Dr. Goff, I think you've put your finger on many of the crucial questions that should concern this conference, from the physiological, as well as the behavioural, point of view. Physiological studies have shown that it is possible to have these late responses even when only the specific system remains and after destruction of most of the non-specific pathways. There is also physiological evidence tending to show that the original conception of colaterals from the rapid sensory systems into the midline is being modified, and that selections of the type of input that goes to these midline structures occurs really very low down in the sensory systems that have longer conduction times. This raises the question of the intensity of your somatic stimulus, for it is the *high threshold* afferents in peripheral nerves which are being conducted to the midline structure, according to some of the analyses, and not the low threshold afferents. In other words, the rapidly conducting fibres may not show any evoked potential in the midline areas, even though you excited all of the rapidly conducting fibres in a given nerve trunk. If you are using a relatively weak stimulus you would not fire these medially conducting afferent pathways. You might excite only the more rapidly conducting fibres of the lemniscal system.

Goff

We have some evidence relevant to this from an intensity study Dr. Burton Rosner and I did (23). We simultaneously obtained subjective absolute thresholds, recorded the median nerve response by averaging from the antecubital fossa, and recorded somative evoked responses from the scalp. We found that if shock intensity was above subjective threshold, we got both early and late responses. There was no differential, suggesting that if primary responses are evoked, ensuing neural activity occurs normally.

Jasper

I remember when we re-examined this notion of anaesthetic action on the brain stem reticular formation instead of on the specific pathways (the original notion of French and Magoun, if you remember) it was our impression that the units of the primary sensory cortex were equally affected by the anaesthetic if you considered only the late responding cells. So that we could get just as good a correlation between levels of anaesthesia by recording cortical responses of single cells of longer latency as occurred in the reticular cells of long latency. It seemed that the effect of anaesthesia was simply on the later components or the later responding cells, whether they be cortex or brain stem. Your results would seem to suggest therefore that anaesthetic action is possibly merely a blockage of the multisynaptic proliferation of the response rather than the primary afferents.

Goff

Our interpretation that the blocking was at the brain stem level now appears incorrect.

Jasper

It blocks also at cortical level.

Grey Walter

As a corollary to your point about electrical stimulation, one must realize that with an electrical stimulus you activate a whole bunch of fibres. This is not physiological—an electric shock feels like an electric shock and nothing else, and this is obviously a peculiar experience for the central nervous system. One wonders to what extent one can regard this as truly physiological.

Goff

We have been working in the auditory system using a physiological stimulus—a click or a tone—and we are getting evoked response waveforms which seem similar to the somatosensory components. Homologies have yet to be established, I grant you, but this physiological stimulus evokes similar potentials. There are obvious differences using physiological visual stimuli however.

Jasper

There's an obvious difference in the somatic system too. We have studied this in man with cortical electrodes during surgery. Stimulating the nerve, very often we were getting rather small evoked potentials directly from the cortex in man, while if we stimulated a spot on the skin with a mild touch stimulus we produced a much larger response than if we stimulated the nerve with maximum voltage. This made us believe that in the somatic field we are probably dealing with a situation similar to the visual system where focal stimulation, as Dr. Cobb has beautifully shown this morning, is much more effective when giving regular evoked potentials of high amplitude than a massive flash of light. Of course, we know that in the somatic system we have the same type of organization, areas of inhibition surrounding the focal areas of excitation. I imagine you would find much more effective responses if you were to take this into consideration.

Goff

We have used index finger stimulation as well as median nerve to see if there was any difference in distribution and there was not.

Jasper

I suppose if you are recording the mass response through the skull this difference wouldn't be so obvious, as when you are recording more local small areas of cortex directly.

Goff

We got essentially the same waveform, but the amplitudes of components were smaller for index finger stimulation.

Jasper

You tend to blur out the localization.

Goff

The physical intensity was the same for both stimulus sites.

Berlyne

I think that some of Dr. Goff's points reinforce suggestions I was making yesterday. So many of the operations that are used with the expectation of manipulating attention while observing electro-cortical phenomena seem to be ones that strengthen particular stimulus-response associations, as shown by either increased probability or increased amplitude of a specific response. And when that is the case, a concept of attention is *not* necessary and is therefore *not* justifiable.

If on the other hand, particular electro-cortical phenomena occur when several different stimulus-response associations are strengthened, then some concept of arousal or activation might be justifiable. The concept of attention seems to be needed in two kinds of situation. One is a situation in which several stimuli are competing for control of behaviour. And electro-cortical phenomena, such as evoked potentials, can surely help us there only if they take different forms or occupy different locations, according to which of the competing stimuli is winning the competition. On the other hand, if you have some operation that strengthens or weakens a large number of stimulus-response associations simultaneously then an intensive concept of attention or attentiveness is applicable and justifiable. Anaesthesia and injection of barbiturates would be cases in point. Dr. Goff, do you agree?

Goff

Yes, I think so. These components certainly do have different distributions which could be used. Instead of attention I think the word *contingency* comes to mind. One thing I didn't mention in my paper is CNV, and I wonder if Dr. Grey Walter would say a word about the relationship between all these late components, which after all immediately precede the CNV, and the CNV itself.

Grey Walter

When we started, we weren't looking for the CNV; we were studying evoked responses and we were struck by this long delayed surface-negative wave. There seems to be some degree of independence between the CNV and the size of the evoked response. In some cases where we are distracting subjects—or as Dr. Berlyne says, when there's competition between stimuli—with buzzes or tones between trials, the evoked response with its complexity both in the early and late components, disappears but leaves the CNV intact. So the cortex is still getting the one bit of information from the conditional stimulus, although there's no sign of this in the form of an evoked response proper. It's easy to separate both the early components, the late components and the CNV by various procedures such as introducing competing stimuli. So different cortical processes are at work for different situations.

Goff

Has anyone ever examined correlation between changes in the amplitude of the vertex potential and simultaneous changes in the CNV?

Grey Walter

Well this is one case where the vertex potential disappears in interference, but the CNV does not, and conversely, of course, the CNV can be attenuated by equivocation and the evoked response remains, so again they're independent. I think the CNV is a very wide-spreading depolarization of the cortex following the conditional stimuli.

Beck

Would you care, Dr. Grey Walter, to speculate on what is the source of the CNV?

Grey Walter

In fact one can only speculate at this stage. We suppose that it is depolarization of the apical dendrites, in patches, over a wide area. It's certainly on the surface of the cortex as far as we can see from derivations with cortical electrodes.

McCallum

Were you using a.c. coupled amplifiers Dr. Goff? It seemed to me that in one of your traces, in which as I recollect a tone was associated with a shock stimulus, there appeared the beginning of a CNV. There was an undoubted negative rise, but I was uncertain from your illustration, which showed only the first 500 ms, how it developed.

Goff

We only seemed to see that in that one subject. We were doing a perception experiment and the subject did have to make a response. He reacted to 10 ms tone pips—there were six of them. The frequencies differed by 100 Hz and he had to identify them. We were suppressing the vertex potential by preceding it at various time intervals with a shock, and then seeing what psychological variables we could alter. In other words, could we interfere with discrimination of these tones by deliberately suppressing the vertex potential? The subject did have to press a button. He had to wait 500 ms while we were sampling the evoked response and then within the next 2 s to press the key which he thought correctly identified the tone. This is a CNV paradigm.

McCallum

As Dr. Grey Walter indicated, the vertex potential and the CNV can vary quite independently. Nevertheless, there are frequent instances in which they appear to be varying in the same way. For example, in the experiment which I illustrated yesterday, where subjects were asked to make a special effort of concentration on certain trials in which two stimuli were associated and a motor response made to the second, both the vertex potential to the first stimulus and the CNV showed similar increases in amplitude.

Beck

The onset of the CNV is about 200–250 ms?

McCallum

The latency to onset after the first stimulus is about 300–350 ms.

Beck

About 20 μV and how long?

McCallum

We have seen it quite clearly with inter-stimulus intervals ranging from 0·5 to 13 s.

Grey Walter

We tried one at 30.

Beck

Dr. Goff's conclusions are supported by the work of Bignall, Imbert and Buser. With almost complete destruction of the diencephalon, that is, thalamus and other subcortical structures except the optic afferent system, photically evoked responses appeared in the visual cortex, in 'association' areas of suprasylvian and anterior marginal gyri. As evoked responses persisted in these areas after extensive destruction they are not dependent on structures medial to the lateral geniculate body.

REFERENCES

1. Abrahamian, H. A., Allison, T., Goff, W. R. and Rosner, B. S. (1963) *Anesthesiology*, **24**: 650
2. Allison, T. (1962) *Electroenceph. clin. Neurophysiol.* **14**: 331
3. Allison, T., Goff, W. R., Abrahamian, H. A. and Rosner, B. S. (1963) In: (Ed. by Hernández-Peón). 'The Physiological Basis of Mental Activity'. *Electroenceph. clin. Neurophysiol.* Suppl. **24**: 68
4. Davis, H., Davis, P. A., Loomis, A. L., Harvey, E. N. and Hobart, G. (1939) *J. Neurophysiol.* **2**: 500
5. Davis, H., Mast, T., Yoshie, N. and Zerlin, S. (1966) *Electroenceph. clin. Neurophysiol.* **21**: 105
6. Dement, W. and Kleitman, N. (1957) *Electroenceph. clin. Neurophysiol.* **9**: 673
7. Domino, E. F., Matsuoka, E. F., Waltz, J. and Cooper, I. S. (1965) *Electroenceph. clin. Neurophysiol.* **19**: 127
8. Gazzaniga, M. S. (1967) *Scient. Am.* **217**: 24
9. Gazzaniga, M. S., Bogen, J. E. and Sperry, R. W. (1965) *Brain*, **88**: 221
10. Goff, W. R., Allison, T., Shapiro, A. and Rosner, B. S. (1966) *Electroenceph. clin. Neurophysiol.* **21**: 1
11. Goff, W. R., Rosner, B. S., and Allison, T. (1962) *Electroenceph. clin. Neurophysiol.* **14**: 697
12. Jasper, H. H. (1960). In (Ed. by Field) *Handbook of Physiology*. Section 1, Vol. II. American Physiological Society, Washington, D.C.: p. 1307
13. Jasper, H. H. (1966). In: (Ed. by Eccles). *Brain and Conscious Experience*. Springer-Verlag: New York, p. 256
14. Liberson, W. T. (1966) *Am. J. physiol. Med.* **45**: 125
15. Liberson, W. T. and Scott, D. (1964) *Electroenceph. clin. Neurophysiol.* **17**: 456
16. Lindsley, D. B. (1960). In: (Ed. by Field), *Handbook of Physiology*. Section 1, Vol. III. American Physiological Society, Washington, D.C.: p. 1553
17. Osborn, A. G., Bunker, J. P., Cooper, L. M., Frank, G. S. and Hilgard, E. R. (1967) *Science, N.Y.* **157**: 574
18. Satterfield, J. N. (1965) *Electroenceph. clin. Neurophysiol.* **19**: 470
19. Toglia, J. U. (1961) *Dis. nerv. Syst.* **22**: 428
20. Toglia, J. U. and Lapayowker, M. S. (1966) *Conf. Neurol.* **27**: 345
21. Wilkinson, R. T. and Morlock, H. C. (1967) *Electroenceph. clin. Neurophysiol.* **23**: 50

22. Williams, H. L., Hammack, J. T., Daly, R. L., Dement, W. C. and Lubin, A. (1964) *Electroenceph. clin. Neurophysiol.* **16**:269

Discussion

23. Rosner, B. S. and Goff, W. R. (1969) (Ed. by Neff), Vol. 2. Contribution to Sensory Physiology. New York: Academic Press, p. 169

Spectral Analysis of EEG Data from Animals and Man During Alerting, Orienting and Discriminative Responses

W. R. Adey

Space Biology Laboratory, Brain Research Institute, University of California, Los Angeles, California

INTRODUCTION

The search for correlates between brain function and behavioural processes may follow several routes. To this time, the classic trinity of ablation, stimulation and electrophysiological recording have been only modestly successful in unravelling the broad pattern of cerebral interrelations that are presumably consistent concomitants of particular behavioural performances.

To the extent that the brain's own electrical activity is a faithful mirror of transactional processes in any cortical domain, it might be anticipated that our examination of electroencephalographic data would reveal these patterned substrates of behavioural performance. It is only too well known that here lies the weakest link in our investigative chain. The very complexity of EEG data has traditionally limited their evaluation to the simplest parameters, and the correlates with behaviour to the broadest classifications. Since mental processes are rapid and changing, our windows on these processes must be provided by techniques that are equivalently refined and effective with epochs of data only a second or so in duration—far shorter than segments of record that can be analysed reliably with most analogue computations. They must also provide details of interrelations between different cerebral domains, certainly in terms of linear relations, and hopefully in the near future in subtle but important non-linear relations.

There is the added requirement that our methods should allow evaluation of the intimate relations between the electric activity of a single neuronal element and the domain of neurones of which it is part. In this way, we would seek an understanding of processes that might determine interactions between a population of such elements and not necessarily restricted to classic synaptic interactions.

The following view of cerebral system physiology and neuronal organization is a collected account of work in which I have participated with my colleagues, Anatol Costin, Zeev Elazar, Rafael Elul, Raymond Kado, David Lindsley, Mica Radulovacki and Donald Walter. Our attention has been directed to the role that allocortical systems might play in establishment of a 'set' of attention necessary for orienting and discriminative responses, and changes that occur at different levels of learning. We have studied in animals EEG patterns relating temporal lobe structures with diencephalic and midbrain regions during these responses. Our evidence indicates that the hippocampus modulates ascending streams of activity in the non-specific systems of the rostral midbrain and subthalamic areas during specific states of attention necessary for visual discrimination, and that this influence is manifested in turn in sensory cortical areas.

We have directed much effort to the elucidation of patterns in scalp EEG records in man that might be similarly correlated with states of focused attention and discriminative performance. Necessarily, quantitative criteria have been established by extensive computational estimates of EEG spectral parameters, since their evaluation even by simple automated techniques, such as analogue methods of frequency analysis, are quite inadequate for this purpose. In turn, these parameters from digital spectral analysis have been used in further automated pattern recognition studies, in which the computer has been assigned the task of selecting parameters that best classify epochs of EEG records in individual subjects in particular test situations, including correct and incorrect decision-making.

ANATOMICAL AND PHYSIOLOGICAL ORGANIZATION OF THE LIMBIC SYSTEM IN ESTABLISHMENT OF THE 'SET' OF ATTENTION

Anatomical Substrates

Interrelations of allocortical structures, including hippocampus and amygdala, with subcortical zones in diencephalon and midbrain have long been matters of speculation in both structural and functional schemes. Elliot Smith (21) in his Arris and Gale lectures noted the antiquity of hippocampal cortex in vertebrate evolution, and the essentially minimal changes in its location and structure in the vast metamorphoses that have occurred in the remainder of the hemisphere. He emphasized the role of the 'paraterminal body' (septum) in the genesis of a major band of fibres that appear in the reptilian brain and pass directly into motor co-ordination centres in the midbrain tegmentum, and thus constitute the first major efferent pathway from the cerebral hemisphere. Later investigators, including Herrick (31) and Papez (45) stressed the possible role of the hippocampal system in interrelating afferent volleys from environmental stimuli with internal, visceral inputs. These hypotheses were based primarily on anatomical considerations, including cortico-cortical connections between neo- and allocortical structures. Papez proposed the series of pathways that bear his name in the 'Papez circuit', which emphasizes re-entrant or cyclic aspects of hippocampo-diencephalic paths through mammillo-thalamo-cingulate afferent and hippocampo-fornico-mammillary efferent paths.

More recent reviews have emphasized alternate paths within this system on the basis of electrophysiological mapping of evoked potentials and EEG activity combined with focal lesions (1, 12, 29). These studies have indicated that the septum is a pre-eminent way station on afferent, as well as efferent hippocampal paths. Septal lesions are followed by loss of characteristic EEG theta rhythms in the hippocampus. Moreover, both hippocampus and amygdala are connected with the caudal diencephalon and rostral midbrain by paths that traverse dorsal hypothalamic and subthalamic areas (1). From the hippocampus, these connections may be directed through the overlying hippocampal (entorhinal) cortex. Experimental lesions in this cortex first disclosed monosynaptic pathways to periaqueductal grey matter and dorsal tegmentum of the rostral midbrain in the marsupial (10). Electrophysiological evidence (11) and later anatomical studies (43) have indicated that in eutherian mammals, including primates, these descending hippocampal pathways are predominantly polysynaptic (*Figure 1*).

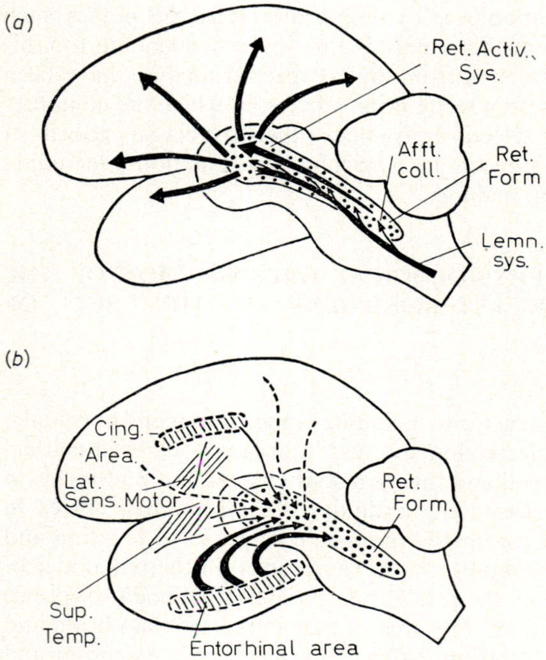

Figure 1. (a) Diagrammatic representation of ascending reticular system. Although the diagram shows inputs primarily from the lemnisci, activation also occurs substantially from spinothalamic and other non-lemniscal paths. (b) Arrangement of main cortical areas (lateral sensorimotor, superior temporal gyrus, cingulate and entorhinal) capable of modifying ascending reticular conduction. Corticifugal projections from the entorhinal area exercised a major influence on ascending reticular conduction

(From Adey *et al.*, 1956, by courtesy The Royal Australasian College of Physicians)

In summary, our model of hippocampal relations with subcortical structures must consider it as set astride the diencephalon at two levels, rostrally at the level of perifornical and mammillary tissue, and caudally at the level of the rostral midbrain. Stimulation studies have suggested that these two zones may represent partly autonomous areas in behavioural mechanisms (33). At the caudal, midbrain level, descending hippocampal influences strongly influence ascending conduction in dorsal reticular and periaqueductal pathways (11).

Investigations of functional organization in these hippocampo-subcortical systems has emphasized the role of the junctional zone between thalamus and hypothalamus (9, 38, 39), in both afferent and efferent limbs of hippocampal connections. We have applied the term 'subthalamus' to this area, but its limits extend beyond the classic subthalamic nucleus of Luys, which it envelops. Stereotaxic lesions in this region in cat and monkey induced sharp but temporary changes in states of attention. Spontaneous movements were much reduced, and the gaze was fixed ahead for long periods. Food held before the face of the unfed animal was not accepted, but food placed in the mouth was vigorously masticated and swallowed. Our electrophysiological studies have indicated afferent subthalamic paths from the anterolateral columns of the spinal cord (38). This subthalamic region clearly influences centres both caudal and rostral to itself, and fine analysis of its EEG activity cited below (19, 20) favours the view that these effects involve it as an integrating centre, rather than merely by fibres of passage arising elsewhere.

Stereotaxic lesions in this zone were associated with diminished excitability of units in the rostral midbrain reticular formation. Effects of these lesions on EEG correlates with behaviour were striking. We have tested discriminative ability for food rewards in a modified T-box with visual cues, and in delayed response performance, also based on visual cues (15). In T-box testing after a unilateral subthalamic lesion, performances were at or below chance levels, with a marked incapacity to respond to visual cues in the visual field contralateral to the lesion. Concomitantly, hippocampal EEG activity was reduced in amplitude and lacked the regular trains of theta waves characteristic of the alert and performing animal (5, 27). Regular trains of theta waves at 6 Hz that accompany both T-maze performance and delayed response tasks (14, 32) no longer occurred (*Figure 2*).

These performance decrements were temporary, and improvement after 4–7 days was contemporaneous with reappearance of 6 Hz theta activity during task performance. A second stereotaxic lesion in the contralateral subthalamus caused decrements in T-box discrimination and delayed response performance, again associated with discriminanda presented in the contralateral visual fields, and with reduced hippocampal theta activity. Recovery of performance and of hippocampal theta occurred more slowly after the second lesion.

HIPPOCAMPAL THETA PATTERNS IN ATTENTION, ORIENTING AND DISCRIMINATION

These initial studies of hippocampo-subcortical interrelations have led us to more detailed investigations, with attempts to quantify subtle aspects of intrinsic hippocampal EEG patterns, as well as their relations with subcortical and cortical EEG patterns accompanying specific behavioural situations (19, 20, 47).

Distinctive patterns in the cats' hippocampal EEG activity consistently accompanied certain categories of attentive behaviour. These studies initially involved a simple averaging of the EEG, with onset of averaging time-locked to presentation of behavioural cues. Parameters which survived as deflections in these averages were considered consistent components of EEG activity in behavioural performances. Later studies have confirmed these findings, by digital spectral analysis, including calculations of coherence as a measure of linear interrelatedness between different recording sites. Our results have confirmed that theta rhythm patterns in the cat cannot be interpreted as a stereotype, relating simply to a general class of responses, such as the orienting reflex (27, 28). We have recognized patterns separately associated with alerting, orienting and discriminative responses.

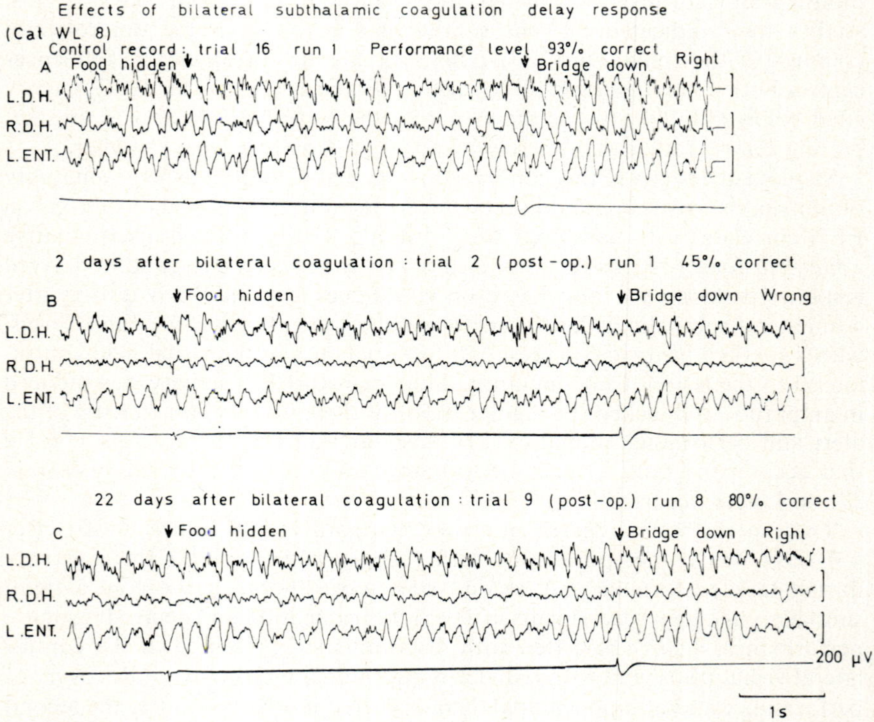

Figure 2. Correlation of recovery of delayed response performance following bilateral subthalamic lesions, with reappearance of normal 6 Hz *bursts during approach. Two days after lesion, this activity was essentially abolished* (*B*), *but it returned after* 22 days (*C*). (*L.D.H., left dorsal hippocampus; R.D.H., right dorsal hippocampus; L.ENT., left entorhinal cortex*)

(From Adey *et al.*, 1962, by courtesy of the Editor, *A.M.A. Archs. Neurol.*

As described in detail elsewhere (13, 47), approach to food in a T-maze on the basis of visual cues by the fully trained animal was accompanied by a regular 'burst' of hippocampal EEG waves at 6 Hz, beginning about 300 ms after opening of the test-box doors and lasting 1·0–1·5 s, but terminating prior to attainment of food (5, 47). The regularity of this theta rhythm in the computed average increased between chance levels of performance and the 80 per cent level of correctness. Further training led to truncation of the terminal cycles in the computed average, but the initial portion of the burst remained unaltered, even after many thousands of trials in over-training (46).

Computed averages in simple orienting responses were markedly different. These were evoked by the presentation of reinforcing cues, such as a tone, or the opportunity to observe the reward area in the T-box, but without the opportunity to approach the goal. Averages were much less regular, and exhibited slower dominant frequencies around 4 Hz. In simple alerted states, a broader spectrum and varying dominant frequencies were noted, both in computed averages and in spectral analyses.

Hallucinogenic and psychotomimetic drugs, such as LSD, psilocin and psilocybin (3) and cyclohexamine drugs and their derivatives (4) produced both acute and long term effects (47). These drugs induced paroxysmal slow wave discharges, lasting 5–20 s and spreading widely in diencephalon and midbrain from an apparently focal origin in hippocampal tissue. They were associated with disruption of learned behaviour and with 'absences' and staring, resembling psychomotor epilepsy. Recovery from the acute effects of these drugs after 12–24 h was followed by a period of exalted orienting behaviour. This lasted 2–4 days after cyclohexamine drugs and as long as 20 days after a single dose of LSD. Consistent changes in computed averages of hippocampal theta trains accompanied both orienting and discriminative responses during this recovery period. Computed averages 4–7 days after LSD showed enhanced amplitude and duration of 6 Hz activity during discrimination in comparison with pre-drug averages. This enhancement declined thereafter (47). By contrast, averages during orienting responses progressively incremented in amplitude and regularity at a later time, from approximately the seventh to eleventh days, and at lower frequencies (4–5 Hz) than during discrimination. This enhancement declined with diminished orienting behaviour after the twentieth day.

EEG PATTERNS IN CORTICO-SUBCORTICAL SYSTEMS IN ALERTING, ORIENTING AND DISCRIMINATION

These findings in simple averages of the hippocampal EEG suggested that a finer analysis, as by spectral analysis, might reveal more clearly not only distinctive patterns in hippocampal activity relating to a repertoire of behavioural performances, but also modes of interaction in cortico-subcortical systems. It was recognized that a highly desirable objective in such studies would be reliable recognition of patterns in EEG epochs substantially shorter than the 10 or 15 s epochs customarily used in spectral analysis.

From our earliest applications of digital spectral analysis to EEG data (13, 50, 51), it was apparent that, despite a high and reliable resolution down to the lowest spectral frequency around 1 Hz, power densities so calculated for each frequency represented a mean covering the whole epoch analysed. These methods necessarily took no account of EEG patterns characterizing states of consciousness that might last one second or less, as in rapid decision making. Evidence to be presented below on shifting amplitude distributions in the EEG of man clearly indicates a mathematical stationarity lasting less than 5 s (7, 23). Yet the reduction of a single data epoch to 1·0 s would offer no solution, since due regard must be paid in such a statistical analysis to the number of degrees of freedom involved in such parameters as epoch length, bandwidth and sampling rates. Where the data sample is short, digital filter bandwidths necessarily become broad, so that only a crude analysis can be validated, and adjacent filters are several Hz wide.

For these reasons, our colleague Dr. D. O. Walter encouraged us in the development and application of a method we have termed 'evoked spectral

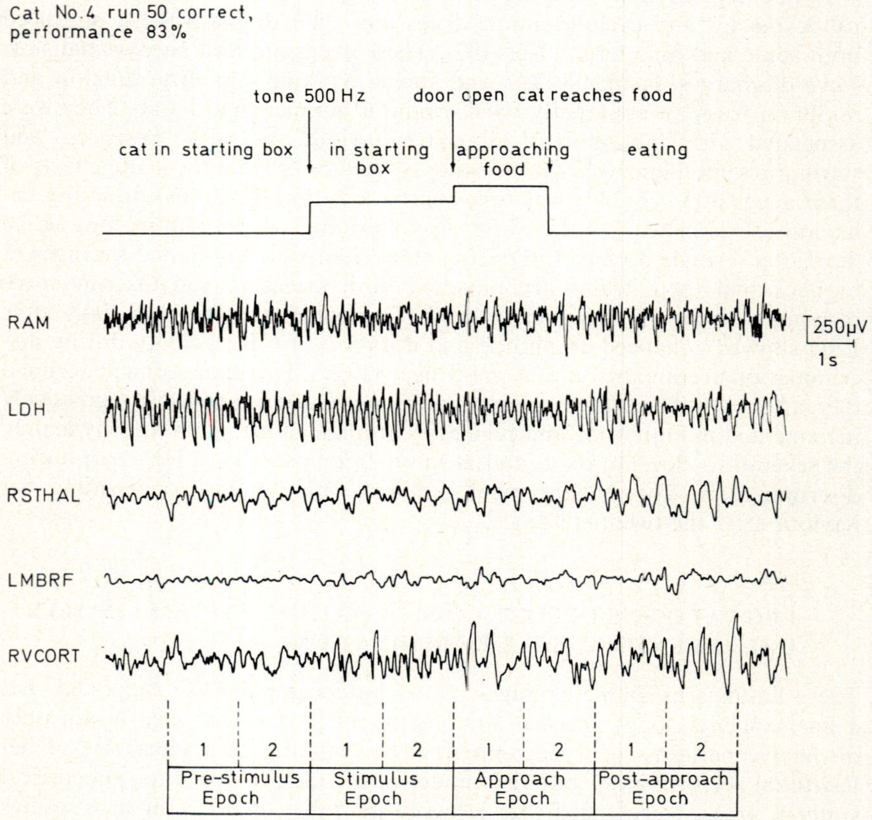

Figure 3. Segments of EEG records during one trial which was recorded simultaneously on paper and on magnetic tape. The upper stepwise line shows the event marker, and the sequence of behavioural states. Division of the record into successive segments for spectral analysis is shown below. In each behavioural epoch, two periods (1, 2) *each* 1–6 s *long were analysed. RAM, right amygdala; LDH, left dorsal hippocampus; RSTHAL, right subthalamus; LMBRF, left midbrain reticular formation; RVCORT, right visual cortex*

(From Elazar and Adey, 1967a, by courtesy of the Editor, *E.E.G. Journal*)

analysis'. Essentially, this involves selection of epochs recorded during particular behavioural responses, with durations between 0·5 and 1·5 s. These separate short epochs were culled from the magnetic data tapes after their conversion to digital form, and a composite of 10 or 20 such episodes collated into a single data file from 10 or 20 repetitions of a particular behavioural response (19, 20). In this way it was possible to preserve a fine grained analysis, with the filter only 1·0 Hz wide across the spectrum from 0 to 25 Hz for data epochs lasting 1·6 s.

This evoked spectral method was applied to data from cats performing in a modified T-box similar to that already described (47), but with facilities for recording a sequence of accurately timed epochs, each 1·6 s long beginning with two 'pre-stimulus' epochs, followed by two 'stimulus' epochs, when a 500 Hz tone was presented, then two 'approach' epochs covering the visual discriminative performance and approach to the food reward, and finally, two 'post-approach' epochs associated with eating (*Figure 3*).

(*a*) Evoked spectral analyses of hippocampal EEG data

When applied to hippocampal EEG data in early training, the spectral contours were inconsistent, with highest peaks varying between 4 and 6 Hz at each phase of the performance. Although the shift to a spectral peak at 6 Hz during the approach epoch appeared early in training, increased regularity of the waves, as expressed in the heightening and sharpening of the spectral peaks at each phase of the performance, appeared much more slowly. In advanced stages of learning (*Figure 4*), with performance levels beyond 80 per cent, a clear sequence of shifting spectral peaks emerged in correct responses, with 4 Hz characterizing the pre-stimulus period, 5 Hz occurring during the warning tone, and 6 Hz during approach (19). By contrast, a well developed 6 Hz peak did not occur during incorrect responses, even though the somatomotor pattern of movement toward the windows persisted. Distinct differences in hippocampal EEG patterns accompanying incorrect responses indicate that 'attention sets' immediately preceding incorrect performances were substantially different from those in correct performances.

We have interpreted these findings as indicating that the 6 Hz peak is not correlated with the goal directed somatomotor response, but with the connection between this pattern and a conditional stimulus which gives biological significance to the somatomotor pattern. The consistency of the regular shift in frequencies and sharpness of peaks in the approach period at advanced stages of learning suggests that they may be associated with the consolidation processes of learned behaviour.

(*b*) Evoked spectral correlates of learning in cortico-subcortical systems

In view of the clear sensitivity of the hippocampal EEG to changing states of focused attention, we have examined concurrent patterns in subcortical regions that might provide indices of afferent and efferent hippocampal systems, in accordance with schemes described above.

Spectral analysis was applied to cat EEG records from hippocampus, amygdala, subthalamus, midbrain reticular formation and visual cortex, during the T-box sequence of behavioural responses described above (20). Much variability was found between single behavioural epochs and between different daily runs. This variability decreased in advanced training. A common feature in the autospectra of all structures was a general shift toward frequencies in the theta range in the late stimulus and approach periods. This

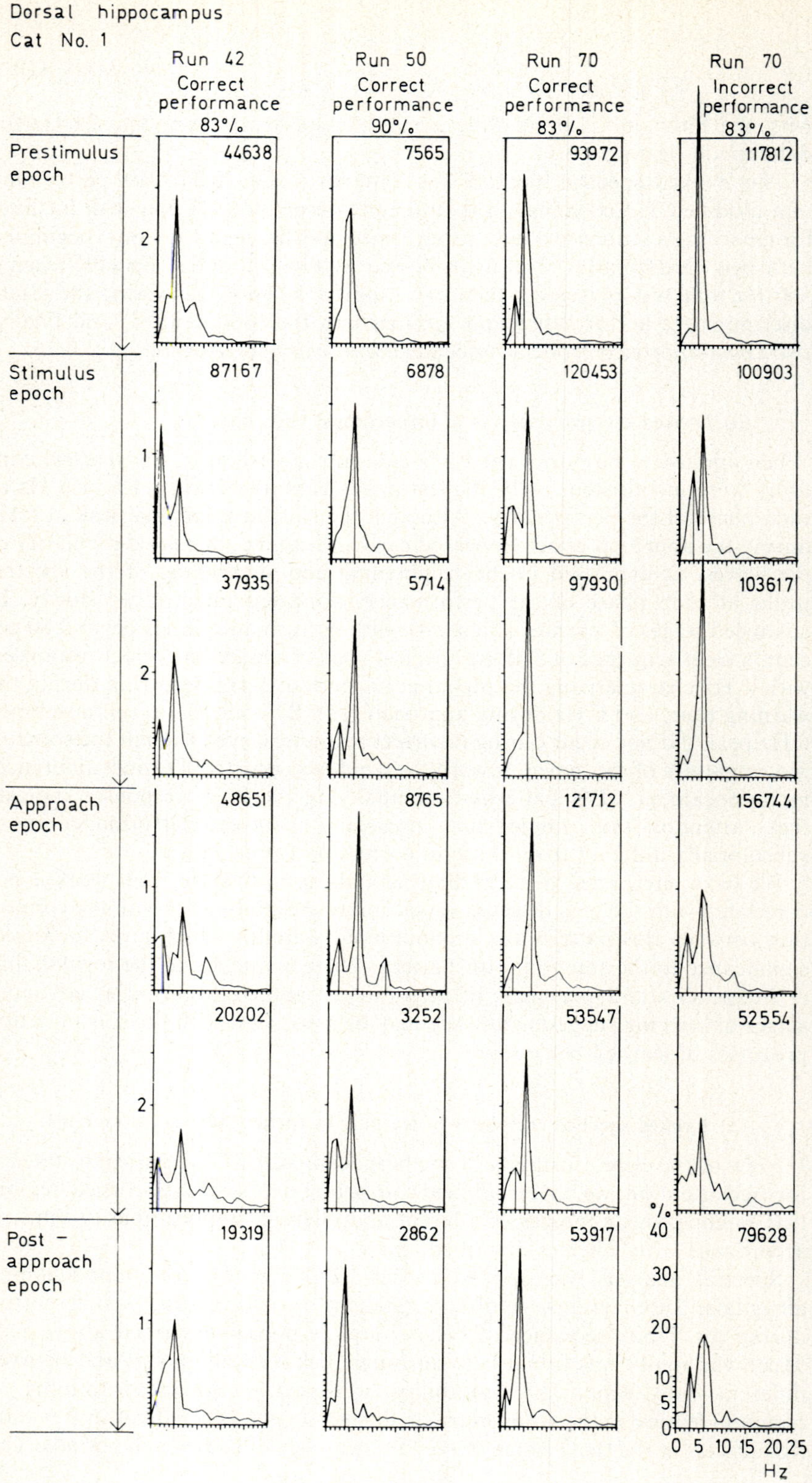
Dorsal hippocampus
Cat No. 1
Run 42
Correct performance
83%
Run 50
Correct performance
90%
Run 70
Correct performance
83%
Run 70
Incorrect performance
83%
Prestimulus epoch
Stimulus epoch
Approach epoch
Post-approach epoch
44638
7565
93972
117812
87167
6878
120453
100903
37935
5714
97930
103617
48651
8765
121712
156744
20202
3252
53547
52554
19319
2862
53917
79628
%
Hz

tendency appeared in some cases as a broadening of the spectral contour toward these frequencies, without appearance of new peaks, or in other cases as new single or multiple peaks. Spectral peaks at 6 Hz occurred mainly in the approach epoch, but were also seen in the subthalamus in the pre-stimulus epoch.

Livanov and Poliakov (41) first observed that, during formation of a conditioned reflex, similar wave patterns appear in different areas of the cortex. Later studies by Livanov and his colleagues (37, 40) with more elaborate analyses indicated a widespread synchronization in the initial stage of generalization of a conditioned reflex. With consolidation, the synchronization was limited to a few cortical points. A similar evolution of intercortical relations during conditioning has been observed in evoked potentials (35, 36, 48).

We have therefore examined interrelations between cortical and subcortical regions at each phase of the behavioural paradigm, and at different levels of learning. We have estimated the cross-spectra (50), with calculation of the coherence function as a measure of predictability of linear interrelations between pairs of structures. The coherence function has values between 0 and 1, and a probability level can be calculated for each level of coherence, taking account of the degrees of freedom associated with length of data epoch, digital filter bandwidths, data bandwidths and sampling rates. Plots of coherence can thus reveal levels of linear interrelations that lie above or below chance levels of probability in their statistical significance.

Autospectral densities in extrahippocampal structures, such as amygdala, subthalamus, midbrain reticular formation and visual cortex, all showed high spectral peaks at 1, 2 and 3 Hz during the visual discrimination and indicated increased regularity in slow rhythms. They also showed peaks in the theta range at 5 to 6 Hz, but these theta peaks were smaller and more variable than in the hippocampus.

Coherence calculations from the cross-spectra revealed high and consistent coherences between symmetric hippocampal points. Although more variable, high coherences also occurred between hippocampus and subthalamus, midbrain reticular formation and visual cortex, mainly in the theta range, during the approach epoch, and in the advanced stage of learning. *Figure 5* exemplifies these patterns, and particularly the hippocampo-subthalamic relations in the theta range detected in our earlier studies (15) described above.

Moreover, there were clearly defined simultaneous coherences between the hippocampus and subcortical structures, such as the subthalamus, in a certain

---

Figure 4. Spectral density plots of dorsal hippocampal EEG. A 'run' involves 30 consecutive trials given on one day, and different runs thus involve different training days. The sequence of plots should be read from above downward, as indicated by arrows. The number at the top right on each plot is a measure of total power in the spectrum from 0 *to* 25 Hz. *Each point on the plot is a percentage of the total power occurring at that frequency. Peak powers are indicated by vertical lines drawn to the abscissa. Peak powers shifted from* 4 *to* 6 Hz *in the sequence of pre-stimulus, stimulus and approach epochs, with gradual sharpening of the peaks. Power distributions were clearly different in incorrect performances* (*Run* 70)

(From Elazar and Adey, 1967a, by courtesy of the Editor, *E.E.G. Journal*)

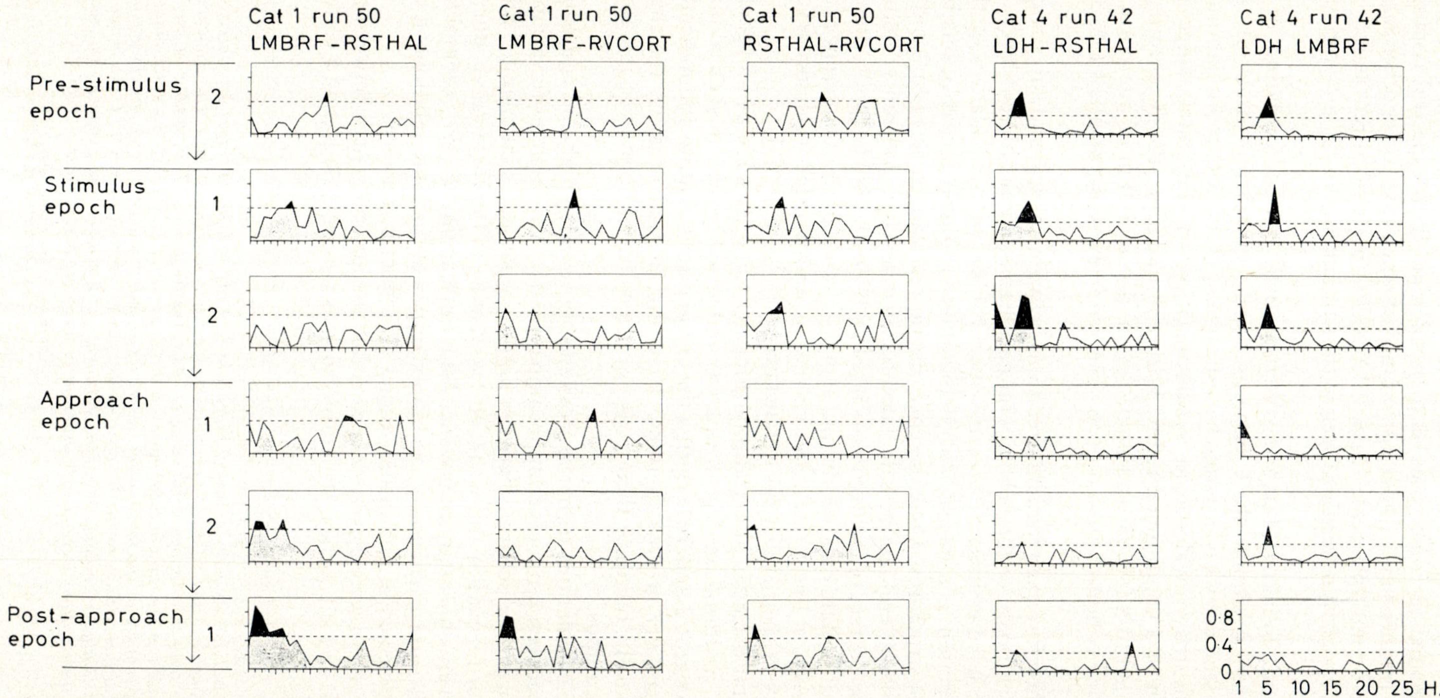

Figure 5. Plots of coherences in pairings of EEG records from mid-brain reticular formation, subthalamus and visual cortex. The sequence of plots should be read from above down, as indicated by the arrows. Each column of plots is for a different pairing. Frequency scale is shown on bottom right map (0 to 25 Hz). *Coherence magnitude between* 0 *and* 1 *is indicated on ordinate, and horizontal dashed line shows* 50 per cent *level of significance*

(From Elazar and Adey, 1967b
by courtesy of the Editor, *E.E.G. Journal*)

pattern of frequencies, and equally high coherences between that same subcortical nucleus and cortical structures at a different set of frequencies. Within the frame of linear interrelations tested by this method, there is an immediate confrontation with important problems. Thus, there is a possibility of parallel processing of different information streams between hippocampus and subthalamus on the one hand, and between subthalamus and visual cortex on the other. There is an even more important question as to the basis of simultaneous appearance of identical bands of frequencies in physically separated brain regions, so closely phase-related that they yield high coherence values. Are they phase-locked on the basis of a simple frequency coded information transfer between regions, or are they the manifestation of strictly local processes in the separate recording sites, with coherence established on the basis of an entrainment by a transfer signal that does not necessarily bear a direct frequency coding? No categorical answer is possible, but the aggregate of the evidence indicates that probably the wave processes are locally engendered, and locally entrained by transfer signals lacking direct frequency coding. This conclusion is suggested from an examination of phase relations at each frequency in cross-spectra from remote regions. Phase angles tend to be uniform across more than an octave of frequencies, a phenomenon noted in our early applications of spectral analysis (13), and difficult to explain other than on the basis of EEG genesis by a local domain or ensemble of generating elements. *A priori*, it would seem unlikely, though by no means impossible, that such a fragile parameter as frequency and phase relation of waves in different cortical domains would be directly sustained on the basis of pulse-coded neuronal firing between them. Certainly, this problem merits close consideration beyond the mere detection of correlated neuronal firing patterns at separate recording sites, and invites specific attention to the pattern of interrelations in the total spectrum of waves arising in each of the separate locations. We shall consider these findings further below in relation to the genesis of the EEG.

(*c*) Role of somatic components in orienting responses

We have considered EEG activity in limbic, subcortical and neocortical systems accompanying the total pattern of behavioural responses in alerting, orienting and visual discrimination. We have further investigated effects of altered bodily habitus on EEG activity in subcortical and cortical structures, with special reference to vestibular and somatic components contributing to EEG reactions to passive changes in body position (16). A series of cats were prepared with electrodes implanted bilaterally in cerebellar nuclei, vestibular nuclei, midbrain reticular formation, putamen, caudate nucleus, anterior hypothalamus, septum, amygdala and hippocampus. Postural effects were tested in unanaesthetized animals with limbs and trunk lightly wrapped. EEG records were taken with the animal successively in prone, dorsal and right and left lateral positions.

Spectral analyses of prone records indicated peaks in density at frequencies between 1 and 4 Hz in all structures. Placing the animal in lateral positions produced new spectral peaks with densities as much as four times higher than in the prone position. In subcortical structures these new peaks were located at the same frequencies as maximum energies in the prone position. In both dorsal and ventral hippocampus, however, the new peaks appeared at substantially higher frequencies, and usually in a narrow band around 5 Hz (*Figure 6*).

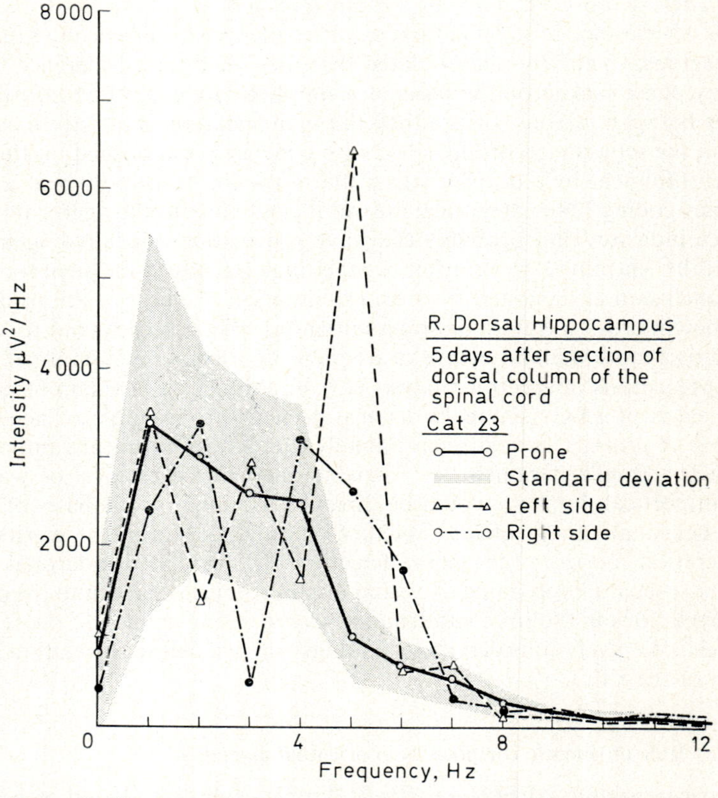

Figure 6. Hippocampal EEG spectral density changes induced by shift in body position from prone to right and left lateral and to dorsal positions, after section of dorsal columns of spinal cord. Shaded band around graph of control data shows variance (1 *O.S.D.*)

(From Costin *et al.*, 1967, by courtesy of the Editor, *The Physiologist*)

The relative significance of proprioceptive and vestibular afferents in initiating these responses to postural alterations was tested by separately sectioning the dorsal columns of the spinal cord in the upper cervical region, and bilateral vestibular nerve section. Section of the dorsal columns did not abolish the augmented low frequency peaks. The dorsal hippocampal response was reduced in amplitude but at a higher frequency than pre-operatively. Section of the vestibular nerves abolished all statistically significant responses. It is also relevant, in a consideration of these parameters in orienting responses, that contour plots of serial spectral analyses frequently showed a waning after the first 10 or 20 s of non-prone posture, followed by a transient recrudescence after a further half minute or more. Maximum evoked shifts always occurred on initiation of the postural change, supporting the view that they related to orienting responses.

RELATED PATTERNS IN SCALP EEG RECORDS OF MAN; PREPARATION OF A NORMATIVE LIBRARY

It is a historical paradox in many fields of biomedical research that substantially more is known about the pathological physiology of disordered systems than about subtleties and nuances of system responses in healthy organisms. For the EEG, complexities of multichannel records, and particularly the differences between subjects, have discouraged fine interpretations of a gamut of behavioural states that might range from quiet wakefulness to extreme focusing of attention. Visual examination of paper records could scarcely provide the basis for such an interpretation. Yet the need for such a comparative baseline is great, as a building block in schemes for direct monitoring of central nervous states in performing man.

We have therefore examined a 'normative library' of EEG and other data, collected from 200 astronaut candidates in a series of behavioural tests presented to each subject under identical conditions (53). These subjects were tested by means of a programming device developed in our laboratory, using magnetic tape to control presentation of one hour of perceptual and learning tasks. All subjects were tested in the Department of Neurophysiology at the Methodist Hospital, Houston, by Drs. P. Kellaway and R. Maulsby, who prepared essentially artefact-free records from all subjects. During the test, the subject was comfortably seated before a projection screen, with a push-button device at his right hand for responses to behavioural tasks. Epochs of physiological stimuli and behavioural tasks were interspersed with 'resting' periods in which the subject sat quietly, with eyes open or closed, as specifically commanded. Behavioural tasks ranged from simple auditory vigilance tests, with a requirement to recognize the presence or absence of three tones in a series presented at 6·0 s intervals, to recognition of the largest of six circles in a matrix. This visual discrimination involved a 3·0 s test epoch in some series, and a 1·0 s epoch in others. The 3·0 s task required moderate concentration, whereas the 1·0 s task was near the limit of performance capability in many subjects.

(*a*) Data acquisition and analysis methods

A wide range of physiological parameters was recorded from each subject, including 18 channels of EEG from leads placed according to a modified international 10–20 schema, together with the EOG, electrocardiogram, respiration and galvanic skin responses. Initial spectral density calculations were made for each of the 18 EEG leads from 50 subjects. To synthesize the data, spectral outputs were averaged over all 50 subjects. A synthetic 'head' was thus constructed, with the average for each scalp region presented as a bar

RESPONSES OF ELECTROENCEPHALOGRAM TO DIFFERING SITUATIONS

TOPO SPECTROGRAPHIC VARIATIONS OF
AVERAGES OVER 50 ASTRONAUT CANDIDATES

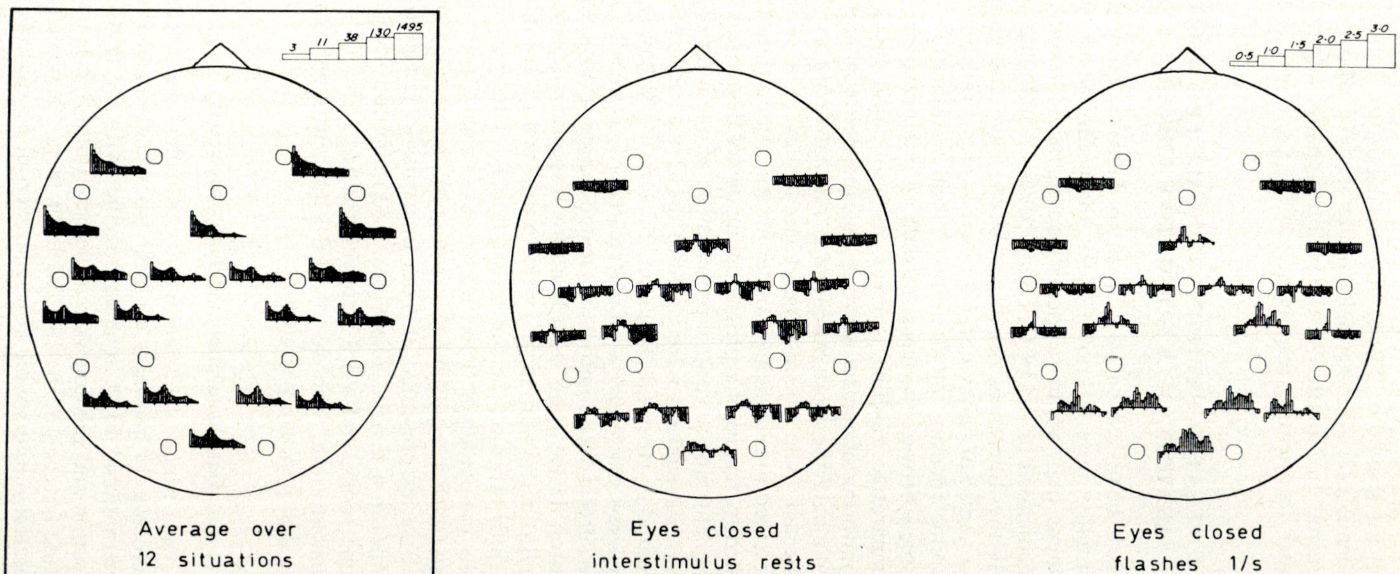

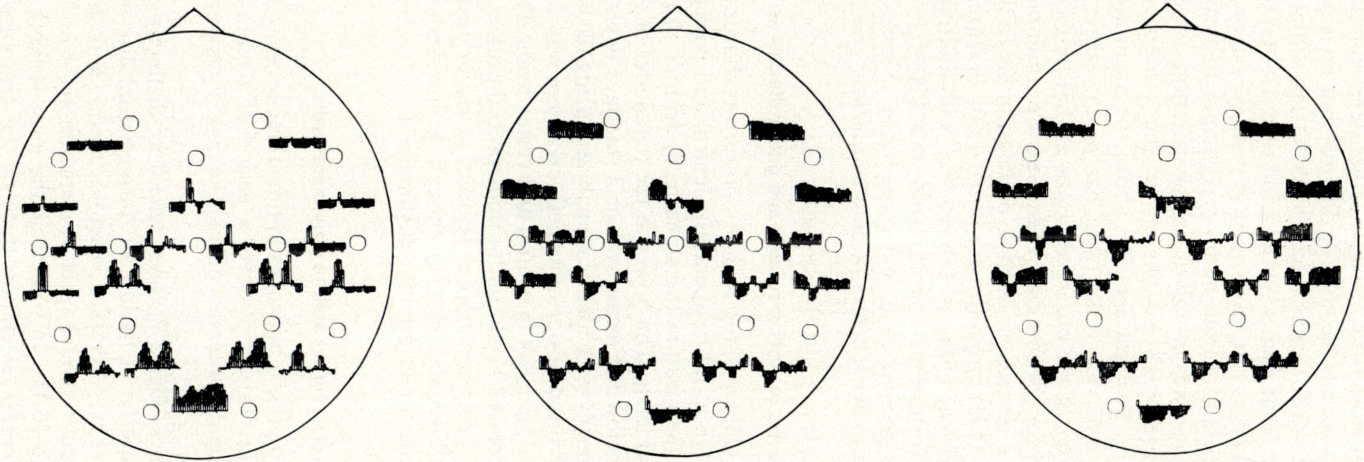

Figure 7. Averaged spectral densities over the range 0 *to* 25 Hz *for a population of* 50 *subjects, with each spectrum presented as a series of bars at* 1 Hz *intervals, and located at the appropriate location on the scalp. The top left figure is an average for all subjects across* 12 *situations (see text). The contour of this average was then used as the mean against which to measure deviations in the succeeding* 5 *situations, with powers at any frequency above the mean shown as a bar above the baseline and vice versa. Calibrations for average over* 12 *situations in* $\mu V^2/s$ per cycle; *for separate situations, in standard deviations*

(From Walter *et al.*, 1967,
by courtesy of the Editor, *Aerospace Medicine*)

graph covering the spectrum from 0 to 25 Hz. Further development of this display technique provided separate 'heads' for different test epochs.

After computation of spectra for each subject over all test situations, an average was prepared for all 50 at each scalp location over 12 situations, including sitting with eyes closed at rest, eyes open at rest, eyes closed during one per second flash stimuli, and during auditory vigilance and visual discrimination tasks. Subsequent heads in *Figure 7* show variations for each test situation about the mean previously established by the contour of these 'lumped' or averaged spectra. Spectral densities above the mean are shown as bars above the baseline and vice versa, with bar height expressing the number of standard deviations of the observed power at that frequency.

This study has established for the first time that there are common factors that clearly categorize EEG records in such a population for a series of specified behavioural states. This comprehensive initial analysis required 25 h of main computation time for each hour of subject data, with multiplications in spectral calculations performed at approximately 500,000/s. From this initial broad baseline, it has become clear that recognition of states of consciousness can be based on simpler schemes of data acquisition and analysis.

(*b*) Averages of EEG spectra in resting records and during flashes at 1/s

With eyes closed and at rest (*Figure 7*, top middle), powers in all frontal leads were below the group average at all frequencies. Central and parietal leads showed a small peak above the mean in the alpha band around 10 Hz. This peak was broader in occipital leads. The power in the remainder of the spectrum from 0 to 25 Hz was below the mean by 1 or 2 standard deviations.

These findings contrast sharply with identical analyses during flashes at 1/s with eyes closed. Although frontal power remained low, there was a marked increase in the power of the alpha peak in central and vertex leads. There was a further broadening and increased height in this mid-spectral peak in parietal areas. In the occipital and occipitotemporal regions, a major increase in spectral power occurred in all frequencies, even though the low flash rate was clearly not responsible for any direct driving phenomenon.

(*c*) Averages of EEG spectra during auditory vigilance tasks

The subjects were required to press a button each time three tones were heard in a succession of tests at intervals of 6 s over a period of 8 min. Unusual groups were presented at irregular intervals during the task. Eyes were closed throughout.

The resulting EEG averages (*Figure 7*, lower left) were distinctly different from those during slow repetition of flashes, or during visual discriminative performances described below. Vertex, parieto-occipital and temporo-occipital leads showed distinct bimodal spectral distributions, with separate peaks centred around 10 and 20 Hz. In summary, the most striking changes were those in the belt of cortex extending through the temporo-parieto-occipital regions.

(*d*) Averages of EEG spectra during visual discrimination performances

The task involved identification of minor size differences in sets of six circles, presented in two series at 3 s and 1 s intervals. The averaged spectral 'heads' were clearly different from either resting or vigilance task spectra.

In the 3 s tests (*Figure 7*, lower middle), frontal activity was sharply incremented over the whole spectrum. This declined gradually toward central and vertex regions, where the alpha peak noted in the previous test situations was now sharply reduced, and powers in the alpha band were reduced below the multi-situational mean. In the bioccipital lead, powers were reduced below the multi-situational mean at all frequencies above 2 Hz.

Trends characterizing the 3·0 s tests were more obvious in those lasting only 1·0 s, but the two 'heads' show obvious similarities (*Figure 7*, lower right). A clear distinction between the two was possible, however, by utilization of coherence measurements of shared activity between different cortical regions. These distinctions were clearly manifested in automated recognition of patterns of parameters that best characterized the two data sets.

(*e*) Application of pattern recognition techniques to EEG characteristics in individual subjects

The disclosure of patterns in these collective 'heads' that clearly differ between behavioural situations, yet represent a congruent pattern across individuals, is indeed challenging. Nevertheless, it is obvious that detection of a population characteristic will be of little value where the variance of an individual may be sizeable in relation to the population mean. Moreover, the constellation of parameters that collectively best characterize the EEG of a single subject may not be identical in all respects with those characteristics of the group mean for the same behavioural situation. That is, each individual may have sets of EEG signatures uniquely his own in their aggregate description of brain states accompanying particular behavioural states and responses.

Our application of pattern recognition methods supports the view that, not only is it feasible in individual subjects to relate with a high reliability a particular behavioural state and concomitant EEG patterns, but also that these patterns are best described by parameters of which some are shared with other subjects, and others are unique to a particular subject (54). Using a stepwise discriminant analysis of parameters in a multivariate matrix, EEG correlates were sought in four subjects for five basic behaviour states, including eyes closed resting, eyes open resting, during performance of the auditory vigilance task, and during the 3·0 and 1·0 s visual discrimination tasks (*Figure 8*). Using only the four best measurements for each subject, between 62 and 69 per cent of the epochs were correctly classified. After selection of 15 parameters, epochs were correctly classified in 95, 93, 96 and 90 per cent of cases for each individual. Since the epochs were only 10–20 s in length, and classification involved only 4 EEG channels, it may be inferred that each individual may have spatially and temporally characteristic EEG 'signatures', that change in an ordered way in relation to specific behavioural states. Detailed discussion of the parameters selected for each pattern is presented elsewhere (54), but it is relevant that the rank ordering of parameters follows closely the accepted

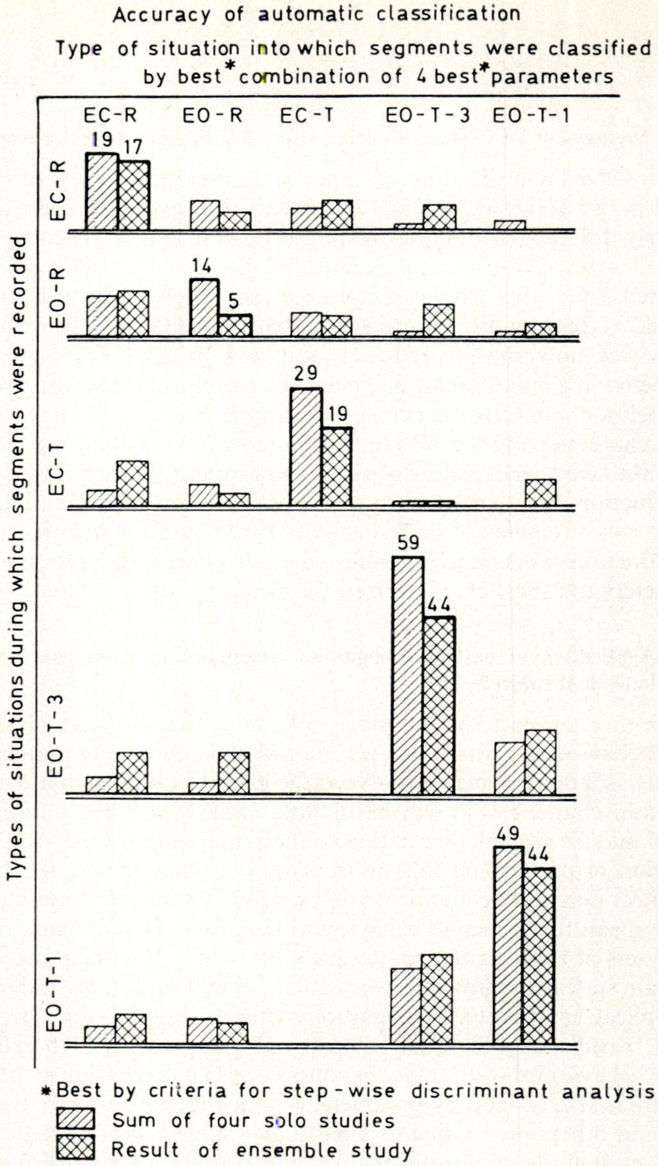

Figure 8. Distribution of EEG segments by an automatic discrimination programme. Segments were recorded in 5 situations: EC-R, eyes closed resting (34 *segments); EO-R, eyes open resting* (32 *segments); EC-T, eyes closed performing auditory vigilance task* (40 *segments); EO-T-3, performing a visual discrimination of material presented for* 3·0 s *for each situation* (80 *segments); EO-T-1, the same visual task with only* 1·0 s *exposure* (78 *segments). Five related studies are summarized: in four 'solo' studies, each subject's records were evaluated separately. In the first analysis, only four parameters were selected that best categorized his records. In the other 'ensemble' study, records from all subjects were treated as if for a single subject. The horizontal rows of bars indicate the situations from which the records were taken, and the columns indicate categorizations by selected parameters. Heavily outlined bars on the diagonal represent correct categorizations. Single hatching indicates summed categorizations in the four solo studies, and cross-hatching indicates the ensemble study. After parameters were selected, records from individual subjects were correctly classified on* 95, 93, 96 *and* 90 *per cent of the tests, compared with only* 60 *per cent for the group as a whole*

(From Walter *et al.*, 1967, by courtesy of the Editor, *E.E.G. Journal*)

schemes of cortico-subcortical interrelations in specific sensory and association pathways. The parameters thus appear to have relation to cortical integrative mechanisms.

A similar study of EEG parameters in chimpanzees playing tic-tac-toe has allowed complete separation of EEG patterns accompanying correct and incorrect decision making (30).

APPLICATION OF 'NORMATIVE LIBRARY' BASELINE TO EEG DATA FROM MAN IN THE WEIGHTLESS STATE

The past decade has seen increasing recognition of the striking impact of environmental factors in determining the performance capability of man. Estimates of performance limits in the laboratory necessarily involve varying degrees of unreality. The physiologist has recognized the need to monitor a gamut of central nervous, autonomic and cardiovascular parameters in man on a non-interference basis, even where task performances may involve circumstances dangerous to life. Developments in bioinstrumentation have materially assisted in achieving these goals.

As a specific application of the baselines developed in the normative library study, 55 h of EEG data from Astronaut F. Borman were analysed (8). The records began in the immediate pre-launch period and ran continuously into the third day of the Gemini GT-7 flight. The flight data were compared with the astronaut's own baselines in the normative library tests, and in tests in a Gemini flight simulator. Electrode and signal conditioning techniques in the flight experiment have been described by Dr. Robert Maulsby (42), under whose supervision the flight preparations were made. Two channels of EEG were recorded for the first 29 h of flight. Inadvertent detachment of an electrode at that time left only a single effective channel up to the termination of recording at 55 h. Electrodes were arranged to give wide bipolar pairs of leads, one pair spanning from vertex to occiput, and the other covering a similar span to the left of the midline. Seven channels of flight data were recorded continuously with a special magnetic tape recorder, with a data bandpass of 0·5–30 Hz.

The prime interest in this experiment was in evaluation of sleep states. Drowsy or sleep records were sampled for two consecutive epochs, each of 20 s, every 2 min. The pre-launch period and first orbit were sampled continuously at this rate, and awake records in the remaining data were sampled for two consecutive epochs, each of 30 s, every 10 min. Contour plots of spectral density and coherence prepared in this way thus have an 'elastic' time base, expanded to present drowsy and brief sleep episodes in greater detail, and sharply contracted at other times.

(*a*) Analyses of baseline data from laboratory tests and Gemini simulator

The subject's resting EEG was clearly normal, with a well developed alpha rhythm at 9–11 Hz, having a power of 300 μV^2/s per cycle. At the same time, and important in comparisons with flight records, powers in the spectrum 2–7 Hz were low (30 μV^2/s per cycle or less). Powers below 2 Hz were higher than in the range 2–7 Hz but typically were lower than simultaneous alpha densities.

By contrast, visual discriminative tasks in the normative library test caused alpha activity at 9–11 Hz to decline sharply in all areas, whereas powers in the

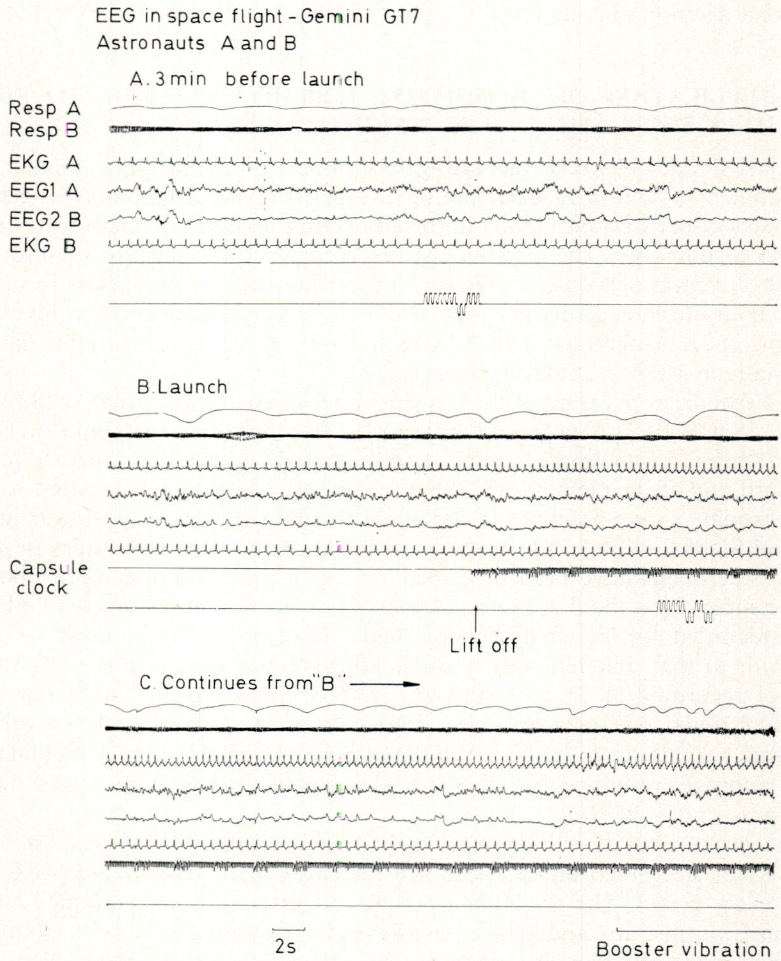

Figure 9. Samples of physiological data from Gemini GT-7 flight, covering epoch immediately prior to launch and during lift-off. RESP A, respiration of F. Borman; RESP B, respiration of J. Lovell; EKG A, EKG from Borman; EEG 1 A and EEG 2 A, EEG channels from Borman; EKG B, EKG from Lovell. Channel 7 shows spacecraft clock (beginning at lift-off)

(From Adey *et al.*, 1967, by courtesy of the Editor, *Aerospace Medicine*)

range 0–3 and above 14 Hz were considerably augmented, in good agreement with the findings for the population of subjects in the normative library. In comparison with his records in space flight presented below, powers in the theta band from 5 to 7 Hz were low in all resting and visual task records. His EEG reactions during the auditory vigilance test showed a recrudescence of high powers in the alpha bands in central, parietal and occipital areas, also in good agreement with the group study.

EEG data were also recorded in a Gemini spacecraft simulator during a thermal vacuum test, using electrode placements, amplifiers and a recorder identical with flight equipment. These records were taken during alerted states, and showed relatively low powers at all frequencies above 5 Hz. Both EEG channels showed a broad and ill-defined alpha peak at 9–13 Hz, but higher powers occurred in the theta band from 4 to 7 Hz. This decreased alpha and augmented theta activity in the altitude chamber test was clearly intermediate between laboratory tests and space flight records.

(*b*) Analysis of Gemini GT-7 flight data at launch, and in early and late orbits

In computation of the flight records (*Figure 9*), a low frequency cut off at 3·0 Hz was arbitrarily designated to minimize contamination of the analysis by movement artefacts, although very few records were significantly affected by movements.

(*i*) *Launch and first orbit*—Theta rhythm activity was increased in the pre-launch period over baseline records. This increased density broadened to include higher frequencies as launch became imminent, and at one minute before launch, the augmented theta band merged with similarly raised powers in alpha and beta bands (*Figure 10*). These findings are interpreted as relating to strongly focused attention and orienting responses in an undoubtedly novel situation, and closely resemble the normative library data during strongly focused attention. At launch, EEG power density was augmented by a factor 10, in a strong neurophysiological arousal reaction. These augmented densities declined slowly, with higher powers in bands above 10 Hz recurring in the first half hour of flight.

Coherence between the two EEG channels was high at theta frequencies up to the time of lift-off, and then fell to low values across the entire spectrum for the next 30 min. Coherence values reached unusually high levels from 3 to 9 Hz in the second half of the first orbit, sustaining almost continuously for about 40 min, and declining slowly thereafter. No comparable coherence was detected either in baseline records or in later flight records. Our laboratory studies (52) have indicated that such a resurgent or 'rebound' coherence between cortical regions frequently follows a psychically challenging experience, and that, as in the present case, it is correlated with a more relaxed attitude following such an experience. This interpretation is also supported by simultaneously diminishing powers in both leads at higher frequencies. Moreover, low coherence at both high and low frequencies during the booster phase is strong evidence against electromechanical artefacts induced by vibration of the head as significant contributors to the spectral density of the EEG.

(*ii*) *Records of 'day' and 'night' sequences in space*—Careful assessment of awake flight records throughout the remainder of the 55 h indicated increased power in the theta band by comparison with laboratory and flight simulator data. Our knowledge of the significance of theta rhythms in normal man has

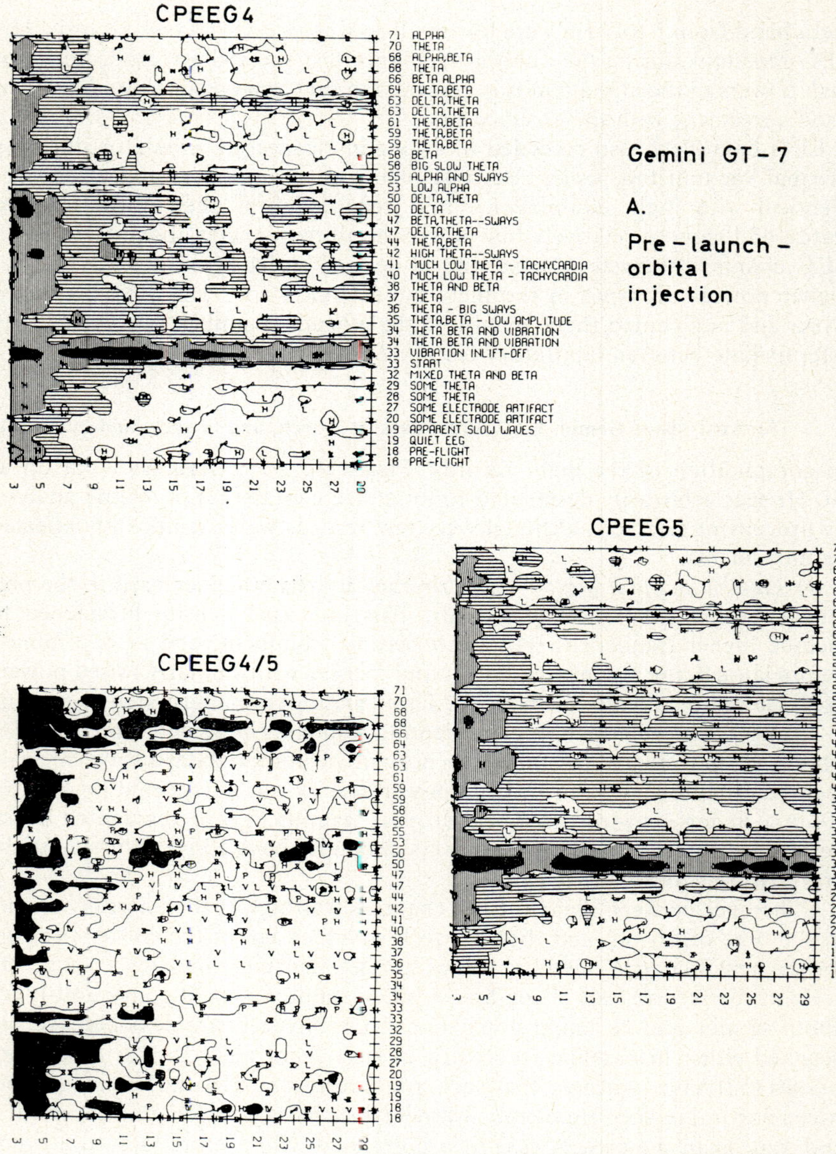

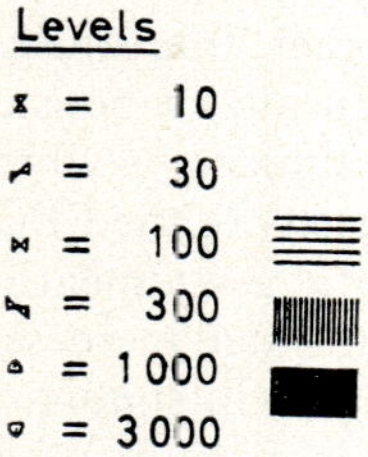

Figure 10. Contour maps of EEG data from F. Borman in pre-launch and launch period of Gemini GT-7 flight. Autospectral densities in two EEG channels (CPEEG4 and CPEEG5) showed enhanced theta activity pre-launch, and great exaltation of many EEG frequencies before and during lift-off (epoch 33). Compared with later awake and sleeping records, coherences between the two channels (CPEEG4/CPEEG5) were low at this time. Spectral densities above 10

B
Continuation of
first orbit

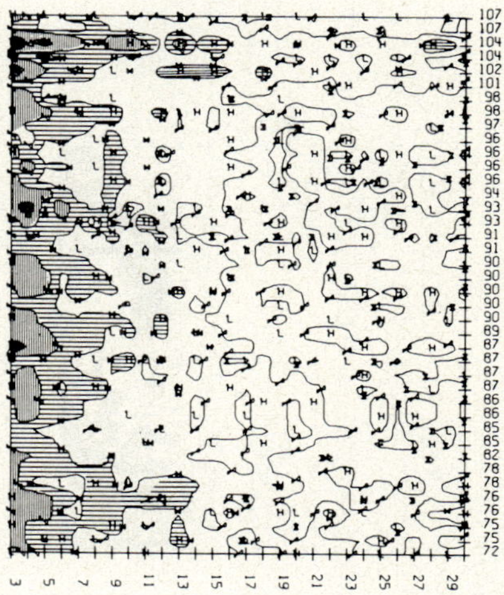

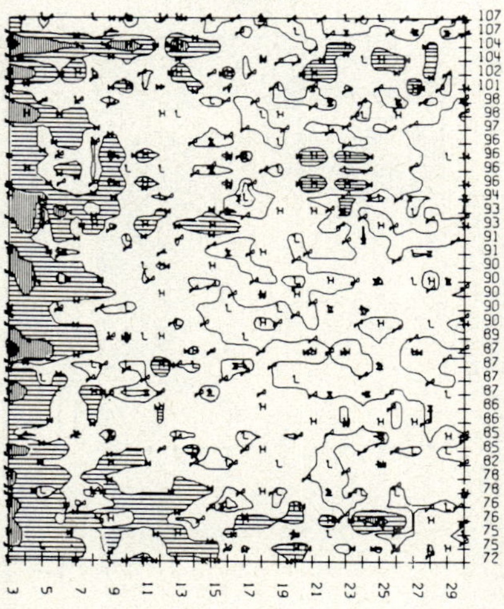

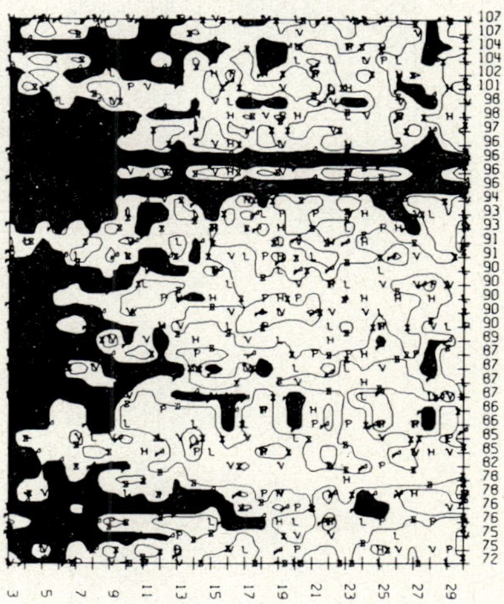

Levels
x = 0·40
↗ = 0·7 ■

Hz *gradually declined in the first* 30 min *of flight*, *but very little alpha activity appeared before* 70 min. *Numerals on ordinates indicate elapsed minutes of record in that analysis frame. Contour markings: horizontal shading*, 100 μV²/s per/cycle; *vertical shading*, 300 μV²/s per cycle; *solid black*, *in excess of* 1000 μV²/s per cycle

(From Adey *et al*., 1967, by courtesy of the Editor, *Aerospace Medicine*)

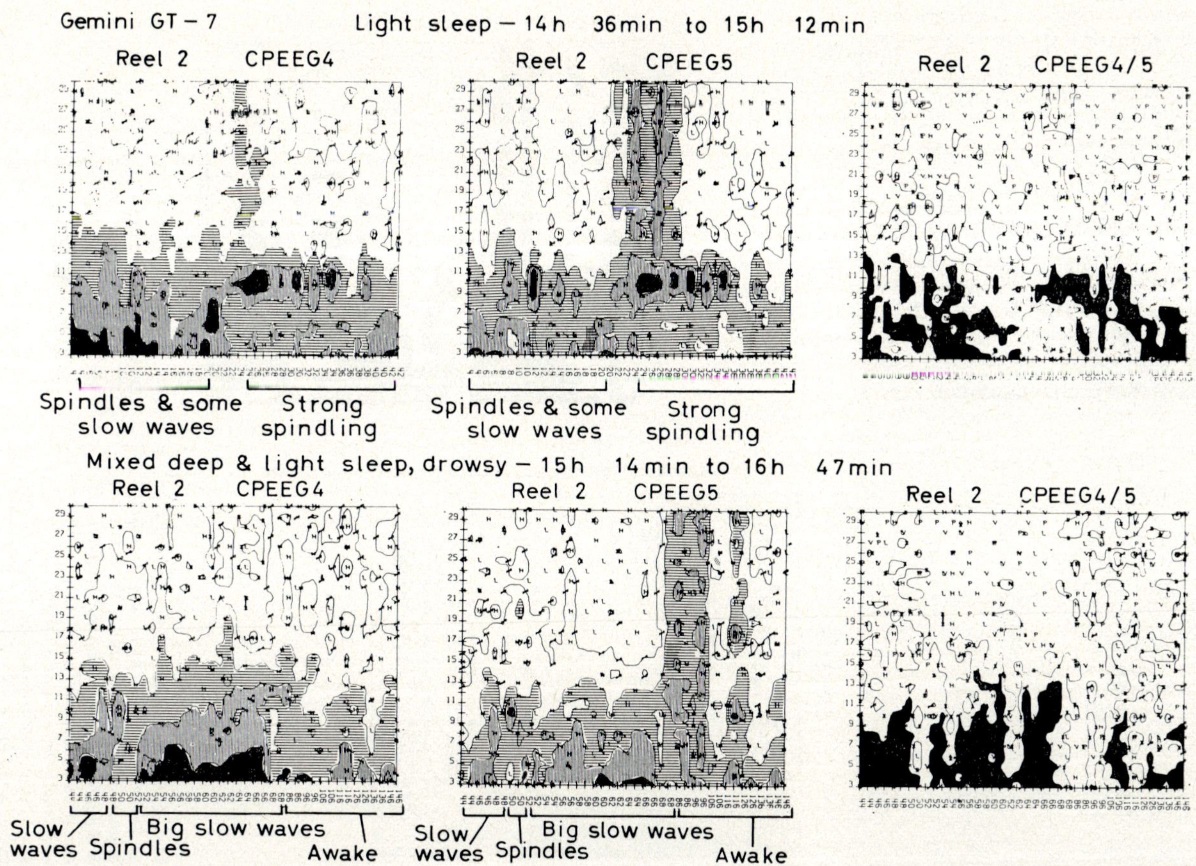
Gemini GT – 7
Light sleep – 14h 36min to 15h 12min
Reel 2 CPEEG4
Reel 2 CPEEG5
Reel 2 CPEEG4/5
Spindles & some slow waves
Strong spindling
Spindles & some slow waves
Strong spindling
Mixed deep & light sleep, drowsy – 15h 14min to 16h 47min
Reel 2 CPEEG4
Reel 2 CPEEG5
Reel 2 CPEEG4/5
Slow waves
Spindles
Big slow waves
Awake
Slow waves
Spindles
Big slow waves
Awake

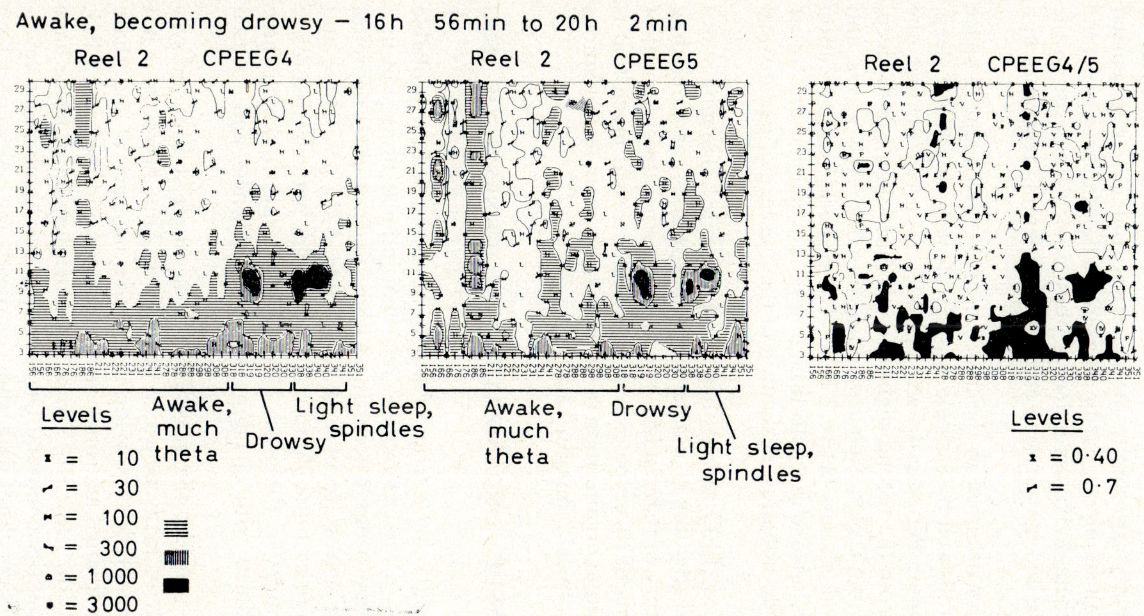

Figure 11. Contour maps of EEG data from 15th to 21st hours of Gemini GT–7 flight. This period covers first 'night' in space, and shows a mixture of states from awake to drowsy and light to medium sleep. This slow wave sleep lasted only about 25 min *(middle row of figures), and was the only deep sleep on the first night. Coherence between channels was sharply increased across the spectrum from* 3 *to* 10 Hz *in this episode (middle row, right figure). Calibrations as in Figure 10*

(From Adey *et al*., 1967,
by courtesy of the Editor, *Aerospace Medicine*)

remained small. Grey Walter (55) has attributed these rhythms in young adults to responses to frustration and disappointment, but such correlates may not describe exclusive relationships. Rather, these rhythms may arise in the physiological substrates of alerting and Pavlovian orienting reflexes to novelty in the environment, as discussed above. Comparable findings in Soviet cosmonauts in early exposure to the space environment are discussed in detail elsewhere (8). In summary, our findings in the flight data support the view that augmented theta activity is indeed a physiological response to the weightless environment, and may arise in augmented orienting responses to a quite unusual experience.

Long drowsy episodes occurred in the first 6 h of the first night in space (*Figure 11*). This was followed by a single episode of deep slow-wave sleep, lasting about a half hour. Periodic drowsy episodes occurred through the second day. Records from the second night showed more than 6 h of sleep, with successive 90–100 min cycles, each descending into deep slow waves, and with brief periods of wakefulness between each. The posterior scalp location of the EEG leads precluded clear detection of eye movements, and other features of the records do not allow evaluation of the presence of paradoxical sleep phases.

The sensitivity of these EEG records has clearly revealed changing states of alertness and focused attention in task performances probably as demanding as any other in the history of man and his machines. Further development of data acquisition systems of this type can be expected to reveal much that is new about the fundamental organization of the human brain, in test situations far removed from the typical laboratory setting.

CEREBRAL TISSUE ORGANIZATION IN GENESIS OF EEG RHYTHMS

It is only in the last 5 years that a series of studies have revealed the probable origin of the EEG, not in the ensemble of neuronal impulse firing, but in a quite different slow wave process occurring within single cerebral neurones (18, 22, 26, 34). These waves are of high amplitude, 5–15 μV. These and related studies have given us a first glimpse of the complex patterns that relate individual wave generators to the gross EEG, and of ways in which these relations may be modified with changing states of alertness and focused attention (7, 24).

It would be easy to consider the gross EEG as a simple sum of the synchronized wave activity of many neuronal generators. This, however, is not the case. Coherence calculations provide a sensitive measure of the phase relations between an individual neuronal wave generator and the gross EEG. Intracellular waves and the gross EEG from the same domain of tissue have quite similar spectral density contours. On the other hand, coherence calculated on the basis of 10 s epochs of data, show extremely low values over long periods (*Figure 12*). These studies (24) have suggested that cortical generators appear to meet the requirements of the central limit theorem of Cramer (17), exhibiting individual frequency distributions which are not linearly related, and possess a mean and a finite standard deviation. The contributions from

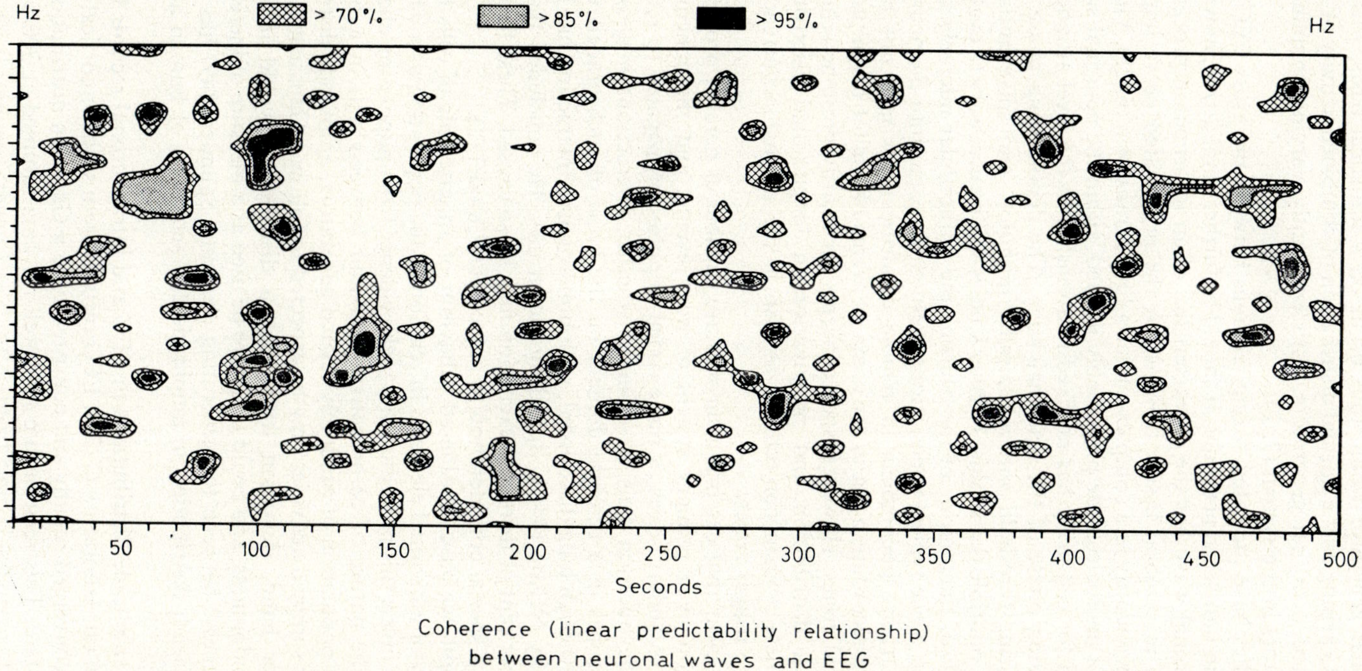

Figure 12. Plot of coherence between intracellular waves and surface EEG for a sequence of 10 s *epochs over period of* 500 s. *Coherence remained below statistically significant levels at all frequencies for major part of analysis. Significant coherences (black) occurred at about chance rates throughout the analysis. The findings are interpreted as indicating origins of the EEG in a population of largely independent neuronal generators*

(From Elul, 1964,
by courtesy of the Editor, *The Physiologist*)

the intracellular generator would thus occur on a statistical basis, with the resulting EEG arising as the normal distribution of combined activity of non-linearly related neuronal generators, with constantly varying coupling to other generators in the system (7, 24).

These studies have indicated strong relations between the gross EEG and wave events in individual neurones. It should be further emphasized, however, that the neuronal wave has the characteristics of an information processing phenomenon in cerebral tissue, from which the propagated nerve impulse arises only secondarily as a complex and non-linear transform (6, 25). Viewed at the tip of a microelectrode presumably located within the neuronal soma, the threshold for cell firing has no fixed relationship to the level attained in the depolarizing phase of the neuronal wave. The depolarizing phase may exceed the firing threshold without impulse initiation, and spike initiation may occur at multiple thresholds. Further studies will be necessary to evaluate by how much of the interior of the neurone the intracellular microelectrode is illuminated. Since typical penetrations presumably occur into the neuronal soma, depolarizing phenomena in remote dendritic fields may not be as visible from a recording point in the soma as events of similar magnitude in more proximal zones. However, such concepts necessarily presume schemes of differential distribution and propagation in longitudinal electrotonic processes. These remain conjectural, as also is the normal occurrence of multiple spike trigger zones.

It appears clear that cerebral neurones are capable of parallel processing of information by a primary system of wave events, and the concurrent genesis of a secondary series of pulse coded impulses propagated to remote locations. The question thus arises as to whether interactions may occur between neurones in a cerebral domain on the basis of such graded wave processes. Although long discounted, such non-synaptic interaction has been investigated by Nelson (44) in spinal motoneurones. He concluded that as much as 10 per cent of threshold excitation may be so caused. These findings would merit a special consideration in any cerebral model (2).

To this point, we have considered behaviour of an ensemble of cells from close scrutiny of individual elements in relation to the population as a whole. We may now invert this procedure and seek evidence from population characteristics on the nature of co-operative behaviour between cells in a single domain. In a population of randomly related generators, amplitude distributions of their summed activities may be expected to follow a Gaussian curve. On the other hand, deviation from such a distribution would indicate decreased randomness, and could therefore be used as a measure of increased 'co-operativeness' between the generating elements. Saunders (49) has described a Rayleigh distribution in amplitude characteristics of human alpha rhythm.

Trends in amplitude distributions in EEG and intraneuronal records have been tested in our laboratory (7, 23). EEG records from scalp or cortical surface in man converged rapidly and consistently to a Gaussian distribution in the waking state. This relationship held well for data sampled at 200/s and 7000/s, and for epochs lasting from 2 to 5 s. Longer epochs of 10–15 s frequently showed declining Gaussian characteristics, relating perhaps to

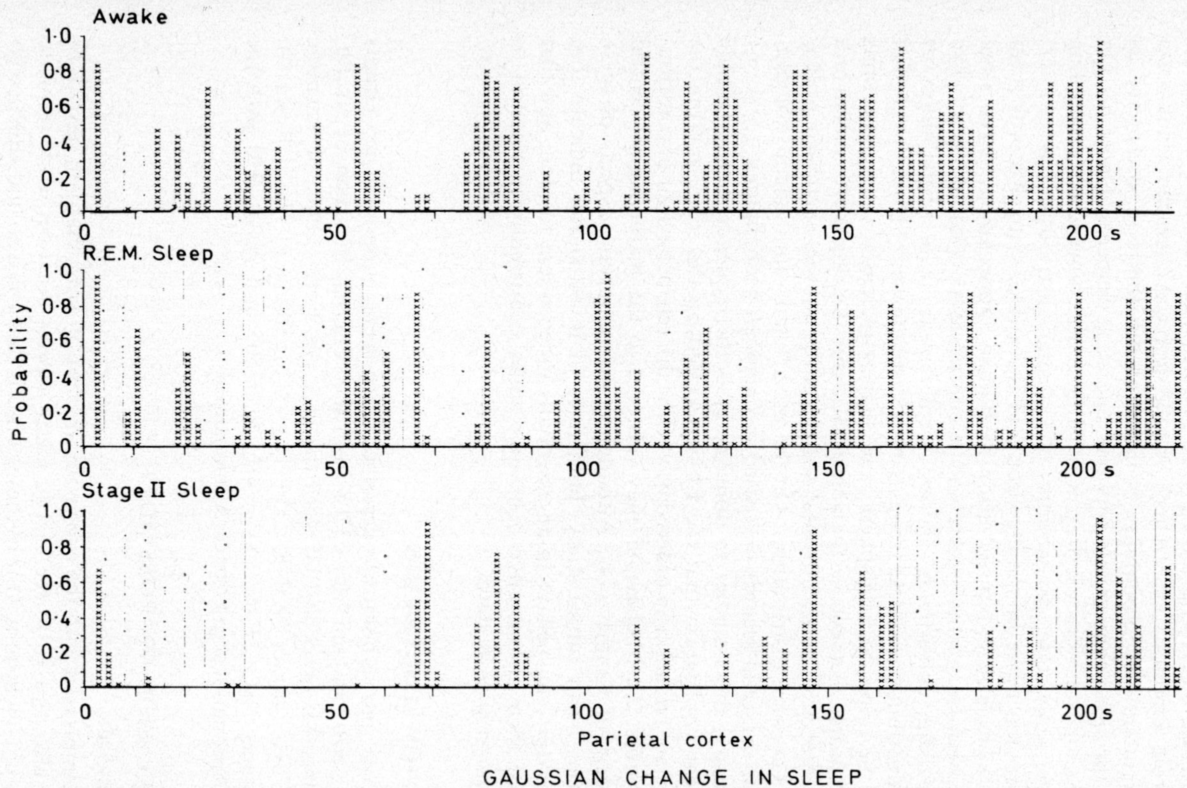

Figure 13. Changes in amplitude distributions in EEG records from parietal cortical screws in man in the awake state, during REM sleep, and in spindle sleep. Successive 2·0 s *epochs were analysed for probability of Gaussian amplitude distributions. The regular spindle trains in Stage II sleep reduced probabilities of Gaussian distributions to a low level*

(From Adey *et al.*, 1966,
by courtesy of The American Electroencephalography Society)

non-stationary behaviour of the data, associated with shifting states of attention. On the other hand, amplitude distributions in intracellular wave records regularly produced skewed, non-Gaussian distributions.

EEG amplitude distributions were studied in normal individuals in spindle and paradoxical sleep, and during task performance, and also in patients suffering from temporal lobe epilepsy. Spindle sleep was characterized in scalp records by diminished probability of Gaussian distributions, interpreted as indicating increased connectivity between generating elements (*Figure 13*). On the other hand, in paradoxical sleep, with EEG records superficially resembling those in the wake state, probabilities of Gaussian distributions were intermediate between those in spindle sleep and the awake state. Records from bilaterally symmetrical bipolar leads in human hippocampal tissue were compared between normal tissue and a contralateral epileptic focus. The abnormal tissue consistently showed trends away from probabilities of Gaussian distributions. These trends clearly distinguished it from the normal region of the opposite hemisphere, and suggest increased interconnections between neuronal elements in the abnormal tissue. During performance of a task in mental arithmetic, bifrontal scalp records showed increased probabilities of Gaussian distributions. On the other hand, simultaneous longitudinal fronto-occipital records decreased in Gaussian probabilities.

In summary, these studies in domains of both frequency and amplitude carry a clear implication that the EEG recorded on the scalp or from the cortical surface relates in an ordered way to the discrete behaviour of contributing generators. This behaviour may conform to the mathematical central limit theorem of Cramer (24). Although many important aspects of these interrelations await future study, both lines of investigation indicate that these generators can behave in an essentially independent fashion, and that the level of their co-operative behaviour shifts consistently in relation to changing physiological states.

ACKNOWLEDGEMENTS

Studies from our laboratory described here were supported by Grants NB-01883, NB-2503 and MH-03708 from the National Institutes of Health, by Contracts AF (49) 638-1387 with the U.S. Air Force Office of Scientific Research, by Contract NONR 233-(91) with the Office of Naval Research, and Contracts NsG 237-62, NsG 502, NsG 505 and NsG 1970 with the National Aeronautics and Space Administration.

DISCUSSION ON ROSS ADEY

Creutzfeldt

I am very impressed with the enormous amount of work that you have done and I am also pleased to see that in many aspects our own work seems to support your conclusions. One of these points is that you also find alpha peaks persisting even in difficult behavioural situations. Where we seem to disagree is over the question of whether EEG-changes are related to the difficulty of the task. I tried to explain in my paper that in our test battery we tried to avoid any emotional or stress component. Furthermore we tried to stay within the same behavioural context. All tasks had to do with the visual-motor behaviour, motor activity and/or abstraction. Your experimental conditions seem to have differed quite considerably: Firstly, in their emotional stress content, and secondly, clearly, in their sensory modality content. This might be the reason for the different EEG patterns, which you found during the performance of the different tasks. Now if you programme your computer to recognize from the frequency content of a curve the performed reaction, you are implying a very restricted definition of the EEG. We were asking a different question: 'Is the incidence of alpha directly related to the degree of mental activity, or to the attention to be invested in solving a certain task?'

The second point concerns the relationship between the EEG and unit activity. I completely agree with you that if you look at one unit you cannot predict the average relationship between this single unit and the surface EEG during a record of, say, 10 or 20 min, or even one minute. There is sufficient evidence available, however, that certain EEG-phenomena are related in a consistent manner to the activity of one single unit or a group of units. This relationship is of course statistical. One such phenomenon showing a consistent relationship of this kind is the occurrence of the spindle waves. Here the rhythmical waves of the EEG are accompanied by rhythmical depolarizations of the membrane potential. This finding is now well established and even confirmed by computation (cross correlation). The question may arise as to whether these spindles produced by slight barbiturate anaesthesia are really the same as the human alpha rhythm. This is arguable. There are other conditions where the two activities are related to each other in a reversed phase relationship, and other periods where they do not seem to be related to each other at all. Our findings indicate, however, that the potential changes of one unit are related in a predictable manner to different EEG-phenomena. If you take a record over 2 min where you have a mixture of such different EEG-phenomena, then you might fail indeed to demonstrate a significant correlation.

Adey

That's not what we did however.

Creutzfeldt

Very well. But if you really stick to a certain phenomenon—let's say rhythmical EEG waves during a spindle or a sharp wave, like a strychnine wave or a metrazol wave—then your relationship becomes consistent. Why these relations between neuronal activity and EEG patterns are different is, of course, a difficult problem which we should not discuss now.

Adey

Taking the first point first. Now the tasks that I showed, and for which I was drawing specific inferences relating to their severity, were only two in number and they related to essentially identical tasks. In one instance 3 s were allowed for its performance, and in the other, only one second. Under these conditions there was a clear transition between the two states which was visible in the collected autospectra, from the heads of the 50 individuals and which, when one took the individuals separately, showed as different sets of signatures that the computer recognized, as I mentioned, with a high accuracy. So we did not mix the modalities as you suggest. It is true that in this study we delineated only two states of the EEG that related to two levels of difficulty. On the other hand, I would make the critical observation in relation to your own work, that you were not able to see a difference in alpha density—or at least the occurrence of alpha in your tests as they related to increasing difficulty. It was clear in our results that the decrement in alpha followed very closely the increase in the difficulty of the tasks for the population as a whole and for the individuals. I repeat that the tasks were purely visual, and we were not mixing modalities. With regard to the EEG and the intracellular waves, that coherence diagram (*Figure 12*), was based on analysis every 10 s over some hundreds of seconds. There were many, many analyses and each analysis stands in its own right as a measure of the statistical level of inter-relationship between unit activity (or neuronal wave) and the surface EEG. If within that epoch of 10 s, there were transitions that were significant, then it is true that we would still be looking at too large a block of data. You might contend that a single EPSP, for example, would then show some aspect of a consistent relationship and be revealed in a 10 s analysis. On the other hand, the fact that successive sampling of a quite considerable population of neurones in different preparations, and by that I mean somewhere between seventy and a hundred, for periods in excess of 100 s each, and the consistent failure of the unanaesthetized preparation to show coherence by this method is, to me, strongly suggestive that in the aggregate as a population characteristic, the units are not contributing synchronously to the EEG as usually recorded. Your contention that one can look at spindles and see the relationship between intracellular waves and the surface process I find unimpressive unless you can show me that you did some correlation analyses that are statistically valid on what, after all, would be very short epochs of data. The spindle burst barely lasts more than one second and unless you have done correlations which aggregated these bursts for a time duration which would be statistically valid, then your results are suggestive but not convincing.

Creutzfeldt

It may interest you to know that we did cross correlation between slow cellular potential changes and spindle waves of the surface EEG. The correlation factors were between 0·5 and 0·8.

Another problem which I would like to raise is the Gaussian and non-Gaussian distribution of EEG and cellular slow potentials. Would you not expect an asymmetric non-Gaussian amplitude distribution of post-synaptic potentials, because their depolarization is limited by their firing threshold?

Adey

Well I think that is a completely irrelevant argument, for some very good reasons. If you have perhaps a hundred synaptic potentials occurring simultaneously on the surface of one neurone, the slope of the individual synaptic potential is not going to determine what you see with a micro-electrode at one particular point in the cell. It will be the aggregate of their activity dispersed in time and space, and will have nothing whatever to do with the shape of the individual synaptic potential. Moreover, the skewing of the Gaussian amplitude distributors can have nothing to do with the rising or falling contours of individual potentials.

Lilly

My only comment is a methodological one. I am delighted to see the EEG power spectra presented, plotting frequency versus time with amplitude preserved. In our work with human speech and with the dolphin emissions, we have had similar problems to these EEG ones. We have found that we must preserve the amplitude data as well as the frequency and time data to do an adequate Fourier analysis. To do a Laplace transform, one may obtain some simplification, but the primary data must still be available. I agree with Dr. Adey that one should preserve amplitude, frequency and time.

Dewan

I was somehow never quite aware of the unhappy coincidence that the period of one orbit is exactly the period that Kleitman found for REM. And I have the feeling this might be a very ominous development if it resulted in selective REM deprivation.

Adey

This is something that Robert Maulsby has investigated in detail and has been unable to convince himself that the orbital period has anything to do with the sleep cycle length as it happened to occur in that one precious night's data.

Dewan

Well, I think it is vital to get at this because if in a manned orbiting laboratory there were conditions in the environment that would tend to cause the man to lose his mind, nothing could be more important.

REFERENCES

1. Adey, W. R. (1959) *Int. Rev. Neurobiol.* **1**:1
2. Adey, W. R. (1969) *Prog. Physiol. Psychol.* Vol. 3.
3. Adey, W. R., Bell, F. R. and Dennis, B. J. (1962) *Neurology*, **12**:591
4. Adey, W. R. and Dunlop, C. W. (1960) *J. Pharmac. exp. Ther.* **130**:418
5. Adey, W. R., Dunlop, C. W. and Hendrix, C. E. (1960) *A.M.A. Archs. Neurol.* **3**:74
6. Adey, W. R. and Elul, R. (1965) *The Physiologist*, **8**:98
7. Adey, W. R., Elul, R., Walter, R. D. and Crandall, P. H. (1966) *Proc. Am. Electroenceph. Soc.* p. 86
8. Adey, W. R., Kado, R. T. and Walter, D. O. (1967) *Aerospace Med.* **38**:345
9. Adey, W. R. and Lindsley, D. F. (1959) *Expl. Neurol.* **1**:407
10. Adey, W. R., Merrillees, N. C. R. and Sunderland, S. (1956) *Brain*, **79**:414

11. Adey, W. R., Segundo, J. P. and Livingston, R. B. (1957) *J. Neurophysiol.* **20**:1
12. Adey, W. R. and Tokizane, T. (1967) *Structure and Function of the Limbic System*. Elsevier, Amsterdam. 489 pp.
13. Adey, W. R. and Walter, D. O. (1963) *Expl. Neurol.* **7**:186
14. Adey, W. R., Walter, D. O. and Hendrix, C. E. (1961) *Expl. Neurol.* **3**:501
15. Adey, W. R., Walter, D. O. and Lindsley, D. F. (1962) *A.M.A. Archs. Neurol.* **6**:194
16. Costin, A., Elazar, Z., Walter, D. O. and Adey, W. R. (1967) *The Physiologist*, **10**:148
17. Cramer, E. G. H. (1955) *The Elements of Probability Theory*. New York: Wiley. 168
18. Creutzfeldt, O. C., Fuster, J. M., Lux, H. D. and Nacimiento, A. (1964) *Naturwissenschaft*. **51**:166
19. Elazar, Z. and Adey, W. R. (1967a) *Electroenceph. clin. Neurophysiol.* **23**:225
20. Elazar, Z. and Adey, W. R. (1967b) *Electroenceph. clin. Neurophysiol.* **23**:306
21. Elliot Smith, G. (1910) *Lancet*, **1**:147
22. Elul, R. (1964) *The Physiologist*, **7**:125
23. Elul, R. (1966) *Proc. Am. Electroenceph. Soc.* 80
24. Elul, R. (1967) In *Progress in Biomedical Engineering* (Ed. by Fogel and George). Washington, D.C., Spartan Books, p. 131
25. Elul, R. and Adey, W. R. (1966) *Nature, Lond.* **212**:1422
26. Fujita, J. and Sato, T. (1964) *J. Neurophysiol.* **27**:1011
27. Grastyan, E., Lissak, K., Madarasz, I. and Donhoffer, H. (1959) *Electroenceph. clin. Neurophysiol.* **11**:409
28. Grastyan, E., Karmos, G., Vereezkey, L. and Kellenyi, L. (1966) *Electroenceph. clin. Neurophysiol.* **21**:34
29. Green, J. D. (1964) *Physiol. Rev.* **44**:561
30. Hanley, J., Rhodes, J. M., Walter, D. O. and Adey, W. R. (1968) *Nature, Lond.* **220**:879
31. Herrick, C. J. (1933) *Proc. natn. Acad. Sci., U.S.A.* **19**:7
32. Holmes, J. E. and Adey, W. R. (1960) *Am. J. Physiol.* **199**:741
33. Hunsperger, R. W. (1956) *Helv. physiol. pharmac. Acta*, **14**:70
34. Jasper, H. H. and Stefanis, C. (1965) *Electroenceph. clin. Neurophysiol.* **18**:541
35. John, E. R. (1962) In *Information Processing in the Nervous System* (Ed. by Gerard and Duyff). Excerpta Medica, Amsterdam, p. 288
36. John, E. R., Ruchkin, D. S., Leiman, A., Sachs, E. and Ahn, H. (1965) *Proc. int. Union physiol. Sci. IV*, Lectures and Symposia, XXIII Congr., Tokyo. Excerpta Medica, Amsterdam, p. 618
37. Knipst, I. N. (1961) *Works. Inst. higher new. Act. Physiol. Series*. Acad. Sci. U.S.S.R., Israel Program Sci. Transl. **5**:1
38. Lindsley, D. F. and Adey, W. R. (1961) *Expl. Neurol.* **4**:358
39. Lindsley, D. F., Zaroodny, T. and Morton, T. H. (1967) *Expl. Neurol.* **17**:210

40. Livanov, M. N. (1962) *Proc. int. Union physiol. Sci., XXII Congr. Leiden.* Excerpta Medica, Amsterdam, Vol. 1, Part 2:899
41. Livanov, M. N. and Poliakov, K. L. (1945) In (In Russian) *Izv. Akad. Nauk U.S.S.R.* (*Otd. Biol. Nauk*), **111**:345
42. Maulsby, R. L. (1966) *Aerospace Med.* **37**:1022
43. Nauta, W. J. H. (1958) *Brain*, **81**:319
44. Nelson, P. G. (1966) *J. Neurophysiol.* **29**:275
45. Papez, J. W. (1937) *Archs. Neurol. Psychiat.* **38**:725
46. Porter, R., Adey, W. R. and Brown, T. S. (1964) *Expl. Neurol.* **10**:216
47. Radulovacki, M. and Adey, W. R. (1965) *Expl. Neurol.* **12**:68
48. Ruchkin, D. S., Villegas, J. and John, E. R. (1964) *Ann. N.Y. Acad. Sci.* **115** (2):799
49. Saunders, M. G. (1963) *Electroenceph. clin. Neurophysiol*, **15**:761
50. Walter, D. O. (1963) *Expl. Neurol.* **8**:155
51. Walter, D. O. and Adey, W. R. (1963) *Expl. Neurol.* **7**:481
52. Walter, D. O., Berkhout, J. and Adey, W. R. (1967) In *Progress in Biomedical Engineering* (Ed. by Fogel and George). Spartan Books, Washington, D.C. p. 125
53. Walter, D. O., Kado, R. T., Rhodes, J. M. and Adey, W. R. (1967) *Aerospace Med.* **38**:371
54. Walter, D. O., Rhodes, J. M. and Adey, W. R. (1967) *Electroenceph. clin. Neurophysiol.* **22**:22
55. Walter, Grey (1959) In *Handbook of Physiology* (Ed. by Field, Magoun and Hall). Am. Physiol. Soc., Washington, p. 279

Novelty, Attention and Habituation

Gabriel Horn
Department of Anatomy, University of Cambridge

RESPONSES TO NOVEL STIMULI

In some circumstances an animal will select a stimulus to which it will respond, in other circumstances a stimulus may determine the response. Among these response-determining stimuli are the distress and mating calls of animals of the same species, a person's own name (30) and a sudden unexpected change in the environment (6). The potency of these response-determining stimuli may vary from one individual to another and, in a given animal from one physiological state to another. Under optimum conditions, however, these stimuli share in common a capacity to elicit a response from the organism or to interfere with ongoing activity which is not possessed by other stimuli.

A behavioural response which is commonly elicited by these response-determining stimuli and which has been extensively studied in recent years (37) is the orientation reaction. This reaction is complex, involving both somatic and vegetative changes and may easily be evoked in the laboratory by sudden unexpected events. These stimuli rapidly lose their capacity to elicit a reaction for if repeatedly presented, the response gradually fails. In some ways, therefore, novel stimuli exercise remarkable control over the various command systems controlling behaviour—a control which is rapidly lost as the stimulus loses its novelty by repetition.

As a first step in understanding the factors controlling the operations of these command systems we need to know whether there exist neurones, the response characteristics of which parallel those of behaviour, i.e. giving a brisk initial response which wanes on repetition of the stimulus.

The responses to repeated stimulation of many cells in the optic pathways of the cat and the rabbit appear to be quite stable though there are some cells both in the lateral geniculate body (1) and the primary visual cortex (42) of rabbit in which the responses gradually wane. Response decrement also occurs in some cells in the auditory cortex (8, 23) and in the visual cortex (25) of the cat, in the optic tectum of frog (29) and rabbit (18, 19) and in cells of the reticular formation of the cat (3, 35). Response decrement does not occur among primary sensory fibres in lumbar dorsal roots of the cat (43) but it does occur among nerve cells in the spinal cord (5, 39, 43). It occurs widely in neurones in the nervous system of invertebrates (15, 21, 22, 32, 44).

Our knowledge of which cells show decremental responses and which do not, is still rudimentary chiefly because the number of regions in the nervous system where the phenomenon has been looked for is small and because we know little of the ways in which various experimental procedures influence the properties of the cells being studied (12, 20, 43). On the basis of our present knowledge, however, it seems, in mammals at least, that some cells in the sensory pathways leading from receptor surfaces to primary cortical areas continue to respond to repeated stimulation whereas others fail. The relative proportion of these two classes of neurones appears to be different between different sensory systems of a given animal and between the same system in different species. These findings make it difficult to extrapolate the relative proportions of the two classes of neurone from, say, the visual to the auditory pathways in the cat, and no less difficult to extrapolate from the lateral geniculate body of the cat to that of the rabbit.

SOME CHARACTERISTICS OF RESPONSE DECREMENT IN NEURONES

The gradual waning of the orientation reaction to a repeated stimulus is known as habituation. The decrement is, with certain qualifications, specific to the stimulus, for another stimulus may be effective in evoking a response. The orientation reaction may also be evoked by the original stimulus if it is withheld for some time before being presented again. Furthermore, an interpolated stimulus is sometimes effective in making the repeated stimulus transiently potent—a phenomenon known as dishabituation.

A question that has interested my colleagues and myself in the past few years is whether some or all of these characteristics can be demonstrated in the cells which show decremental responsiveness. Without this knowledge we are unlikely to gain any fundamental insight into those properties of the nervous system by which novel stimuli are effective and familiar stimuli relatively ineffective in activating the command systems controlling behaviour.

The cells that we have been studying are found in the midbrain of the rabbit (work done in collaboration with Dr. R. M. Hill) and in the brain of an insect, the locust (work done in collaboration with Dr. C. H. F. Rowell). The insect has many advantages over mammals for this sort of work, perhaps the most important of which is that records may be made of the discharge of single neurones for 24 h or more without any difficulty. The two animals come of course from different phyla, but habituation probably occurs universally in animals (14, 40) so that, in our present state of ignorance, similarities between animals are heuristically more important than differences.

Temporal Characteristics

In both animals, units can be found in which the response to a slowly repeated stimulus gradually wanes. If the stimulus is withdrawn for some time and presented again the response can be elicited once more. Usually the number of spikes evoked depends on the interval between successive groups of stimuli. Thus, while recording from a cell in the optic tectum stimuli were presented once every 4 s and it was found that the number of impulses elicited by each stimulus in a group slowly waned (*Figure 1*). The stimulus was withdrawn for some time and then presented again. An interval between groups of stimuli of 11 s was inadequate to restore the response. With longer intervals the response appeared, being greater the longer the inter-group interval.

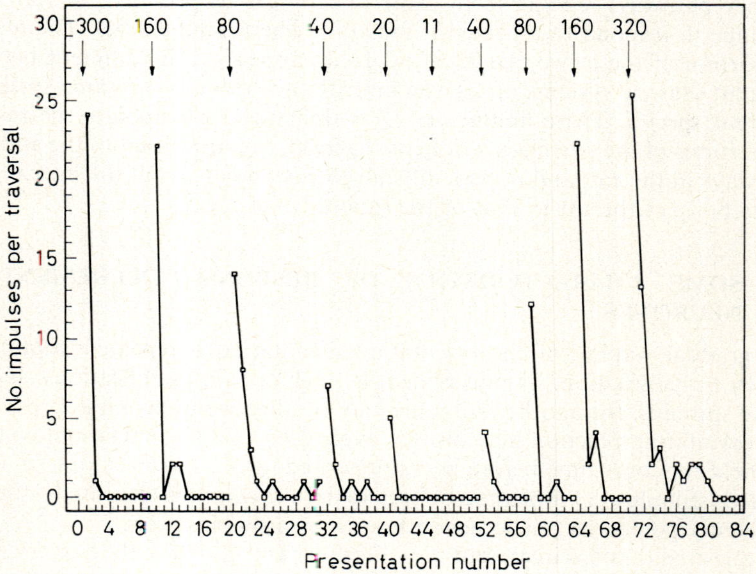

Figure 1. Response decrement to repeated stimulation with recovery by time lapse. Tectal unit of rabbit. Object moved across visual field once every 4 s. *Unit showed no spontaneous activity during the time that data for these curves were recorded. In this and in Figures 2, 3, 5, 7 and 9 the number of impulses present during each traversal (ordinate) are plotted against stimulus presentation number (abscissa). Numbers above arrows give duration of pause in seconds between successive groups of stimuli (inter-group interval). During these intervals no stimuli were delivered. After an inter-group interval stimuli were presented until they failed to evoke a response*

(From Horn and Hill, 1966a, by courtesy of Academic Press)

The effect of varying the inter-group interval can be more conveniently studied in the locust since the recorded cell can be held for long periods of time. While making such studies we (21) noticed some interesting properties of the response curves. In the unit whose responses are plotted in *Figure 2* a small black disc was moved across the visual field once every 10 s. Following an interval of 16 min (*Figure 2*, curve 1) the initial stimulus evoked a brisk burst of impulses—25—followed by a slowly declining response to successive stimuli. Following an inter-group interval of 2 min (*Figure 2*, curve 2), the response (24 impulses) to the initial stimulus was almost as high as that following a 16 min pause, but the subsequent responses declined more rapidly than those following the longer pause. When the inter-group interval was 60 s (*Figure 2*, curve 3), the initial response contained only 14 spikes. There was no significant recovery following a pause of 15 s (*Figure 2*, curve 4).

The rate of response decrement of many neurones is related to the rate of stimulus presentation, as is the case with certain aspects of the orientation

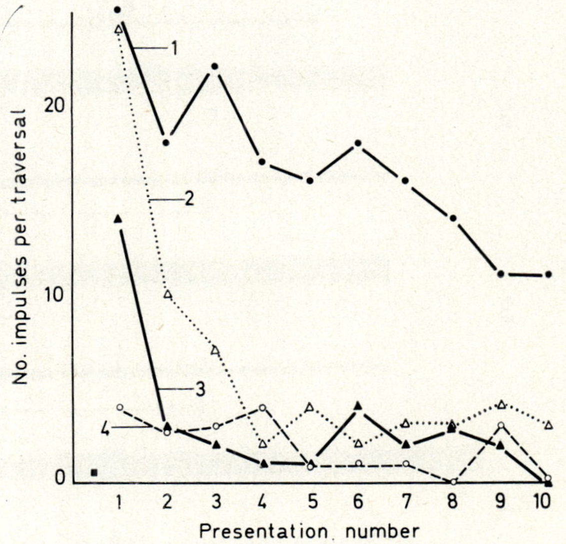

Figure 2. Effect of varying the inter-group interval on the response of a tritocerebral unit in the locust to a black disc moved across the visual field once every 10 s. *The inter-group interval preceding the set of responses plotted in curve 1 was* 960 s, *that before curve 2 was* 120 s, *that before curve 3*, 60 s, *and that before curve 4*, 15 s

(From Horn and Rowell, 1968, by courtesy of Cambridge University Press

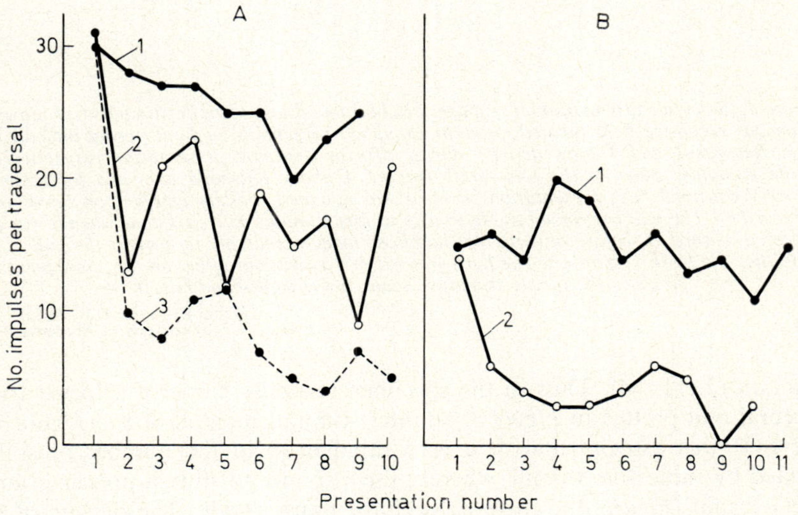

Figure 3. The effect of varying the inter-stimulus interval on the response to movement of a black disc in the visual field. Tritocerebral unit of locust. A. For curves 1, 2 and 3 the inter-stimulus intervals were 120 s, 40 s *and* 5 s *respectively. B. Response of another unit. The inter-stimulus intervals were* 10 min (*curve 1*) *and* 10 s (*curve 2*)

(From Horn and Rowell, 1968, by courtesy of Cambridge University Press)

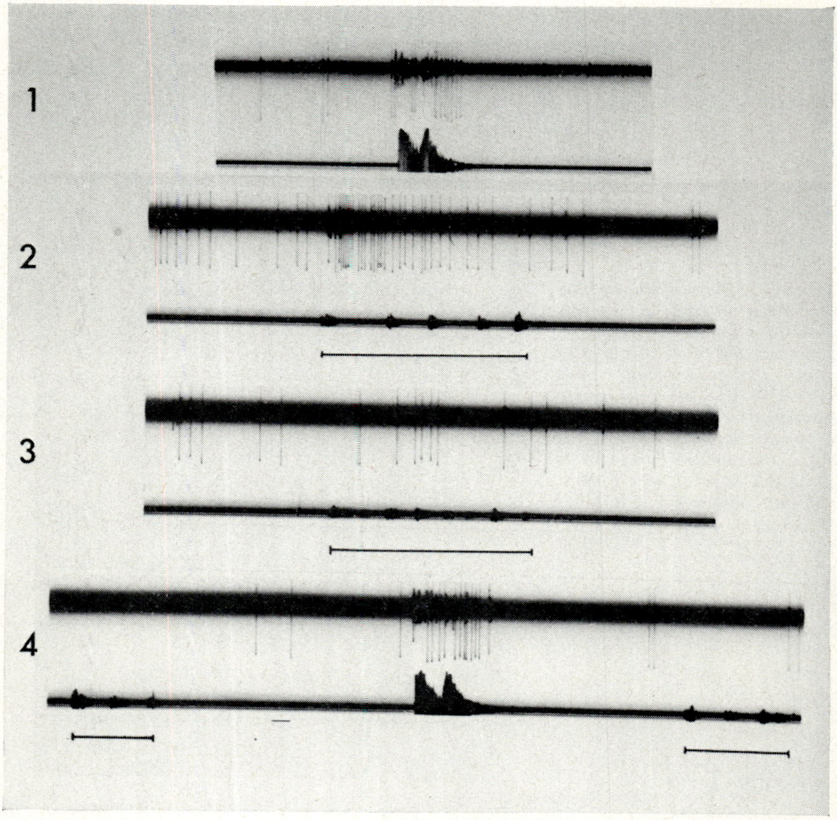

Figure 4. Selective attenuation of response. Rabbit tectal unit. Tactile stimulus (air blown on cutaneous receptive field located on right shoulder) present throughout period indicated by bars. Records 2 and 3 show activity during fifth and seventh presentation respectively to tactile stimulus. Between the forty-sixth (record 4 above left-hand bar) and forty-seventh (above right-hand bar) presentations a horn was sounded (middle artefact on lower trace of record 4). The unit responded as vigorously to this stimulus as it had done when it was first delivered (record 1), before any attempt had been made repeatedly to present the puff of air. Note that the brisk response to the horn (record 4) was without effect on the subsequent response to the tactile stimulus. Scale, lower right-hand bar, 0·5 s

(From Horn and Hill, 1966a, by courtesy of Academic Press

reaction (7, 11, 34). Thus in the specimen response curve of a locust tritocerebral unit plotted in *Figure 3*, an inter-stimulus interval of 5 s (*Figure 3A*, curve 3) was associated with a very rapid drop in the number of spikes elicited by successive stimuli, whereas the response to stimuli presented once every 2 min (*Figure 3A*, curve 1) declined more slowly. The decline of the evoked discharge with inter-stimulus intervals of 40 s (*Figure 3A*, curve 2) was between these two extremes.

The 'sag' in the tail of the response curves could usually be eliminated if the stimuli were presented once every 3–10 min. In *Figure 3B* two extreme forms of response of one unit are shown. When the inter-stimulus interval was 10 s (*Figure 3B*, curve 2) there was a marked decline of the responses, but little or no decrement when the interval between successive stimuli was 10 min (*Figure 3B*, curve 1). Whether the response would have declined had more stimuli been presented is not known, but occasionally even inter-stimulus intervals of 10 min appeared to be inadequate to prevent a decline of the responses in the course of ten presentations.

Specificity

It is possible to evoke a discharge from the recorded cell following attenuation of the response by presenting some other stimulus. This must be done without interfering with the rhythm of stimulus presentation, otherwise the recovery observed may be the result of the lapse of time and not a specific response to the additional stimulus. By taking the necessary precautions it is possible to show that loss of response to the repeated stimulus may be without any effect on the ability of the cell to respond to another stimulus. The unit whose responses are illustrated in *Figure 4* was recorded from the optic tectum of a rabbit. This unit gave a brisk discharge to acoustic and tactile stimulation. The upper trace of record 1 shows the response to the acoustic stimulus (artefact on lower trace) presented before repeatedly applying a tactile stimulus. The response to the fifth presentation of this stimulus is shown in record 2. The response gradually waned and was absent by the 46th trial (record 4, above left-hand line). The acoustic stimulus was then presented and elicited a vigorous response (record 4, above middle artefact) that was no less brisk than the control response (record 1). When, after the acoustic stimulus had been delivered, the tactile stimulus was re-applied, no response was evoked (record 4, above right-hand line).

This specificity can be observed between stimuli presented to the same modality as well as between modalities and can often be quite subtle. Thus the responses in *Figure 5* are of a tectal unit stimulated by a spot of light moved across its receptive field. Control responses to single movements of the spots of different intensities were first plotted. The less intense stimulus (I/10) was then repeatedly presented until the response had waned to a low level (29th trial). The neutral density filter in front of the light source was then removed and, without interrupting the rate of presenting the stimulus (1 per 2 s), the more intense spot (I) was moved across the receptive field. It may be seen that the response to the more intense spot on the 30th trial was not affected by repeated exposure to the less intense spot.

Similar specificity may be seen in locust cells. Thus, when the response has waned to a disc moved repeatedly across one axis of the visual field, a vigorous response may still be elicited by moving the disc in some other part of the visual field (21).

Generalization

In contrast to the specificity of the decremental processes, it is possible to demonstrate, in cells, response characteristics that appear to be similar to

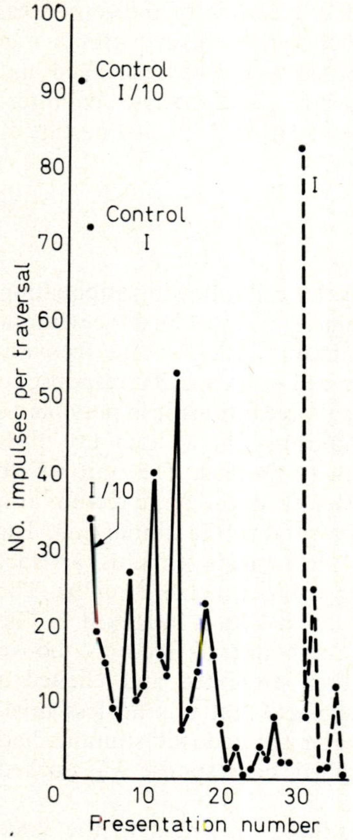

Figure 5. Rabbit tectal unit. Response to a spot of light (2 *degree diameter) moved repeatedly across receptive field. Intensity* (I=21 cd/m²) *controlled by neutral density filter. Reading from left to right, the first point is the initial response to the spot of light intensity* I/10, *and the second the initial response to the spot at full intensity* (I). *An interval of* 20 s *intervened between the presentations of these two control stimuli. Following a further* 20 s *pause the response to the less intense spot was attenuated by repeated presentations at approximately* 2 s *intervals. Between the 29th and 30th presentations the filter was withdrawn, and the unit responded to the more intense spot which initially evoked more spikes than its corresponding control*

(From Horn and Hill, 1966a, by courtesy of Academic Press)

stimulus generalization. For example, a tectal unit (*Figure 6*) responded to a 1000 Hz and to a 950 Hz tone (upper two records) presented several minutes apart. The 1000 Hz was then presented at 2 s intervals and after the response to this stimulus had waned (record 2) alternate 1000 Hz tones were omitted and tones of other frequencies were presented instead. In *Figure 6*, records 2–5 are continuous. It may be seen that though the response to the 1000 Hz tone is weak or absent the unit responded to a 900 Hz tone (records 2 and 4) and gave an 'off' discharge to a 1050 Hz tone (record 5). It did not, however, respond to the interpolated 950 Hz tone (record 2). At the end of this sequence several minutes were allowed to elapse by which time the cell responded once again to the 1000 Hz stimulus.

Dishabituation

Dishabituation is described in single cells in the reticular formation of the unanaesthetized curarized cat (3) but Hill and I have not seen tectal neurones in the rabbit behaving in this way, possibly because our animals were anaesthetized. However, an interesting form of dishabituation can be

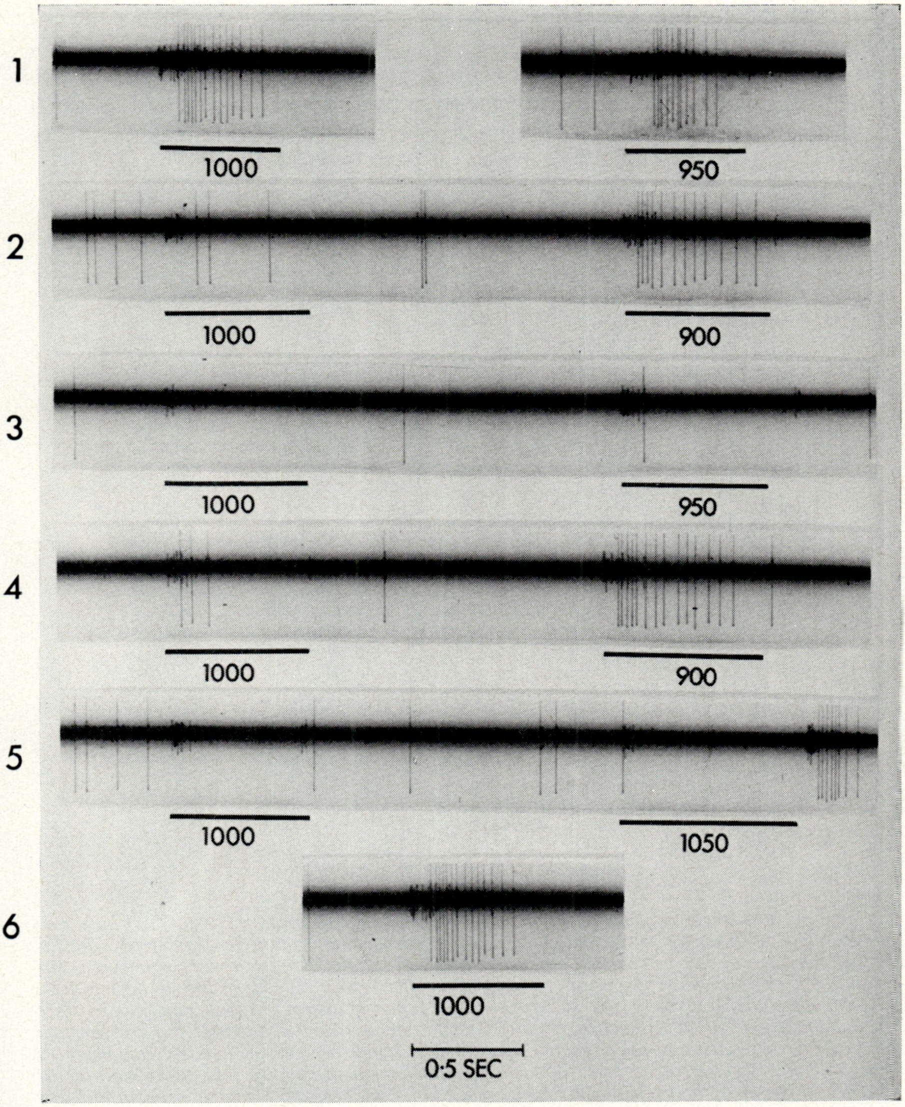

Figure 6. An example of stimulus generalization. Rabbit tectal unit. The unit initially responded briskly to a 1000 Hz *and* 950 Hz *tone (record 1). When the response to the* 1000 Hz *tone had waned, the unit responded to a* 900 Hz *tone (records 2 and 4) and to a* 1050 Hz *tone (record 5), but failed to respond to the* 950 Hz *tone. Several minutes after this experiment had been completed, the unit responded again to the* 1000 Hz *tone (record 6). Scale* 0·5 s

Horn and Hill, unpublished

demonstrated in the locust (33). The cells which we have studied can be activated trans-synaptically by stimulating the neck connective. Following a 10 s train of shocks (0·5 ms duration 50 V amplitude, 36/s) applied to the contralateral connective, the background firing rate of the cell may be elevated, compared with the pre-stimulus rate, for up to 15 min. The effect is quite specific, for shocks applied to the ipsilateral connective influence the cell only weakly or not at all.

In order to study the influence of connective stimulation on the response to visual stimulation, a disc was repeatedly moved across an axis of the visual field at 10 s intervals until each movement elicited only a weak discharge

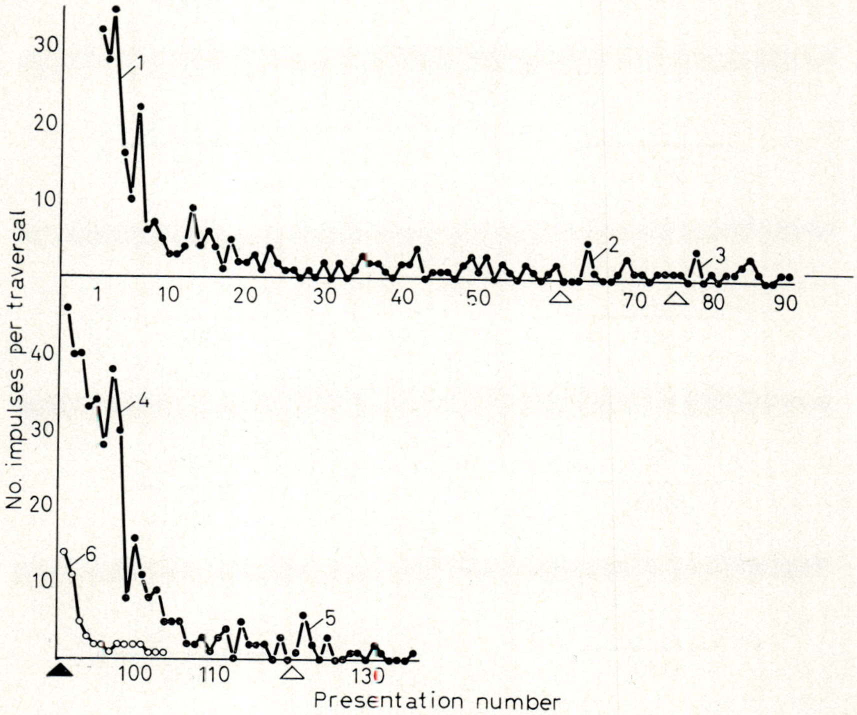

Figure 7. An example of dishabituation. Tritocerebral unit of locust. The unit initially responded vigorously (curve 1) to a black disc moved across the visual field at intervals of 10 s. *In the* 10 s *interval between the 60th and 61st presentation the ipsilateral neck connective was tetanized. There was no recovery of response to the disc (curve 2). This failure to dishabituate was demonstrated (curves 3 and 5) several times (open triangles indicate that shocks were applied to the ipsilateral connective). In the* 10 s *interval between the 90th and 91st presentation of the disc, the contralateral connective was tetanized and there was a massive enhancement (curve 4) of response to the disc. When the contralateral connective was stimulated in the absence of the disc, there was a marked increase in its background firing rate. From this was calculated the number of spikes that would be 'expected' in the time the disc, when present, was in the field of view. These expected values are plotted in curve 6. It can be seen that the observed number of impulses greatly exceeds the expected number*

(From Rowell and Horn, 1968, by courtesy of Cambridge University Press)

(*Figure 7*, curve 1). Between a selected pair of successive movements, a series of shocks was delivered to the ipsilateral cord. There was no recovery of response to the disc (*Figure 7*, curve 2). This manoeuvre was repeated with a similar result (*Figure 7*, curve 3). In the 10 s interval between the 90th and 91st movements, the shocks were delivered to the contralateral connective and were followed by a marked and prolonged enhancement of the response to the moving disc (*Figure 7*, curve 4). Shocks applied once again to the ipsilateral cord remained without any effect on the responsiveness to the disc (*Figure 7*, curve 5).

The increased discharge to the disc which follows stimulation of the connective is a real increase in responsiveness and is not entirely due to an increase in background firing which follows this procedure. Thus, prior to the experiments which generated the responses plotted in *Figure 7*, the effect on the background firing rate was studied following stimulation of the contralateral connective when no visual stimulus was present. From the data provided by these experiments it was possible to calculate the number of impulses which would be expected in 1·4 s—the time the disc, when present, was exposed in the visual field. These numbers are plotted in *Figure 7*, curve 6 and may be compared with the observed number when the disc was moved across the field after stimulating the contralateral connective (*Figure 7*, curve 4). Clearly the observed values greatly exceed the expected values and we conclude that shocks applied to the contralateral connective increased the responsiveness of the cell of the visual stimulus.

NEURAL MECHANISMS

There are a number of neuronal mechanisms that might underlie response decrement of the kinds that are described in the present study. They may be grouped into two classes. On the one hand it is supposed that the gradual waning of response to a set of neurones is due to an active shutting off of an input as the result of activity originating in some other part of the brain. On the other hand, it is supposed that response decrement is brought about by a self-generated depression (SGD) of activity in the neurones mediating the response of the recorded cell. The first explanation is an extension to single neurones of the view which Sokolov (36) proposed in a different context, to account for habituation of the orientation reaction. It was supposed that a neuronal model of a stimulus is built up during its repeated presentation. When complete this model matched the stimulus in all its parameters (e.g. intensity, duration, repetition rate, etc.). When a match occurred between the incoming signal and the model, inhibitory impulses passed to other regions in the nervous system and shut off the sensory inflow to them. If there was a mismatch between model and input, cells in these regions were thought to discharge in proportion to the mismatch. On the basis of this hypothesis we should expect that once the response to a regularly presented stimulus had waned, omission of one of the stimuli from the sequence should be accompanied by a 'response' (36). Although we have repeatedly tested for this effect it was never observed. Furthermore, with very rare exceptions, the response of all cells decreased whether the stimuli were presented at regular or at random intervals. These observations are not consistent with the simple 'active blocking' theory, which is neither necessary nor sufficient to account for the large majority of cases of response decrement that are reported here. This is not, however, to deny that an environment matching system does not exist and can influence the course (but is not a necessary condition for the appearance) of the decremental processes.

A more parsimonious view is that response decrement can take place without reference to 'higher order' neurones. The cells we have studied, both in the mammal and in the insect, are not themselves inhibited since their background firing rate and their ability to respond to other stimuli are not changed even when the repeated stimulus has ceased to evoke a response. It seems, rather, that the ability of the system of neurones to transmit the evoked impulse activity to the recorded cell is depressed as a result of having been active. Such self-generated depression could occur in several ways. It

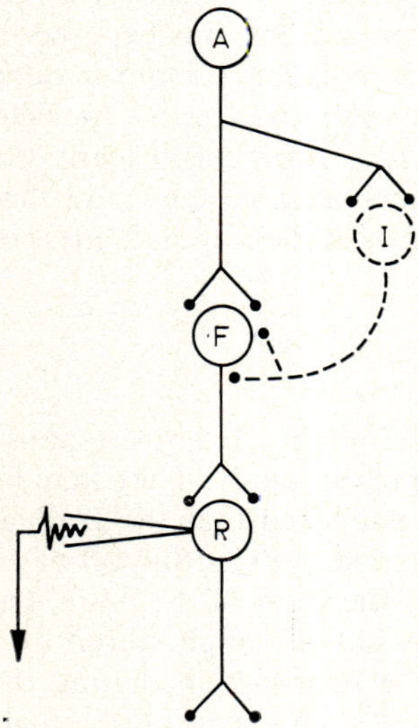

Figure 8. Afferent collateral inhibition. For discussion see text

could be brought about by a gradual build up of some inhibitory processes as for example afferent collateral inhibition (15, 38). Such a system might operate in the following way (*Figure 8*). Cell A is considered to respond without decrement to a slowly repeated stimulus whereas the recorded cell R exhibits decremental behaviour. On the first occasion the stimulus is presented cell R gives a maximum response. The activity in A discharges inhibitory neurone I. The after-effect of this discharge lasts for many seconds (e.g. through repetitive firing of I). If the second stimulus is presented during this time, cells F and R will give a smaller discharge compared with their response to the first stimulus. When the second stimulus is applied, cell I will again be activated. Cell F will be still further inhibited and this effect will continue as the stimulus continues to be repeated. As a result of this sequence of events, the responses of cell R will gradually wane. On the afferent collateral hypothesis therefore we should expect to find cells, like cell F which, in the course of presenting a series of stimuli, become increasingly subject to hyperpolarizing influences. Holmgren and Frenck (15), while making intracellular recordings from cells in the land snail, have indeed shown that as the stimulus is repeatedly applied waves of hyperpolarization may be observed in the cell.

While such a mechanism is inherently plausible and may operate in some systems it will not readily account for all the forms of response decrement that have been described. Strong but not overwhelming evidence against it comes from the findings that response decrement still occurs following administration of strychnine (39) which depresses certain forms of post-synaptic inhibition, and picrotoxin (28, 39) which depresses pre-synaptic inhibition. More particularly, the simple afferent collateral inhibitory theory is not consistent with the findings (*Figure 2*, curves 1 and 2) that the magnitude of the response to the second stimulus of a group is not determined by the magni-

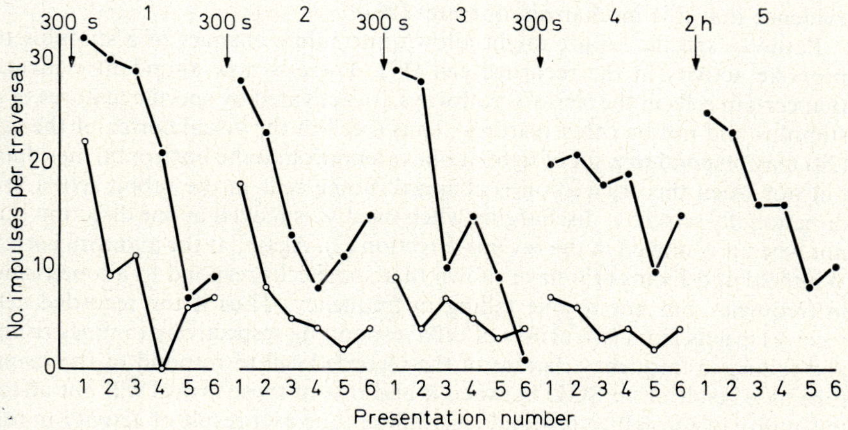

Figure 9. Influence of prior stimulus exposure on responsiveness. The short series (closed circles, curves 1–4) are responses to groups of six movements of the black disc separated by inter-group intervals of 300 s. *The open circles are severally the responses to the first six of a group of sixty movements (long series). The inter-group interval here too was* 300 s. *The long series of presentations followed the short series. After the last of the long series of movements had been made, the stimulus was withdrawn for* 2 h *and presented again: the responses plotted in curve 5 were generated. In all cases the interstimulus interval was* 10 s

(From Horn and Rowell, 1968,
by courtesy of Cambridge University Press)

tude of response to the first stimulus. Furthermore, response decrement has been described in *Aplysia* among neurones thought to be linked monosynaptically to the stimulated afferent fibre (4). This finding, in this preparation, would seem to rule out afferent collateral inhibition, recurrent inhibition and indeed any process mediated by some third neurone. In *Aplysia* response decrement is thought (4) to be an expression of an imbalance between transmitter mobilization and utilization. Such a view would not be incompatible with our findings either in the rabbit or locust; and they would be consistent with the findings (*Figure 9*) that the more often a cell is activated by a stimulus the less effective becomes a given period of rest in restoring the response. It would, however, be premature and probably incorrect to suppose that, even in a particular nervous system, habituation is brought about by only one form of SGD.

If response decrement is due to an SGD in the pathways mediating the response to the recorded cell, how can we account for the various properties of habituation and dishabituation exhibited by such a cell? There is in fact very little difficulty in doing so, at least for many though possibly not all of these properties. Recovery of response following a lapse of time would be due to a gradual restitution of synaptic efficacy—or, on the afferent collateral model, to a falling off of the inhibitory effects that had been built up in the course of presenting the sequence of stimuli. With respect to stimulus specificity it is clear that if some additional stimulus evoked activity in the recorded cell over pathways quite separate from that which carried the response evoked by the repeated stimulus, the recorded cell would respond to the additional stimulus while remaining unresponsive to the repeated one. There is in fact evidence that this mechanism operates (26).

Pathway specific failure might allow quite subtle changes of a stimulus to provoke activity in the recorded cell (17). There is now abundant evidence that certain cells in the sensory pathways are activated by specific features of a stimulus and not by other features. Thus a cell in the visual cortex of the cat (24) may respond to a slit of light at one orientation to the horizontal meridian but not when the orientation is changed. Some cells in the rabbit retina are directionally selective, discharging when the disc is moved in one direction but not when it is moved in the reverse direction (2). Again, at the auditory cortex Whitfield and Evans (45) have shown that some cells respond to a tone rising in frequency but not to one falling in frequency. Thus if the recorded cell receives inputs from two arrays of cells responding respectively to tones rising and falling in frequency, failure of the recorded cell to respond to the rising tone as a result of an SGD between it and one of these arrays will not affect the ability of the cell to respond to a falling tone as a result of activity in the other array. The effect would be that the recorded cell would respond to a tone falling in frequency even though it had ceased to respond to a tone rising in frequency.

There is evidence (10) that some cells are intensity selective. These cells discharge less frequently if the intensity is increased or decreased about a given value. If an array of such cells with a given selectivity ceases to discharge the recorded cell because of an SGD in the connecting links, an adequate change in intensity, either an increase or a decrease, should activate the recorded cell.

Some, though not all of the cells that we have studied behave in these ways (19). That is, although a given change in the parameters of a stimulus might activate one cell the change may be ineffective for another. Failure of response to a stimulus change resulting from prior exposure to the repeated stimulus is an example of stimulus generalization (*Figure 6*). One simple explanation for this effect is that at some region between receptor surface and the recorded cell, the pathways mediating activity evoked by the repeated stimulus and the additional stimulus overlap. If the overlap is complete, stimulus equivalence is complete; if the overlap is incomplete there will be a response to the additional stimulus but the response will be less than maximum.

Some neurones in the mammalian hippocampus show response decrement to repeated stimulation (41). When a stimulus is omitted the cell discharges. This observation is not easy to account for on the basis of the SGD/channel dependent hypothesis alone, largely because the relevant information about the signalling within the nervous system of information about intervals of time is

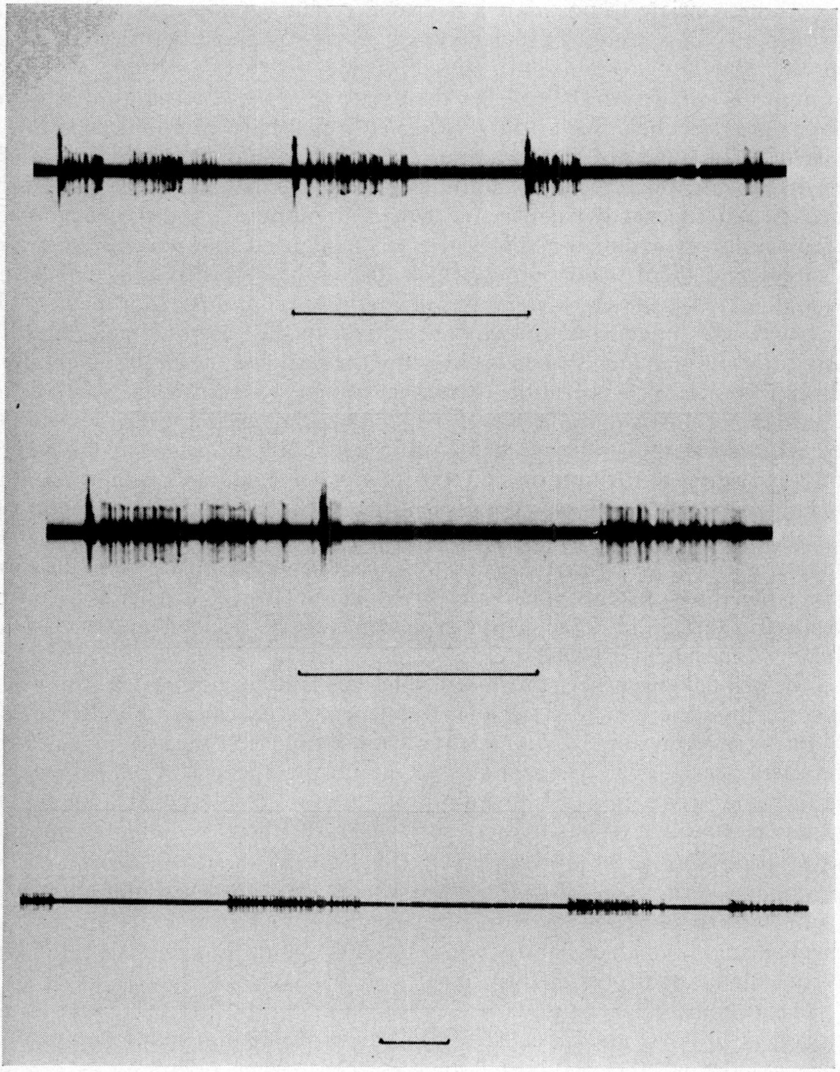

Figure 10. A single unit which responds to diffuse retinal illumination (upper record) and to weak electric shocks applied to the skin over the back (middle record). The cat from which these records were taken was free to move about within its training box. Upper record shows responses to the last three flashes (indicated by artefacts) in the series (of ten flashes) together with a burst which appeared at the time the next flash would have occurred, had the series been continued. A similar discharge (middle record) occurred at the end of the sequence of skin stimuli, the last two of which are indicated by the artefacts, and thereafter (lower record) the unit, no longer driven by an external stimulus, continued to discharge spontaneously at irregular intervals. The last burst of the middle record is the first burst of the lower. Scale 2 s *in all cases*

(From Horn, 1962)

simply not available, at least in the detail required. At the present time it seems appropriate to postulate a neuronal system encoding certain temporal properties of a stimulus. In such a system we may suppose that a cell or group of cells repeatedly responds to a stimulus presented every *t* seconds. When the stimulus is withdrawn the cell, for the next one or two occasions, discharges following *t* seconds. Such cells (*Figure 10*) which have been observed in the visual cortex (16) and hippocampus (41) may be supposed to 'match' the environment. When there is a stimulus present the matching activity and the evoked activity exactly balance and there is no output from the system. When the stimulus is withdrawn the system is unbalanced, the matching cell discharges and there is an output which drives the recorded cell. The result would be that a cell whose responses to a repeated stimulus had waned, would respond when one of the stimuli was omitted. In this case the function of the environment matching system is to restore the response, not to bring about its disappearance. The operation of such a system is discussed in some detail elsewhere (17). It should, however, be stressed that such a model appears to be required in the studies of habituation which are discussed in this paper, only in respect of the unit which 'responds' when a stimulus is omitted. This behaviour is not a characteristic of all neurones showing decremental behaviour for it was not seen in any of the cells in the rabbit tectum nor in the locust brain (21, 31). It may be that it is necessary to postulate other environment matching systems when very prolonged effects of habituation can be demonstrated (9, 13, 27) but this issue can only be decided on the basis of further experimental analysis.

The physiological mechanisms by which response to a repeated stimulus is restored by an interpolated stimulus will depend on the cause of the decrement. If this is brought about by afferent collateral inhibition then response to the repeated stimulus will be restored if the inhibitory influence of cell I (*Figure 8*) is in some way eliminated. If failure of response results from a change in synaptic efficacy, dishabituation will occur if there is some temporary restoration of this efficacy; and how this is brought about will depend on the physiological or chemical basis of the change. Until these variables are more completely understood, it is not particularly constructive to discuss the mechanism by which dishabituation is brought about, though it is not difficult to speculate about possible mechanisms (4, 17, 38).

The responses I have described of single cells, and the behaviour of an organism to novel and repeated stimuli are convergent at many points. One may wonder whether the output of these cells, when operating as novelty detectors, exerts a powerful and sometimes, perhaps, an overriding influence on some of the command systems underlying perceptual and behavioural activity.

ACKNOWLEDGEMENT

The author's work was supported by funds from the Medical Research Council, the Mental Health Research Fund, the United States Public Health Service (Grant NB 04787) and the Wellcome Trust.

REFERENCES

1. Arden, G. B. (1963) *J. Physiol.* **166**:468
2. Barlow, H. B., Hill, R. M. and Levick, W .R. (1964) *J. Physiol.* **173**:377
3. Bell, C., Sierra, G., Buendia, N. and Segundo, J. P. (1964) *J. Neurophysiol.* **27**:961
4. Bruner, J. and Tauc, L. (1966) *Nature, Lond.* **210**:37
5. Buchwald, J. S., Halas, E. S. and Schramm, S. (1965) *J. Neurophysiol.* **28**:200
6. Cherry, E. (1953) *J. acoust. Soc. Am.* **25**:957
7. Coombs, C. H. (1938) *J. exp. Psychol.* **22**:244
8. Evans, E. F. and Whitfield, I. C. (1964) *J. Physiol.* **171**:476
9. Glaser, E. M. and Whittow, G. C. (1957) *J. Physiol.* **136**:98
10. Goldberg, J. M. and Greenwood, D. D. (1966) *J. Neurophysiol.* **29**:72
11. Greer, J. H. (1966) *J. exp. Psychol.* **72**:617
12. Griffin, J. P. and Pearson, J. A. (1967) *Nature, Lond.* **216**:705
13. Hinde, R. A. (1954) *Proc. R. Soc. B,* **142**:306
14. Hinde, R. A. (1966) *Animal Behaviour*. McGraw-Hill: New York
15. Holmgren, B. and Frenck, S. (1961) *Nature, Lond.* **192**:1294
16. Horn, G. (1962) In *Viewpoints in Biology* (Ed. by Carthy and Duddington). Butterworths: London
17. Horn, G. (1967) *Nature, Lond.* **215**:707
18. Horn, G. and Hill, R. M. (1964) *Nature, Lond.* **202**:296
19. Horn, G. and Hill, R. M. (1966a) *Expl. Neurol.* **14**:199
20. Horn, G. and Hill, R. M. (1966b) *Nature, Lond.* **211**:754
21. Horn, G. and Rowell, C. H. F. (1968) *J. exp. Biol.* **49**:143
22. Horridge, G. A., Scholes, J. H., Shaw, S. and Tunstall, J. (1965) In *The Physiology of the Insect Central Nervous System* (Ed. by Treherne and Beament). Academic Press: London
23. Hubel, D. H., Henson, C. O., Rupert, A. and Galambos, R. (1959) *Science*, N.Y. **129**:1279
24. Hubel, D. H. and Wiesel, T. N. (1962) *J. Physiol.* **160**:106
25. Hubel, D. H. and Wiesel, T. N. (1965) *J. Neurophysiol.* **28**:229
26. Kehoe, J. S. and Bruner, J. (1966) *J. Physiol. Paris,* **58**:542
27. Kozak, W., MacFarlane, W. V. and Westerman, R. (1962) *Nature, Lond.* **193**:171
28. Krasne, F. B. and Roberts, A. (1967) *Nature, Lond.* **215**:769
29. Lettvin, J. Y., Maturana, H. R., Pitts, W. H. and McCulloch, W. S. (1961) In *Sensory Communication* (Ed. by Rosenblith), M.I.T. Press: Boston and Wiley: New York
30. Moray, N. (1959) *Q. Jl. exp. Psychol.* **11**:56
31. Palka, J. (1967) *J. Insect. Physiol.* **13**:235
32. Rowell, C. H. F. and Horn, G. (1967) *Nature, Lond.* **216**:702
33. Rowell, C. H. F. and Horn, G. (1968) *J. exp. Biol.* **49**:171
34. Schaub, R. E. (1965) *Psychon. Sci.* **2**:361
35. Scheibel, M. E. and Scheibel, A. B. (1965) *Archs. ital. Biol.* **103**:279
36. Sokolov, E. N. (1960) In *The Central Nervous System and Behaviour* (Ed. by Brazier). Josiah Macy, Jr. Foundation: New York

37. Sokolov, E. N. (1963) *Ann. Rev. Physiol.* **25**: 545
38. Sokolov, E. N. (1965) *Proc. Int. Un. Physiol. Sciences*, **4**: 340
39. Spencer, W. A., Thompson, R. F. and Neilson, D. R. Jr. (1966) *J. Neurophysiol.* **29**: 253
40. Thorpe, W. H. (1963) *Learning and Instinct in Animals.* 2nd edn. Methuen: London
41. Vinogradova, O. S. (1965) Quoted by Sokolov, E. N. (1965) *Proc. Int. Un. Physiol. Sciences*, **4**: 340
42. Vinogradova, O. S. and Lindsley, D. F. (1963) *J. higher nerv. Act.* 13/2 (207)
43. Wall, P. D. (1967) *J. Physiol.* **188**: 403
44. Waterman, T. H. and Wiersma, C. A. G. (1963) *J. cell. comp. Physiol.* **61**: 1
45. Whitfield, I. C. and Evans, E. F. (1965) *J. Neurophysiol.* **28**: 655

The Time Course of the Oculomotor Component of Orienting Reaction

V. Zikmund
Institute of Normal and Pathological Physiology, Bratislava, Czechoslovakia

The oculomotor reaction is considered as an inherent component of a general orienting reaction of the organism to novel stimuli (10). Various data, however, indicate that oculomotor reactions are not only closely related to perceptual changes, but that they accompany various changes in the general level of alertness and certain kinds of mental activity, too.

An increase of eye movement frequency has been described, e.g. during mental multiplication (7) or as a reaction to stressful words in association experiments (3). Much attention has been paid to oculomotor activity during the visual imagery process (1, 2, 5, 6, 8, 9, 12, 13, 14, 15, 16, 17, 19, 20). Several authors, however, consider the increase of oculomotor activity during the above-mentioned processes as being only a common concomitant sign of the simple increase of attention or arousal level (1, 9). Gaarder (4) on the other hand, has described recently the occurrence of fine eye movements during inattention. A particular area of interest is occupied by studies of eye movements during sleep and dreams.

Thus oculomotor reactions are in many ways closely related to changes in the central nervous functions mediating behaviour, and mental processes are sometimes a more convenient indicator of them than many other physiological variables.

CHARACTERISTICS OF OCULOMOTOR REACTIONS

In the present work we should like to stress certain characteristics of the oculomotor reactions to several kinds of acoustic stimuli, these reactions being studies of the general orienting reaction in man.

The present experiments have been carried out in a dark, relatively soundproof cabin on 22 healthy and 22 neurotic subjects, all undergraduates, who sat comfortably in a reclining position in an easy chair. Each subject had previously been instructed to keep his eyes closed and to remain quiet during the study, the purpose of which was an investigation of certain physiological functions especially in a relaxed state.

In 10–15 min intervals of relaxation the following verbal and nonverbal acoustic stimuli were successively applied:

I. A question—without emphasizing the significance of either a positive or a negative answer—as to whether the subject had slept.

II. A short disturbing noise without any particular signal meaning.

III. Each subject was asked to solve a concrete problem involving mental multiplication of binary numbers and to state the result.

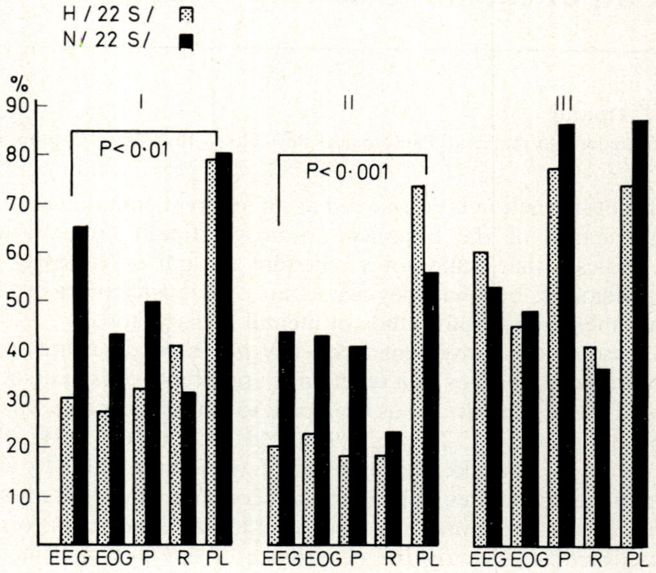

Figure 1. Intra-individually significative changes in the percentage time alpha in the EEG frequency of the horizontal ocular movements (EOG), heart rate (P), respiration rate (R) and plethysmogram (Pl) in healthy subjects (H) and neurotics (N) after (I) *a simple verbal contact,* (II) *a short disturbing noise and* (III) *mental multiplications*

The following physiological parameters were continuously recorded in each subject during the 2 min periods before and after each stimulus.

(*a*) Percentage time alpha in the EEG
(*b*) Frequency of ocular movements from the horizontal EOG recording (on the electrocardiogram channels of the Visiograph Supra Alvar)
(*c*) Heart rate (from the R-R intervals of the ECG)
(*d*) Frequency of respiratory cycles (from the mechanical pneumogram)
(*e*) Vasomotor reactions of the peripheral blood vessels of the right hand (as recorded by plethysmography and evaluated planimetrically).

The 2 min values before and after each stimulus were divided into 10 s intervals for measurement of the reaction described above.

As shown in *Figure 1* the occurrence of intra-individually significant changes (healthy subjects dotted columns, neurotics black columns) varies both for each concrete stimulus and for the parameter evaluated. The most distinct reactions may be found for both groups in response to mental multiplication, which, in fact, represents a stress situation that does not enable a separation to be made of the orientation component from the emotional load of the whole situation. This stimulus was incorporated into

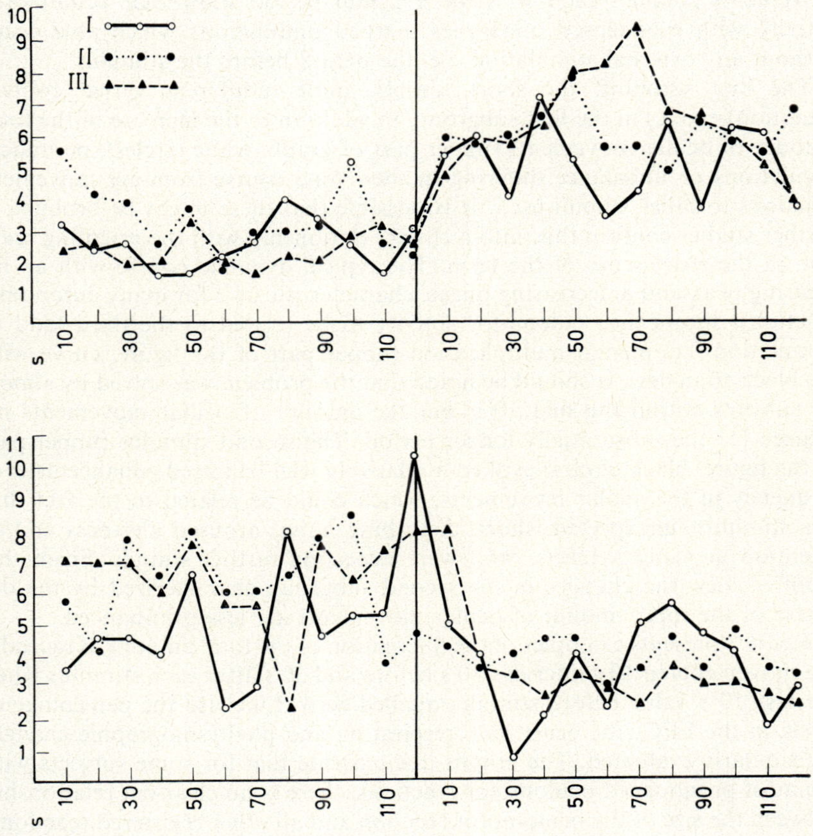

Figure 2. Mean frequencies of the ocular movements as measured in 10 s *intervals for a* 2 min *period before and after the stimuli used, in the subjects with intra-individually significant reactions. White circles*—I. *stimulus, black circles*—II. *stimulus, black triangles*—III. *stimulus*

the experiments for purposes of comparison only. The weakest changes were found to be in response to the second, non-verbal stimulus (the sudden disturbing noise without any signal significance). There are, however, also marked changes in the occurrence of significant reactions in both groups under study especially after the first and the second stimulus.

As may be seen from the same figure the significant eye movement reaction when compared to other physiological variables under study is, in general, one of the less pronounced reactions which is in agreement with the older findings of Voronin *et al.* (11).

Figure 2 shows that the intra-individually significative oculomotor reactions to the stimuli mentioned may be divided into two subgroups. In the first, the larger of them (upper part of the figure) is the total amount of saccadic eye movements after the stimuli increased; in the second one (lower part of the figure) a marked decrease may be observed. The mean values of saccadic eye movements counted each 10 s for a 2 min period show that oculomotor activity with eyes closed undergoes marked undulations which also occur without any external stimulation, i.e. the period before the stimulus.

The first stimulus, the short, simple, quite indifferent verbal contact (question) evokes in the first subgroup, in addition to the increase of the total amount of ocular movements (upper part of graph, white circles), periodical oscillations or imbalance differing in their time course from eye movement reactions to other stimuli used. It is possible that these might be brought, if further studies confirm this, into a closer relationship with the orienting reaction in the strict sense of the term. The typical dynamic course with an increasing peak and a decreasing phase, characteristic also for many autonomic reactions, shows the oculomotor activity to be related to the third kind of stimulation, i.e. mental multiplication (upper part of the figure, curve with the black triangles). It should be noted that the problem was solved by almost all subjects within the first 10 s, but the amount of ocular movements increased for the substantially longer period. The second stimulus (upper part of the figure, black circles) evoked a relatively well balanced enhancement of frequency in the ocular movements, which could be related to the fact that this stimulus (unexpected, short, disturbing noise) aroused alertness and/or attention in many subjects, who were expecting further stimuli. From this point of view the changes in the second subgroup characterized by the decrease of the total amount of ocular movements are less pronounced.

Figure 3 presents examples of the fluctuation of the total amount of saccadic eye movements in 10 s intervals 30 s before and 30 s after each stimulus when the last 10 s value before stimuli equalled zero. Likewise the percentage of alpha in the EEG, the heart rate, respiration and plethysmographic changes are similarly evaluated. The graphs demonstrate that for some subjects with the most pronounced oculomotor reactions, there is no clear-cut relationship between the size of the oculomotor reaction and all other registered reactions. The marked oculomotor reactions may be accompanied by only small changes in the EEG and/or in other functions, and vice versa.

RESULTS IN RELATION TO EEG CHANGES

These findings are also illustrated in relation to EEG changes by some original recordings (*Figures 4–7*). As may be seen the spontaneous saccadic eye movements are characterized in some subjects in the relaxed state by rather irregular outbursts or series of small and frequent ocular movements, these being changed immediately after the stimuli into larger movements of lower frequencies, similar to the voluntary saccadic or scanning eye movements. In several subjects these larger movements change in between the second half of

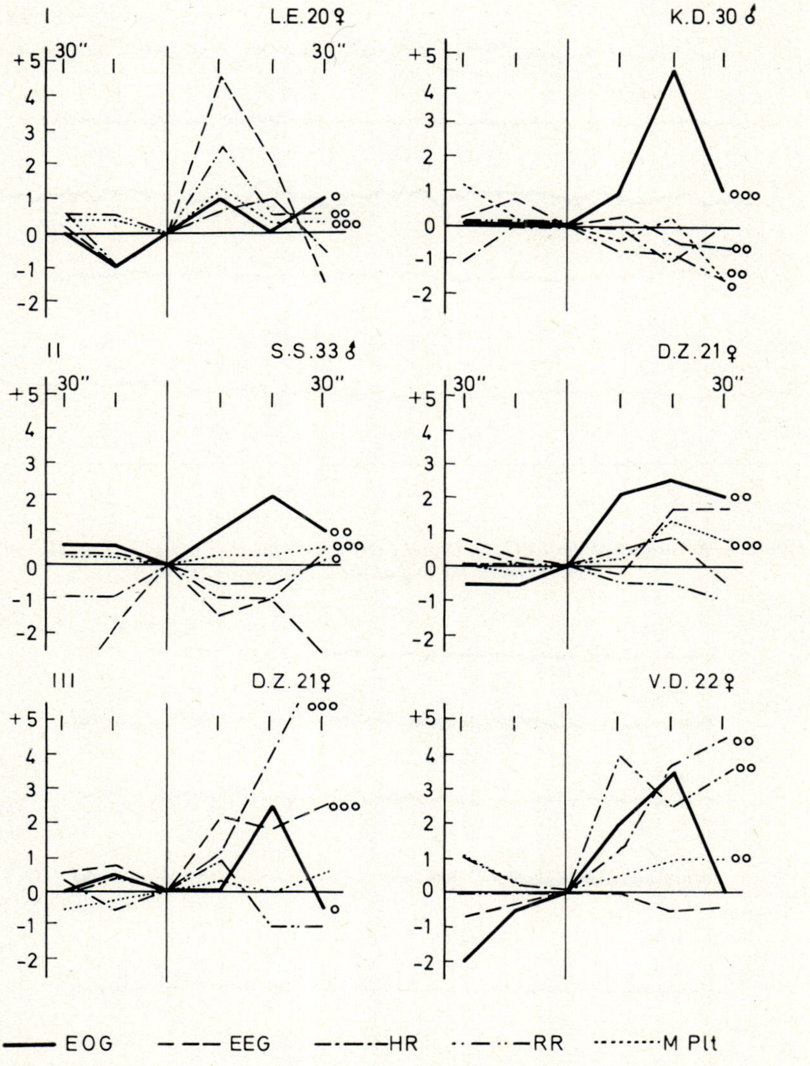

Figure 3. Changes in physiological reactions before and after stimulation in six subjects. Measurements were taken at 10 s *intervals. The value occurring* 10 s *before stimulation was arbitrarily, zero base-line. The vertical axis represents relative values of the physiological reaction. Horizontal axis is time* (t). I—*A simple question.* II—*Brief noise.* III—*Mental multiplication.*

the 2 min period into outbursts of small faster saccades again, especially following the first two stimuli used in this case. During the longer time periods of the resting state, before or after particular stimuli the saccadic movements vanish in some subjects, these being replaced by slow undulatory drifts characteristic of the drowsy state or of falling asleep (*Figure 7*). *Figures 6–8* show that the occurrence of the outbursts of faster saccadic

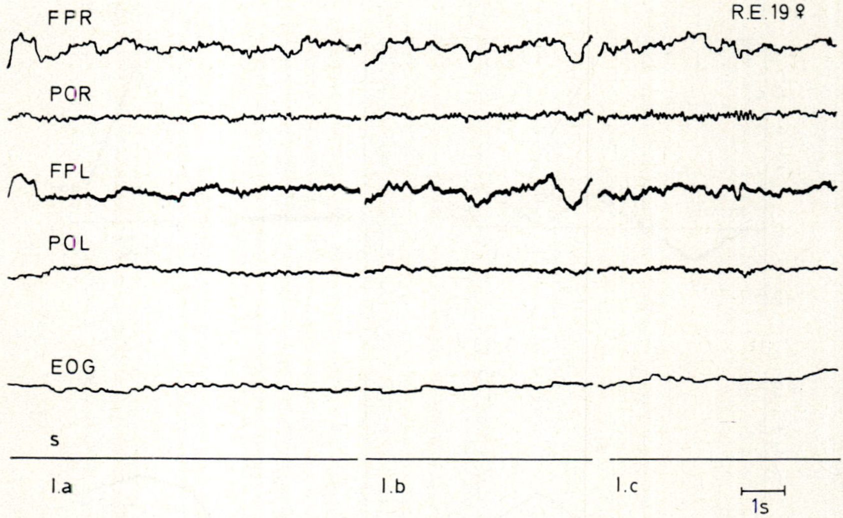

Figure 4. Recording of the EEG, ECG and EOG before the first stimulus (I.*a*), 10 s *after it* (I.*b*) *and at the end of the* 2 min *period* (I.*c*)

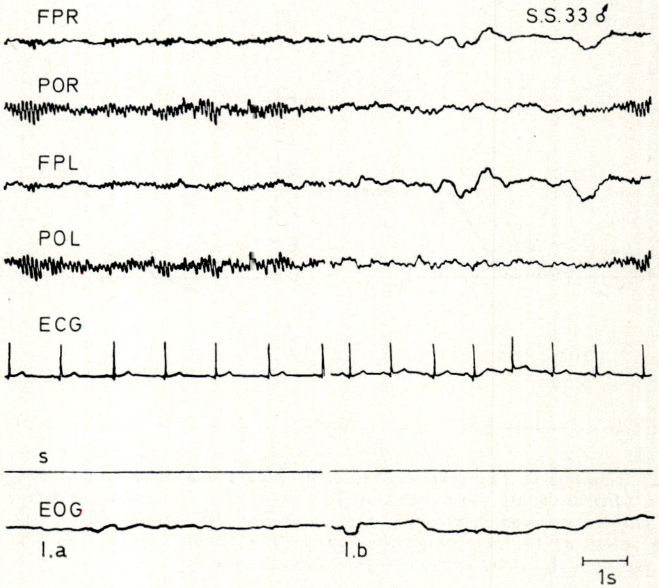

Figure 5. Description as in Figure 4

movements does not produce any change in the total amount of the alpha activity, while on the other hand the less frequent ones affect it to a different degree.

Examples of the above mentioned changes in saccadic eye movements, especially of their time courses and their duration are given in *Figures 8–9*. On

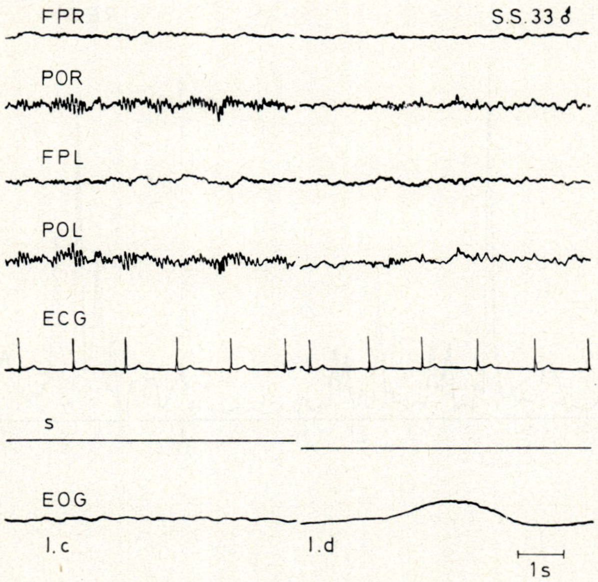

Figure 6. Description as in Figure 4. I.*d recording* 10 min *after the first stimulus*

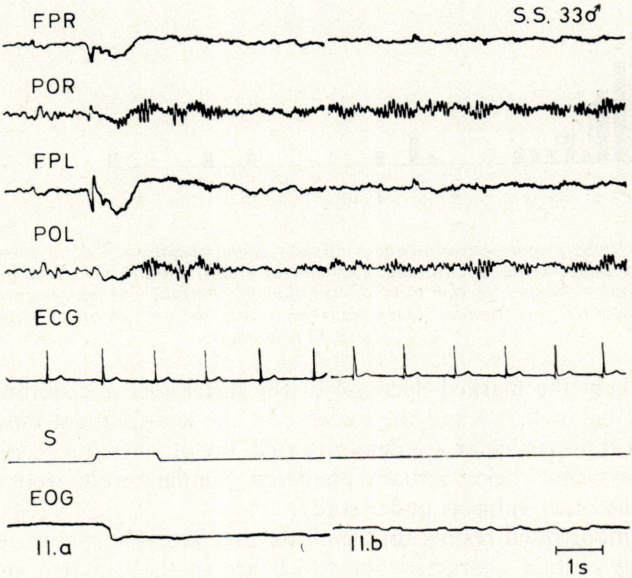

Figure 7. Recording of EEG, ECG and EOG immediately before and during the second stimulus (II.*a*) *and after* 20 s (II.*b*)

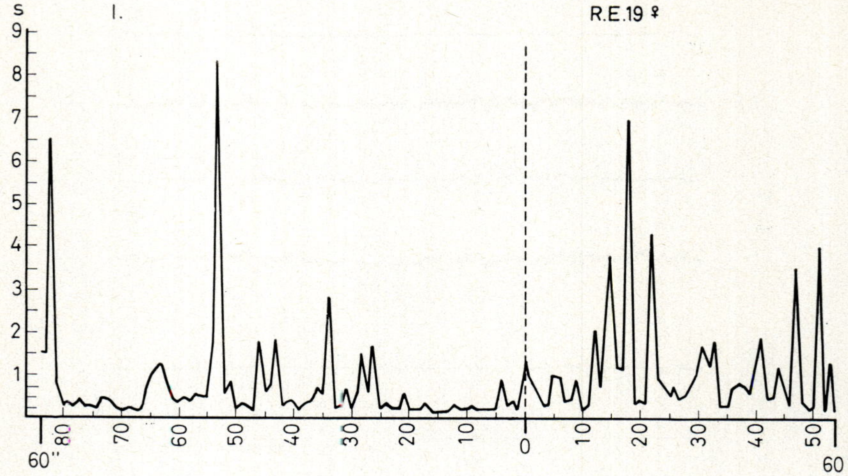

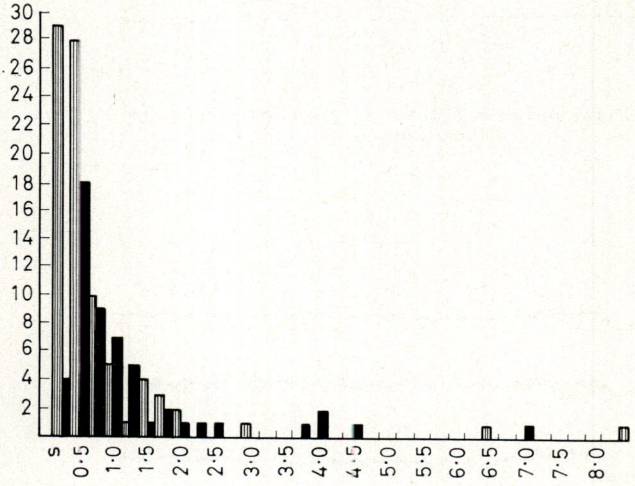

Figure 8. Upper graph—duration and continuity of particular saccadic eye movements in 1 min *period before the first stimulus* (*left side of graph*) *and after it* (*right side of graph*) *Lower graph—changes in the total distribution of saccadic eye movements of various duration after the first stimulus* (*black columns*) *in comparison with the pre-stimulus period* (*striped columns*)

both of them the marked decrease of the faster saccadic outbursts (of the frequency below 0·5 s) and the increase of the less frequent movements (of about 1·0 s) in particular are demonstrated, the other frequencies of saccadic ocular movements being actually unaltered. Similar results were obtained in most of the other subjects under study.

The demonstrated results thus indicate that saccadic eye movement reactions show certain characteristics which are in their course and duration comparable to other physiological changes under study, however, being in general one of the lesser expressed reactions. A clear cut relationship between

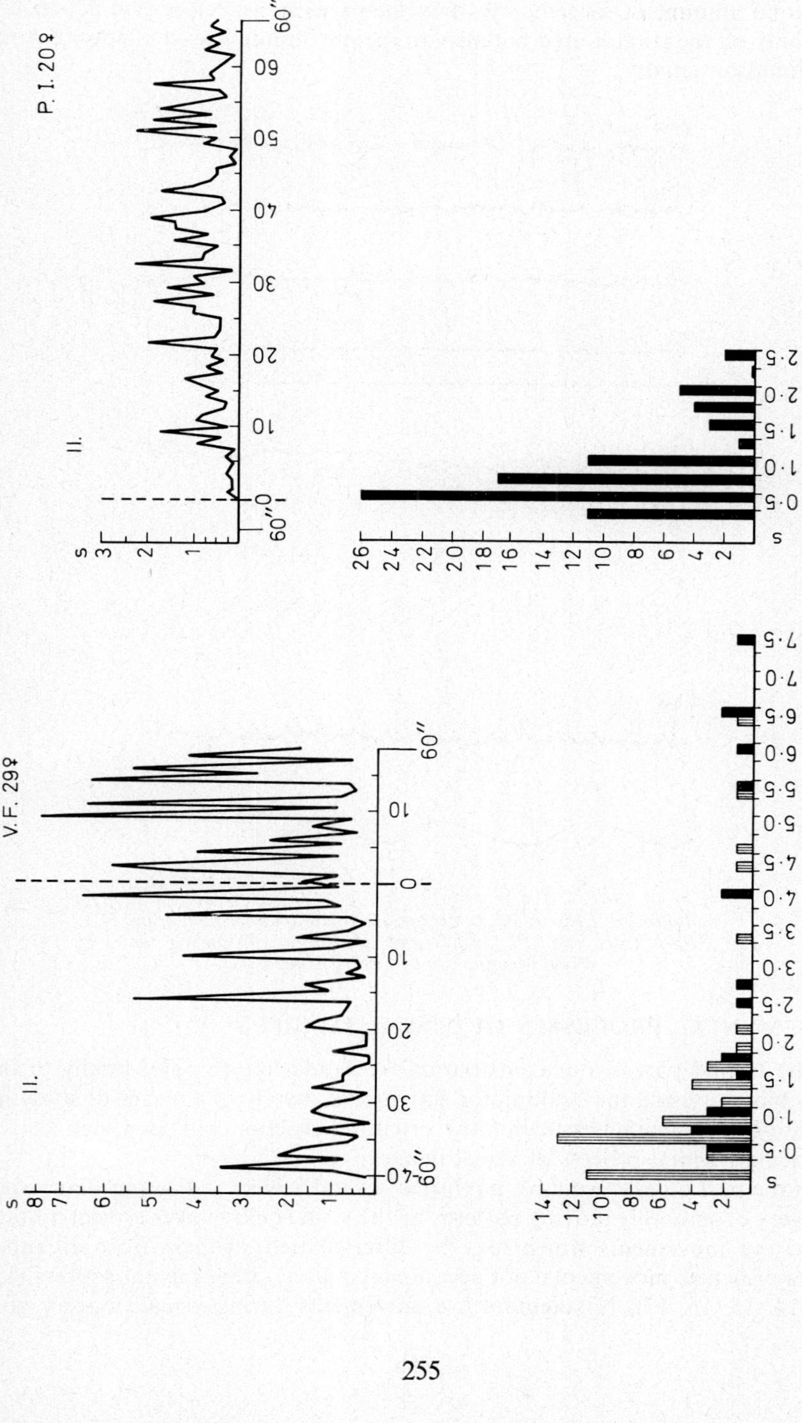

Figure 9. Description as in Figure 8. Second stimulus, two subjects

the size of the oculomotor reaction to stimuli used and the size of the other physiological reactions could not be found. It was shown that the time course and general direction of the oculomotor reaction (increase and/or decrease of the total amount of saccadic eye movements with eyes closed) is dependent not only on the stimuli used but also on the pre-stimulus level of spontaneous oculomotor activity.

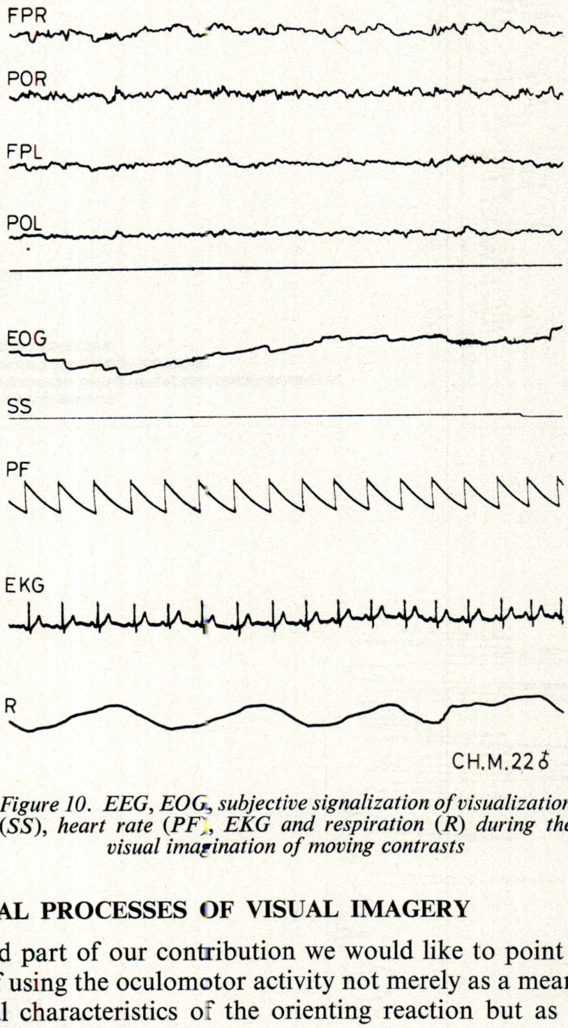

Figure 10. EEG, EOG, subjective signalization of visualization (SS), heart rate (PF), EKG and respiration (R) during the visual imagination of moving contrasts

MENTAL PROCESSES OF VISUAL IMAGERY

In the second part of our contribution we would like to point briefly to the possibility of using the oculomotor activity not merely as a means of studying some general characteristics of the orienting reaction but as a sign of the particular mental process of visual imagery.

In the past 5 years we have published several papers indicating that visual imagery of smoothly moving patterns in the visual field evokes typical nystagmoid eye movements which may be differentiated strongly from all other types of ocular movements, not accompanying any other mental process (12, 13, 14, 15, 16, 17). Nystagmoid eye movements during visual imagery of a moving stimulus pattern are analogous to optokinetic nystagmus, being however, in most cases of a slower frequency and amplitude—as may be seen in *Figure 10*. It has been shown that with repeated training for accurate reporting of the subjective sensory experience during visual imagery, the percentage of agreement between subjective signals of visualization (key-pressing) and the appearance of nystagmoid eye movements significantly increases (14, 17). A significant improvement in visualizing ability was noted in individual cases following administration of the psychostimulant drugs Dexphenmetrazin and Centredin as compared with a placebo (13, 17, 18). Visual imagery of the sensory stimulus used to evoke optokinetic nystagmus as a conditioned reflex, was found to help the elaboration of those conditioned responses which follow immediately after visualization (19, 20). Neither the simple fact of increasing attention nor the other mental processes show a similar facilitating effect.

REFERENCES

1. Amadeo, M. and Shagass, C. (1963) *J. nerv. ment. Dis.* **136**:139
2. Antrobus, J. S. and Singer, J. L. (1964) *J. abnorm. soc. Psychol.* **69**:244
3. Asher, J. and Ort, J. S. (1951) *J. gen. Psychol.* **45**:209
4. Gaarder, K. (1966) *Nature, Lond.* **209**:83
5. Jeannerod, M. and Mouret, J. (1962) *C. r. Soc. Biol.* (Paris), **156**:1407
6. Jeannerod, M. and Mouret, J. (1963) *J. Physiol. Paris*, **55**:268
7. Lorens, S. A., Jr. and Darrow, Ch. W. (1962) *Electroenceph. clin. Neurophysiol.* **14**:739
8. Mouret, J. (1964) *Les mouvements oculaires au cours du sommeil paradoxal.* Tixier: Lyon
9. Singer, J. L. and Antrobus, J. S. (1965) *Archs. gen. Psychiat.* **12**:71
10. Sokolov, E. N. (1963) *Perception and the Conditioned Reflex.* Pergamon Press: Oxford, London, New York, Paris, 309 pp.
11. Voronin, L. G., Sokolov, E. N. and U Bao Chua (1959) *Vop. psikhol.* **6**:73
12. Zikmund, V. (1963) *Activas nerv. super.* **6**:64
13. Zikmund, V. (1964) *Activas nerv. super.* **6**:207
14. Zikmund, V. (1965a) *Activis nerv. super.* **7**:64
15. Zikmund, V. (1965b) 'Mechanisms of the Orienting Reaction in Man'. *International Colloquium*, Smolenice, September 1965. SAV, Bratislava, 1967, p. 93
16. Zikmund, V. (1965c) 'Some Psychophysiological Characteristics of the Visual Imagery of a Moving Stimulus Pattern', *Proc. 8th Int. Congr. Neurology*, Vienna Ix, 1965, Suppl. 260
17. Zikmund, V. (1966) *Studia psychol.* **8**:4, 54
18. Zikmund, V. (1967 in press) 'The Effect of Centedrin on Visual Imaginative Capacity'. *Activas nerv. super.*
19. Zikmund, V. and Visser, P. (1964) *Physiol. bohemoslov.* **13**:202
20. Zikmund, V. and Visser, P. (1966) *Intn. Symp. über kortiko-viszerale Physiologie und Pathologie*, Akademie Verlag: Berlin, pp. 435–439

Relationships between Visual Evoked Responses and some Psychological Processes

E. García-Austt and W. Bûno (Jr.)
Medical School of Montevideo, Uruguay

Since 1959 (7) our group at the Medical School of Montevideo has been working on the visual evoked response in relation to certain psychological responses, using various summation or averaging techniques in man, in the cat, and in the albino rat. We have published a number of papers on the characteristics of visual evoked responses and the changes provoked by attention (1), habituation (6), the state of awareness (5), and maturation (14, 17).

Previously there had been general agreement that the scalp visual evoked response was simply the expression of the cortical visual inflow, and there was considerable confusion regarding its relationship with other waves recorded from the scalp. During a symposium (1964) on evoked responses Walter noted: 'The least satisfactory feature of this admirable symposium is that no two workers used both the same technique and the same experimental conditions. The technical discrepancy and diversity in this field must surely be unique in the history of science; never in the history of human progress have so many paid so little attention to one another' (16).

Now there is conclusive evidence that when the retina of man is stimulated by a short-duration flash, the potentials recorded on the scalp are different in the occipital region as compared to the vertex. The visual evoked response and, when the eyes are moving, the lambda wave are recorded from the occipital region. From the vertex several short-lasting waves and a slow contingent negative variation (CNV) described by Walter *et al.* (15) are recorded.

In our experiments the visual stimuli were delivered with a CRO (Tektronix 502) with a short persistence screen. In general, a 1 mm diameter low-luminosity spot, was on for a period of 10 ms. There were two alternatives: (1) the spot appeared always in the centre of the screen or (2) the spot appeared alternatively on the right and on the left of the screen. Sometimes a horizontal or oblique line was used. The luminance of the stimuli, about 5–10 microlamberts, was near the photopic threshold. Clicks used as auditory stimuli had a low intensity of about 5–10 db above the background noise.

The subject was seated in a dark room, his head 40 cm from the oscilloscope screen. Eye movements were evaluated in all the experiments. When eyes were not permanently fixated on the point indicated, or when they failed to move according to instructions, the data were discarded. Motor performances recorded were the acts of pressing a key or following a skipping spot with the eyes.

The EEG was also recorded throughout the experiment in order to monitor the state of awareness. Electrodes were placed on the scalp according to the 10/20 system. Bioelectrical responses and eye movements were summated with a computer of average transients (CAT 400B). The histogram of the finger movement reaction time was also done by the same apparatus concomitantly with the sums of the responses. From 50 to 250 epochs of 500, 1000, or 2000 ms were summated. The EEG, ocular movements, summated responses and the histograms were recorded by a Grass polygraph, using time constants of 0·65 or 0·80 s.

Until not very long ago we believed that the visual evoked potential recorded on the occipital region of the scalp was associated with the arrival at the occipital cortex of synchronized retinal impulses and that the changes provoked by modifications in attention were due mainly to the gating of the messages on their way from retina to cortex, according to the significance of the stimulus. This gating actually exists and has been demonstrated at receptor level by Hernández-Peón, in the retina of the cat (10) and by us in the guinea-pig cochlea (2). Now there is evidence that the scalp visual evoked response is more closely associated with the processing of information than with the mere arrival of information.

This raises a problem, namely, the relationship of the waveform of evoked responses with the information content of the stimulus. Schiller and Chorover (12) studying metacontrast and evoked potentials, found that the waveforms and amplitudes were nearly identical when stimulating with a disc or a ring. On the other hand, John *et al.* (9) demonstrated that visual evoked potentials are different for different geometric forms of equal area and are similar for versions of the same geometrical form of unequal area. What is the cause of these discrepancies? We do not know their exact nature but our experimental data support the findings of John and co-workers and might also account for such disagreements.

THE OCCIPITAL VISUAL EVOKED RESPONSE (OVER)

In *Figure 1* two experiments on two subjects are illustrated. In the first case (W.B.) the stimulus was a 1 mm diameter spot (indicated by a circle in the inset on the left). Only the right eye was stimulated. The fixation point (indicated by a cross) was also in the centre of the screen and did not change during the whole experiment. Also in every figure the frequency of the stimulation and the number of summated responses are indicated. Response number 1 was obtained with the spot fixed on the centre of the screen and response number 2 with the spot appearing once a second on each border of the screen, alternately on the right and on the left, the gaze being always fixated on the centre. Clearly, the responses are very different as observed in the superposition of both responses on the right (*1 & 2*). This may be explained by the fact that in the second case the spot was displaced 6 degrees away from the centre, toward either side and, as a result, different receptors were stimulated, or, to the fact that the information content was very different in both circumstances.

In the second case (M.B.) the same stimulus was given on the first presentation but both eyes were stimulated. After 100 presentations the spot was

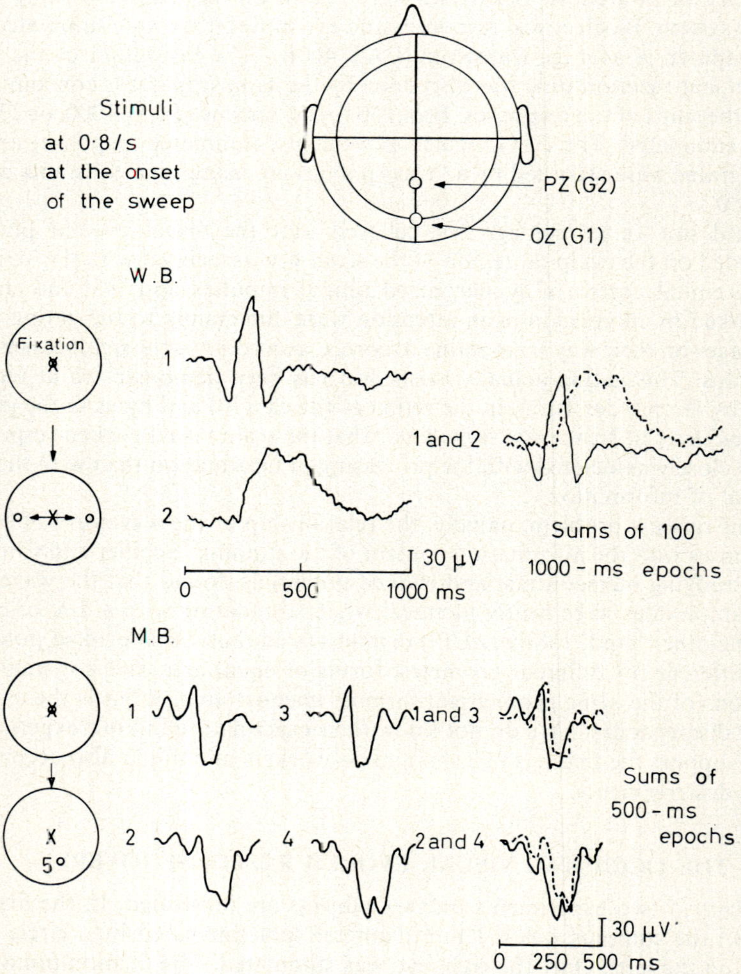

Figure 1. Changes of occipital visual evoked response (OVER) determined by changes in the stimulus pattern

substituted by a horizontal line 5 degrees wide and response number *2* was obtained. The same procedure was repeated once and response *3* and *4* were recorded. A clear cut difference in the response waveform upon changing the shape of the stimulus is observed. On the other hand, the waveform for the same stimulus is very similar as shown in the superpositions on the right (*1 & 3*, and *2 & 4*).

Again in this case it can be postulated that the change is due to stimulation of different receptors, and as the retina has a point-to-point projection in the primary visual area, the changes observed in the scalp may be the result of a different spatial distribution of the areas activated in the brain cortex. Moreover, the horizontal line obviously stimulates a larger number of receptors because their surface was greater than that of the spot.

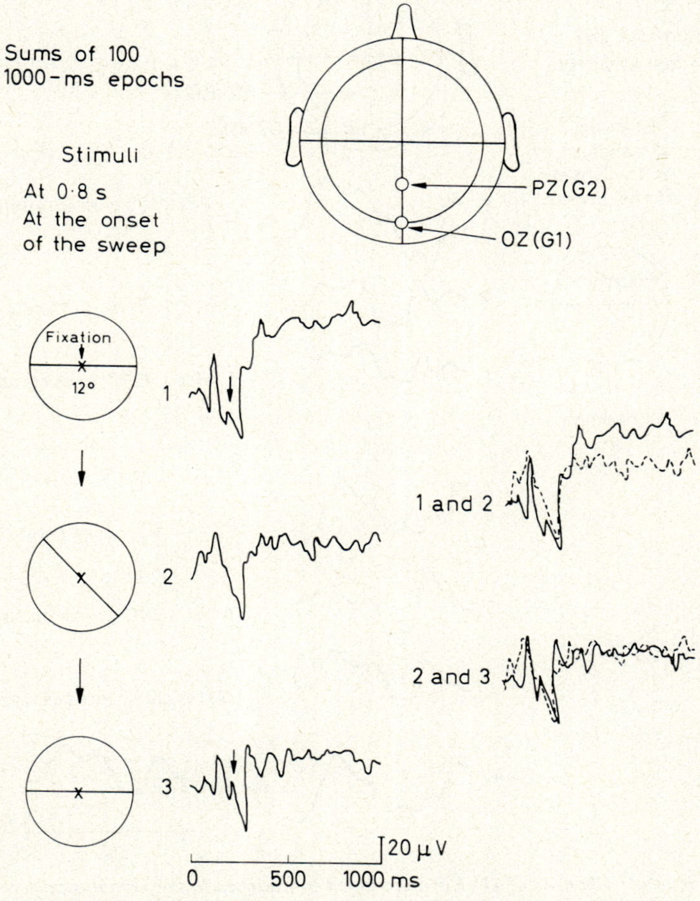

Figure 2. Changes of OVER determined by changes in the stimulus pattern

Figure 2 shows another experiment in which the stimuli—a horizontal line and an oblique one—had the same intensity and the same area. Response number 1 was obtained by stimulating with a horizontal line and number *2* by stimulating immediately after with an oblique one as shown in the inserts on the left. There are differences in the response in the first 300 ms of the epoch as seen in the superposition on the right (*1 & 2*). The notch in the positive wave—indicated by the arrow when the stimulus was the horizontal line—does not appear in the other instance. Moreover, in the last 700 ms of the epoch a steady negative shift is observed in the first response but not in the second one. However, upon further stimulation with a horizontal line, response number *3* shows the same notch indicated by the arrow but the slow negative variation has disappeared—as observed in the superposition on the

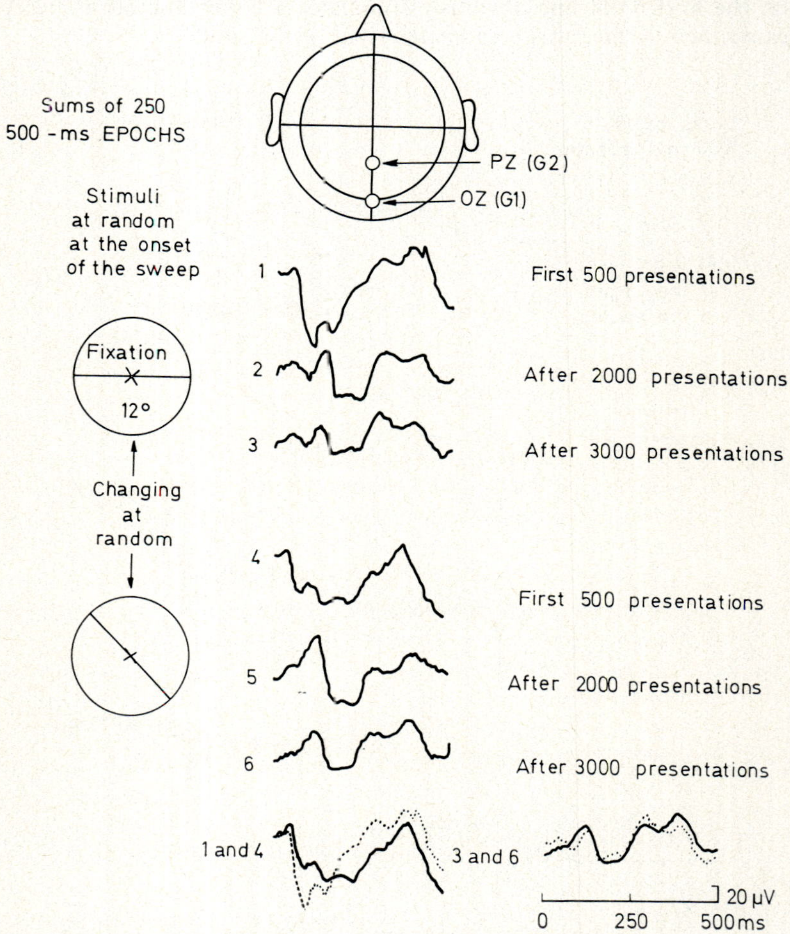

Figure 3. Changes of OVER determined by changes in the stimulus pattern

right (*2 & 3*). We believe that the slow negative variation recorded on the occipital region (which is evidently not determined by the stimulus pattern because it disappeared after 100 presentations) corresponds with the CNV despite the fact that the recording is from the occipital region and that no 'imperative' stimuli were used (15). As for the notch in the positive wave, it cannot be ascribed to a change in the area or intensity of the stimulus, because they are identical. However, it may still be postulated that these differences are the result of stimulation of different receptors.

Figure 3 gives the results of another more complex experiment in which both eyes were stimulated with horizontal and oblique lines changing at random. The responses to horizontal and oblique lines were summated at the same time in two different channels. In the first 500 presentations, response number *1* to the horizontal line, and response number *4* to the diagonal line were very different as seen in superposition *1 & 4* at the bottom of the figure. After 2000 presentations the occipital responses to both stimuli change in shape considerably, and are far more alike. After 3000 presentations there are practically no waveform differences between response *3* and *6* as observed in superposition *3 & 6* at the bottom.

A number of findings and comments are derived from this experiment. 1) The waveform of OVER in the first presentation is very different for different geometric patterns of the visual stimulus with identical intensity and area. 2) Both OVERs change markedly in shape and amplitude as a function of time when both stimuli are monotonously repeated at random. The first positive wave was partially substituted by a negative one followed by a positive wave of longer latency. The general amplitude of the responses are progressively reduced, and after 3000 presentations both responses are similar. 3) At any moment the subject can tell one stimulus from the other, whether the occipital responses are different or similar, i.e. every time he was asked about the type of stimulus his answer was correct. 4) The EEG did not show any important changes and the eyes did not close (except for occasional blinking) or become displaced during the whole experiment. Hence the changes in the response may not be explained by somnolence or eye displacements. 5) The subject reported that he was very attentive to the first stimuli and that he was interested in determining whether the succession of changes followed any definite order. Later on, he realized that the prediction of the class to which the next stimulus belonged was entirely inaccurate and that the probability for any of them was 0·5. When he learned this, he lost interest in the stimuli but he was at any moment able to differentiate one from the other perfectly well. 6) Hence it is probable, that the waveform of the OVER depends upon different variables; the information content of the stimulus, the degree of attention and the subjective probability.

These dependencies of the OVER may well account for its great variability in an individual under varying experimental conditions and also for the inter-individual differences. We do not know whether the very first components of the OVER elicited by low-intensity stimuli, which we never studied in detail, are related to the crude sensory inflow of visual messages but we feel certain that most of the components of the occipital response are linked to the processing of visual information. Undoubtedly this aspect will prove far more stimulating for future investigations than if the OVER had a fixed pattern determined by the arrival of visual input.

The OVER also showed important changes during intensity discrimination tests. In *Figure 4* the changes of OVER are shown when the subject is comparing the luminance of one spot with the luminance of a previous one. Response number 1 was obtained when the subject was instructed to press a key each time he saw the spot which always appeared in the centre of the screen on which the subject had his gaze fixed and at a regular frequency of

1/s. Response *2* was recorded when the subject was requested to press the key only when the luminance of the spot changed. Hence he had to compare the luminance of spot N with that of spot N-1. Twenty per cent of the spots delivered at random had a different intensity, but all the responses were summated. The OVER changed considerably in these circumstances, the amplitude of the first negative wave diminished, a second, also negative, wave of a shorter duration appeared, and the last negative wave present in the first

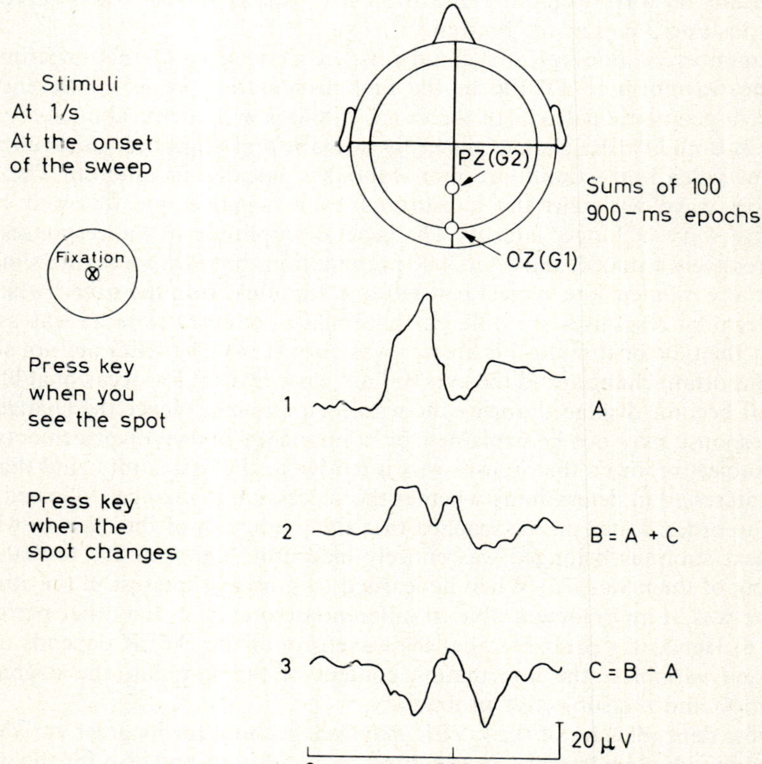

Figure 4. Changes of OVER during luminance discrimination

response was substituted by a positive one. It seems logical to assume that to the first wave A was summated another wave C, which appeared following the operation of comparing between the luminance of points N and N-1—its result being response number *2*. This summated wave C, as assumed, may be obtained in the computer by subtracting B – A. This operation yielded the last response, consisting primarily of two positive waves, whose peak latencies were 360 and 620 ms, respectively, and therefore separated by 260 ms. This wave would hence be a concomitant of the process by comparing luminance between two successive spots and has a latency as short as the one of response number *1*. This finding supports the idea that the OVER recorded in our experimental conditions is fully related or at least concomitant with the processing of visual information. These changes in OVER with intensity discrimination were observed in all the subjects studied as will be seen in some of the forthcoming figures.

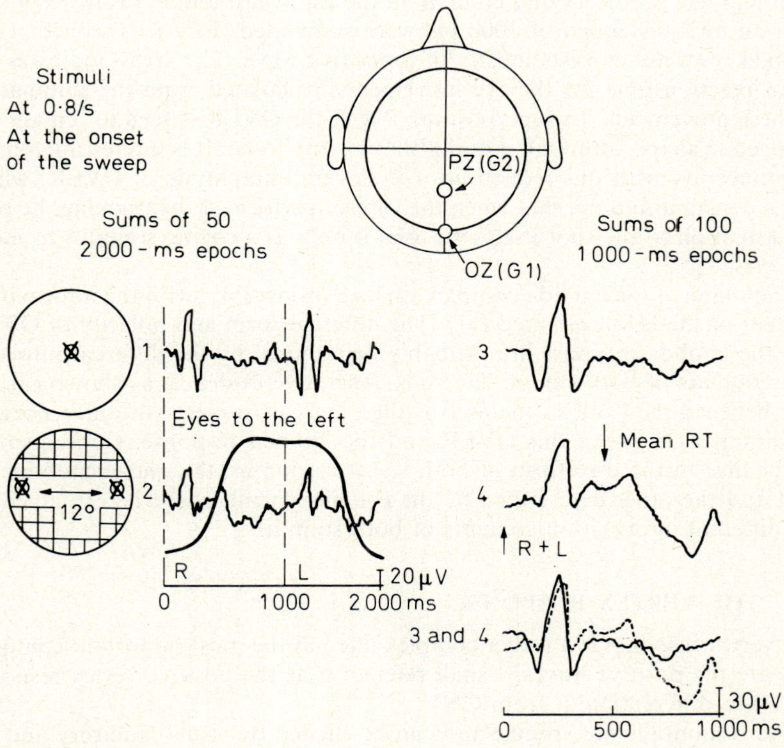

Figure 5. OVER and lambda response

THE LAMBDA RESPONSE

The potential appearing when eyes move (3, 4, 8) is a very well known one. It is not a simple wave and is better called the lambda complex (11) or lambda response. It is observed as a consequence of eye movements (voluntary or involuntary) but does not appear in darkness. It is certainly provoked by the visual stimuli upon moving the eyes, and its shape varies considerably on changing the exposed field (11).

In *Figure 5* an experiment in which the occipital response to a spot appearing on the right and on the left every second, and the lambda response provoked by the eye movements, is shown. The graticule of the oscilloscope screen was illuminated with a scarcely perceptible red light. In *1* and *2* the sums of two responses are seen in the same sweep. In *2* can be seen the responses to the spots that would appear on the right and on the left, successively.

In addition, the averaging of the correctly performed ocular tracking is superposed. The subject was instructed to move his eyes to the left on seeing the spot on the *right*, to wait for the appearance of the spot on the *left*, after which he was to direct his gaze toward the right and so on. It is apparent that both responses are different, with no important differences, however, between right and left. In the second column all the rightward and leftward responses with an analysis epoch of 1000 ms were summated. In *4* it is seen that the lambda response consists mostly of a positive wave. The arrow indicates the mean reaction time for the eye movements, calculated from the summation of these movements. In superposition *3 & 4*, the OVER is seen to remain unchanged in shape, although it diminishes in amplitude. It is interesting to note in connection with this reduction of the overall amplitude of OVER, which was a constant finding, that when the subject performed the tracking, he paid less attention to the spot itself and used it only as a cueing stimulus to move the eyes.

The shape of the lambda complex varies considerably, with the information content of the field presented (11). The different form and polarity of OVER and the lambda response are probably determined by the different information contents of both stimuli. There is conclusive evidence, as shown earlier, that between the OVERs themselves there may exist clear-cut differences, as important as between this OVER and the lambda response. Consequently, the change in the waveform of both visual responses, the one elicited by the spot and the other determined by the eye movement, could be explained by the different information contents of both stimuli.

THE VERTEX RESPONSE

The vertex response is a rather complex one but the most important components are the positive waves. I shall refer to it as the positive vertex response (PVR) to differentiate it from CNV.

The response is not specific and can be elicited by visual, auditory and somatic stimuli. In *Figure 6* can be seen vertex and occipital responses to light and sound stimuli and the corresponding histograms of the reaction times with three different tasks, on the left and on the right, respectively. The first task was merely to perceive the stimulus. The spot determines an OVER in the occipital region (*1*) and the click does not provoke any response in the same region (*4*). On the other hand, both stimuli provoke a similar positive wave at the vertex (*2 & 5*). The peak latency of the vertex response of the visual stimulus is 350 ms and for the click, 260 ms. The difference is due to the shorter auditory latency. The second task was to press a key every time the stimulus was perceived. The OVER is similar to the first one (*7*). There is now a low amplitude response to the click in the occipital region (*10*). Both PVR are now seen to be considerably increased in amplitude and there is the same difference in the peak latency between the visual (430 ms) and the auditory (300 ms) response (*8* and *11* respectively). The motor performance for the click (*12*) shows a much lower dispersion than for the spot (*9*), and coinciding with this fact the PVR for the spot (*8*) has a longer duration than for the click (*11*). With the third task, which consisted in appreciating differences in the intensity of the stimuli, the OVER (*13*) changes as usual, a second negative peak appearing. The PVR becomes wider in both circumstances (*14* and *17*) coinciding with a greater dispersion of the reaction time (*15* and *18*).

This experiment demonstrates that (1) the occipital visual evoked response is specific for photic stimulation; (2) the PVR is similar though different modalities be stimulated; (3) the amplitude of this PVR bears relation to the complexity of the programme; (4) its duration may be related to the dispersion of the motor performance.

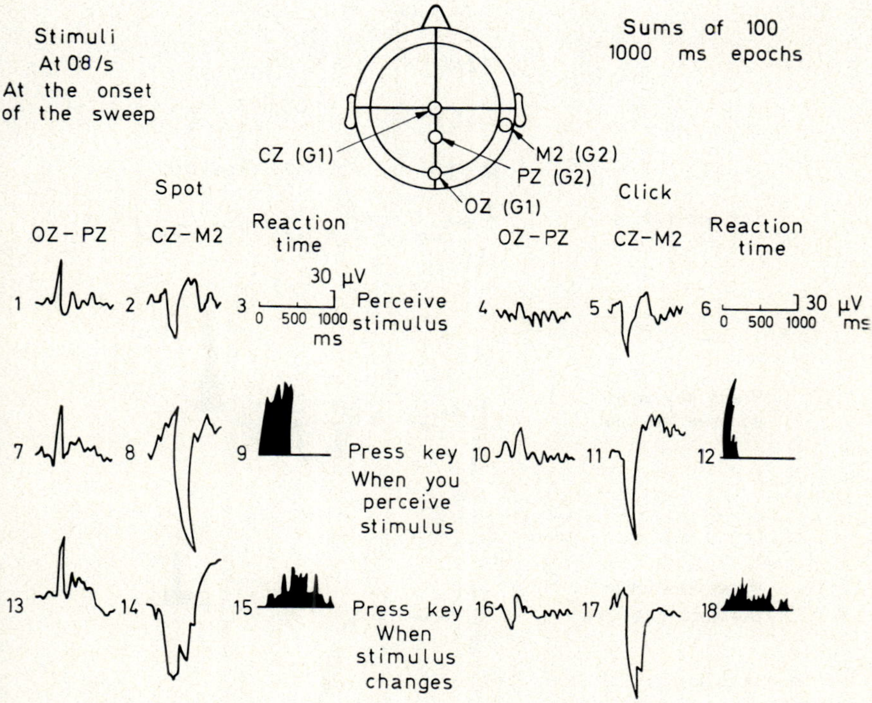

Figure 6. OVER and unspecific vertex responses

The vertex response is not simply determined by a stimulus, for it can be elicited by *negative* stimuli. The responses to two successive spots separated by 500 ms were recorded in the occipital region and in the vertex, and three successive tasks were employed. In the first one the subject was required to see the spot; in the second he had to press a key when he perceived the second spot. Responses in the occipital region did not change significantly, but at the vertex the positive waves increased in amplitude. In addition, in both circumstances, a slow negative variation could clearly be seen. The third task consisted of pressing the key just when the second spot should have appeared—though it actually never did. In the occipital region the response to the second spot was not present and a slow negative variation developed. In the vertex region a positive wave appeared approximately at the time the second spot should have gone on. This wave was of lower amplitude and was delayed a few milliseconds with respect to the equivalent wave of the previous task.

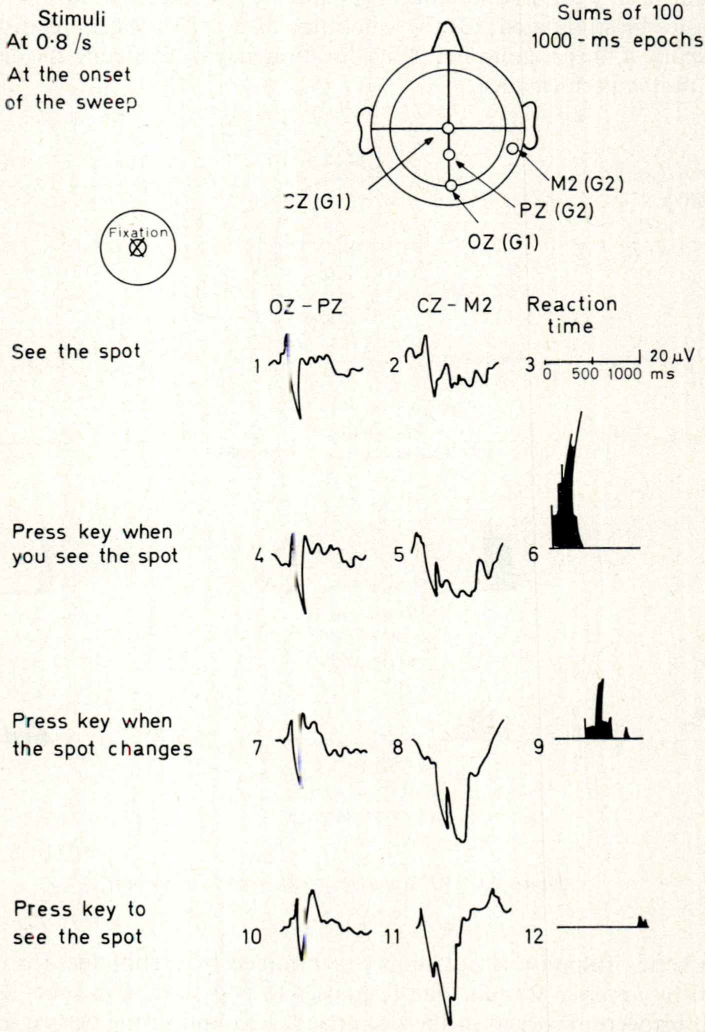

Figure 7. Changes of OVER and vertex response with different programmes

These results are in good agreement with those obtained by other authors (13) working under similar conditions with auditory stimuli, and it may be concluded that in the vertex there appear responses even with negative stimuli, though this does not happen in the occipital region. Moreover it is interesting to note that in this latter area, under certain circumstances, it is also possible to observe a slow negative variation seen in the vertex—as in this case it was of a much longer duration.

In *Figure 7* the changes of evoked responses provoked by light and the histograms of the reaction times with different kinds of tasks are shown. The first again consisted simply in perceiving the stimulus. The spot determines an OVER (*1*) and a rather complex PVR of low amplitude (*2*). With the second task which consisted in pressing the key whenever the spot was seen, a similar OVER is observed (*4*). On the other hand, the PVR increases in amplitude (*5*). The third task consisted in pressing the key when the luminosity of the spot changed. The OVER presented an additional negative peak as described previously (*7*). The PVR again increases in amplitude and presents two peaks (*8*). The fourth task consisted in pressing the key to turn the spot on. The PVR (*11*) is of high amplitude and short duration, and the latency of the peak is shorter. The last histogram (*12*) shows that the subject occasionally pressed the key before the end of the epoch.

This experiment demonstrates that changes of the PVR are dependent on the complexity and duration of the tasks and, consequently, on the complexity and duration of the operation the subject has to perform. On the other hand, the occipital response has a constant latency with every task.

In *Figure 8* changes of evoked responses provoked by light, and the histogram of the reaction times with different kinds of tasks are shown. As in the previous experiment, the first task was simply to perceive the stimulus. The spot determines a typical OVER (*1*). A PVR is observed in the vertex (*2*). The positive wave is followed by a slow negative variation which subsides before the end of the epoch. After this task the subject was trained to press the key every time he saw the spot delivered at a regular frequency of 0·8/s. After 5 min the second sum was done. The OVER does not show any changes (*4*). The PVR (*5*) increased slightly in amplitude and the latency was the same as the previous one. The slow negative variation disappeared. The motor performance was correct, with small dispersion (*6*). The third task was carried out immediately after the previous one and consisted in pressing the key every time the spot was perceived, but on this occasion, it was delivered at random. The OVER (*7*) does not change in shape or in amplitude, but the PVR (*8*) appears later and is inscribed in a slow negative variation. The increase of latency with respect to the two previous responses is very clear, and the histogram of the motor performance (*9*) shows a considerable delay when compared to the previous one (*6*).

The fourth task was to press the key when the spot changed in luminosity and at this time the stimulus was delivered at a regular frequency of 0·8/s The OVER shows the characteristic duplication indicated by the arrow (*10*). The PVR (*11*) has a latency similar to that of the previous programme and a slow negative variation is still observed. The histogram (*12*) shows the characteristic delay and an increased dispersion. The final task was similar to the former but in this case the stimulus was delivered at random. Both responses are similar (*13 & 14*) to the previous ones but the delay of the histogram increases again (*15*).

This experiment indicates that the PVR is not directly related to motor performance because in the last three instances the final positive peak did not

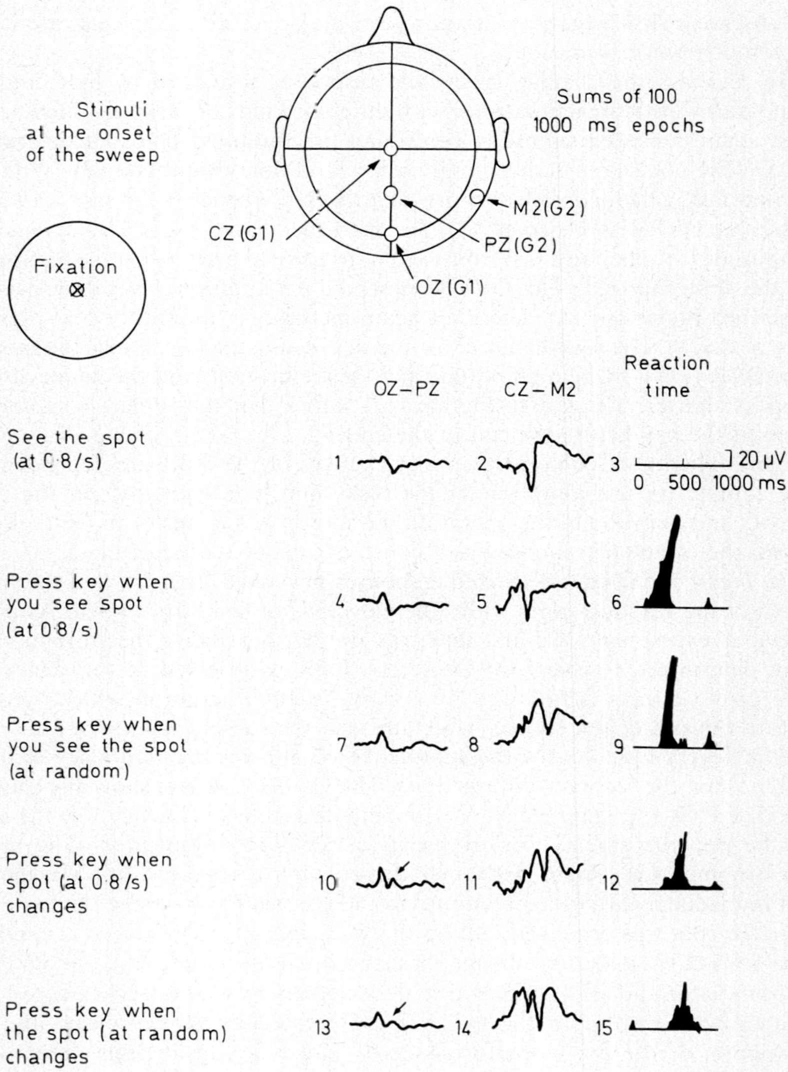

Figure 8. Changes of OVER and vertex response with different programmes

change its latency while, on the other hand, the latency of the motor performance was progressively increased.

In *Figure 9* we see the changes of OVER, PVR and reaction times when the individual is performing different tasks. In all the tracings two responses are summated in the same sweep. When the subject is only seeing the spot the OVER (*1*) is simple and biphasic, the vertex responses (*2*) being more complex but of low amplitude. The second task was to press the key every other spot. The OVER (*4*) does not show any important changes but both PVR (*5*) increase in amplitude. The PVR corresponding to the spot that elicited motor performance (*6*) is more complex, with two successive peaks. The third task was to press the key when the luminance of the spot changed. This luminance was changed only for the second spot as shown by the corresponding histogram (9). Both OVER (*7*) change in the same way, a second negative peak appears (arrows) and the positive wave disappears. Also both PVR change in a similar way showing a double peak (*8*).

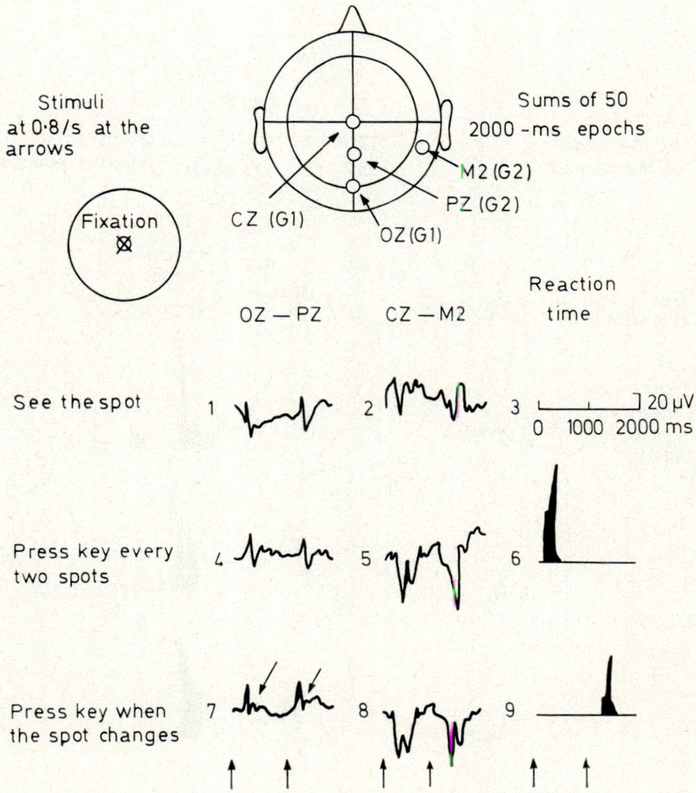

Figure 9. Changes of OVER and vertex response with different programmes

This experiment demonstrates again that 1) when the subject is comparing the brightness of spot N with that of spot N-1 the OVER changes in shape and a second negative wave appears; 2) the OVER changes are not due to the changes in brightness, because it has the same waveform in both responses and only the second stimulus changed; 3) OVER does not change either with or without a motor performance, and when the subject is not required to make comparisons with his previous experience; 4) the PVR becomes more complex and wider with increased complexity of the programme and these changes on the other hand are not specifically related to the comparison of the experience.

In *Figure 10* the changes of evoked responses and of the reaction time when the subject is performing the same task during a determined period are shown. In this experiment the responses were recorded from three different regions: occipital, vertex and frontal. The first task was to see the spot without any other action. It may be observed that the responses from the occipital (*1*) and the frontal region (*3*) are very similar. However, the subsequent evolution of both demonstrates a very different pattern. The second task was to press the key each time the spot was perceived, and here there is a striking change in the

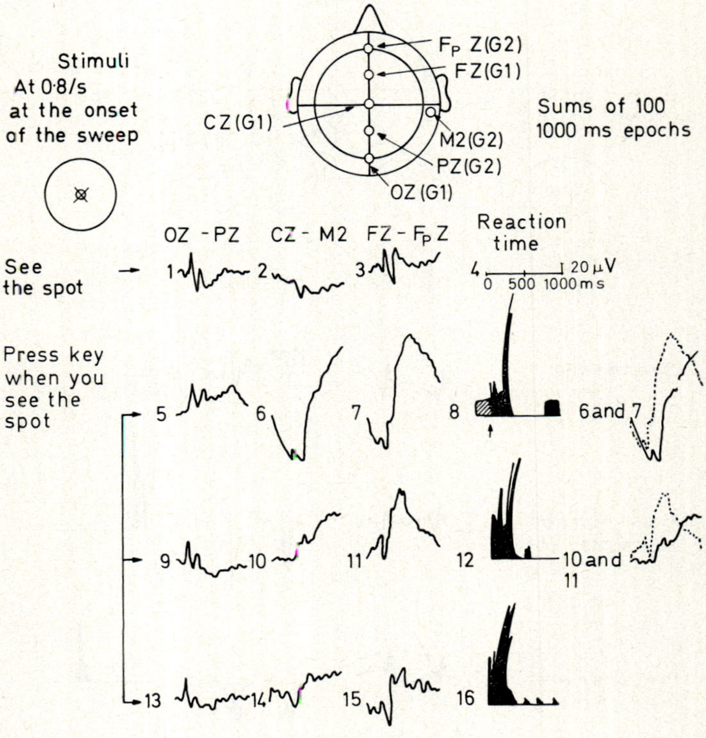

Figure 10. Changes of vertex and frontal responses when the same programme is repeated

frontal response (*7*), with the development of a slow negative variation of high amplitude reaching a maximum of about 600 ms, subsiding before the end of the epoch. This slow variation is preceded by a short latency positive wave. The baseline for each response is represented by the base level of the histogram. In the central region a similar response (*6*) is observed but with some delay, as seen in superposition *6 & 7* on the right side. The OVER (*5*) also shows some changes, the positive peak disappears and a slow negative variation of low amplitude is likewise observed. The histogram of the reaction time (*8*) shows that the motor performance of the subject was not efficient because he pressed the key many times before the appearance of the stimulus. These responses were observed at the end of the epoch and were re-designed on the left side before the beginning of the epoch, appearing as a striped area in the illustration. As the subject continued with the same task his motor performance kept on improving as shown by the successive histograms (*12 & 16*) despite the fact that he always pressed the key before the appearance of the OVER. Under these circumstances the responses continue to change gradually. The slow negative variation in the frontal region kept on decreasing in amplitude and the peak latencies of the positive waves progressively increase until they reach the initial latency (*11* and *15*). Superposition *10 & 11* indicates that there is always a delay in the slow negative variations.

This experiment demonstrates that 1) the occipital and frontal responses generally vary considerably; 2) when the subject is learning a task, a slow negative variation appears in the anterior regions, with a delay between the frontal and central regions—an indication that the recording in the central region is not the result of electrotonic propagation; 3) the positive wave, the peak of which appears approximately within 300 ms, varies in latency when the subject changes the timing of his motor performance; 4) the visual response in the occipital region is relatively constant in shape and latency, an indication that its nature is altogether different from that of other recorded responses.

Figure 11 shows the scalp responses determined by both light and sound stimuli and by eye movements. In trace number *1* two OVERs to a spot projected on to the centre of the screen are being observed, with the gaze also fixed on the centre. Record number *2* is the sum of fifty epochs at the vertex and does not show any significant response. Record number *3* was obtained with a spot appearing every second on the right and on the left. The screen graticule of the oscilloscope was illuminated with a barely perceptible red light. As in the previous record two responses are seen in the same sweep, first, the response to the right spot and second, the response to the left one. The sum of the eyes tracking is superimposed indicating that the subject performs the following task satisfactorily: 'move the eyes to the left on seeing the spot on the right, wait for the appearance of the spot on the left and when you see it at once direct your gaze to the right and so on'. It is apparent that both responses are similar. In trace number *4* obtained from the vertex, a response is seen. It is a positive wave followed by a negative one but obviously, comparing this response to the superimposed recording of the tracking the latency is shorter than that of the lambda complex recorded on the occipital region.

The same motor performance was carried out with auditory stimuli. Record number *5* is the sum of fifty epochs with the eyes centred on the screen. Low-amplitude short-latency responses were observed in the occipital region and in the vertex. In record number *7* the individual was given the following instructions: 'fix your gaze on the right side of the screen, and when you hear the click, move your eyes to the opposite side and wait for the next click, after which move your eyes to the right and so on'. The short-latency response to the click increases in amplitude in the occipital region and is followed by a lambda complex. At the vertex, response number *8* to the click also increases and a positive wave is present with the peak about 300 ms after the stimulus. This response is very similar to responses *3* and *7* determined by eye movements. However, we believe that the positive response recorded in the vertex region is the common wave recorded following the end of any task, and is not a lambda complex which is not usually found on this region.

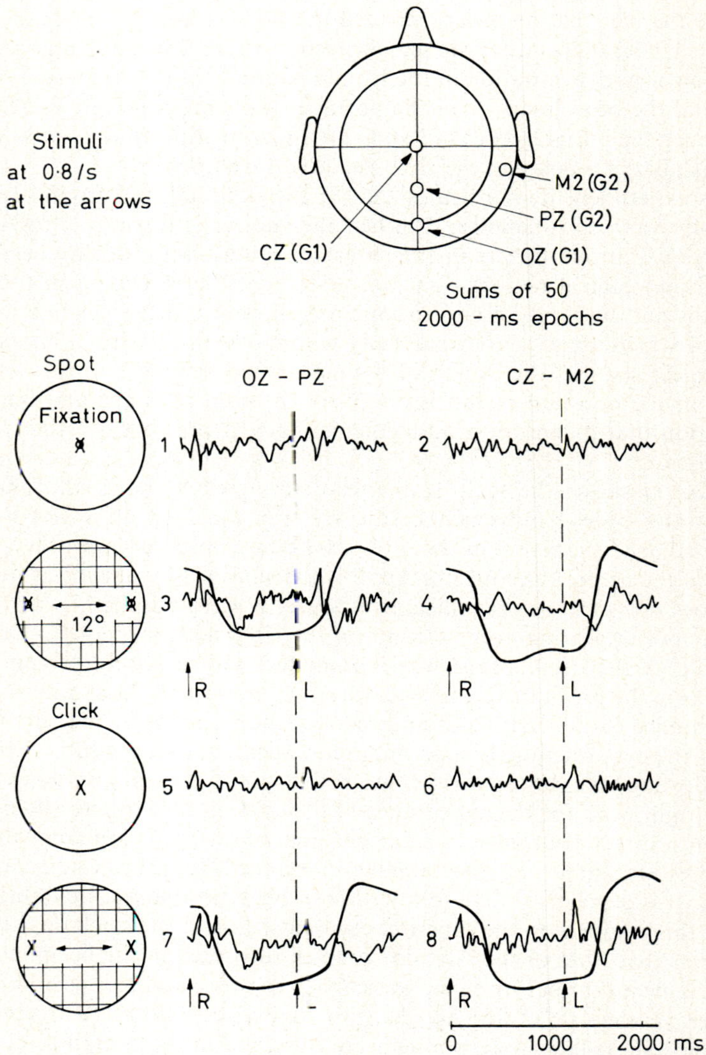

Figure 11. Scalp responses determined by light and sound stimuli and by eye movements

CONCLUSIONS

In our set of experiments which dealt fundamentally with visual stimulation of low intensity and with both scalp occipital-parietal and vertico-mastoid recordings in man, three electrical variations determined by the stimulus were studied in relation to psychological variables.

1) The Occipital Visual Evoked Response (OVER) is concomitant with some processing of specific visual information, the designation of 'visual response' being therefore pertinent. In the occipital regions this wave is not replicated by auditory stimuli. Its waveform varies on changing the processing of visual information. The spatial characteristics of the stimuli influence the waveform of the response. However, we showed that to maintain the differences determined by two different stimuli it is necessary that the interest and attention of the subject is maintained. When the individual is comparing the relative luminance of two successive stimuli, a double-peak positive process develops with almost the same latency as that of the evoked response. This process changes the waveform of the response in a characteristic pattern.

2) The positive vertex response (PVR) is always present with any type of programme, no matter how simple, and is independent of the stimulus modality. This wave appears with a relatively constant latency once the programme or task has been completed. It augments considerably with uncertainty and, in general, with the increased complexity of the programme. Under certain circumstances the PVR can be very similar to the OVER but its variations with the task changes are very different. It is the wave of the accomplished task, and might be the expression of the reset of the subjective programme.

3) Finally, the contingent negative variation has been recorded in the vertex and in the frontal region but also occasionally in the occipital region. In these regions it may have a different temporal course. This wave is observed at the beginning of the performance with whatever task, even the simplest. The existence of an 'imperative' stimulus is not indispensable. We are not sure whether this wave could be always correlated with expectancy. In some instances the negative variation subsides before the beginning of the next task, when expectancy should be maximal.

ACKNOWLEDGEMENT

This work was performed in the Seccion de Neurologia Experimental, Instituto de Neurologia, Facultad de Medicina, Hospital de Clinicas, Montevideo, Uraguay and was supported by the National Institutes of Health under Grant NB–04382–05.

DISCUSSION ON GARCIA–AUSTT

McCallum

I would like to make three points. Firstly, I was most interested to hear about your findings of a second negative component. We too have frequently seen such a late negative component in similar circumstances. However, I was a little more concerned by your positive component. Could you say what controls were made for eye movements? One can see quite a large eyeblink component in responses recorded from the vertex, and the latency of your positive response looked to be of an order similar to that which one might expect from such a blink response.

García–Austt

Well, eye movement control is always present. Subjects do not always maintain their gaze on the fixation point. This is a very important point, and often not considered. Then, in all the cases I have shown, the gaze was fixed in the centre of the screen.

McCallum

Yes, this would eliminate gross eye movements, but I was thinking rather of a reflex blink evoked by the stimulus. Would this not be important?

García–Austt

No. I don't believe it would.

McCallum

The third point relates to the difference in latency which you found on one trace between the responses seen at the frontal and vertex electrodes. It is our experience and that of Bickford and others that there are slow potential changes due to ocular movement which are very similar in their general form to the contingent negative variation, but which of course have a more anterior distribution. I wonder whether it is possible that the shorter latency observed at the frontal electrode was due to the addition of such a component.

García–Austt

I don't think so because the blinking does not synchronize with it. Blinking is not determined by this very low intensity stimulus. Blinking is spontaneous, though it can be determined by an intense flash but not in our experimental conditions.

Grey Walter

In relation to what Dr. McCallum said about contamination of these records by eye movements there's one technical limitation of averaging procedures which is not always appreciated. If you take, say, 100 samples at a particular gain which is far below the saturation of either your amplifiers or your analogue-digital converter, a single very large transient may produce in the average a component as large as the summated responses you are looking for. The very small, say 5 or 10 μV, responses which you are averaging add up to a certain height in your average presentation, but just one event which is 100 times larger than these will produce an equally large component in the average write-out. For this reason Dr. Offner suggested some years ago that one should always use these machines, averagers or computers, almost at saturation for the components that you are looking for, so that the response to a single contaminant can never be much larger than its original size. As a practical example, a single blink in 100 trials, which might be a millivolt or so in amplitude, as compared with the few microvolts of your responses, would look like a 10 μV response but in fact it would have been one event. One way round this is to collect a small number of samples for each average, and then superimpose the averages. Another, more elaborate procedure is to compute the median rather than the mean—as suggested by Borda and Frost.

Horn

Wouldn't that be expected in the occipital one as well though?

Grey Walter

Not so much. The field is from front to back, whatever it is due to. But a single large blink at the right time will produce a component as large as the authentic responses.

García–Austt

But the changes are very consistent and very regular, for example, the changes in the latency of the responses in one of the last illustrations I showed. There is a consistent change very difficult to explain for one blink or for a few blinks.

Grey Walter

Yes, of course it is, especially in analysing your detailed record, but this latency of 200 ms does seem very long for a large component of this type, and it is about the right latency for the peak of the blink responses.

Mulholland

I'm referring to the experiment where you compared the evoked potential from an oblique and a horizontal line. And as I remember, as the sample that went into your evoked potential increased, the averages converged. As the sample size increased they became similar. How do you decide when the difference between two evoked potentials is such that it would be maintained with further sampling? If you're drawing samples from the same population, you might start out with a difference, but as your sample gets bigger the averages converge. Is this a problem in general, in deciding when two average evoked potentials are in fact different?

García–Austt

It is not possible to calculate with the CAT computer the significance of the difference between two averages because we don't know the variance in question. But our sums did not converge. To compare responses the same amount of epochs were summated.

Horn

My question also relates to the quantitative aspect of this work. You just mentioned one difficulty about averaging techniques and that is the difficulty of obtaining variances in order to compare two groups of data. If you were to repeat the experiment several times with the same subject, you would obtain differences between the mean responses and could calculate the variance. Did you do this and to what extent are the average responses producible in one subject?

What population of subjects did you use? Did you for example use 100 subjects and reject 10 of them or perhaps 20? It would be very nice to have some idea of the size of the population from which this sample was drawn.

García–Austt

Well, we used about 15 subjects most of whom were personnel from the laboratory.

Horn

Did you reject any of these?

García-Austt

No. We performed many experiments with them and results were consistent provided the same experimental conditions were maintained.

Horn

But you haven't in fact analysed them yet. Have you the data in some quantitative form so that you can say what the variance, the standard deviation, and so on are?

García-Austt

No. As I already stated, I do not know the variance.

Beck

I can answer that if you don't mind. The early work of my colleague Robert Dustman concerned this problem. He ran 10 subjects separated by 10 min, monthly and in some instances there was a year's separation. Coefficients of correlation, if you want numbers, indicated that the first 250 ms of the response was quite reliable. Test-retest correlations ranged from 0·77 to 0·98. So the response is quite stable until you get out to the very late components, those beyond 300 ms.

Horn

But it's the late components that we are discussing.

Beck

Yes, but speaking of our data, most of these occur from 100–300 ms. We have little response after 300 ms save the 'ringing response' or a general d.c. negative shift. In many people there is a reliable response up to 300 ms.

Goff

What Dr. Grey Walter says about blinks and other types of artefact is very true. One way we attempt to get around this, I think successfully, is to take a series of smaller numbers of summation—48, 64 or some convenient number, and then take six or eight of these averages and superimpose them. A deviant response is very apparent for it is very improbable that, for example, a blink would occur at the same time with the same latency in six or seven different averages. This also, as was pointed out, gives you some variance which you can then use in a statistical analysis. I think you can probably use parametric statistics, although we tend to use non-parametric tests, and then you can take a measure of amplitude, do a statistical analysis and find out whether or not these potentials are indeed different. I think that the averages that you presented were probably typical and that you did more than one response at the time.

Beck

A brief comment about a simple approach to this problem. You can run an experiment instructing subjects to blink to the light pulse and compare the averaged evoked response of this procedure with one when no blinking occurred. There will be a positive component occurring at about 200 ms but this can be distinguished from the blink artefact as the amplitude of artefact is many many times that of an evoked response component, and incidentally has other characteristics that identify it.

Stark

It's very simple to measure the blink of a subject or his eye movements while you're doing these experiments, and then you can tell whether in fact he has blinked or not. It's a very trivial bit of apparatus; a clarex, some red light and you can measure blinks and eye movements (18).

García–Austt

I am quite familiar with the artefact of blinking and you can recognize this in the record quite easily.

Jasper

Just another point about these averages. The information we lose in the temporal course of changes in the evoked potentials always impressed me as being a very serious loss. In the course of recording successive evoked potentials in monkeys and conditioning experiments, the variations within a series of, say, ten evoked potentials can be quite significant in relation to motor responses and maybe only two of the evoked potentials just preceding a motor response may be reduced to as much as a tenth of their original value. The average however would show this only as a slight reduction in the overall response to the stimulus. This detailed information, at least from the point of view of understanding, seems to me to be very significant. In these studies we found that the visual evoked potential was consistently affected by a motor response which was associated with it. Always prior to a conditioned motor response there was a very striking change in the visual evoked potential which was the conditioning stimulus for this motor response. In other words there was a feedback on the input if a movement was associated with that particular sensory input but the averages would not pick this up because it was a matter of only a couple of evoked potentials out of a series which signalled the beginning of the motor responses. I think that, when you're trying to see in your averages, effects of movement in your experiments you may have very significant effects on two or three evoked potentials in the series which physiologically have a great significance but which would be lost in the average.

Beck

I am familiar with this study, so may I comment briefly on how averaging could be applied in this situation. This outstanding work of Jasper, Ricci and Doane suggested an idea to us. We conditioned cats to respond with a foot flexion to a train of five light pulses separated by one second each. Each of the five flashes could be independently summed, that is, all the first flashes throughout one or more sessions, all the second flashes, etc. Thus, if a certain flash in the series, say the flash occurring just before the response, changed differentially in its relationship to others this would be accentuated through the averaging process. At this point, Dr. Sakai who is doing the study, has noted that the response to the first flash is potentiated. As yet data analysis has not been done on the other flashes, but it is our impression that a gradual attenuation of each successive flash up to the time of flexion occurs. I believe this is essentially what Jasper, Ricci and Doane reported. So, in averaging, data may be salvaged if you are willing to go to a lot of trouble.

REFERENCES

1. Bogacz, J., Vanzulli, A., Handler, P. and García Austt, E. (1960) *Acta Neurol. Latinoam.* **6**:353
2. Buño, W. (Jr.), Velluti, R., Handler, P. and García Austt, E. (1966) *Physiology and Behaviour*, **1**:23
3. Evans, C. C. (1952) *Electroenceph. clin. Neurophysiol.* **4**:111
4. Evans, C. C. (1952) *Electroenceph. clin. Neurophysiol.* **4**:371
5. García Austt, E. (1963) *Electroenceph. clin. Neurophysiol.*, Supp. **24**:76
6. García Austt, E., Bogacz, J. and Vanzulli, A. (1964) *Electroenceph. clin. Neurophysiol.* **17**:136
7. García Austt, E., Bogacz, J. and Vanzulli, A. (1961) (In (Ed. by Delafresnaye) *Brain Mechanisms and Learning*. Oxford: Blackwell, p. 603
8. Gastaut, Y. (1951) *Revue Neurol.* **34**:640
9. John, E. R., Herrington, R. N. and Sutton, S. (1967) *Science, N.Y.* **155**: 1439
10. Palestini, M., Davidovich, A. and Hernández-Peón, R. (1959) *Acta Neurol. Latinoam.* **5**:113
11. Rémond, A., Lesevre, N. and Torres, F. (1965) *Revue Neurol.* **113**:193
12. Schiller, P. H. and Chorover, S. L. (1966) *Science, N.Y.* **153**:1398
13. Sutton, S. Braren, M., Zubin, J. and John, E. R. (1965) *Science, N.Y.* **150**:1187
14. Vanzulli, A., Wilson, E. and García Austt, E. (1964) *Acta Neurol. Latinoam.* **10**:129
15. Walter, W. G., Cooper, R., Alcridge, V. J. and McCallum, W. C. (1964) *Nature, Lond.* **203**:380
16. Walter, W. G. (1965) *Acta oto-Lar.*, Supp. 206:162
17. Wilson, E., Bogacz, J. and García Austt, E. (1966) *Acta Neurol. Latinoam.* **12**:91

Discussion

18. Stark, L., Vossius, G. and Young, L. R. (1962) *Trans. Inst. Radio Engrs.* HFE-3:**52**-57

Saccadic Suppression: A Product of the Saccadic Anticipatory Signal

L. Stark*, J. A. Michael and B. L. Zuber

Bioengineering Laboratory, University of Illinois, and Department of Biomedical Engineering, Presbyterian-St. Luke's Hospital, Chicago

INTRODUCTION

Saccadic suppression (46) is momentary elevation of the visual threshold associated with a saccadic eye movement. It is one example of the integrative activity concerned with information flow in and between the visual and oculomotor systems. A suppression of vision during saccades was first observed by Dodge (11, 12) and Holt (20, 21) early in this century. These observations were not pursued further, although hypotheses considering the possible nature of such phenomena were independently advanced by von Holst and Mittelstaedt (19) and Sperry (35) in 1950. The next experimental demonstration of interactions between oculomotor activity and vision was described by Ditchburn (10) who observed that subjects were unable to see the displacement of an oscilloscope trace when the displacement was triggered by the subject's microsaccades. Lettvin (28) also described a similar experiment; a subject in a totally dark room is unable to read when the stroboscope providing the light for reading is triggered by the subject's eye movements, although a second observer is able to read under these conditions.

More recent studies have attempted to quantify this change in visual perception. Volkmann (41) investigated visual functioning during saccadic eye movements using three standard psychophysical tests, and found that, even under conditions of foveal excitation and with no appreciable retinal blurring, there was a small but significant depression of vision. Latour (25, 26) measured the temporal relationship between suppression and the accompanying saccade, and made the significant observation that changes in threshold occur *before* the eye actually begins to move.

It was at this point that work was begun in the Neurology Section, Electronics Systems Laboratory, Massachusetts Institute of Technology (and continued at the Bioengineering Laboratory, Department of Information Engineering, University of Illinois at Chicago Circle and the Department of Biomedical Engineering, Presbyterian-St. Luke's Hospital, Chicago) to confirm and extend these observations. This work will be discussed in some detail in

*Present address: Department of Physiological Optics, University of California, Berkeley.

the next section, but in brief, an attempt was made to establish the extent to which suppression is associated with *all* saccades regardless of their nature, and to determine whether there are physiological correlates of this phenomenon that can be monitored. The results obtained in our laboratory have, in general, been confirmed by recent independent work by other investigators (3, 5, 16, 24, 27) and this work will be mentioned at the appropriate point in the Discussion.

On the basis of our results, the findings of others on suppression, and the growing literature on the organization and function of the oculomotor system a model has been proposed to explain saccadic suppression. It has been postulated that a centre integrating and controlling saccadic eye movements generates a signal, which we have called the saccadic anticipatory signal, whose function is the initiation of the recomputation of the visual frame of reference so that proper processing of visual inputs can proceed in the face of a moving retinal image. It is thought that as a consequence of this computation an elevation of the visual threshold takes place.

METHODS AND RESULTS

Voluntary Saccades (9, 45, 47)

The initial experiments in our laboratory were carried out to determine the effect of voluntary horizontal saccades on vision. These experiments were similar to those reported by Latour (25). A dark-adapted subject seated in front of a perimeter was instructed to maintain fixation on the one of two

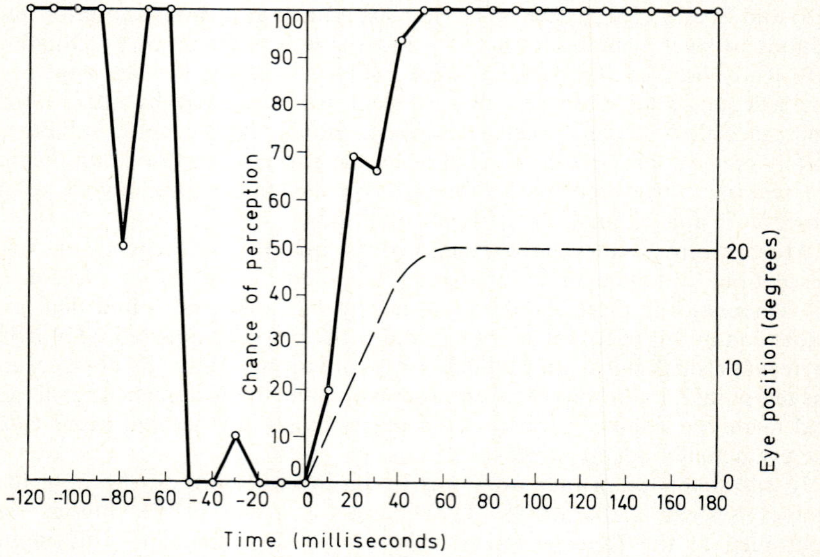

Figure 1. Time course of saccadic suppression for dark-adapted subject making 20 *degree voluntary saccades. Times to the left of* 0 *represent the period before the beginning of the eye movement*

(From Zuber and Stark, 1966, by courtesy of Academic Press)

fixation lights that was lit. Switching these lights on and off then resulted in the subjects producing a pattern of horizontal saccades. Dim flashes from a stroboscope, previously adjusted to be just threshold with the eyes not moving, were presented to the subjects at various times before, during, and after a saccadic eye movement. The subject was able to indicate the perception of the flash by depressing a switch. By examining the record of the subject's eye movements, the time of occurrence of the flash, and the presence of perception, it was possible to obtain data from which could be plotted the probability of perception of the flash as a function of the time of its occurrence (*Figure 1*). Our initial results thus confirmed Latour's finding (25) that suppression occurs before the eye physically begins to move.

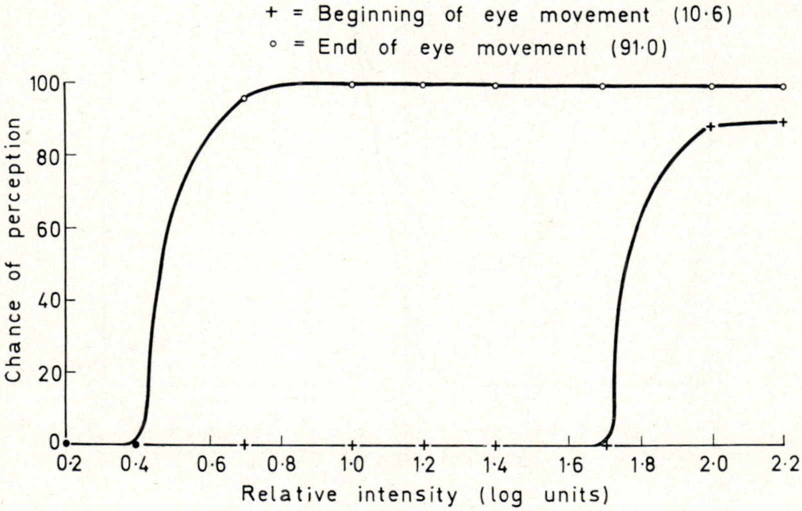

Figure 2. Probability of perception as a function of relative intensity of test flashes presented at two fixed times with respect to the eye movement. At a 50 per cent probability of perception the magnitude of suppression is approximately 1 log unit

(From Zuber and Stark, 1966, by courtesy of Academic Press)

By varying the intensity of the flashes at two fixed times during the eye movement, it was possible to determine the amount of suppression occurring. When the experiment was carried out with flashes presented at two fixed times in relation to the eye movement but with varying flash intensity, at a 50 per cent probability of the flash being seen, the magnitude of suppression was 1 log unit or perhaps a little more (*Figure 2*). The general effect of varying flash intensity on the probability of perception is schematically illustrated in *Figure 3*.

Microsaccades (45, 46)

Ditchburn (10) was the first to report that visual changes occur in conjunction with microsaccades. Using techniques similar to those described above for voluntary saccades, experiments were carried out to determine whether horizontal microsaccades produce a suppression of vision similar to that observed with voluntary horizontal saccades. The subjects maintained fixation on a special target and the appropriate microsaccades were monitored and used to trigger delayed flashes from the stroboscope. By varying the delay between the beginning of the eye movement and flash presentation, it was possible

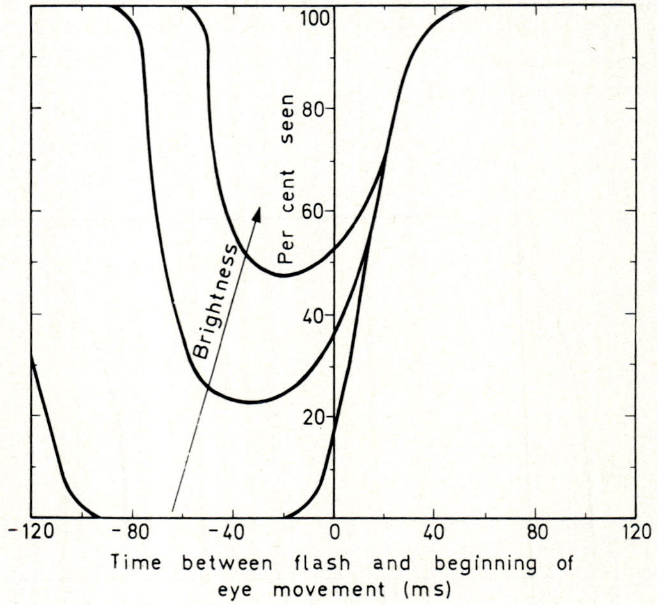

Figure 3. Schematic representation of the effect of flash intensity on the saccadic suppression curve

(From Zuber and Stark, 1966, by courtesy of Academic Press)

to present flashes during and after the just occurring microsaccade or before the next saccade. Although an exact time course was not determined, it was found that during a period from 25 ms before the beginning of the movement to 25 ms after the movement began there was complete suppression of vision. When, however, flashes were presented non-synchronously in relation to the eye movement, essentially every flash was seen.

Vestibular Nystagmus (45, 49)

Because vestibular nystagmus is a pattern of eye movements produced by a non-visual input (although, of course, vestibular nystagmus does make use of some of the same final common pathways and related oculomotor structures as other eye movements), it was considered important to determine whether the involuntary, saccadic fast phase of nystagmus also produced visual suppression. These experiments again utilized techniques similar to those used for examining the effects of voluntary saccades, except that eye movements were produced by spinning the subject on a stool and then bringing him to a stop as quickly as possible; stimulation of the vestibular apparatus then produced a post-rotary nystagmus. Stroboscope flashes could be triggered by the saccades with a variable delay. Suppression was reported by the subjects, and its time course was determined (since only a limited number of test flashes could be presented, the data for relatively long time intervals had to be pooled, resulting in a time resolution more coarse than that obtained for voluntary saccades). Although complete suppression (the probability of flash perception being zero) was not reported, the time course of suppression under these conditions was similar to that seen with voluntary saccades.

The Direct Pupil Reflex (29, 48)

Having established that changes in the visual threshold are associated with all the saccadic eye movements tested, it was next asked whether there is a

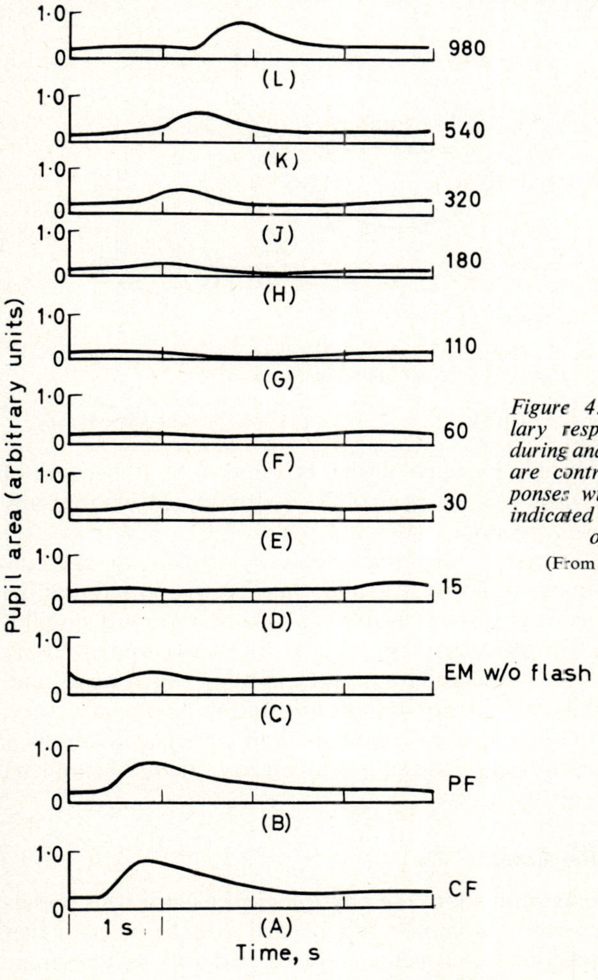

Figure 4. Computer-averaged pupillary responses to flashes presented during and after a saccade. Curves A-C are controls and D-K are the responses with flashes occurring at the indicated times (m ms) after the start of the eye movement

(From Zuber, Stark and Torber, 1966, by courtesy of Academic Press)

corresponding suppression of the direct pupil response to the test flash. A direct-reading infra-red pupillometer was used to monitor the pupillary responses of subjects experiencing saccadic suppression. It was found that the average change in pupil area (determined by on-line averaging of the responses with a digital computer) produced by the test flashes was a function of the time of occurrence of the flash during the eye movement (*Figure 4*). Pupillary suppression has a time course similar to, but not identical with, that of saccadic suppression (*Figure 5*). When pupillary suppression is maximal the average

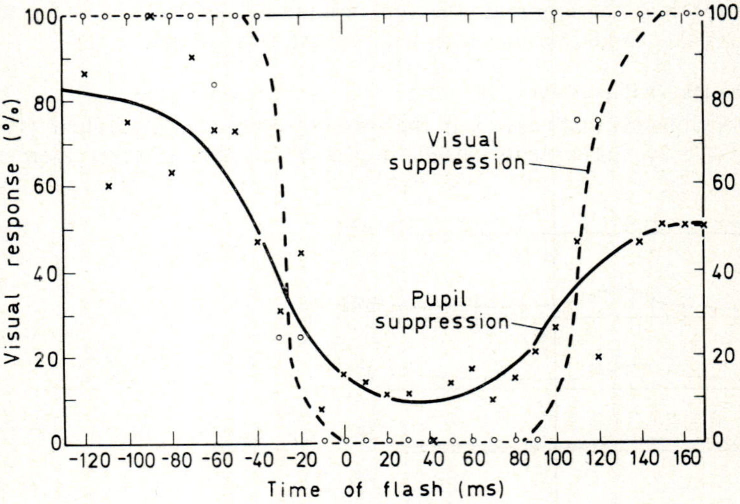

Figure 5. Composite plot showing time course of visual saccadic suppression and pupillary saccadic suppression

(From Zuber, Stark and Lorber, 1966, by courtesy of Academic Press)

change in pupil area produced by control flashes is found to be 10 per cent of the average area change produced by control flashes presented without any accompanying saccadic suppression.

A subjective change in vision phenomenologically similar to saccadic suppression occurs during binocular rivalry, and it has been reported that the pupillary reflex is also reduced during rivalry (2). As the infra-red pupillometer provides an objective measure of pupil area, these experiments were repeated in conjunction with the saccadic suppression pupil experiments, and it was verified that pupillary suppression also occurs during binocular rivalry. The possible analogy between saccadic suppression and binocular rivalry was used to assist in deducing the neural structures involved in these phenomena (see Discussion).

The Visual Evoked Response (31, 32)

In an attempt to learn something about the physiological mechanisms underlying the interaction between oculomotor activity and visual processes that leads to saccadic suppression, experiments were carried out to determine whether suppression is accompanied by any change in the normal visual evoked response (VER) recorded in the electroencephalogram. Subjects were placed in a situation in which saccadic suppression could be produced, and the VER was examined with an average response computer under three conditions: suppression occurring with stimuli presented before the beginning

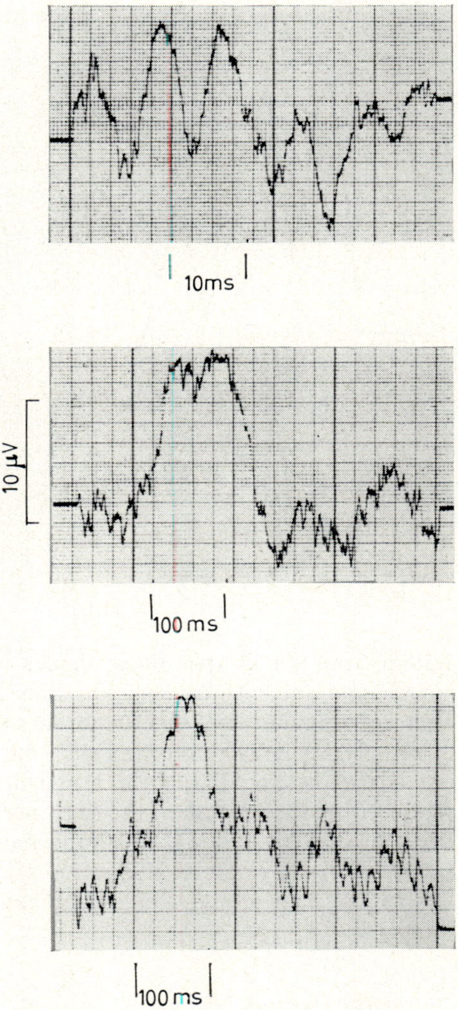

Figure 6. Averaging computer write-outs of visual evoked responses in the EEG from an experiment in which flashes were delivered to a subject 50–60 ms *before the beginning of an eye movement:* (*top*) *VER produced by marginally suprathreshold flashes with the eyes stationary;* (*middle*) *response to flashes occurring in conjunction with a saccade* (*subject reports that suppression occurs*)*; and* (*bottom*) *response to changing fixation lights only* (*no flashes presented*)

(From Michael and Stark, 1967, by courtesy of Academic Press)

of the eye movement, suppression occurring with stimuli presented synchronously with the beginning of the eye movement, and with the intensity of the flashes adjusted so that suppression no longer occurs (*Figure 6*). In all cases where suppression was reported by the subject, obvious concurrent changes in the wave shape of the VER were observed (*Figure 7*). When no suppression occurred, the VER was essentially unchanged within normal limits of variation.

The human observer is generally unavailable for neurophysiological studies beyond superficial clinical physiological approaches. For this reason it was felt that further efforts to uncover the nature of the physiological mechanisms involved with interactions between eye movements and vision would be better carried out on experimental animals. Thus, experiments were carried out on cats to determine the location in the eye movement control system from which arise those signals involved in interactions between eye movements and vision. This was done by producing eye movements with electrical stimulation of different brain sites and comparing the VER produced by test flashes delivered during such movements with the VER produced by test flashes delivered with the eyes still. It was found that eye movements produced by stimulation of the third nerve nucleus produce no change in the VER (*Figure 8*), although eye movements produced by stimulation of the eighth nerve do result in alterations of the wave shape of the VER that resemble the changes seen in the human VER during saccadic suppression (*Figure 9*).

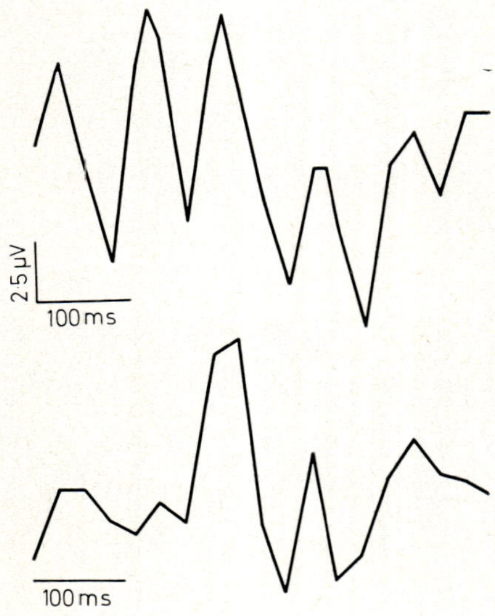

Figure 7. The flash-only response (top) reconstructed by plotting every fourth point of the original response (Figure 6, top) and the corrected-flash response (bottom); the eye-movement-only response (Figure 6, bottom) has been subtracted from the flash-eye-movement response (middle). Note the change in the wave shape of the corrected-flash response (saccadic suppression response)

(From Michael and Stark, 1967, by courtesy of Academic Press)

DISCUSSION

The results of our experiments studying saccadic suppression represent a beginning of attempts to uncover the physiological mechanisms involved in this phenomenon. In summary, we have found that an elevation of the visual threshold occurs slightly before and during such fast saccadic eye movements as voluntary horizontal saccades, horizontal microsaccades, and the fast phase of vestibular nystagmus. It is of particular significance that suppression

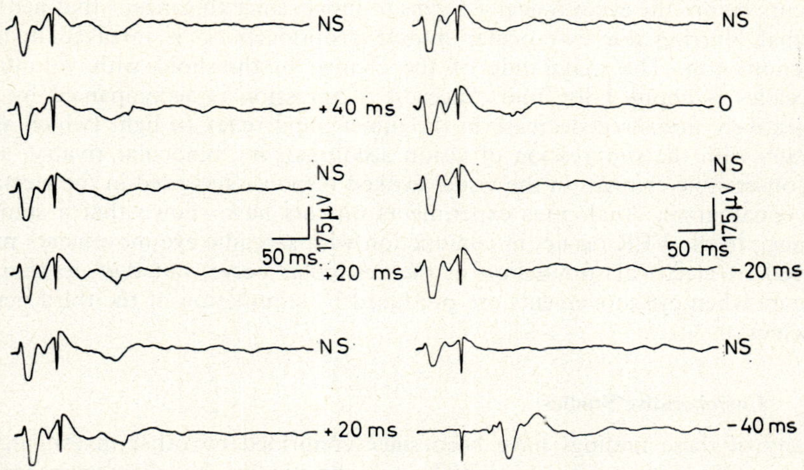

Figure 8. The cortical evoked response elicited by flashes delivered at various times before (+ times) and during (− times) electrical stimulation of the third nerve nucleus; saccadic eye movements resulted from the electrical stimulation. Control responses (NS = no stimulation and hence no eye movements) were in no way different from responses obtained during eye movements

(From Michael and Stark, 1966, by courtesy of the Editor, *E.E.G. Journal*)

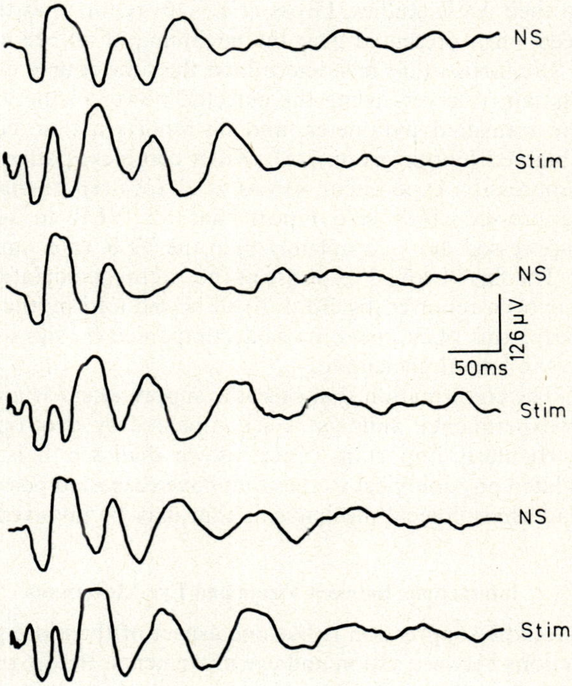

Figure 9. VERs obtained with (Stim) and without (NS) accompanying electrical stimulation of the vestibular nerve; pronounced and consistent changes in the waveshape of the response are evident. Vestibular nerve stimulation produced a nystagmoid pattern of eye movements rich in fast, saccadic movements

(From Michael and Stark, 1966, by courtesy of the Editor, *E.E.G. Journal*)

occurs before the eye physically begins to move, since this means that neither retinal blurring nor extraocular muscle proprioception is involved in this phenomenon. The magnitude of the change in threshold with voluntary saccades is about 1 log unit. Saccadic suppression is accompanied by an objectively measured decrease in the direct pupil reflex to light (which also occurs with the suppression of vision accompanying binocular rivalry) and by observable changes in the visual evoked response recorded in the electroencephalogram. Analogous experiments on cats have shown that a similar change in the VER occurs in conjunction with saccadic eye movements produced by electrical stimulation of the vestibular nerve, but that no change occurs when eye movements are produced by stimulation of the third nerve nucleus.

Corroborative Studies

Many of these findings have been since confirmed by other investigators. Latour (27) has extended his initial work and reports that the occurrence of saccadic suppression is independent of the location of the test flash on the retina, the colour of the flash, and the intensity of the background, but is dependent on the size of the eye movement (or perhaps the speed of the movement, since this is a function of saccade size). Beeler (3) utilized a contact lens technique for recording eye movements and stimulus presentation and found a time course of suppression for microsaccades that extends the one earlier reported by our group (45). A suppression of detection of small target movements associated with microsaccades was also observed. Incidental to their VER studies, Gross *et al.* (16) report that their test stimuli were not seen when presented near the beginning of an eye movement.

Lombrossi (24) has determined the time course of saccadic suppression in human observers using the detected power in the visual evoked response as the measured parameter, and he reports a time course similar to the one obtained by our group (45). When complex, patterned stimuli are used in a suppression experiment, Gross *et al.* (16) report that the VER is completely suppressed; they also report that the VER to such patterned stimuli is suppressed more completely than the VER to a simple light flash.

The flash-evoked response at the lateral geniculate and the cortex of the cat has been reported by Bizzi (5) to be reduced by the rapid eye movements of sleep; this phenomenon was accompanied by signs of presynaptic inhibition at the lateral geniculate.

The confirmation that saccadic suppression (or an analogous phenomenon in experimental animals) is accompanied by electrophysiological correlates is particularly important, since, as we shall see, it is the results of these and related physiological studies that have suggested possible anatomical locations and physiological mechanisms that may be involved here.

Interactions Between Vision and Eye Movements

Saccadic suppression is but one aspect of the more general problem of interactions between vision and eye movements. Before proceeding to a discussion of the anatomy and physiology of suppression it would be well to place this phenomenon in its proper perspective by briefly considering this larger issue.

Visual data processing has two general functions of supreme importance to the organism. First is pattern recognition, the ability to respond appropriately to objects in the environment on the basis of their perceived shape. Second, this must be done in the context of a frame of reference such that the organism knows where objects are located with relation to its body. The proper carrying out of these functions is made complicated by the fact that the organism is mobile—i.e. moves through its environment—and possesses visual receptive organs, the eyes, that are independently mobile. This is significant because it means that internally generated movements of any kind can produce a changing pattern of retinal excitation that must be distinguished from changing retinal excitation produced by changes in the environment. To be able to make this distinction it is necessary that information be available to the visual system about the occurrence of internally generated movements. This is the basis of interactions between oculomotor processes and vision.

Proper visual perception then seems to call for the availability of a signal able to indicate the occurrence, or the anticipated occurrence, of an internally generated movement. It is this signal, which when generated in conjunction with a saccadic eye movement, we have called the saccadic anticipatory signal. Its function is to initiate the recomputation of the visual frame of reference to be put into effect upon the occurrence of the saccadic eye movement. In this way, the movement of the image of an object on to a different area of the retina will not be falsely interpreted as the sudden appearance of a new object in the visual field or as the movement of the object. This might then account for the phenomenon of shape constancy in the face of scanning eye movements. Thus, the anticipatory signal permits a moving retinal image to be interpreted correctly under a variety of conditions. This function would seem to call for the anticipatory signal to bring about the altering of the set of spatial coordinates assigned to a particular pattern of retinal activity.

Another general concept that may be relevant in discussions of saccadic suppression is that of the intermittency of the saccadic eye movement control system. Intermittency refers to the fact that saccades cannot occur more often than every 200 ms. This was demonstrated to be the case for voluntary saccades generated in visual tracking tasks by Stark and colleagues (37, 43, 44); a retinal error signal well above threshold for corrective-saccade production cannot be acted upon for 200 ms following the preceding saccade. A similar intermittency seems to operate during vestibular and optokinetic nystagmus (36), and the same mechanism probably produces intermittency for both voluntary and involuntary saccades since it has been shown that following an optokinetically induced saccade a voluntary saccade cannot be made for a 200 ms period (Stark, unpublished observation). Since convergence eye movements utilize the same final common pathway as saccadic movements and since there seems to be no intermittency in convergence movements (36), it is unlikely that intermittency is a property of the final common pathway. And since intermittency occurs with vestibular induced eye movements as well as with visually directed eye movements, it is unlikely that intermittency is strictly a visual, sensory phenomenon.

The possible relationship between saccadic suppression and intermittency is that both seem to arise from an eye movement controller somewhere between the sensory and final motor ends of the oculomotor system. This will be discussed further at the appropriate place.

The physiological mechanisms underlying saccadic suppression and intermittency and their anatomical locations cannot now be specified with any certainty. However, as we will see, it is possible to eliminate certain central nervous system locations from consideration.

Localization of the Site of Suppression

The information currently available about saccadic suppression makes it possible and profitable to ask where in the central nervous system the relevant neural centres are located and the nature of the physiological mechanisms at work. In asking the first question, we are actually asking about the location of two centres, first, the site at which interactions take place, and second, the locus from which arises the controller signals producing the interaction.

In considering the location in the visual system at which interactions take place, the retina might seem to be an obvious choice—particularly since suppression of the pupillary reflex occurs along with perceptual changes in both saccadic suppression and binocular rivalry (29, 48). In fact, it has been suggested that one function a centrifugal pathway to the retina might have is to mediate such suppression (18). However, there are many reasons for believing that suppression does not occur at the retina. First, during binocular rivalry the visual evoked response recorded from the scalp is reduced in amplitude but with no concomitant reduction in the magnitude of the electroretinogram (1). Bizzi (5) has observed that flash-evoked responses recorded at the lateral geniculate and the cortex are reduced in amplitude by eye movements, but no changes are seen in the responses of the optic tracts. These findings thus suggest that information is being transmitted from the retina with no modification and that interactions are occurring at some higher level. If saccadic suppression is in some way linked to the computation of the frame of reference, as we will discuss below, then the amount of data processing needed to transform the spatial co-ordinates of retinal activity would call for a major centrifugal pathway to the retina. In fact, although such a pathway undoubtedly exists in some animals (18), there is no good evidence that it is present in man or other mammals (8). For these reasons it seems unlikely that the retina is the site of interactions between vision and eye movements.

The lateral geniculate nucleus has been implicated as a possible site for oculomotor-visual interactions by neurophysiological studies carried out by a number of workers. Bizzi (5) has demonstrated that during the REM-phase of sleep in the cat there are signs of pre-synaptic inhibition at the LGB. This is accompanied by the reduction of the flash evoked response recorded at the geniculate and the cortex. Pre-synaptic inhibition has also been seen at the geniculate level during eye movements (22), although this was observed to occur with spontaneous smooth tracking movements in pre-trigeminal sectioned cats. Feldman and Cohen (15) report that evoked potentials associated with rapid, saccadic-like eye movements can be recorded from the geniculate in the monkey. Such saccade-associated responses at the LGB may be mediated by a pathway from oculomotor centres in the pons and mesencephalon to the geniculate (7). Thus, the lateral geniculate nucleus would appear to be one possible site at which saccadic suppression could take place.

One possible argument *against* the LGB being the site of suppression is the fact that only half of the visual field is represented at each geniculate. This is relevant since during binocular rivalry, for instance, vision in *one* eye is suppressed while vision in the other eye remains intact. A simple demonstration can show that the recomputation of the frame of reference that occurs with an eye movement involves vision in *one* eye, not in hemi-fields; if one eye is restrained from moving, and a voluntary eye movement is then attempted, the field in the restrained eye will appear to jump, although the other monocular field will not. This can be explained by the fact that the representation of the field of the non-moving eye was computed on the basis of the eye movement that was anticipated but that never occurred because the eye was physically prevented from moving. If these mechanisms act at a geniculate level they would have to selectively influence those fibres or cells corresponding to the two individual hemi-fields. It can be argued (32, 48) that this selectivity is unlikely, and that one should look for a more central site at which monocular data processing is possible (this is not to say that we expect to find a discrete, localizable representation of each monocular field somewhere in the cortex, but that at a cortical level the mechanism needed for such data processing would have more access to the neural elements involved).

Knowledge about the visual data processing mechanisms in the primary sensory receiving area (the striate cortex, area 17) is growing rapidly. The possible involvement of this area in suppression phenomena is suggested by the reports of Evarts (14) and Valleala (40) that units in the primary visual cortex of the unanaesthetized cat (the lateral gyrus) fire in a definite temporal relationship to the REM's of sleep. However, in spite of the accumulating information on the role of the striate cortex in visual processes little, if any, information is available about the physiological basis for visual *perceptual* mechanisms, and we cannot even point to likely areas where investigation might be particularly fruitful. Perhaps single unit studies in association areas where visual and other sensory inputs appear to be manipulated would be of value.

Localization of the Origin of the Saccadic Anticipatory Signal

The possible site of origin of the saccadic signal giving rise to interactions with vision will now be considered. It would seem that there are only two mechanisms that could provide the kind of signal needed. One possibility is a system in which signals from the motor controller are sent to sensory and/or perceptual centres to make possible the correct processing of incoming data in the face of the disturbing motor act. This idea was first seriously proposed by Helmholtz in the nineteenth century (17), and the corollary discharge theory of Sperry (35) and the reafference theory of von Holst and Mittelstaedt (19) are obviously modern restatements of this idea. Teuber (38, 39) has commented on the possible role of a corollary discharge in perceptual phenomena.

The other possible mechanism for signalling an eye movement (or any other motor act for that matter) would involve feedback from the proprioceptive system in the extraocular muscles. This was first proposed by Sherrington (34), who did much to establish the function and importance of such proprioception in the control of skeletal muscles.

However, the most important characteristic of saccadic suppression is the fact that its time course begins well before the eye physically begins to move (3, 27, 45), and thus at a time when there can be no change in the length of the extraocular muscles. This would seem to eliminate proprioception as a possible mechanism. This conclusion is strengthened by the persistent inability of investigators to conclusively demonstrate any oculomotor function for the proprioceptive signals arising in the extraocular muscles (42). Then, if proprioception is not the mechanism involved, we must look for a saccadic signal arising from the oculomotor system.

The oculomotor system has components located at many different levels of the central nervous system, and the organization of these components is complex and not completely understood. Nevertheless, it is possible to distinguish components operating at different functional levels. At the lowest level is the final common motor pathway, which for eye movements of all kinds, must be neurones in the cranial nerve nuclei, since the motor nerves to the extraocular muscles arise here. These nuclei are found in the brain stem. At this same anatomical level are found the vestibular nuclei which are well known to have a powerful influence on the generation of eye movements, and certain oculomotor control centres whose exact anatomical location are still unclear, but whose function seems to be the integration and control of all eye movements. Finally, at a cortical level are found two areas which have classically been thought to be involved with eye movements; the frontal motor field is thought to be involved in some way with the voluntary initiation of eye movements, and the occipital area is felt to be involved with visually induced reflex eye movements (see reference 33 for a discussion of the entire question of cortical 'centres' for eye movements). The reasons for believing that the controller is at a brain stem site will be presented in a moment.

There are at least two reasons for thinking that it is unlikely that the saccadic anticipatory signal arises from the cranial nerve nuclei. First, the experiments of Michael and Stark (31) demonstrated that saccades produced by stimulation of the third nerve nucleus do not lead to alterations of the visual evoked response, although similar eye movements produced by vestibular stimulation do show signs of interacting with visual processes. (It should be borne in mind that electrical stimulation of a complex structure such as the third nerve nucleus is not very physiological, and if the pattern of stimulation is important for activating the saccadic anticipatory signal mechanism, it is perhaps not surprising that no sign of this mechanism was seen in these experiments. These findings are thus only suggestive.) Then too, the motoneurones of the cranial nerve nuclei are the final common pathway for *all* eye movements, not just saccades. However, there is no evidence that suppression occurs with any other kind of oculomotor activity, and the kinds of interactions between vision and other eye movements would certainly have different temporal characteristics. For these reasons, it is possible to de-emphasize the cranial nerve nuclei as the site of origin of the saccadic anticipatory signal.

Possible Models of Visual-Oculomotor Interaction

However, elimination of the cranial nerve nuclei from consideration still leaves a vast area of the central nervous system for exploration. As an aid in

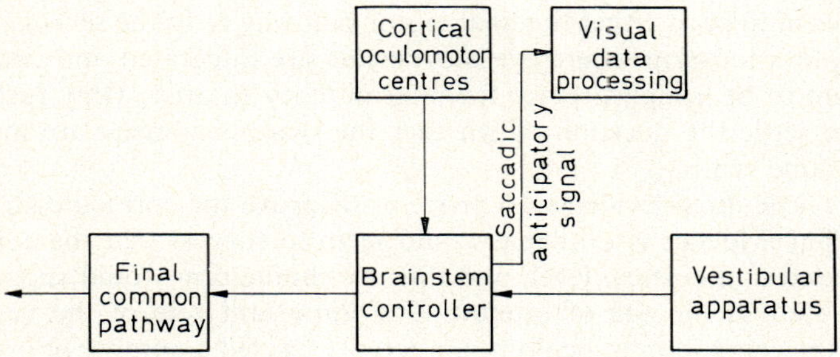

Controller for saccadic eye movements and anticipatory signal at a brainstem level

(a)

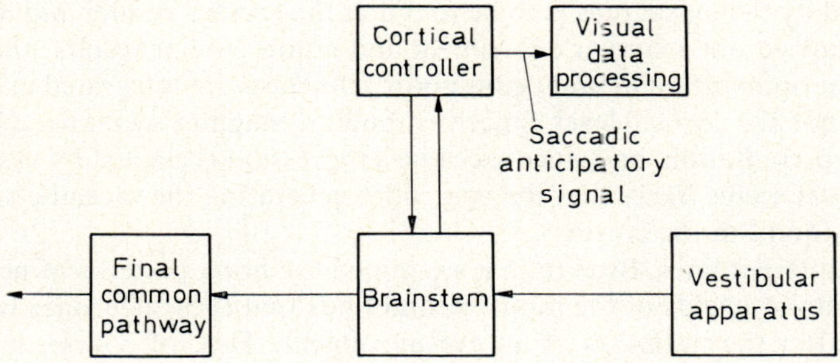

Controller at a cortical level

(b)

Figure 10. Diagrams of two possible models to explain the origin of saccadic suppression (see text)

this search we would like to propose two models around which the known facts can be organized, and which can hopefully lead to further fruitful investigation. The first of these models (*Figure 10a*) proposes that cortical and vestibular influences impinge on a controller in the brain stem which serves to generate integrated eye movements and an anticipatory signal which 'warns' the visual system of the impending saccadic eye movement. The origin of this saccadic signal is thus to be found at a brain stem level according to this model. The second model (*Figure 10b*) proposes that cortical and vestibular influences are finally integrated at a cortical level from which signals return to the brain stem and hence to the final common pathway. It is then this cortical controller that gives rise to the saccadic anticipatory signal.

The postulated controller that gives rise to the saccadic anticipatory signal may also be the centre from which arises the intermittency phenomenon previously discussed. It has been mentioned that the intermittency operator does not seem to be at either the final motor pathway or in the sensory system as such, and a location where eye movements are integrated and controlled would seem to be a logical place for intermittency to arise. Only further research can settle the question of whether the two phenomena are mediated from the same centre.

There is little direct evidence to prove or disprove the correctness of either model. What evidence is available would seem to suggest that the controller is not located at a cortical level, and thus by elimination would suggest that the brain stem is the site of interest. The important role of the vestibular system in the generation of oculomotor activity has led a number of investigators to examine the phenomenon of vestibular nystagmus and to ask where optovestibular and oculomotor influences are integrated. Kornhuber and daFonseca (23) have found that although there is rhythmic, nystagmoid activity recordable from single units in the striate cortex during vestibular nystagmus, this activity disappears in the dark; thus, there is no cortical representation of vestibular nystagmus as such, but only the visual activity modulated by the nystagmus is to be found at the cortex. Bender and Shanzer (4) have carried out a similar experiment and report similar results; they concluded that optovestibular and oculomotor influences are integrated at a brain stem site, not at a cortical level. Since vestibular nystagmus seems to lack a true cortical representation, and since saccadic suppression is elicited by vestibular nystagmus it seems likely that the controller generating the saccadic signal is not to be found in the cortex.

Along different lines, Bizzi (6) has examined the firing patterns of neurones in the frontal eyefields of the monkey, and finds that these neurones begin to fire only after the initiation of an eye movement. This, of course, makes it unlikely that these cells are initiating the eye movement, but it does not mean that neurones in other areas of the cortex might not be initiating and controlling eye movements. However, at least one area long thought to be concerned with control of eye movements has been shown to be *not* involved in the initiation of oculomotor activity.

The results just described would suggest that the cortex is not the locus from which arises the saccadic anticipatory signal. However, the experiments of Dumont and Dell (13) demonstrating an inhibitory influence on vestibular nystagmus arising in the cortex would suggest that the cortex might in fact play a role in the control of eye movements.

In summary, the little evidence that is available would seem to point to the cortex *not* being the site of the saccadic eye movement controller, but rather, that the brain stem is a more probable locus for this centre.

Function and Origin of Saccadic Suppression

We have proposed a model in which a brain stem controller integrates and generates saccadic eye movements. In addition, this controller provides a signal for the visual system whose function is the initiation of a recomputation of the visual frame of reference so that the saccadic eye movement about to occur will not disrupt ongoing processing of visual inputs. We would further postulate that saccadic suppression is a non-functional consequence of this neural computation; the change in visual threshold that is observed in fact has no useful function. This would then solve the problem of explaining the function of a change in threshold which all investigators agree amounts to no more than 1 log unit. Certainly one cannot claim that under real-life conditions such a threshold change can serve to eliminate the retinal blur produced by a saccade; considering the normal range of light intensities over which the visual system can function, 1 log unit represents a literally inconsequential decrease in threshold under normal conditions.

Assuming that suppression has no function as such, we still must explain the mechanism that causes it. Perhaps, in the most general terms, during recomputation of the frame of reference preceding a saccade the visual data processing centres are so 'busy' that a change in the threshold for detection of incoming signals occurs. This might be because the frame of reference computation acts as added noise in the signal detection system, thus decreasing the signal/noise ratio and making detection less likely. Or, it may be that the involvement of many neural elements in the frame of reference computation produces a reduction in the signal processing gain, again leading to a reduced probability of detection.

If, as we have postulated, saccadic suppression is without a true physiological function but is simply the consequence of other ongoing activity, this would not mean that suppression is an uninteresting or non-valuable phenomenon for further study. Rather, it would seem to be useful as a behavioural measure of important interactions occurring between oculomotor and visual processes.

CONCLUSIONS

Saccadic suppression is a momentary elevation of the visual threshold that occurs in conjunction with all saccadic eye movements. It is accompanied by an objectively measured decrease in the amplitude of the pupillary reflex to light and by alterations or reductions in amplitude of the visually evoked response. It has been postulated that suppression is a functionless consequence of the recomputation of the visual frame of reference initiated by a saccadic anticipatory signal sent to the visual system by a brain stem controller for eye movements.

ACKNOWLEDGEMENT

We wish to acknowledge partial support from the W. Clement and Jesse V. Stone Foundation, from National Institutes of Health grants (NB-06197, NB-06487 and GM-01436) and the Office of Naval Research (N-0014).

DISCUSSION ON STARK

Lombroso

We have some electrophysiological data relevant to the beautiful demonstration by Dr. Stark and his colleagues. They support the concept of a central mechanism for visual suppression during voluntary saccades. We were concerned in our study to eliminate as far as possible retinal blurring and displacement—two questions left unanswered in previous studies. Also, we were interested to observe what took place just before, as well as during and after, a voluntary saccade. Our subjects sat in front of an oscilloscope with a short persistence screen and the horizontal movement of a spot was controlled by a d.c. voltage produced by the eye movement itself. Thus, it was possible to

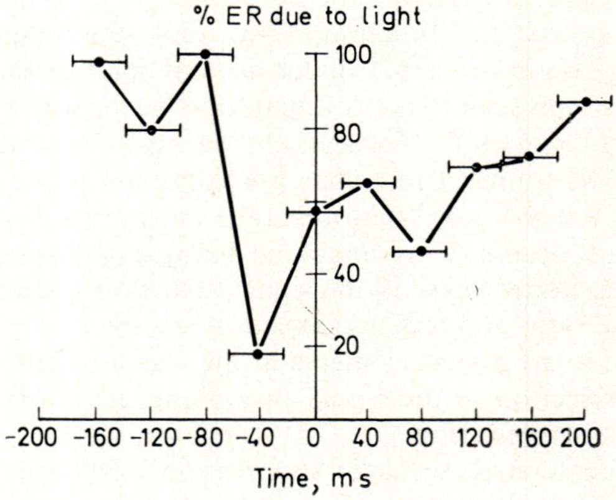

Figure 11.

have the spot persist at the point of central fixation during small horizontal eye movements. Each voluntary eye movement subtended 20 degrees. After calibrating the system, the spot intensity was reduced, its reappearance as a flash 5 ms in duration, about 2 log units above threshold, occurring at random. The EEG was obtained from an orthogonal pair of electrodes on the left occiput. Subjects were asked to move their eyes from the left to the extreme right of the screen. The EEG pulses, which corresponded to the onset of eye movements, as well as pulses corresponding to the onset of flashes, were recorded on FM tape and later converted into digital form. A programme was developed to measure the relationship of each flash to the nearest eye movement and to construct a histogram of the number of flashes recorded within each interval of 40 ms. It was possible to examine periods of time just prior to the onset of eye movement without having to alert the subject. We were assured that flashes were distributed in a random fashion. Then, we computed the average evoked response for each interval of the histogram. The magnitude of the visual evoked response of each interval which was attributable to the light flash was then approximated by the arithmetic difference between the average power of the ER for an interval during which a flash had occurred, and a comparable interval during which a flash did not occur. In other words, we were thus substracting components due to noise, eye movements *per se* and so on. Power was approximated by the function: $I_p = Tv(t)^2 dt$. In *Figure 11* the average power due to the ER by light was plotted against time of the saccade. It was expressed as a percentage of the average power due to light at an interval sufficiently distant from the onset of the saccade (over 500 ms) not to be affected by the eye movement. In this diagram, you can see a profound suppression of the ER at between 60–40 ms prior to the onset of eye movement, and it takes something like between 150–200 ms to recuperate. There is also this interesting dip (at about 150 ms) which I observed in your first slide also Dr. Stark, and I would like to have your comment since we are confirming what you saw in your perception study.

Stark

Latour has also shown data indicating great variability of suppression. I've discussed it with him and he doesn't have a clue and I don't have a clue as to what it's all about.

Lombroso

Finally, I wanted to ask you, in terms of concepts, whether this profound suppression of ER to brief flashes may be due to the mechanism that prevents visual blurring during voluntary saccades and whether it may be a mechanism which briefly decouples the visual input from the eye tracking servo loop permitting smooth voluntary saccades. We are interested in seeing how this mechanism could work in some pathological conditions, where smooth saccades appear to be defective. I would like to ask you Dr. Stark, how you think this mechanism operates and whether you consider it to be central or peripheral?

Stark

As I said before an awful lot of computation has to go on to compute the frame-of-reference for an eye movement that has been commanded; and this computation is presumably going on a hundred to zero milliseconds before the eye movement. Perhaps the poor brain is too busy to keep attending to further information as well as doing this frame-of-reference computation.

Jung

Well this brings us to the mechanism of the suppression in the brain. I think it's very significant that Dr. Stark has found the beginning of suppression occurring well before the eye movement. Some information about the intended ocular movement must reach the visual system and this must cause inhibition in preparation of each ocular saccade.

Stark

It must be going on in the visual cortex or the frontal cortex.

Grey Walter

I was fascinated by this suggestion which Dr. Stark made about the importance of computation preceding a voluntary eye movement. There are two experiments one can do on oneself to show how important this is. One is to look at oneself in a mirror; look first at one eye and then at the other. Of course your eyes have just moved quite a long way, but you never see them actually moving. In other words you can prove to yourself that during a period of voluntary change of gaze there is complete suppression of perception but there is no gap in your perception! This is interesting for not only does somewhere in the cortex compute and predict what's going to happen when you move your eyes, but it also maintains a memory of what's happened so that there's no perceptual gap. You don't black out in this period. The other experiment of course is to move your eye by physical means. If you close one eye and move the other with your finger the whole world does move in a very disagreeable way. If you keep the other eye open, you can see the movement, but it doesn't worry you, you keep fixating with the steady eye even if you push your dominant eye. The unmoved eye holds the picture while the other image moves. So there is an extraordinarily elaborate computation. One wonders going back to the story of eye movements and alpha rhythms, whether this isn't such a large task for the brain computer that it really can't do anything else. It's taking up a lot of the core in doing this job.

Stark

Well we originally got into this because of what is called the *sample-data* theory for eye movements. Engineers, especially when they're using digital computers as part of a control system, have been employing sample-data control systems. These are systems which have *intermittent* action. Often they will take in some burst of sensory information, then shut off this intake, do computations, make control movements, and then return to take in more sensory information. There are two very interesting neurological control systems, one for positional control of the eye and the other for the movement of the hand—both intermittent systems, and in these the intermittency is not in the sensory end. The intermittency is in the central control portion for the eye movements; for the hand movements the intermittency seems to be in the spinal cord where the postural control system and the voluntary control system are alternating control of the final common pathway. Both of these systems can be modelled in a very powerful way by the engineering mathematics of sample data control systems (51, 52).

Byford

I understood Dr. Stark to say the amplitude of the saccadic wave at the time was at its maximum. I think that several people have shown that the rise time of involuntary saccades is a constant regardless of the amplitude of the movement. If this is so then a period of 100 ms before the saccade, would include a substantial part of the whole of the slow phase of nystagmus. The corneo-retinal potential would form a very unfortunate choice of mechanism for stabilizing a retinal image. This potential is so non-linear and contains so many unwanted signals that it is most unlikely that it could be used successfully to drive a target spot on an oscilloscope; the spot would not in fact appear in the same retinal location with the precision required for a stabilized image.

Stark

The rise time of the saccade *is* a function of its amplitude. We used special infra-red photocell eye movement monitors (50) so as to remove the problems Dr. Byford correctly points out are associated with the EOG.

Kennedy

I understand that Dr. Bender has a few points relevant to this topic that he would like to make. This might be a suitable way to round off the discussion.

Kennedy

I understand that Dr. Bender plans to take up some points relevant to this topic in his own contribution. This would seem tc be a suitable moment to invite him to present his paper.

REFERENCES

1. Balen, A. Th. M. van (1964) *Documenta ophth.* **18**:440 (1964)
2. Barany, E. H. and Hallden, U. (1948) *J. Neurophysiol.* **11**:25
3. Beeler, G. W., Jr. (1967) *Vision Res.* **7**:769
4. Bender, M. B. and Shanzer, S. (1964) In *The Oculomotor System* (Ed. by Bender). New York: Harper & Row
5. Bizzi, E. (1966) *J. Neurophysiol.* **29**:861
6. Bizzi, E. (1967) *Science, N.Y.* **157**:1588
7. Bizzi, E. and Brooks, D. (1963) *Archs. ital. Biol.* **101**:666
8. Brindley, G. S. (1960) *Physiology of the Retina and the Visual Pathway.* London: Edward Arnold (Publishers) Ltd.
9. Davis, J. (1962) Unpublished experiments. Neurology Section, MIT
10. Ditchburn, R. W. (1955) *Optica Acta* **1**:171
11. Dodge, R. (1900) *Psychol. Rev.* **7**:454
12. Dodge, R. (1905) *Psychol. Bull.* **2**:193
13. Dumont-Tyc, S. and Dell, P. (1962) *J. Physiol. Paris* **54**:331
14. Evarts, E. V. (1962) *J. Neurophysiol.* **25**:812–816
15. Feldman, M. and Cohen, B. (1967) *Proc. a. Conf. Engr. Med. Biol.* Vol. 9, p. 2.3
16. Gross, E. G., Vaughan, H. G., Jr. and Valenstein, E. (1967) *Electroenceph. clin. Neurophysiol.* **22**:204
17. Helmholtz, H. von. (1963) *Handbuch der Physiologischen Optiks.* Leipzig: Voss (1867). English trans. from edn. 3, 1925. J. C. P. Southall, *A Treatise on Physiological Optics.* New York: Dover
18. Holden, A. L. (1966) *Nature, Lond.* **212**:837
19. Holst, E. V. and Mittelstaedt, H. (1950) *Naturwissenschaften* **37**:464
20. Holt, E. B. (1903) *Harv. psychol. Stud.* **1**:3
21. Holt, E. B. (1906) *Psychol. Bull.* **3**:68
22. Kawamura, H. and Marchiafava, P. L. (1966) *Brain Res.* **2**:213

23. Kornhuber, H. H. and daFonseca, J. S. (1964). In *The Oculomotor System* (Ed. by Bender). New York: Harper & Row
24. Lambrossi, C. (1968) Personal communications
25. Latour, P. L. (1962) *Vision Res.* **2**:261
26. Latour, P. L. (1963) In *Nerve, Brain and Memory Models.* (Eds. Wiener and Schade.) Amsterdam: Elsevier, pp. 30–36
27. Latour, P. L. (1966) *Cortical Control of Eye Movements.* Institute for Perception RVO-TNO, Sosterberg, The Netherlands
28. Lettvin, J. Y. (1960) 'Human vision'. *QPR 58, Research Laboratory of Electronics, MIT*, pp. 254–258
29. Lorber, M., Zuber, B. L. and Stark, L. (1965) *Nature, Lond.* **208**:558
30. Michael, J. A. (1967) *Nature, Lond.* **214**:1008
31. Michael, J. A. and Stark, L. (1966) *Electroenceph. clin. Neurophysiol.* **21**:478
32. Michael, J. A. and Stark, L. (1967) *Expl. Neurol.* **17**:233
33. Pasik, P. and Pasik, T. (1964). In *The Oculomotor System* (Ed. by Bender). New York: Harper & Row
34. Sherrington, C. S. (1918) *Brain* **41**:332
35. Sperry, R. W. (1950) *J. comp. physiol. Psychol.* **43**:482
36. Stark, L. (1963) 'Physiology of Visual Control System: The Intermittency Operator in the Control System for Eye Movements'. *NASA Progress Report No. 2* on NAS2-1328
37. Stark, L., Vossius, G. and Young, L. R. (1962) *IRE Trans. hum. Factors Electron. HFE-3:* 52
38. Teuber, H. L. (1960) Perception. In *Handbook of Physiology, Sect. 1: Neurophysiology Vol. III.* Washington, D.C.: Am. Physiol. Soc.
39. Teuber, H. L. (1961) *J. Nerv. Ment. Dis.* **132**:32
40. Valleala, P. (1967) *Archs. ital. Biol.* **105**:1
41. Volkmann, F. C. (1962) *J. opt. Soc. Am.* **52**:571
42. Whitteridge, D. (1960) Central control of eye movements. In *Handbook of Physiology, Sect. 1: Neurophysiology Vol. II.* Washington, D.C.: Am. Physiol. Soc.
43. Young, L. R. and Stark, L. (1963) *IEEE Trans. hum. Factors in Electron. HFE*-**4**:38
44. Young, L. R. and Stark, L. (1963) *IEEE Trans. Military Electron. MIL*-**7**:113
45. Zuber, B. L. and Stark, L. (1966) *Expl. Neurol.* **16**:65–79
46. Zuber, B. L., Crider, A. and Stark, L. (1964) *QPR 74, Research Laboratory of Electronics, MIT*, p. 244
47. Zuber, B. L., Michael, J. A. and Stark, L. (1963) *QPR 73, Research Laboratory of Electronics, MIT*, p. 217
48. Zuber, B. L., Stark, L. and Lorber, M. (1966) *Expl. Neurol.* **14**:351
49. Zuber, B. L., Horrocks, A., Lorber, M. and Stark, L. (1963) *QPR 73, Research Laboratory of Electronics, MIT*, p. 221
50. Stark, L., Vossius, G. and Young, L. R. (1962) *Trans. Inst. Radio Engrs.* HFE-**3**:52–57

Discussion

51. Stark, L. (1968) *Neurological Control Systems*, Section IV: 'The Eye. Studies in Bioengineering'. Plenum Press: New York
52. Stark, L. (1968) *Neurological Control Systems*, Section V: 'The Hand. Studies in Bioengineering'. Plenum Press: New York

The Oculomotor System and the Alpha Rhythm

Morris B. Bender
Department of Neurology, Mount Sinai School of Medicine, New York

One of the questions raised at this conference is whether the oculomotor system is a contributory source to the alpha rhythm. Mulholland and Evans have shown that this rhythm appears to be most prominent when the eyes are partly elevated and the lids partly closed (4, 5). It has also been said that some degree of alertness is an important feature; so-called 'relaxed' wakefulness has been considered essential for maintenance of regular alpha rhythm. The phenomena of alerting and changes in states of consciousness have been found in experiments on the reticular formation of the brain stem. This region, when electrically stimulated, activates various systems including that of the oculomotor. Thus electric stimulation or lesions within this zone invariably cause alterations in eye movements, but it is not generally appreciated that the reticular activating system and the oculomotor pathway are located in the same zone of the brain stem tegmentum.

OCULOMOTOR DECUSSATION AND PATHWAYS IN THE BRAIN STEM

Experiments in our laboratories have revealed that electrical stimulation on one side of the brain stem in the alert monkey, results in a variety of oculomotor responses mostly in the form of conjugate eye movements. The reactive area was situated on both sides of the midline through the tegmentum of the brain stem, from the pretectum caudally to the level of the vestibular nuclei. Analyses of these eye movements revealed that most of them were conjugate and the direction was horizontal, less frequently oblique and occasionally (when the stimulation was in the midline) vertical. Further analysis disclosed that the direction of the horizontal or oblique movement, depended upon the level of the brain stem stimulated; if above the level of the oculomotor-trochlear nuclei, the conjugate deviation was contralateral, and if below, ipsilateral. This pattern of crossing which was found on repeated occasions, suggested that there must be decussation in the reactive area in the tegmentum of the brain stem at the level of these nuclei.

Investigation with restricted electrolytic lesions in the brain stem revealed a similar distribution. The areas in the brain stem in which small lesions yielded impairment of conjugate horizontal gaze, extended from the pretectum downward to the level of the vestibular nuclei and situated in the paramedian zones of the tegmentum of the pons and midbrain. Lesions above the level of the oculomotor and trochlear nuclei caused impairment of contralateral gaze, while below this level there were defects or paralysis of ipsilateral horizontal gaze. On correlating the experimental data it became apparent that the 'oculomotor pathway', defined by electric stimulation or lesion, must cross at the level of the oculomotor-trochlear nuclei (1).

Other experiments in our laboratory have further indicated that the so-called 'oculomotor pathway' extends caudally to the level of the abducens and vestibular nuclei. For instance, a 1 mm lesion in the left paramedian zone at the level of the abducens nucleus, abolished all conjugate horizontal eye movements to the left, such as those ordinarily induced by electric stimulation of either the right frontal, occipital or pretectal regions; the lack of ocular responsiveness to electric stimuli in regions rostral to the brain stem indicates a blocking of oculomotor impulses transmitted caudally to the level of the vestibular nucleus. Translated in other words, optic stimuli such as moving stripes activating oculomotor impulses, resulting in optokinetic nystagmus must be transmitted 'downward' or caudally in the tegmentum below the decussation, and then project upward to reach the oculomotor nuclei.

On the other hand, the same lesion in the left paramedian zone will also block impulses originating at the vestibular level. Unilateral vestibular stimulation activates the oculomotor system, resulting in tonic deviation of the eyes to one side and nystagmus to the opposite side—depending upon the type of stimulation. This effect can be induced by electrical stimuli through electrodes implanted within the horizontal semicircular canal, by thermal stimulation of the ear, or by various types of motion of the head or body. In these situations the impulses originate at the vestibular level and are mediated 'upward' or rostrally to the ocular nuclei.

EYE MOVEMENTS ASSOCIATED WITH LID CLOSURE

There is other experimental evidence showing that impulses originating in the lower brain stem activate vertical eye movements. Thus electric stimulation in the ventral paramedian regions of the tegmentum of the pons on one side, will induce the following: patterned bilateral eye movements, eyelid closure, upward rolling of the eyes and pupillary constriction (6). (*Figure 1.*) This response is another example in which electrically induced oculomotor impulses are transmitted 'upward' or rostrally. However, it should be noted that the impulses for associated movements of eyelid closure, upward rolling of the eyes and pupillary constriction which occur in the normal subject may not be analogous to the experimental situation. It is only an assumption that such impulses may originate or pass through the tegmentum of the brain stem.

VERTICAL EYE MOVEMENTS

Most of the foregoing experiments were concerned with conjugate horizontal eye movements. There were also some with vertical eye movements which occur during eyelid closure. Eye movements in the vertical plane were induced by simultaneous electric stimulation of homotopic points within the right and left cerebrum, unilateral electric stimulation of the semicircular canals (anterior induced upward, while posterior induced downward movements) and by bilateral thermal stimulations of the ears. Paralysis of upward and to a

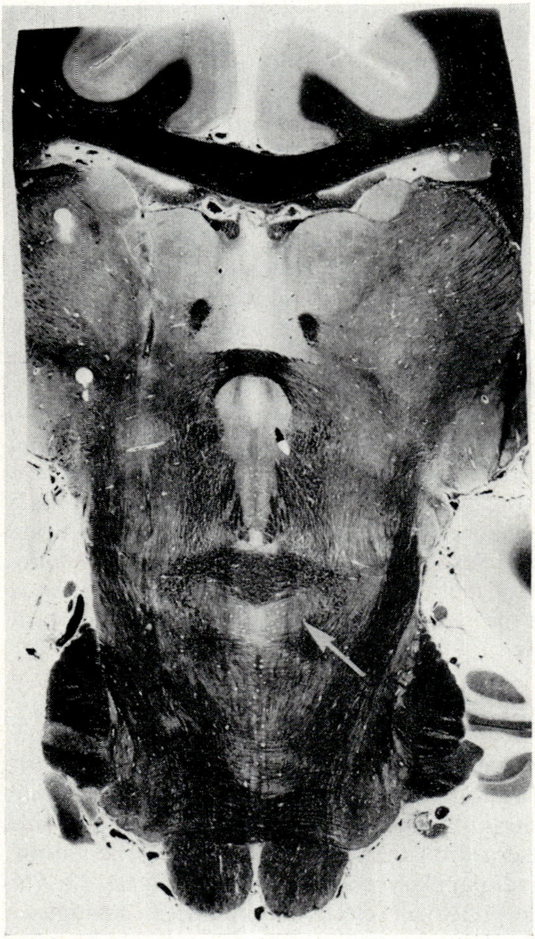

Figure 1. Coronal section of brain stem of a monkey 2 mm *anterior to the zero plane of stereotaxic co-ordinates, extending through the corpus callosum and posterior commissure in the dorsal region and pyramidal tracts of the medulla in the most ventral region of the section. Electric stimulation at tip of arrow* (2·5 mm *laterally to the median and ventro caudally to the brachium conjunctivum in the region of the reticulo olivary tract) induced bilateral closure of the eyelids, upward rolling of the eyes and pupillary constriction. The responses were asymmetrical and always greater on the side of the stimulation*

lesser degree of downward gaze with preservation of horizontal gaze, was obtained with bilateral lesions in the pretectum or with one small lesion within the posterior commissure. Paralysis of gaze in all directions of gaze including convergence was produced by bilateral lesions within the paramedian zone of the tegmentum of the pons involving the median line at the level of the abducens nuclei. (*Figure 2*) These findings imply that all types of eye movements are mediated through the tegmentum of the pons; the median raphe separates the vectors for right and left eye movements while vertical eye movements derive impulses from both sides and the midline. There is no clear indication as to the direction of movement in the vertical plane. Whereas lesions of the posterior commissure abolish upward movement there is as yet no clearly localized lesion which will cause downward paralysis.

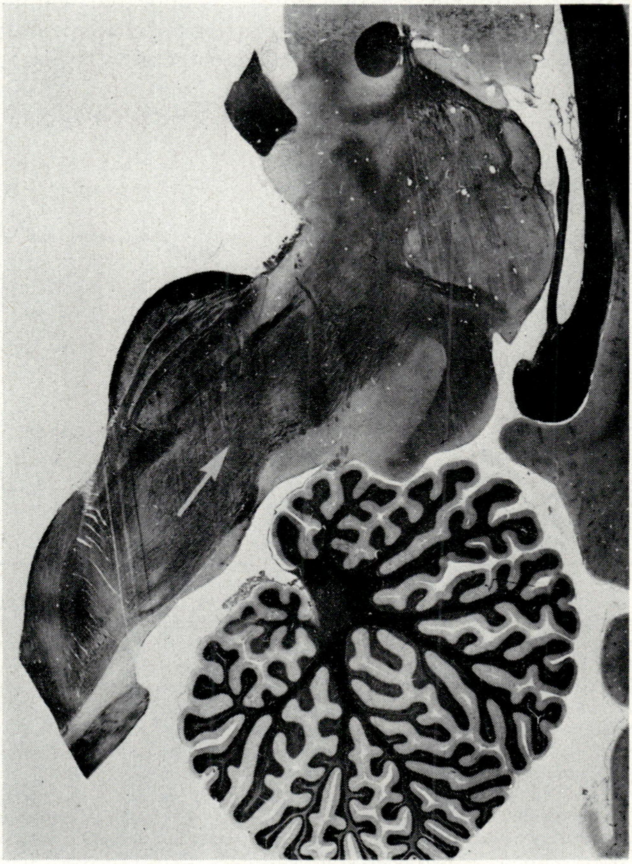

Figure 2. Sagittal section of monkey brain-stem and cerebellum 2 mm *from midline showing depth in the tegmentum of the pons which on electric stimulation at tip of arrow produced the same response as described in the protocol or coronal section*

ELECTRICAL ACTIVITY IN THE LATERAL GENICULATE AND OCCIPITAL AREAS DURING EYE MOVEMENTS

Investigators in our laboratories have also demonstrated that electrical activity can be evoked in the cerebrum and lateral geniculate body during rapid eye movements in light and darkness (1, 2, 3). Electrical recordings within the brain stem, lateral geniculate body, occipital and frontal lobe of the normal alert monkey during vestibular optic or spontaneous movement revealed that action potentials appeared in the tegmentum, the lateral geniculate and occipital lobe. These electrical deflections were almost synchronous with the beat of the quick component of the induced nystagmus. For each saccade or jerk there was a concomitant electrical deflection in the pons, lateral geniculate, and occipital lobe. These action potentials appeared within the pons, lateral geniculate body and calcarine cortex during eye movements in ordinary illumination. Such potentials also occurred in the dark; they were better in the L.G.B. but much less in the calcarine cortex. Each rapid eye movement in

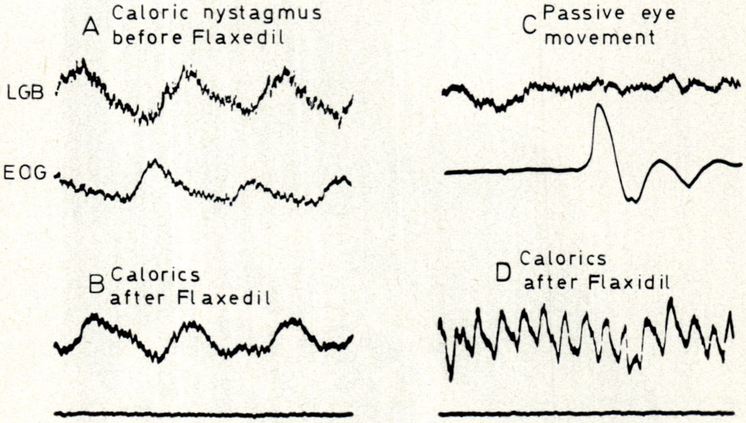

Figure 3. Electric recordings from left lateral geniculate body (L.G.B.) during caloric irrigation in darkness before and after paralysis by gallamine triethiodide. Top trace L.G.B. is monopolar recording in lateral geniculate body. Lower trace is electro-oculogram of left eye RC-coupled with a 1 s *time constant*

A. Calorically induced nystagmus with quick phases to right at frequency of 3 beats/s

B. After gallamine the eyes were motionless, and the E.O.G. was flat. However, the L.G.B. action potentials were similar to those in A

C. Left eye was moved rapidly to the right by forceps as shown by deflection in E.O.G. but during this rapid eye movement there was no response in the L.G.B.

D. Same as B but at slower time base. Time base is shown by the horizontal bar under D. (it equals 200 ms *for A, B, C, and* 1 s *for D). Vertical bar represents* 10 μV *for L.G.B. recordings and* 30 *degrees of horizontal deviation for E.O.G. Negative is up for L.G.B. recordings*

(Erom Feldman and Cohen, 1968, by courtesy of American Physiological Society)

the dark whether a spontaneous saccade or quick phase of vestibular nystagmus or of the optokinetic after nystagmus, induced 100 μV monophasic potentials in the lateral geniculate body, 35–40 ms after the onset of eye movement. These potentials were bilateral for any one type of conjugate eye movement and not direction dependent. Such potentials were changed but persisted when the retinae of both eyes were eviscerated.

The potentials persisted when the eye movements were abolished by Flaxedil (Gallamine) and the 'electric nystagmus' seemed to continue (*Figure 3*). Potentials elicited in the pontine reticular formation associated with eye movement preceded potentials evoked simultaneously in the eye muscles by

5–10 ms and in the L.G.B. and calcarine cortex by 40–50 ms. Electric stimulation of the pontine reticular formation evoked potentials in the L.G.B. and very little in the calcarine cortex (54). The potentials were not present in the L.G.B. during passive movement of the eye.

Since it is claimed that alpha rhythm becomes more evident during upward rolling of eyes with partial eyelid closure and since such an ocular syndrome could be activated in the brain stem it is conceivable that some of the electrical activity in the occipital lobe may originate in the brain stem. Some of the experiments showed that most of the oculomotor impulses were transmitted through the brain stem and then projected rostrally toward the lateral geniculate body and occipital lobe. It is conceivable that the alpha rhythm, which can be induced by upward movement of the globes (a bilateral effect), may originate in part at the brain stem level and that the electrical activity of this phenomenon is reflected in the cerebral cortex.

REFERENCES

1. Bender, M. B. and Shanzer, S. (1964) The Oculomotor System (Ed. by Bender). P. B. Hoeber: New York. pp. 81–140
2. Cohen, B. and Feldman, M. (1968) *J. Neurophysiol.* **34**:806
3. Feldman, M. and Cohen, B. (1968) *J. Neurophysiol.* **34**:455
4. Mulholland, T. B. and Evans, C. R. (1965) *Nature, Lond.* **207**:36
5. Mulholland, T. B. and Evans, C. R. (1966) *Nature, Lond.* **211**:1278
6. Weinstein, E. A. and Bender, M. B. (1943) *Archs. Neurol. Psychiat.* **50**:34

Studies of Paroxysmal EEG Phenomena and Background EEG in Relation to Impaired Attention

Allan F. Mirsky*
Division of Psychiatry, Boston University Medical Center, Mass.

During the course of the past 15 years a group of colleagues and I have been engaged in studies of the central nervous system in impaired attention. Early in this work we discovered the usefulness of the EEG as a method of adding precision, predictive power and understanding to our behavioural observations. The EEG has been useful in adding precision to patient selection and subdivision of clinical groups (7, 14), as an aid in classifying the effects of agents which impair attention (13), as a way of predicting behaviour under specified conditions, sometimes with a high degree of accuracy (1, 20) and hopefully, as a method of comprehending the nature of the changes within the central nervous system in various states of impaired attentiveness. The intent of this communication is to review some of this research.

THE CONTINUOUS PERFORMANCE TEST

A considerable part of the work has involved a particular technique for studying sustained attentive behaviour; therefore it is well to describe it at this juncture. Rosvold and I, who with some other colleagues designed the first device (18) wanted a method for detecting brief lapses or losses of consciousness. We devised a relatively continuous prolonged visual discrimination task in which the subject fixates a visual display (letters of the alphabet, one at a time) and responds, or not, depending upon the stimulus. This is the Continuous Performance Test or CPT.

The subject is seated and watches the letter display through a small aperture. The task is to press a response key when a specified stimulus, i.e. the letter X appears. Subjects work at this task for periods of 10 min or longer. Another version of this task is modified to permit simultaneous EEG study. The subject watches the letters as they are projected on to a screen. The stimuli

* Recipient of a Research Scientist Award K5-MH-14,915 from the National Institute of Mental Health, Public Health Service.

shown to the subject and his responses are recorded simultaneously with and on the same tracing as the scalp EEG.

STUDIES WITH EPILEPTIC PATIENTS

In one of the first investigations with this method (14), we compared the performance of several groups of epileptic patients subdivided on the basis of both clinical and EEG criteria. The EEG in this first study was used as a diagnostic instrument to divide the pool of epileptics and was not recorded simultaneously with behaviour. The results indicated that the *petit mal* group (with their distinctive and characteristic spike-and-wave EEG pattern) performed more poorly on this test of attention than the other populations. Focal epileptic groups, frontal and temporal, do not differ by and large from normal persons. This particular finding has been replicated at least three times (2, 5, 7) with three different populations of epileptic subjects of varying ages. Patients were matched where possible on relevant variables such as age, I.Q. and duration of illness.

This investigation led to a study of *petit mal* patients in which the EEG was recorded simultaneously with the performance. This work has been reported in detail elsewhere (17). Only a few of the results will be abstracted for the purposes of discussion. Although *petit mal* patients make errors on this test even in the absence of spike-and-wave patterns, and occasionally, moreover, are capable of performance during spike-and-wave, it is nevertheless true that the correlation between errors of omission (failure to press for the X) and the appearance of spike-wave patterns is very high, for a psychophysiological relationship. On the average, in the group of 18 patients that was studied, the chance of an error of omission occurring to an X falling within a spike-and-wave episode was at least 75 per cent. In some patients the proportion was as high as 97 per cent. *Figure 1* shows a typical relation between performance and EEG.

Table 1. Comparison of Burst Characteristics Group 'A' *v.* Group 'B'*
Mean values by patient

Characteristic	*Group 'A'*	*Group 'B'*	*P*
Voltage (μV)	186·1	130·3	<0·05
% S and W or PS and W†	89·4	50·8	<0·01
% Max. Fr-cent‡	91·0	66·3	<0·01
% Well organized	70·2	16·1	<0·001
% Normal background§	61·2	0	<0·000 01
Frequency (Hz)	3·15	3·21	NS
Burst length (s)	11·5	6·1	NS
% Errors non-burst	14·9	15·2	NS
Total Bursts	182	406	—
Total Stimuli	517	577	—

* Group 'A' patients are those with many CPT errors in bursts; 'B' type make fewer errors.
† Percentage of time burst had spike and wave or polyspike and wave character.
‡ Percentage of time burst was maximal in voltage F_3–C_3 or F_4–C_4 locations.
§ Normal background was defined in terms of frequency.

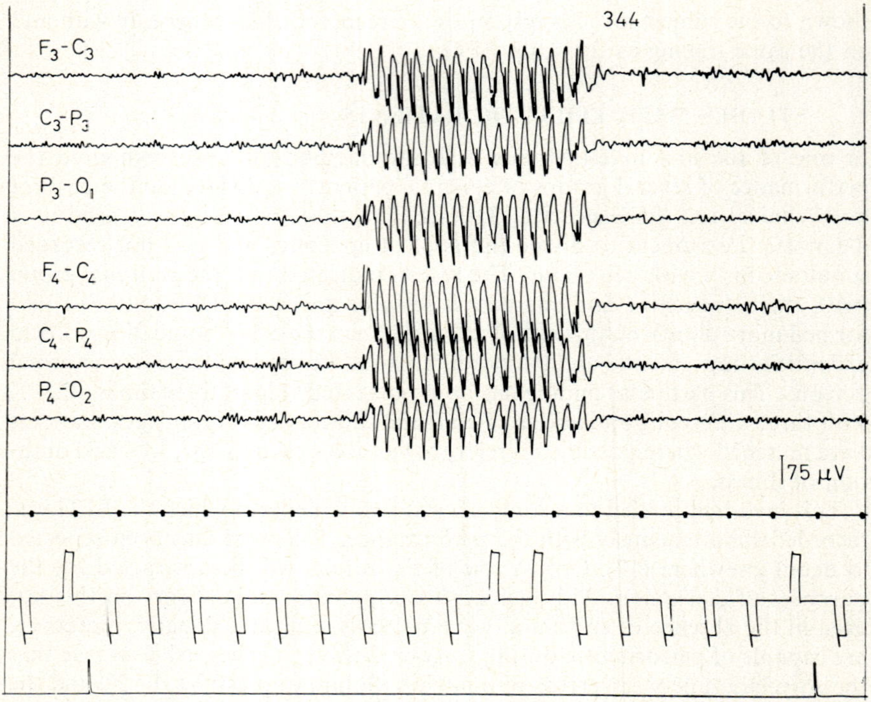

Figure 1. A sample EEG-CPT recording in a petit mal patient is shown. The six EEG channels, according to the '10–20' system, are F_3-C_3, C_3-P_3, P_3-O_1, F_4-C_4, C_4-P_4, P_4-O_2. The seventh channel is a 1 s *time mark and the eighth channel a representation of the stimulus duration, interval and quality. Positive (X) stimuli appear as upward deflections from the base line of this channel, and non-X stimuli as downward deflections. Responses are shown as brief upward deflections on the bottom (ninth) channel. In this* 20 s *sample, the patient has responded correctly to X's at the beginning and end of the tracing, but has failed to respond to the two X's which fell within the* 5 s *burst of spike-and-wave activity*

There were some patients who could respond to a reasonably large proportion of 'correct' stimuli during their bursts and some who were virtually completely impaired. Electrographically, these two groups (i.e. those unimpaired during bursts *v.* those who are impaired) were quite distinct. *Figure 2* shows samples of the kinds of bursts that tend to be associated with complete or nearly complete absence of performance as contrasted with those in which some performance is maintained. Note that the bursts shown here on the left, associated with zero or near-zero performance differ in terms of character (spike-and-waveness, if you will), organization, maximal locus of seizure discharge voltage and voltage itself, from the bursts shown on the right. The latter tend to be associated with less severe performance deficits. Table 1 presents some of these contrasts statistically.

Two additional results occur from that study; Table 2 shows the relationship between the EEG variables, voltage and length of burst, and the tendency to commit omission errors in the burst. There is a regular, systematic and statistically significant relationship between performance and these characteristics of the EEG.

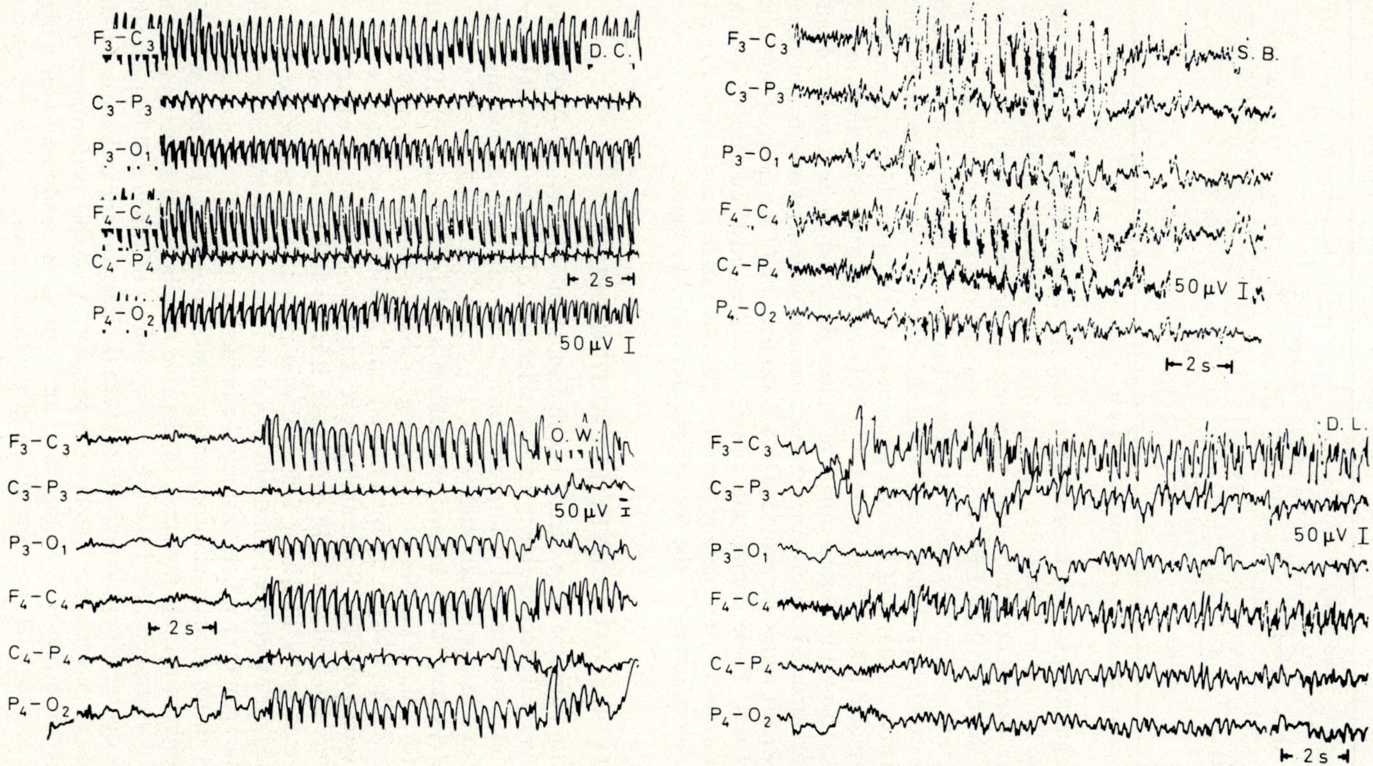

Figure 2. The EEG channels in these four tracings represent the same recording configuration as described in Figure 1. The samples shown on the left were associated with poorer CPT performance than those on the right

Table 2. Relation of Voltage and Length of Burst to CPT Performance

Voltage			*Burst length*		
Range (μV)	% *Errors**	*No. of stimuli*	*Length* (s)	% *Errors**	*No. of stimuli*
25–74	45·8	155	1–4	57·2	362
75–124	67·4	328	5–8	63·8	174
125–174	72·4	225	9–15	77·6	183
175⁺	86·3	386	16⁺	87·2	375

* The percentage array differs from chance at $p < 0{\cdot}0001$.

An argument may be raised concerning the relationships that have been shown so far, namely, that they have involved convulsive phenomena and may therefore represent some extreme condition or end-state of the nervous system; therefore these EEG-behaviour associations do not necessarily apply to other non-epileptic states. Our view is that the phenomena demonstrated here are examples of a neuropsychological relationship which underlies other and possibly all states of reduced awareness. The relationship in a somewhat oversimplified form is this: increases in EEG voltage or modifications in EEG frequency from the normal range of values will tend to be associated with altered or impaired performance. Some instances where this may be illustrated in non-seizure states are provided by a consideration of the effect of centrally-acting drugs. These compounds provide an alternative method of altering attentive capacity and thereby provide additional techniques which may add to the understanding of the central changes that take place.

STUDIES WITH CENTRALLY ACTING DRUGS

In speaking of these drug effects, I omit discussion of behaviour-EEG studies in man and will proceed directly to a consideration of some monkey data using secobarbital sodium and chlorpromazine. These compounds have also been investigated in our human studies (12, 15).

Monkeys have been trained to perform a CPT-like task (10). They are taught to make a simple lever-pressing or button-pushing response to one colour stimulus (red) in a visual display and to withhold responses to other stimuli (green or blue). Some animals are taught to perform for water reward and others to avoid shock. In either case, it is possible to test the animals on such a task every half-hour for a 10 min period. This procedure can be continued for experiment durations of 4 h or longer. This provides a usefully long time span over which to study the induction, peak effect and waning of effect of drugs such as chlorpromazine and secobarbital. Further, working with monkeys permits the study of the role of subcortical structures in this kind of performance. In this work, a few samples of which will be presented here, the degree of correspondence between the electrographic changes and the measures of performance has been impressive. A sample performance-EEG is presented in *Figure 3*. In this case, an experiment conducted with a high dose of secobarbital, the percentage of correct responses is plotted against the counts of the number of waves in several frequency categories. Each EEG performance measure is derived from an 11 min time sample at the specified intervals. This is from a single cortical placement, a frontal-parietal derivation. The analysis of frequency counts is based upon a digital computer analysis (20) which yields β_2 (33–21 Hz), β_1 (20–14 Hz), α (13–8 Hz), θ (7–4 Hz), and Δ (1–3 Hz). Only β_2 and θ are presented in the figure. The correlations between the EEG measures and performance are very high indeed, indicating again how sensitive a measure of attentive behaviour the EEG is in this instance.

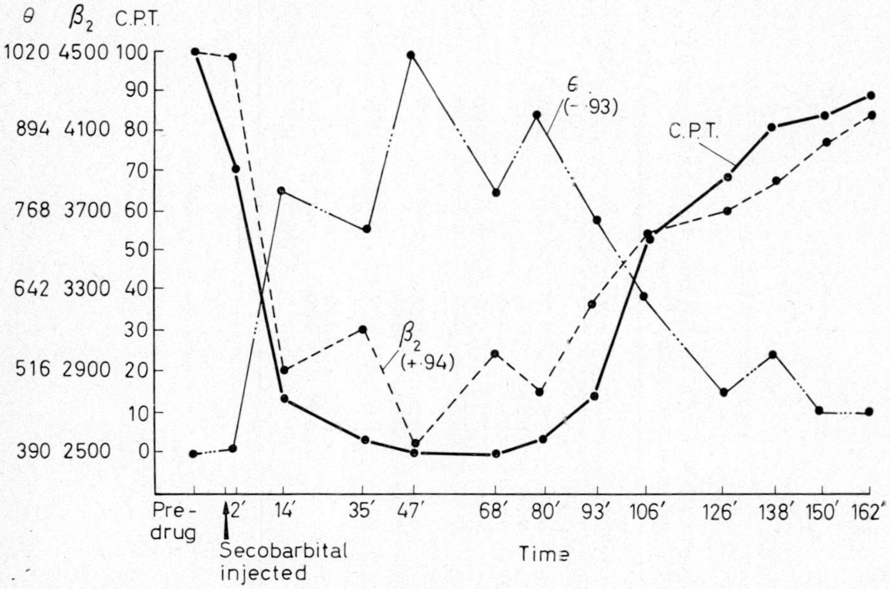

Figure 3. The relation is shown between the percentage of correct responses on the CPT and counts of waves falling within the β_2 *frequency band* (33–21 Hz) *and the* θ *band* (7–4 Hz). *The CPT scores and wave counts from a fronto-parietal bipolar derivation appear on the ordinate, time on the abscissa. In this study the monkey was tested for* 162 min *post injection and the analysis revealed a correlation of* 0·94 *between CPT and* β_2 *counts and* −0·93 *between CPT and* θ *counts*

(From Mirsky and Bakay Pragay, 1967 (*Sleep and Altered States of Consciousness*)
Reprinted with permission from the Williams and Wilkins Company, Baltimore, Maryland)

Such measurements have been done on a number of trained animals, with concomitant recording from a variety of cortical and subcortical placements. The latter are mostly in the area of the midline thalamus and upper brain stem. The results of such a correlation analysis of the data from several animals, comparing the drugs secobarbital and chlorpromazine, has suggested the following relationship: with secobarbital there tend to be more significant performance-EEG correlations involving cortical than subcortical points. With chlorpromazine, on the other hand, the clustering of significant correlations is reversed, i.e. the relationship between behaviour change and EEG frequency is more impressive in subcortical than cortical points. A great deal of such data is summarized in Table 3.

These observations suggest that secobarbital's behavioural effects are

Table 3. Correlations between EEG Frequency Analyses and CPT Scores for Various Brain Loci, as Influenced by Secobarbital and Chlorpromazine—Computer-Aided Analyses

		Monkey G					Monkey D				
		β^2	β^1	α	θ	Δ	β^2	β^1	α	θ	Δ
Fronto-central dura	S	*0·95*	*0·85*	*−0·89*	*−0·95*	−0·30	*0·80*	*0·95*	−0·36	**−0·60**	−0·30
	C	*0·79*	0·38	**−0·58**	**−0·52**	−0·12					
Central-occipital dura	S	*0·78*	0·27	*−0·85*	*−0·93*	*−0·86*					
		0·85	*0·80*	**0·61**	*−0·92*	*0·80*	*0·90*	*0·70*	0·13	*−0·75*	−0·50
	C	*0·95*	−0·10	0·19	**0·53**	−0·22	*0·93*	*0·96*	*−0·71*	*−0·91*	*−0·72*
Midline thalamus*	S	*0·65*	0·02	−0·32	−0·36	−0·56					
		−0·45	*−0·75*	*−0·88*	−0·30	*0·70*					
	C	**−0·61**	*−0·66*	*−0·69*	*−0·65*	*0·74*	0·10	*−0·79*	*−0·93*	*−0·82*	*0·86*
Medial lemniscus*	S						**0·60**	**−0·60**	*−0·63*	−0·40	0·10
Central gray substance*	S	0·54	−0·29	*−0·79*	*−0·75*	−0·30					
	C	*−0·66*	*−0·66*	*−0·65*	**−0·59**	*0·70*	*−0·95*	*−0·96*	*−0·95*	*−0·93*	*0·89*
Caudate nucleus*	S						0·50	−0·50	−0·47	0·30	−0·15

Italic figures = $p < 0{\cdot}01$
Bold figures = $p < 0{\cdot}05$
Others, nonsignificant
* The animals were alive at the time of this presentation; consequently these are the intended locations of the placements.

mediated largely (although not exclusively) via cortical changes and that chlorpromazine's effects occur via some kind of subcortical (midline thalamic, brain stem) alterations.

This is supported by a summary analysis of the correlation array such as is seen in Table 4. Note the preponderance of significant EEG-CPT correlations in cortical leads with secobarbital and the clustering of significant EEG-CPT correlations in subcortical leads with chlorpromazine.

Table 4. Summary of Significant and Nonsignificant Correlations in CPT Frequency Analysis (Table 3)

	Secobarbital			*Chlorpromazine*	
	Significant	*Nonsignificant*		*Significant*	*Nonsignificant*
Cortical	18	7	Cortical	10	5
Subcortical	9	16	Subcortical	19	1

A further note concerning secobarbital: not only does the correlational analysis show a relation between EEG amplitude and frequency and omission errors, but there is a significant predictive association involving commission errors, as well. These errors (pressing for the wrong stimulus) are common with secobarbital and tend to occur when the amplitude increase begins to diminish and the frequency has returned to normal (1). This is clearly seen in *Figure 4*.

This molar analysis of behaviour-EEG relations (overall performance in a 10 or 11 min segment of performance *v.* summed EEG frequencies) has been

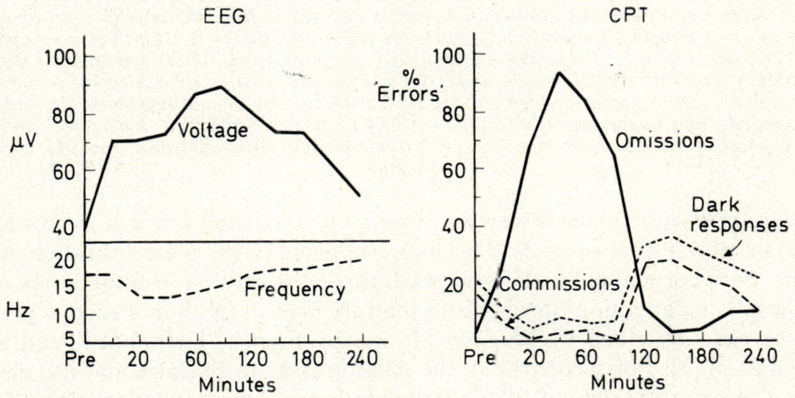

Figure 4. Pooled data based on secobarbital administration to three monkeys; hand-analysis of EEG voltage and overall frequency appears on the left, and CPT performance on the right. Plotted separately on the right are percentages of errors of omission (failure to press for red) errors of commission (pressing to blue or green) and dark responses (pressing when no stimulus is in view). The period of maximal commission errors (150 min *post drug*) *coincides with a drop in the maximum voltage seen with the drug* (90 min *post*). *At the same time* (*i.e.*, 150 min *post*) *the frequency has returned to the pre-drug level and is even beginning to exceed the pre-drug values. The EEG is recorded from a fronto-parietal bipolar derivation. The drug dose was* 20 mg/kg

paralleled by a more molecular analysis. Here the goal is to examine specifically the changes, if any, in the EEG segments which immediately precede erroneous performance, as contrasted with correct performance. A similar approach has been employed by Elazer and Adey (3, 4) in analysis of the role of the hippocampal region and other subcortical areas in conditioning in cats. The analysis of behaviour-EEG relations under drugs has involved primarily chlorpromazine, secobarbital and chlorambucil. The latter compound is discussed in Mirsky, Bloch-Rojas and McNary (11) and will not be described here. The data analysed so far are based on cortical placements. In the case of chlorpromazine, there is a pronounced, highly significant tendency for errors of omission (failure to press the response key) to be preceded by higher voltage EEG than that preceding correct responses (*Figure 5*). The increase in

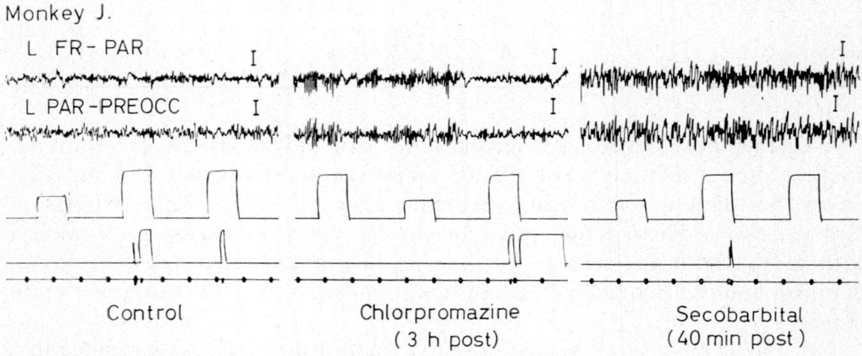

Figure 5. These are sample EEG and behaviour tracings obtained under control, chlorpromazine and secobarbital conditions from a monkey. The two top channels (calibration = 100 μV) *are EEG from fronto-parietal and parietal-preoccipital placements respectively. The third channel represents the stimuli: the smaller deflections from the baseline are negative (blue or green stimuli), the larger deflections are positive (red). On the fourth channel appear the bar press responses. Below this are indicated one-second time marks and (smaller deflections) reinforcements. In the control sample* (0·6 mg/kg), *an omission error occurred to the first red stimulus, accompanied by high voltage EEG activity. The animal responded correctly to the second red stimulus (note lower voltage EEG) and was reinforced. Under secobarbital* (20 mg/kg), *the omission error cannot be distinguished electrographically from the correct response*

voltage prior to errors is somewhat more evident in fronto-parietal derivations than in more posterior leads. This may be related to the more robust connections, both corticifugal and corticipetal, that exist between frontal cortex and diencephalic 'attention' mechanisms than are present for more caudal regions. These were described by Magoun (8) and by French, Hernández-Peón and Livingston (6). For secobarbital, the relation is more complex and less clear-cut. There are a variety of EEG changes following administration of the drug (1, 9), but the period at which there are many omission errors is characterized by sustained high voltage. During this time, the antecedents of omission errors and correct responses are indistinguishable electrographically (*Figure 5*). This is in contrast with the greater variability in voltage seen with chlorpromazine (9, 19).

This failure of secobarbital omission errors to be preceded by and predicted from cortical EEG changes is intriguing, and the question is under study in our laboratory.

Currently being examined are such factors as the existence of weak, partial movements in the presence of the positive stimuli; these may be of insufficient strength to produce a key press, but may nevertheless reflect some muscular response to the stimulus. If it can be shown that responding (of this partial nature) does not disappear completely during secobarbital administration, this would suggest that the nature of the omission error may differ depending upon the agent producing it; in turn, this presumably is related to the locus of action of the agent within the brain. Subthreshold 'responding' might also be evident in some EEG change during the appearance of the positive stimulus. Studies of the brain wave changes following the stimulus are also planned, as well as detailed studies of subcortical EEG changes associated with attentive performance decrement.

This discussion has served to underscore an important difference between the drugs. From the EEG point of view, chlorpromazine tends to produce an obtunding of attention; responses drop out in association with high voltage bursts, similar to that which occurs in sleepy or sleep deprived subjects (12). The chlorpromazine effects are not identical with those of sleep loss (for example, frequency changes are unremarkable in chlorpromazine and marked in sleep deprivation) but they are in contrast with the secobarbital effects. Here there are major changes in frequency and amplitude. Several behavioural effects probably characterize secobarbital; clearly, one of these is confusion and/or excitement. The animal acts behaviourally (and possibly electrographically) as though it cannot distinguish among the stimuli. The clarification and neuropsychological identification of these and other drug effects are some of the goals of our research effort.

ACKNOWLEDGEMENT

The collaborators in these studies have been many; I wish to acknowledge especially the contributions of Drs. H. Enger Rosvold, Cosimo Ajmone Marsan, John M. Van Buren, Janice R. Stevens, Joseph J. Tecce and Eva Bakay Pragay. The recent studies reported here have been supported by grant MH 12568 from the National Institute of Mental Health, Public Health Service.

DISCUSSION ON MIRSKY

Grey Walter

There is a simple corollary to Dr. Mirsky's extremely interesting observations on *petit mal* wave-spike. A patient studied by my colleague, Mr. Winter, for about 15 years has the unusual and useful property of producing wave-spike discharges to a single flash of light or a pair of flashes. Quite recently he studied this in relation to CNV using the flash of light for the conditional stimulus which produced a wave and spike discharge and was also a conditional signal for the patient to respond to the subsequent stimulus. What he found in effect was that in spite of the very large size of this wave and spike discharge, the patient was still able to produce a CNV and perform correctly, even while the spike and wave discharge was going on. The point about this was that this particular spike and wave discharge was in the posterior regions mainly, and the wave component never reached the frontal lobes. This is quite unusual and what is perhaps the most interesting point of all; the patient had never been greatly disturbed by the discharges and in fact medication had been withdrawn because the patient did not mind having the minor seizures. This suggests that the frontal mechanisms which are involved in the CNV were intact, although the whole of the posterior part of the brain was engaged in this relatively enormous discharge. This may be related to what you were saying about your different categories of wave and spike response—this unusual location.

Mirsky

It does seem to be true that if the discharge is not maximal in the frontal region, then it is likely to be less deleterious as far as performance is concerned. In these drug studies, as well, the frontal parietal placement seems to be more highly correlated with performance. What we are doing now is observing smaller and smaller areas within this massive frontal surface, and looking for a more precise kind of relationship. Although we have not studied CNV in *petit mal* patients, we have been working with a number of patients trying to see whether or not there are visual evoked potentials during the spike and wave episode. I think we have encountered every artefact in the book and some that haven't yet been described. But, nevertheless, I think it is true that one can see a visual evoked response in the spike and wave discharge if it is present. The question is whether the augmentation in the maximum-minimum amplitude that we see is real or represents some kind of entrainment with high-voltage components of the discharge. This is something that we are trying to investigate now with the help of a LINC-8 computer, where we can ask all kinds of questions. If one were to subdivide the total attentive process into stimulus input, central decision, and response—then one would say that so far as the spike and wave discharge is concerned, the information in some form seems to be getting back, at least as far as the striate cortex. Whether or not the form is one that the organism can use is another question, but information *is* getting back. It is not, as was shown by one of yesterday's speakers, a complete blanking out of the evoked potential. Rodin in Detroit also showed this.

Cobb

I thought this was an excellent presentation and feel a little ashamed to make some fairly critical remarks about it.

Mirsky

Please do!

Cobb

Dr. Mirsky told us that 'CPT' stands for continuous performance technique. But that is just what it is not. It is discontinuous and the proper word is 'continual'. Now you may think this is trivial, but the discontinuity is, I think, a major defect. It is very difficult to design a truly continuous technique in this case. You say that the interval between stimuli is 1 s, and that 'X' occurs once in every four or five stimuli.

Mirsky

Yes.

Cobb

So you have a definition of only something in the order of 5 s. Now, if you can make your task more nearly continuous, I think you will find that quite a lot more of the *petit mal* attacks are associated with disturbance of consciousness. Furthermore, quite often a disturbance comes late in, and may not persist throughout, the whole of the spike and wave episode. There is great variability from subject to subject but certainly some can respond, for example, for the first 3 s, cannot for the next 5 s, and may be able to respond again before the spikes and waves finish. One more point, a proportion of patients with absences are prevented from having an absence by a visual task.

Mirsky

We attempted to develop a version of our task—continuous or continual—that was closer to continual, by requiring the subjects simply to execute the motor response by pressing the key at a comfortable repetitive rate. We did this task in precisely the same subjects during the same experimental sessions in which we investigated the more complicated function, and we saw exactly what you described: the patient would press into the first 3 s of the burst, stop and then pick up responding again 3 or 4 s before the burst ended. Moreover, this same patient, if you required him to make this motor response to a *stimulus* presented at this time, could not respond. Thus, adding this discriminative act makes the task more sensitive, at least with the group of patients that we studied. We developed a further refinement in hopes of getting at just the input side, which ultimately is what led to our study of the evoked potential. We presented a stimulus to the subject and required him to make no response until some later time. We were attempting to dissociate the responding from the input during the burst. We found that this task also is less sensitive than the complex in which you are required to watch the display, decide whether or not this is the one that requires a response and finally respond. These simpler tasks are not quite so sensitive as measures. Would you remind me of your final comment?

Cobb

Quite a lot of patients with *petit mal* do not have attacks, or very rarely have attacks when they are concentrating on a visual task. I suppose if the task is non-visual it is that much better.

Mirsky

We have a version of this task in which the subject was required to listen to a loudspeaker and to respond when he heard the letter 'X', and the results are by and large similar to what we reported with the visual task. I think if the visual stimulus comes on against a background of blackness, then you do block a spike and wave discharge. But what happens here is that stimuli are appearing quite repetitively at the rate of one per second and there is a degree of habituation which takes place to the visual stimulus. The work of Tizard and Margerison, I believe, has shown the inverse relationship between the difficulty of the task and the number of bursts.

REFERENCES

1. Bakay Pragay, E. and Mirsky, A. F. (1967) *Proc. 75th Annual Convention Am. Psychol. Ass.* **2**:71
2. Campanelli, P. A. (1967) (*Ph.D. Thesis*, University of Maryland) Ann Arbor, Mich.: University Microfilms, No. 67-11294
3. Elezar, Z. and Adey, W. R. (1967a) *Electroenceph. clin. Neurophysiol.* **23**:225
4. Elezar, Z. and Adey, W. R. (1967b) *Electroenceph. clin. Neurophysiol.* **23**:306
5. Fedio, P. (1964) (*Ph.D. Thesis*, Catholic University of America) Ann Arbor, Mich.: University Microfilms, No. 64-11081
6. French, J. D., Hernández-Peón, R. and Livingston, R. B. (1955) *J. Neurophysiol.* **18**:44
7. Lansdell, H. and Mirsky, A. F. (1964) *Expl. Neurol.* **9**:463
8. Magoun, H. W. (1954) In (Ed. by Delafresnaye) *Brain Mechanisms and Consciousness*. Springfield, Ill.: Thomas, p. 1
9. Mirsky, A. F. and Bakay Pragay, E. (in press) In (Ed. by Mostofsky) *Attention: A Behavioral Analysis*. New York: Appleton
10. Mirsky, A. F. and Bloch, S. (1967) *Psychopharmacologia*, **10**:388
11. Mirsky, A. F., Bloch-Rojas, S. and McNary, W. (1966) *Acta Biol. Exp.* (Warsaw), **26**:55
12. Mirsky, A. F. and Cardon, P. V., Jr. (1962) *Electroenceph. clin. Neurophysiol.* **14**:1
13. Mirsky, A. F. and Kornetsky, C. (1964) *Psychopharmacologia*, **5**:161
14. Mirsky, A. F., Primac, D. W., Ajmone Marsan, C., Rosvold, H. E. and Stevens, J. R. (1960) *Expl. Neurol.* **2**:75
15. Mirsky, A. F. and Rosvold, H. E. (1963) In *Psychopharmacological Methods*. New York: Pergamon, p. 302
16. Mirsky, A. F. and Tecce, J. J. (1968) *Proc. C.I.N.P.* 638–645
17. Mirsky, A. F. and Van Buren, J. M. (1965) *Electroenceph. clin. Neurophysiol.* **18**:334
18. Rosvold, H. E., Mirsky, A. F., Sarason, I., Bransome, E. D., Jr. and Beck, L. H. (1956) *J. Consult. Psychol.* **20**:343
19. Rothballer, A. B. (1955) McGill University, Thesis
20. Tecce, J. J. and Mirsky, A. F. (1967) *IEEE Trans. Bio-Med. Engn.* **14**:202

Cybernetics and Attention

Edmond M. Dewan

Data Sciences Laboratory, Air Force Cambridge Research Laboratories, Bedford, Mass.

INTRODUCTION

The main purpose of this paper is to discuss certain aspects of self-regulating systems in engineering technology which appear to have relevance to the concept of attention in psychology and physiology. This discussion will be strictly at the qualitative level, and at the outset it should be emphasized that we will present no explicit models for attention or associated physiological activity here. Instead, we shall present information which seems to be potentially useful for the building of such models.

The requirements for precision in aerospace technology have increased in an unprecedented manner because of what has become known as 'the space age'. This need for better control systems which can function in vastly differing environments has given a great impetus to the development of such fields as adaptive control theory, optimum filtering, and the associated problems of process identification. These concepts will be described in some detail below. It will be shown that they have given rise to devices exhibiting a certain degree of 'artificial intelligence' which are of interest in regard to 'attention'.

Another important concept for understanding certain aspects of brain waves (and biology in general) is 'mutual entrainment'. This will be illustrated by an example in technology and the relevance of this phenomenon to the general problem of self-organization will be explored.

There will be some discussion concerning a 'computer model' for the function of sleep which suggests that there may possibly be very important connections between acquisition of certain stages of sleep and the ability to 'pay attention' in a certain appropriate manner.

The concluding part of the paper will describe a cinematic demonstration, presented at this conference, which concerns the relation between eye position, occipital alpha activity and attention.

The sequence of the discussions of these topics will be 1) optimum filtering, 2) attention and REM sleep, 3) entrainment and self-organization, 4) adaptive control and the concept of 'testing signals' for 'process identification', and 5) the demonstration of eye position effects on EEG.

OPTIMUM FILTERING

The concept of optimum filtering arose during World War II in connection with problems involving radar detection of targets and the control of antiaircraft guns. Two problems were involved. The first was to filter out the noise from the data coming from the radar, and the second was to anticipate the future position of the target (the problem of 'prediction'). The theory of optimum filtering, which is an attempt to obtain the 'best possible' filter under certain assumptions, succeeds in solving both of these problems simultaneously and has general relevance to the field of communication. It has been widely applied by communication engineers to a large variety of problems involving the separation of message from noise.

The theory of optimum filtering was discovered independently by Wiener in the U.S.A. and Kolmogorov in the U.S.S.R. It assumes such things as linearity and stationarity; that is to say, it assumes that the class of filters to be considered is a particularly simple one and that the data to be treated has statistics which remain constant in time. In recent years this theory has been generalized along certain lines. The most recent of these generalizations, by Kalman and Bucy, will concern us here for reasons which will soon become apparent. Before discussing this, however, it will be instructive to consider qualitatively some of the underlying mathematics.

As can be seen from Ross Adey's paper at this conference, certain mathematical functions play a particularly important role in the computer analysis of EEG data. (*a*) the power spectrum, (*b*) the cross-power spectrum, (*c*) the auto-correlation function, and (*d*) the cross-correlation function. It may be interesting to note that these same 'mathematical objects' play a crucial role in the Wiener-Kolmogorov optimal filtering theory. They serve the purpose of characterizing the message and noise properties of the signal in a *statistical* manner. Optimum filter theory is based fundamentally on probability theory, and the mathematical equation which specifies the optimum filter is a famous one called the Wiener-Hopf equation. Suffice it to say here that it specifies the filter in terms of the so-called 'transfer function' or frequency response, and that it is optimal in the sense that it is the linear filter which has the smallest root mean square error. In other words, the error between the message as filtered from the noisy signal, and the message as actually *sent* is minimized by this filter.

The Kalman–Bucy filter, or as some call it, the Wiener–Kalman filter, is mathematically equivalent to the original Wiener filter in some sense. On the other hand, the mathematics and the conceptual picture for this filter provides a rather different point of view. The formalism involving the power spectra, cross-correlations, etc., is replaced by other statistical entities while the transfer function is replaced by the original differential equations from which it came. This mathematical framework is far more complicated for humans to use, but it is much more suited for the computer. The general availability and reduced size of the latter have made this an advantageous choice. In addition, the new formalism (often referred to as the 'state space approach' or 'modern control theory') allows one to make less severe assumptions (e.g. the condition of non-stationarity can be removed). Such powerful generalizations are very useful in space vehicle control system problems. An additional bonus is that it leads to a simple basic idea which is very suggestive when one considers the concept of attention; it will soon become obvious why the concept of optimum filtering is relevant to present at a conference on attention.

To describe the Wiener–Kalman filter let us consider its present application to the Apollo project. In order to take a man to the moon, make a soft landing, take him back to earth, and land again in a specific geographical location, one must continually provide the control system with precise information concerning its attitude, position, trajectory, etc. This is done inside the vehicle using such instruments as accelerometers, gyroscopes, etc. These give the measurements to an internal computer which generates the necessary information. Unfortunately, however, all these measurements are unavoidably mixed with noise. The great requirements for accuracy force one to use the best possible filter to get rid of this noise, hence, the need for the Wiener–Kalman optimum filter. This consists, in part, of a computer aboard the vehicle which is programmed to solve the differential equations, obtained from Newton's laws, which model the dynamic behaviour of the vehicle. That is to say, *the vehicle contains within itself a model of its own dynamic behaviour.* As long as the measurement information agrees with the behaviour of the model, the computer ignores it; that is, the computed behaviour is compared with the measured behaviour, and as long as they agree, the computation is not altered. Whenever there is a discrepancy, it is fed into the computer and incorporated into the problem solved by the computer. The discrepancy, however, is not inserted in a 'raw' form but is weighted by a factor which is essentially the ratio of 1) the estimated uncertainty in the calculated values of the parameters describing the vehicle's state to, 2) the estimate of the uncertainty in the measurements. In other words, the computer also has a programme which allows it to 'judge' the probable accuracy of the information coming into it. The output of the entire computation is the 'best' estimate of the vehicle's dynamic behaviour. This information is then used to guide the vehicle. The method to 'judge' the validity of the incoming information regarding the 'discrepancy' involves a computation which also involves the dynamic characteristics of the vehicle and, furthermore, the 'discrepancies' are smoothed out by the model before the estimates are made. In summary, the filter uses a model of the vehicle to do two things: (*a*) to estimate the reliability of information regarding the discrepancy between the modelled behaviour and measured behaviour, and (*b*) to 'smooth' out the changes in the measurement to be compatible with the vehicle's capabilities. Stated more simply but less accurately, the filter (*a*) ignores information which is already being modelled, (*b*) it ignores measurements which do not agree with what the model says is possible, and (*c*) evaluates the 'likelihood' of certain aspects of the measurement coming to it.

In connection with the above it is interesting to note that Sokolov (15) introduced the concept of the 'neuronal model of the stimulus' to explain the concept of selective attention. He says, 'By neuronal model is meant a certain cell system whereby the information is stored concerning the properties of a stimulus which has been applied many times. The term suggests that the nervous system produces an exact model of the properties of external objects acting on the sense organs. . . . The fact that the orienting reaction only disappears for the stimulus which had previously been used in the experiment, suggests the selective blocking of impulses from the afferent system to effector mechanisms for the orienting reaction. The neuronal model would, therefore, function as a selective filter. . . . The greater the difference between the model and the stimulus delivered, the larger the reaction . . . the neuronal model . . . acts as a compound filter, arresting or blocking the reaction to the definite combination of properties which is characteristic of the given stimulus. . . . The ability of the nervous system to "model" compound stimuli, in the course of the extinction of the orienting reaction, compels us to recognize that the cortex is the most important element in these processes. Extinction of the orienting reaction is impossible in decorticate dogs.' It should be clear from such remarks as these that there is at least a superficial resemblance between the phenomenon of selective attention and the most sophisticated engineering attempts to 'filter' information.

The generality of the Wiener–Kalman filter further suggests that it can also be used as an *adaptive* optimum filter. It appears that a form of this filter could be designed to detect changes either in its environment or in the dynamics of the vehicle (in the above example) and automatically adjust itself to take all these into account. If this were done, the computer would change its 'model' for the vehicle if, for example, a part were jettisoned; and would change its evaluation of environmental influences as it travelled from deep space to a planetary atmosphere.

I am not sure whether or not such adaptive optimum filters actually have been built, but this is the sort of thing which is currently being actively developed. At this conference Ross Adey has already mentioned a similar type of 'self-improving' or 'learning' computer programme in connection with pattern recognition.

The role of feedback in the Wiener–Kalman filter is clearly in evidence and it would appear that some of the above considerations apply to that task of the brain which involves filtering the enormous amount of information coming in through the senses. 'Attention' seems to involve, at least in part, certain complex 'programmes' of filtering in the brain which far exceed in versatility what man can accomplish artificially at this time. Nevertheless, it is perhaps not too much to hope that techniques someday used in technology may lead us to mathematical models to assist us in our understanding of this concept.

ATTENTION AND A POSSIBLE CONNECTION WITH THE NEED FOR CERTAIN STAGES OF SLEEP

As a human being, or any higher organism, goes through life, his brain must store information for at least three important purposes: memory, potential behavioural patterns, and potential tendencies for selective attention. Various economic and optimality considerations lead naturally to the hypothesis that all three types of information must be stored in a manner which changes as the needs and environment of the organism are altered. We shall use the computer analogy of 'stored programmes' which can be altered as needs arise. Assuming that these stored programmes are being altered (perhaps reminiscent of adaptive control processes to be discussed below), the question arises as to when such programming or re-programming takes place. The answer that suggests itself for several reasons is that it happens during sleep.

The word 'programme' in the present context will mean very nearly what other authors intend to say when they use the words 'functional pathways in the brain'. In other words 'programme' will refer to the functional neuronal structures which give rise to the storage of information responsible for memory, potential motor acts, and potential acts of selective attention.

Elsewhere (5) we have investigated the hypothesis that *REM sleep*, the particular stage of sleep which correlates highly with time of dreaming, is the necessary and sufficient condition for the 'reprogramming' to occur. Data and references to the literature were cited there which supported this notion and various tests were proposed. Space does not permit much discussion of this model here but we might mention that several other authors have proposed models for sleep which involve 'programming', in particular Newman and Evans (13), and also Greenberg and Leiderman (9), Gaarder (7) and Breger (3). Our main idea was that the reprogramming hypothesis for REM sleep would explain the ontogenetic and phylogenetic development of REM as well as many experimental observations including dream content.

Turning back to the problem of attention we recall that it involves a multiple input system which selects according to pre-established rules (or by means of a 'model' as was previously discussed), only that information which has a high probability of being useful and relevant to the needs of the organism. Since we have also hypothesized that such a 'programme' would be changed as needs change, we now consider the possibility that an important aspect of this occurs during sleep. More precisely we have elsewhere examined the possibility that REM is the necessary and sufficient condition for the organization of the 'attention programmes'. The question we wish to raise here is, how could such a hypothesis be tested experimentally?

The work on the 'expectancy wave' or CNV which was reported at this conference and elsewhere (21) suggests an interesting practical test to ascertain the necessity of any particular stage of sleep for the establishment of 'attention programmes' in the brain. In particular, the following will illustrate the type of experiment that we wish to propose. The purpose of the experiment is to test whether Stage 1 REM is necessary for the acquisition of 'attention programmes' in some sense. (Actually, any particular stage of sleep including any combination of stages up to 'total sleep' could be tested in the following way.)

Two groups of subjects would be exposed to a complicated learning situation involving complex shifts of attention and anticipations. CNV's would be recorded to ascertain patterns of attention and expectancy. One group would be REM sleep-deprived on the following night while the other group would be awakened only during non-REM sleep. The effect of this during the subsequent day's exposure to the task on the CNV patterns as well as performance would be measured and the whole process repeated again if necessary. The hypothesis would imply that the REM deprived subjects would not develop efficient attention and anticipation patterns in contrast to the control group.

More particularly we suggest that the subjects consist of trainees learning to fly aircraft and that the task consists of a session in a cockpit simulator in which a set of 'emergencies' are presented to the subject. Ordinarily a good simulator will mimic reality to such an extent that the subject will soon react as if he were in a 'real' situation and in fact respond physiologically in the appropriate way. GSR and CNV patterns could be measured during the simulation and the effects of REM and non-REM sleep deprivation on these during a similar simulation on the subsequent day could be ascertained. The theory proposed in reference (5) predicts that the subjects who are not REM deprived will show evidence of a more efficient mode of attention shifts and anticipation with a lowering of affect. In contrast the REM deprived subjects should not show this sort of adaptation of attention 'programmes' but should have a comparatively lower average performance accompanied with inefficient shifts of attention and little lowering of affect as shown in CNV and GSR measurements. If indeed such findings did result from such an experiment they would not only give us important clues to the purpose of REM, but would also give us very practical information concerning learning in general and the training of pilots and astronauts in particular.

ENTRAINMENT

Grey Walter (20) and Norbert Wiener (23) have independently suggested that 'mutual entrainment' a phenomenon involving synchronization to be described in some detail below, plays a fundamental role in the activity of the central nervous system as reflected in the EEG. Perhaps the very fact that one can obtain periodic potential changes between widely spaced scalp electrodes suggests some sort of partial synchronization or 'coherence' in the EEG, especially since it is known that a detailed study of local oscillations reveals a marked difference between the oscillations of one part of the cortex and another even when they are closely spaced (19).

Grey Walter (18) found that under certain conditions he could induce epileptiform spike-wave patterns by stimulating certain subjects with a particular sequence of periodic photic stimulations. He explained his results in terms of (*a*) the entrainment of certain frequencies in the EEG to the photic stimulation and (*b*) the subsequent mutual entrainment or synchronization of the EEG. Norbert Wiener, on the other hand, used the concept of mutual entrainment to explain the existence of the 'alpha rhythm'. Of course there is no such thing as '*the* alpha rhythm' since, as was mentioned, the EEG differs markedly from location to location. On the other hand, both in the normal EEG, and more clearly in the EEG of epileptic activity, there seems to be a certain partial degree of synchronization involved. Other implications of entrainment in EEG power spectra will be mentioned below. Implications for biological oscillations in general will also be mentioned, but first the concept must be explained.

To explain 'entrainment', or more specifically 'mutual entrainment', we shall now turn to the elegant example given by Norbert Wiener (23). This example, while giving a clear insight into the nature of the phenomenon, nevertheless avoids the underlying mathematical theory. It consists of the now famous electric power generator network. The components of this system consist of alternating current generators which are commonly rotated by steam engines. Each generator is equipped with a device called a governor which regulates its speed. Historically it was Watt's invention of the 'fly-ball governor' which made the steam engine a useful device, since only by such means will the engine be kept rotating at a constant speed despite changes in the load. Although each generator has such a governor, no single generator is accurate enough to control a typical synchronous electric motor-driven clock. When, however, a large number of such generators are connected together (in parallel) a very interesting phenomenon takes place. As will be seen, their behaviour becomes quite different. But before going on one must first realize that the faster the generator rotates (i.e. the higher the frequency) the higher the voltage becomes. Such a connection between frequency and amplitude implies that the oscillator is 'non-linear' and 'interaction' is to be expected to take place when such oscillators are coupled together (in contrast, there would *not* be interaction in a 'linear' situation). It was also mentioned in this connection by Wiener that after-discharges in the brain often display such a connection between amplitude and frequency.

We now consider the situation in which many generators are coupled together. Let us suppose that, for some reason, one of the generators starts to run slightly slower or lag in phase behind the others. In this case, energy from neighbouring generators would immediately flow into it and cause it to speed up and get back in step. The lagging generator would thus be treated by the rest of the network somewhat as if it were a motor causing it to regain its proper frequency. But now suppose that a generator slightly exceeds the speed of the others. In this case it would not only be carrying more than its share of the load but it would be losing energy to all the other generators. As a result it would experience a retarding force which would tend to slow it down until it was 'in step' again. Since all the generators interact with the network in this fashion, they all tend to 'pull together in frequency'. This is obviously a form of 'self-organization', for the system of oscillators now acts as a *unit*. This situation can be regarded as the emergence of a 'virtual governor' which does not have a physical existence in the usual sense but is created as a property of the system as a whole by means of the interaction. Perhaps the first instance of mutual entrainment of man-made oscillators was that observed by Huygens who noted that pendulum clocks hung together on a thin wall, would lock together in synchronism.

The reader will probably recognize that there are many instances in biology in which the interaction of many units gives rise to a system that behaves as one super unit. An organism consists of a system of organs while the latter consist of systems of cells. There are some indications that entrainment, or a process similar to entrainment, plays a role in this self-organization.

Returning to the alpha activity of the EEG, Wiener pointed out that the dominance of the 10 Hz rhythm might be due to a 'pulling together of frequencies' or 'entrainment' of oscillations in the brain. As was already mentioned, after-discharges display the nonlinear property of an amplitude frequency relationship, and this already suggests the possibility that entrainment can take place. But if indeed the alpha activity involved entrainment, then this would imply the existence of a certain property in the brainwave power spectra. Theoretically, the 'pulling together' of frequencies in the brain at the alpha frequency could cause a dearth of oscillations to exist at frequencies near the main one. Thus, if the oscillations in the brain were actually pulling toward each other in frequency, the power spectrum should show a *dip* on each side of the main alpha peak. *Figure 1* gives an illustration of how such

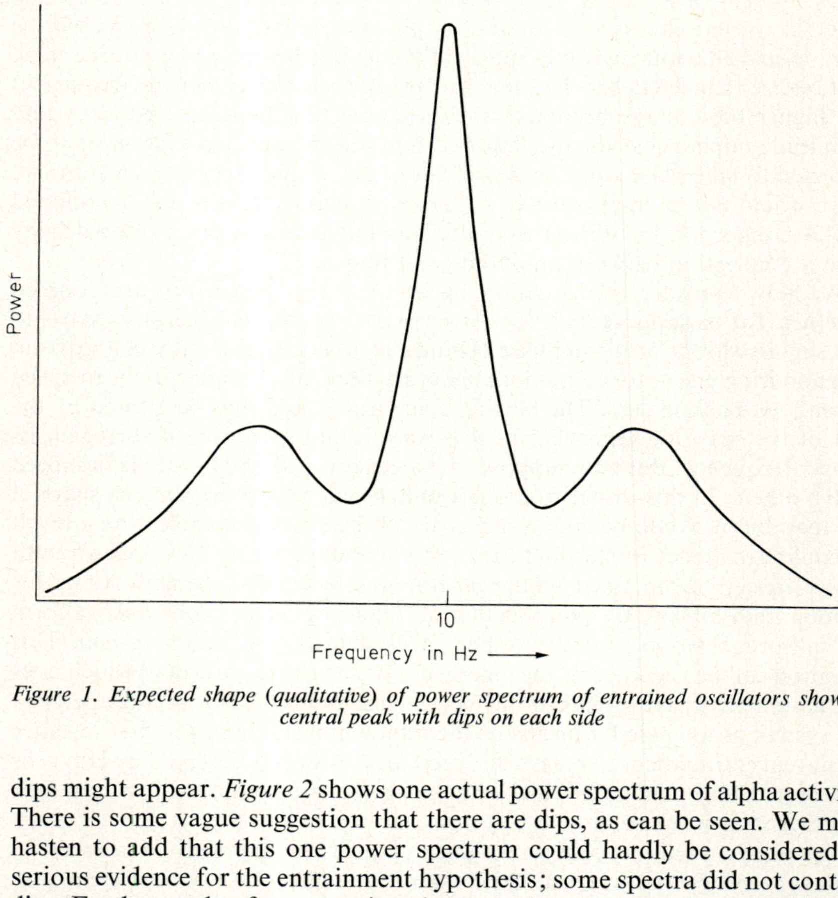

Figure 1. Expected shape (qualitative) of power spectrum of entrained oscillators showing central peak with dips on each side

dips might appear. *Figure 2* shows one actual power spectrum of alpha activity. There is some vague suggestion that there are dips, as can be seen. We must hasten to add that this one power spectrum could hardly be considered as serious evidence for the entrainment hypothesis; some spectra did not contain dips. Further study of a comparison between the power spectra of oscillators in various degrees of mutual entrainment and power spectra of EEG, however, might be an interesting research project.

While the possible role of entrainment in EEG is not very clear at present, there is no doubt that it plays a very important role in biological rhythms, particularly in the circadian rhythms. Pittendrigh and Bruce (4, 14), for example, in trying to 'explain' some of their data, hypothesized the existence of a 'master oscillator' in the body which can be entrained by the earth's light-dark cycle. In addition they hypothesized the existence of a number of secondary oscillators (as evidenced by circadian rhythms in temperature, sleep, activity ,and chemical changes found in the blood, urine, pineal gland, etc.). These secondary oscillators are phase locked to the master oscillator in their model but, under the influence of alterations of lighting schedules, these secondary oscillations can be knocked temporarily out of synchronism with it.

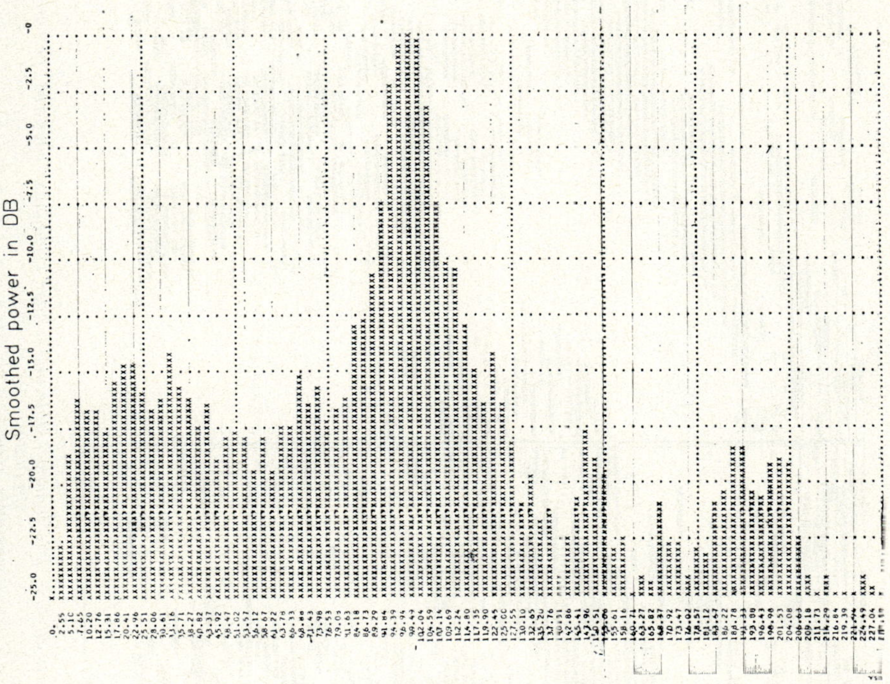

Figure 2. Digital power spectrum of an EEG. The peak is at 10 Hz *and the minima of the dips are at approximately* 5·6 Hz *and* 16 Hz

Figure 3 shows an example of this. Here one can see that the oscillator for the animal's activity cycle starts off in a 'free running' state and then is eventually 'captured' or entrained by the master oscillator.

Entrainment also seems to play an important role in the temporal organization in cell biochemistry as has been pointed out by Brian Goodwin of the University of Sussex (8). This investigator has actually entrained the division cycles of populations of bacteria by periodic chemical pulses.

The current research in circadian rhythms indicates that the discomfort one feels after a long jet trip in the East–West direction is due partly to the temporary desynchronization of one's biological rhythms both to the environment and to each other. Perhaps going by boat has certain advantages in view of these considerations.

Many other examples of entrainment of biological rhythms exist. The action of the heart involves entrainment of the contraction oscillations of cells.

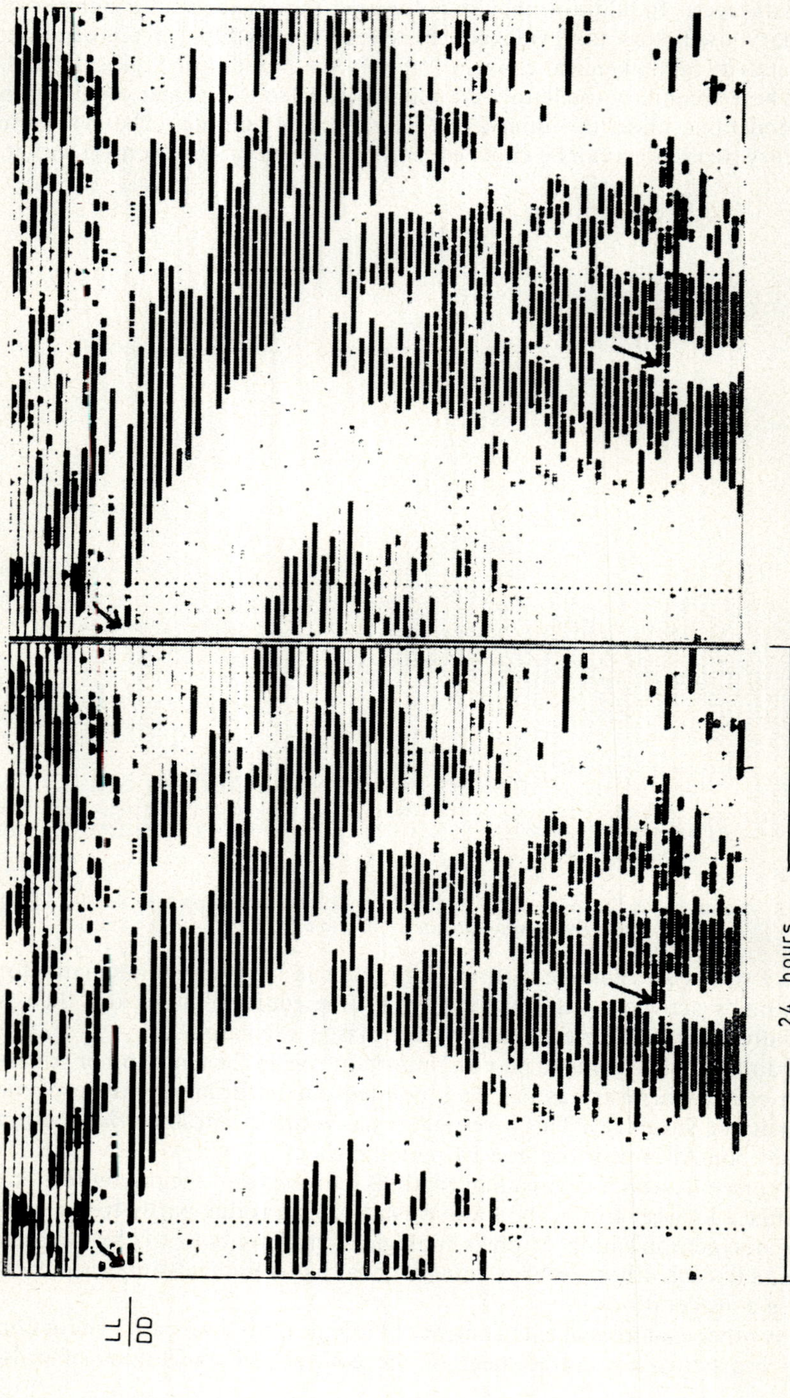

Figure 3. Illustrating that the oscillator starts off in a 'free running' state and is then eventually 'captured' or entrained by the master oscillator

(From Pittendrigh, 1960, by courtesy of the Publisher, Long Island Biological Association, Cold Spring Harbour, L.I., N.Y.)

Breakdown of this entrainment results in fibrillation. The treatment for this condition consists of the resynchronization of the oscillations by means of externally applied electrical impulses. Another famous example of mutual entrainment is the synchronous flashing of fireflies in S.E. Asia (23).

In a recent study, A. Winfree (24) has simulated the entrainment of 100 non-linear oscillators on a large scale digital computer. *Figures 4–7* show clearly

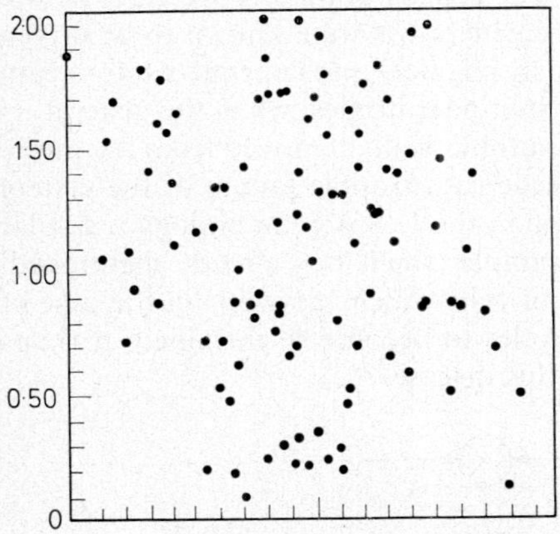

Figure 4

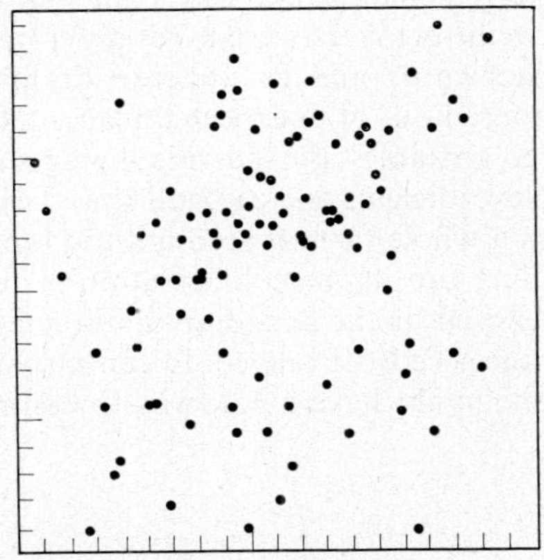

Figure 5

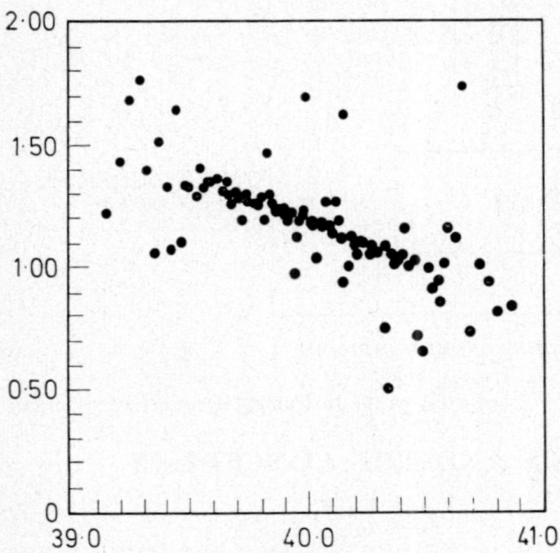

Figure 6

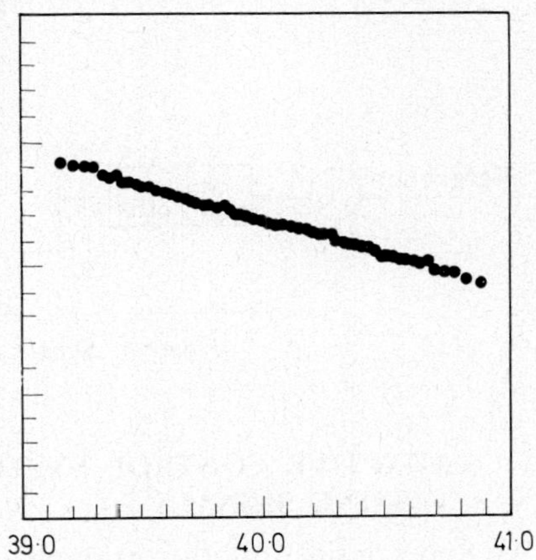

Figure 7

Figures 4–7. Development of coherence from initial chaos among 100 oscillators with free running periods plotted on abscissa and phases on ordinate (selected snapshots from computer experiment). Vertical scale is in units of π, *from* 0 *to* 2·0 π.

(From Winfree, 1967, by courtesy of Academic Press)

the 'pulling together' of these oscillators as they finally become mutually entrained. In these figures the ordinate shows the relative phases of the oscillators and the abscissa shows their 'free running' periods, i.e. their periods when there is no interaction between them. Each dot represents one of the 100 oscillators (there are 100 dots in each figure). As we go from the first to the last of these figures it can be seen that as the oscillators interact the mutual entrainment causes a 'condensation of order out of chaos'. We thus have a rather dramatic visual display of this form of self-organization. Notice that in the final figure the faster oscillators all lead the slower oscillators in phase; hence, the line slopes downward to the right.

As mentioned above, power generators behave in a similar way; however, we also know that under certain conditions the generators can break loose from each other and can 'fight' one another instead of pulling together. Before circuit breakers were used, un-entrained generators were known to be driven backwards and to undergo dramatic mechanical movements while steam poured out of unexpected places. Now in modern times, when the generators 'go unstable', the individual plants are automatically removed from the mesh. Nevertheless, one can still have large scale catastrophic failure of the system as a whole (as was seen not too long ago in the U.S.A.). In biology a similar thing can happen. Fibrillation is an example which was already mentioned. Another is the circadian division cycle of cells which 'go wild' in the case of cancer cells. If one could cause these cycles to become re-entrained, perhaps one might have a new way to control this disease.

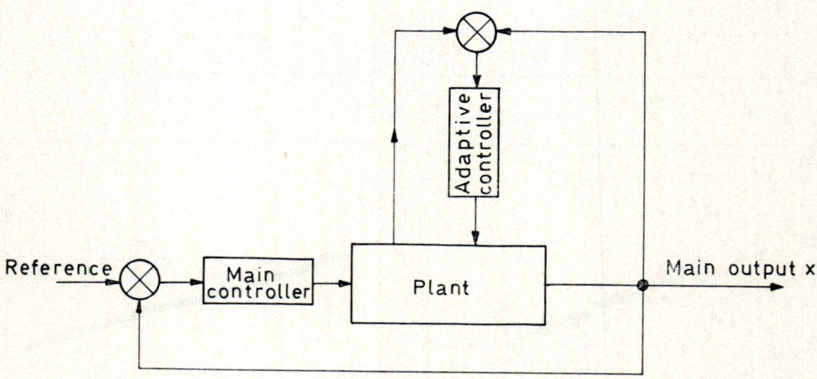

Figure 8. Simple adaptive control system

(From Tarjan, 1962, by courtesy of the Instrumental Society of America)

ADAPTIVE CONTROL SYSTEMS AND THE CONCEPT OF TESTING SIGNALS

We now consider another example of the ingenuity of space age technology which seems to provide insight into various aspects of the concept of attention. This consists of the concept of 'adaptive control' and, in particular, the associated concept of 'testing signals'. H. S. Tsien (17) attributes the idea of adaptive control to workers in the field of 'artificial intelligence' who were attempting to understand the 'ultrastability' as seen in organisms. Ross Ashby's (2) work was explicitly mentioned and it is interesting that a concept which may have originated in the attempt to explain the self-regulatory aspects of biology was subsequently applied to engineering technology with success and the latter results are now being applied back to biology.

The fundamental idea behind adaptive control is illustrated in *Figure 8* from a paper by Tarjan (16). In this figure is shown the usual feedback loop between output and input. In addition there is another block and feedback loop connected to the 'control' block which can change the system as needs demand. An ordinary feedback control system would be effective only over a finite range of demands placed on it, but the additional feedback loop which can cause a modification of the properties of the control system will allow it to 'cope' with a much larger range of environments.

In order to further increase the range over which the system can 'adapt' to external demands, adaptive control loops can be cascaded as shown in *Figure 9*. As can be seen, this method of cascading, can, in principle, be continued

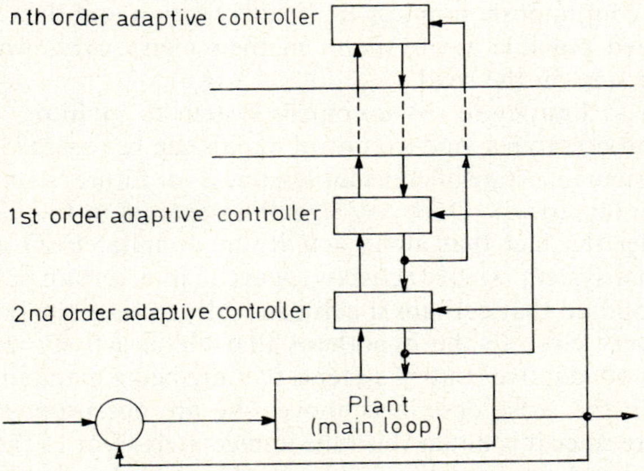

Figure 9. nth order adaptive controller
(From Tarjan, 1962,
by courtesy of the Instrumental Society of America)

without limit. No doubt something like adaptive control takes place in biological systems and we therefore consider some specific technological cases to illustrate this concept.

An especially interesting system which actually exists and which illustrates an important idea is the 'Minneapolis-Honeywell Adaptive Autopilot'. It is used in, for example, the X-15 experimental aircraft which is a vehicle which must cope with greatly differing environments. Near the ground it controls its orientation and trajectory by using the reaction of the air impinging on its control surfaces; but, as it penetrates to higher altitudes in the atmosphere, these surfaces must move over an increasing range to achieve the same effects. At highest altitudes the effects of these control surfaces become negligible and must be enhanced, and, at extreme altitudes, replaced by the effects of small rockets. These rockets orient this vehicle in a manner entirely similar to the standard methods used in space vehicles.

A crucial aspect of adaptive control is that of 'process identification' and 'testing signals'. The adaptive control system of the X-15 must somehow adjust itself in an appropriate manner for the environment and must therefore continually test the effectiveness of the control surfaces. It does this by oscillating them up and down at about 4 Hz and sensing the vehicle's reaction to this procedure. The amplitude of the vehicle's reaction is automatically held constant (using feedback) by increasing or decreasing the amplitude of the oscillations of the control surface. In the thin air of high altitude, the amplitude of the control surface oscillations will of necessity be much larger than at lower altitudes. These oscillations are a 'built-in' property of the system composed of the vehicle and its control system, and occur spontaneously. Since the amplitude is constant, this oscillation is a so-called 'limit cycle'. The system actually alters its so-called 'gain' to maintain constant amplitude.

The basic idea embodied by the limit-cycle autopilot resembles a procedure reputedly employed by racing car drivers. They are said to continually oscillate their steering wheels back and forth slightly while they drive. In this way they can continuously monitor the road's friction and thus ascertain the maximum speed possible at any point in their manoeuvres which will still allow them to stay on the road.

This use of a 'limit-cycle' by a control system to monitor 'gain' or the 'ability' of a process to respond to control signals can be regarded as a method to keep the system in a state of constant readiness for future 'commands'. The question naturally arises: 'Does any signalling process in the CNS resemble this?' Consider the fact that alpha activity in occipital EEG appears only when the visual system is not 'actively engaged' in a certain sense. In 1934, Adrian (1) proposed that perhaps it exists in order to keep the brain 'awake'. This idea is very close to the hypothesis that alpha activity may represent testing signals of adaptive control systems that are being maintained in a state of readiness in the sense described above. We are not proposing this as a hypothesis here since it is not at this time immediately clear to the writer what consequences could be tested experimentally. On the other hand, it illustrates that such analogies may be useful for suggesting possible hypotheses.

A more general type of testing signal will now be discussed. *Figure 10* shows a block diagram of the general adaptive control system. In this diagram the input is first processed by a filter—the sort of filter already discussed above under optimum filtering. Note also that the device which adaptively alters the 'process' must have feedback information in the form of information concerning the parameters describing the process being controlled. In the diagram one can see that a testing signal can be used to ascertain these parameters. In the case of 'linear systems' one can use white noise to 'see' the process parameters and to know how the system will respond to any given input. This method is based on the classic work of N. Wiener and Y. W. Lee (10) who used an approach involving power spectra, cross-correlations, etc. For the theory to work, the 'white noise' must consist of random signals having an entirely flat power spectrum (i.e., having a constant power) for all frequencies relevant to the system's behaviour. Lee and Wiener showed that one can quickly obtain a mathematical description of the dynamics of the 'process' by feeding noise in at the input and by making a 'cross correlation' between input and resulting output. The beauty of this technique is that the system can continue to perform its functions even as the testing procedure is in progress and that this activity does not interfere with the test results nor vice versa.

The use of white noise as a testing signal in the above case of linear systems would appear at first sight to have no bearing on the activity of the CNS since the latter is obviously non-linear in the extreme sense. On the other hand, white noise can also be employed as a testing function for a very general class of non-linear systems as shown by Wiener. His theoretical work has been developed further by a number of workers for several years, but it has not yet been employed in practical systems. This is due to the high level of complexity and number of calculations involved in this approach which still require, at this time, further developments in the computer field before becoming in any way a practical procedure. Nevertheless, since we are now considering processes that may be taking place in the brain (that is to say processes which are generally assumed to be infinitely more sophisticated than anything

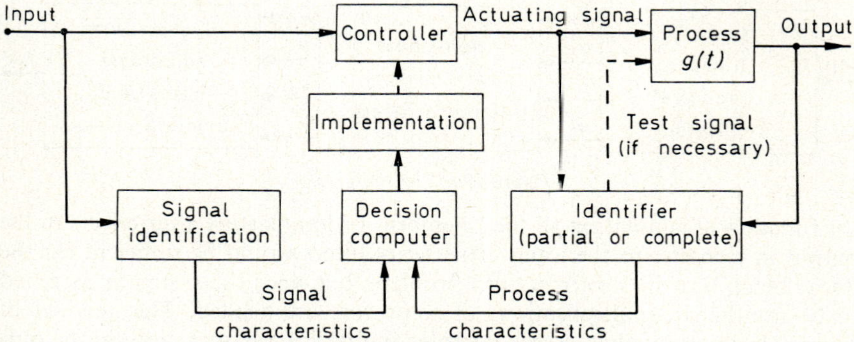

Figure 10. General adaptive control system

(From Mishkin and Braun, *Adaptive Control Systems*, by courtesy of McGraw-Hill Book Co.)

accomplished in the computer field) it is perhaps appropriate to consider this advanced theory of Wiener to see if any analogies are useful. After describing Wiener's theory of the synthesis of nonlinear systems we shall see that Gaussian white noise again appears to be a crucial testing function.

First we recall that we are considering a specific part of the adaptive control system which involves the problem of 'process identification' or the identification of the dynamics of the system. This can be regarded as the problem of obtaining a representation of what is going on in a 'black box' (the usual engineering terminology used to refer to a system with unknown properties). The only way to ascertain what goes on in the black box is to put in various inputs and watch to see what happens to the output. The solution to this problem as thus stated consists of what might be called a 'white box' which has an output as identical as possible to that of the black box for any arbitrary input given to both boxes. In the linear case one merely observes the response of the black box to a series of sine wave signals which cover the range of frequencies relevant to the system. From this frequency response information a 'white box' can be constructed which will respond to *any arbitrary input* in the same manner as does the black box. This procedure works only in the linear case (for reasons relating to the mathematical definition of the word 'linear system'). The Wiener–Lee theory mentioned above represents a procedure based on these principles. In the non-linear case the principle of 'superposition' so crucial in linear calculations, ceases to apply and the linear theories are valid no longer.

One of Wiener's great achievements was to show that white Gaussian noise can approximate, given sufficient time, all possible input functions to the black box. Actually the mathematics involves something called 'differential function space' and plays a fundamental role in certain parts of the mathematics of probability theory. Thus the white box–black box problem will be solved if one can make the white box respond to white noise as does the black box to the same white noise. His technique for doing this is shown in block diagram form in *Figure 10a*. The first set of filters (labelled Laguerre in reference to the so-called 'Laguerre polynomials' they represent mathematically) serve

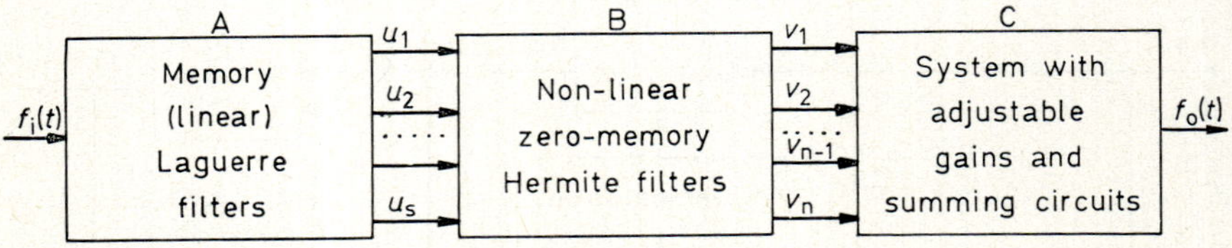

Figure 10a. The white box

the function of simulating all the 'memory' or long lasting after-effects in the output in response to the input. This set of filters would be sufficient for the linear black box or 'identification' problem, but since the system is assumed to be non-linear, multiplications of some sort are needed. The next set of filters (Hermite) involves a very complex multiplicative operation on the output from the Laguerre filters which theoretically can simulate all possible non-linearities in the system but has 'zero memory' (since the 'memory' aspect has already been taken into account by the first filters). Wiener shows that a straightforward process exists for setting the values of the 'gains' of all these filters (involving a comparison between white box and black box outputs) which will solve the problem of making the two boxes essentially identical in behaviour to all inputs. Theoretically, an infinity of calculations must be done in order to do this and one does not know how many are actually needed in practice.

Of course this procedure would not work for all non-linear systems. Consider, for example, a black box that puts out a signal caused by an influence occurring at a time in the infinitely remote past. Since the Laguerre filter would not be able to simulate this, the above technique would fail. It would also fail if the black box had a 'spontaneous' oscillation or a limit cycle 'kicked off' by some influence in the remote past. For this reason one must assume with Wiener that the system has 'a certain deadbeat character in which the output is asymptotically independent of the remote-past input' (22).

We repeat that the main point of bringing this concept of non-linear system identification to the reader's attention is to show that, just as in the linear case, *noise* can be used as a testing function to diagnose the dynamics of the internal mechanism of a particular system. Such a process must take place in any adaptive control system. Since adaptive systems would seem to play an important role in the CNS, one can seriously consider the hypothesis that *noise* might play a role in 'process identification'. We hasten to add that many other practical methods of process identification exist; however, none seem to have quite the generality of the method we have been considering.

MOVING PICTURE DEMONSTRATION OF THE EFFECT OF EYE POSITION ON ALPHA ACTIVITY

The purpose of this demonstration was to show that a subject can exhibit a maximum amount of occipital alpha activity during a period of maximum attention.

Historically, Berger's original hypothesis was that the presence of alpha activity indicated a state of relaxation while a 'blocking' would occur whenever the subject was paying *attention*. Lord Adrian, however, showed in 1934 (1) that one could pay attention without blocking alpha activity provided that there was no visual fixation or 'visual attention'.

In the picture, the subject's task was to send Morse code messages by means of voluntarily controlling her occipital alpha rhythm. The EEG signals were fed to a computer which displayed the resulting alphanumeric symbols on a CR tube and typed them out on a teleprinter. To block alpha the subject was instructed to fixate her eyes on her hand as shown in *Figure 11*. To turn on

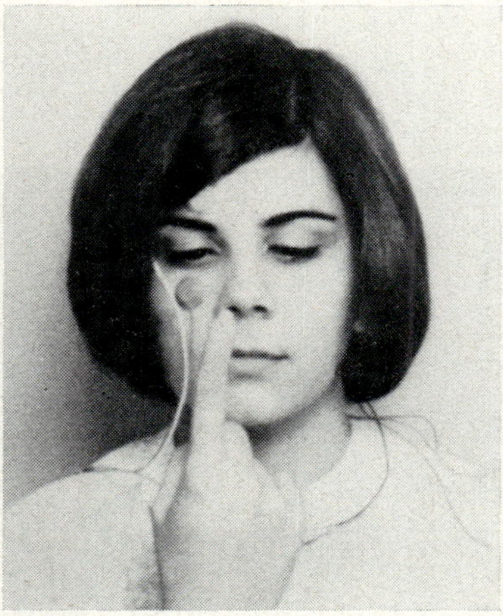

Figure 11. Subject focusing open eyes on finger placed close to face. Note d.c. eye position electrodes

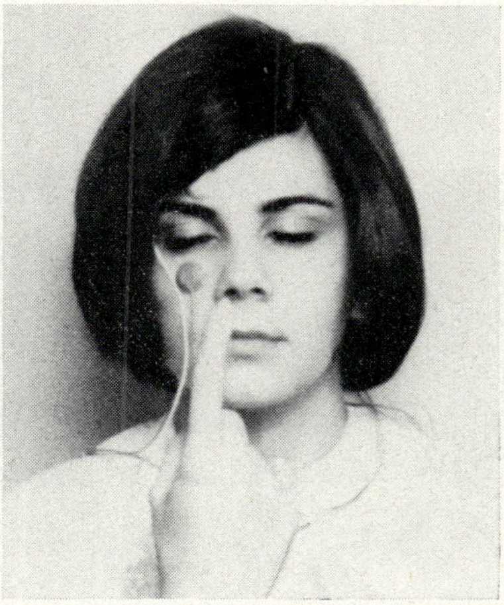

Figure 12. Subject with eyes up and closed sending a 'dot' or short burst of alpha activity

alpha she closed her eyes and turned them upwards as shown in *Figure 12*. It was important that the subject should not try to 'look' up or fixate or converge her eyes when in this position, for this would tend to block alpha. When sending a dash (a long burst of alpha), this subject found it necessary to let her lids flutter (*Figure 13*) since her eyes were turned to the maximum upward position while the lids were kept relaxed. Not all subjects found it necessary to do this nor did they all need to open their eyes in order to fixate and block alpha.

Figure 14 shows the EEG machine's output and one can clearly see the correlation between the upward deflection of the pen showing eye position (corresponding to eyes up) and maximum alpha in the raw and filtered EEG. The computer output (the pen labelled 'COMPUTER') registers no alpha

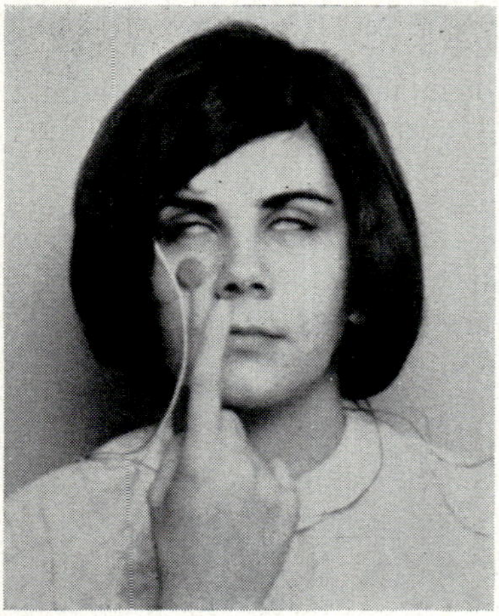

Figure 13. Subject with eyes up to highest position to send an unbroken 'dash' or long train of alpha activity

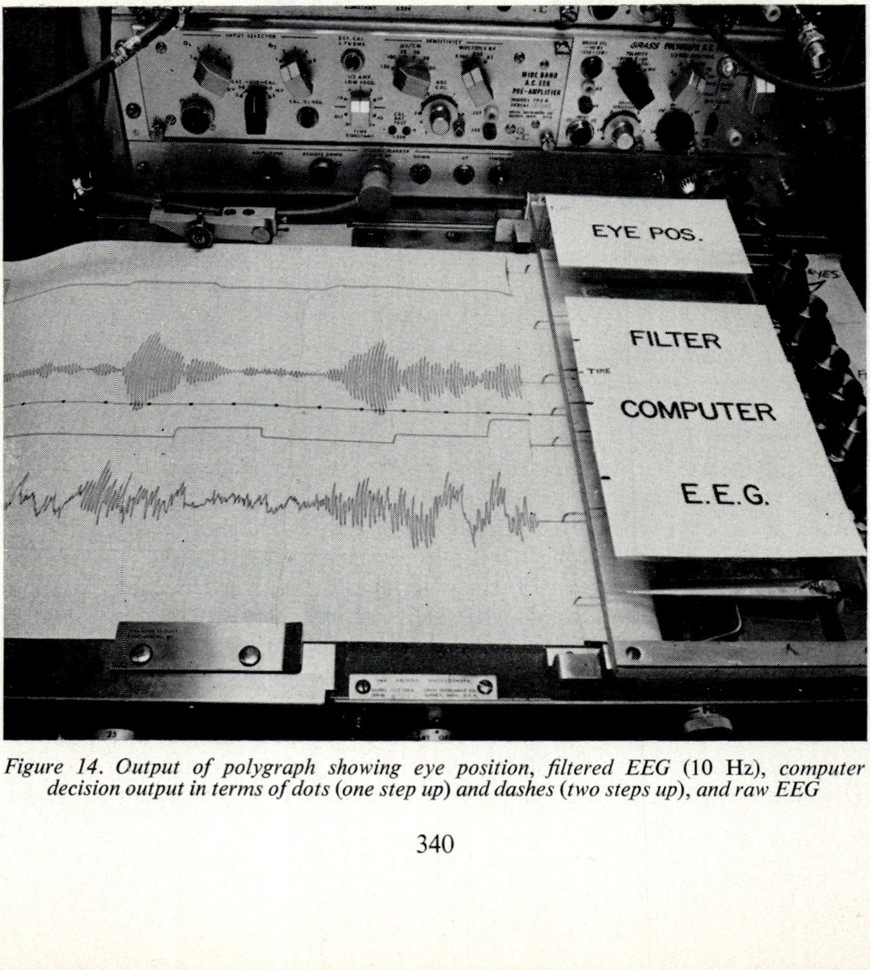

Figure 14. Output of polygraph showing eye position, filtered EEG (10 Hz), *computer decision output in terms of dots* (*one step up*) *and dashes* (*two steps up*), *and raw EEG*

(lowest pen position), alpha burst long enough for a Morse code 'dot' (middle position), and alpha burst sufficiently extended to be a dash (maximum upward deflection). The subject heard a tone whose pitch went up and down with the pen deflections. This signal provided her with the feedback necessary to send the code. The main point was that, while waiting to hear the registration of a dash, she had to (*1*) attend intently to the signal, (*2*) remember what she was in the middle of sending, and (*3*) remember what to do next. Thus maximum attention (from subjective account) was in evidence during maximum alpha.

Figure 15 shows the display on the computer's CR tube while the subject

Figure 15. C.R. tube output of computer during sending an 'S'

Figure 16. C.R. tube output of computer after subject stopped 'sending'

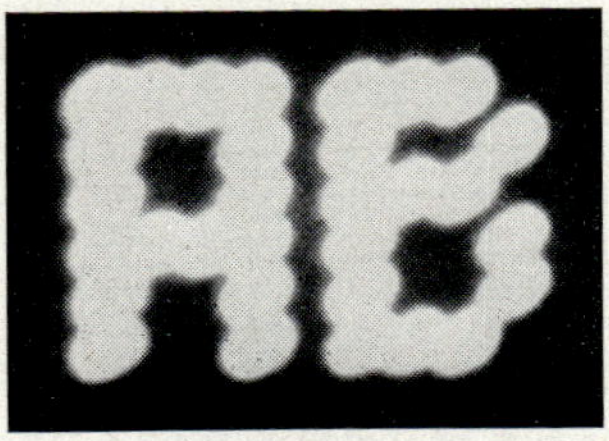

Figure 17. Computer output during sending 'C'

Figure 18. Response of computer (see text)

was sending three dots. The sound track of the picture carried the signal heard by the subject coming from the computer. *Figure 16* shows the registration of the letter sent. Similarly, *Figure 17* shows the sequence of letters already sent while the subject was sending, the letter C. The sound track carried the signal for dot-dash-dot-dash but this was followed by a brief dot by mistake. *Figure 18* shows the result. This was to illustrate that the subject cannot maintain perfect voluntary control of the EEG in this experiment. There seems to be a random uncontrollable aspect to occipital EEG under these conditions.

Cyclomydril which tends to prevent accommodation had no effect on voluntary control of alpha activity. The question clearly raised by all these considerations is: 'Does fixation alone block alpha activity or is it *visual attention*?' (The reader will recall that the subject was engaged in *auditory* attention but not fixating eyes during maximum alpha activity.) One way to test this would be to examine the effect of attending to the behaviour of an after-image while eyes are in the maximum upward position. This experiment was actually performed by Mulholland and Evans (11, 12). Their results showed that one could visually attend to an after-image while exhibiting large amounts of alpha activity. This experiment should be repeated, but if this were to be verified, it would clearly indicate that visual attention alone need not block alpha activity.

Interesting possibilities exist if 'the *attempt* to fixate' turned out to be the necessary and sufficient condition for blocking the alpha-activity. This would

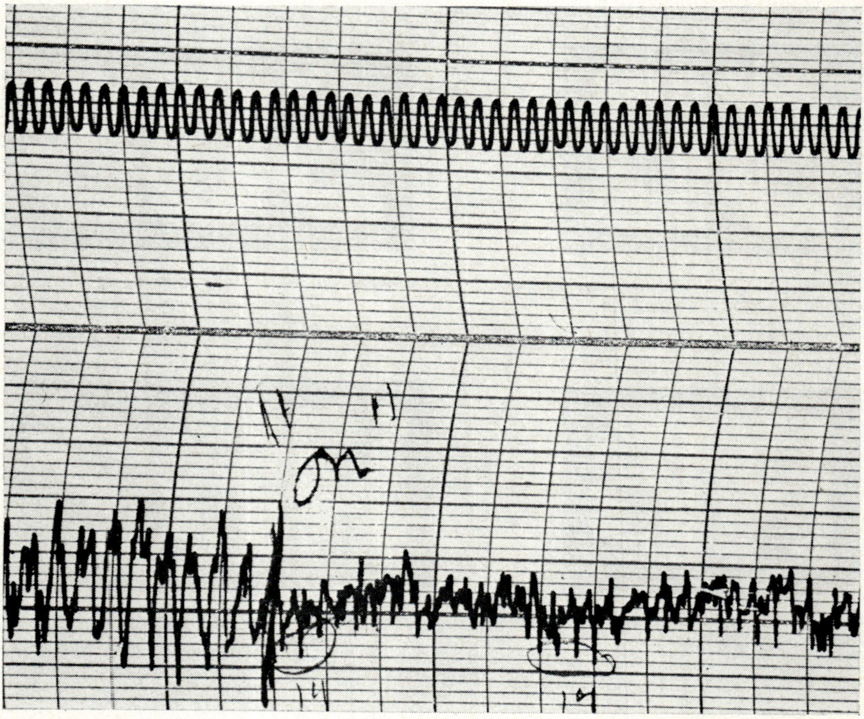

Figure 19. Upper trace corresponds to photic stimulator trigger circuit while lower trace shows EEG record with alpha on and off

not only have the various basic implications but would also have implications for prosthesis. Severe conditions of akinetic mutism, either induced by drugs (curare) or by automobile accidents, could perhaps be helped by this method of communication. Of course, if the subject could even move his eyebrow enough to give an EMG signal, he could use the same programme on the computer (provided by my colleague S. Michaels for the LINC computer) to type out letters to his friends and family. This has not been actually done yet for a paralysed subject; however, a normal subject was able to send Morse code by eyebrow EMG at about eight times the rate that he could do it by EEG. The latter takes about 35 seconds per letter on the average.

Voluntary control of the EEG gives one further bonus. It allows one to examine the effects of 'flicker' (i.e. periodic flashes from a stroboscope) on occipital EEG both during alpha and during alpha blocking. In addition, this made it possible to examine the subjective difference in visual effects under these two circumstances. *Figure 19* shows the EEG during flicker with alpha on and alpha off. When the alpha was turned off by having the subject fixate his eyes behind the lids during photic stimulation, intense colours were seen and the centre of the pattern *remained stationary*. The EEG clearly showed evoked potentials. When the alpha was turned on (eyes up) the amount of colour went down dramatically (perhaps due to the decrease in intensity of light admitted to the eyes when they were up). In addition, the pattern changed character and there was no longer a fixed centre. Also, the evoked potentials were not as easy to see. The significance of these observations is not known at present but at least it demonstrates that a new technique has become available for the study of so-called flicker effects.

DISCUSSION ON DEWAN

Grey Walter

I have just one comment on this last point of yours about 'noise' in the nervous system, which obviously must interest us as we are always dealing with it—usually trying to get rid of it in fact. There's one function in which noise seems to have some useful effect, described by Van der Tweel and his colleagues in Amsterdam. This is concerned with the response of the visual system to sinusoidally modulated light, which produces a rather sinusoidal response in the visual cortex, which in itself is rather surprising. Now if you stimulate the retina with light modulated at only 5 per cent, which is scarcely perceptible as modulation, at say 5 Hz, the response in the visual cortex contains the fundamental and second harmonic, that is a 5 Hz component and a 10 Hz component, indicating non-linearity in the visual system. If now you superimpose on the 5 Hz modulation of the light, a *noise* component, that is irregular modulation as well, then the second harmonic disappears. In other words you've linearized the system by adding noise. Now this could be a valuable property in any system which must contain non-linear threshold components.

Dewan

On this point I would just like to mention that engineers frequently use noise of high frequency to alter non-linear characteristics. The 'bias signal' in tape recorders, for example, 'linearizes' the characteristic to get rid of hysteresis. The engineers sometimes call it 'dithering'. I think that the observation you mentioned is very significant because it shows that 'dithering' can occur in biology.

Jasper

I was just saying to Adrian before you started that one new element in our understanding of the brain is this—the ideas we get from computer technology which we didn't have when we were working at it originally. Of the two questions I had in mind, one was your interpretation of the function of the alpha rhythm, which is a fascinating one. The thing that impresses me about the alpha rhythm, however, analysing it into what we think may be a representative of the alpha rhythm, with micro electrodes, is that it has a very powerful gating action, not only of an activating function; there is a large amount of inhibition between cycles. This is such a powerful inhibition that it may prevent conduction in synaptic pathways in the cortex that otherwise would go through, and which are then facilitated in the other phase of the cycle. This is not a new idea of course; Dr. Grey Walter and others have suggested this scanning and gating hypotheses, but I think we now have the micro electrode evidence that this is a very real phenomenon. What is the functional significance of this gating action of alpha-like rhythms?

Grey Walter

I wondered if it mightn't be better defined if we thought of the rhythms like alpha rhythms as both *controlling* signals, and controlled *by them*. This seems to be one of our difficulties; we saw yesterday that you can potentiate or promote alpha rhythms even with a single flash of light. You can also drive or entrain the alpha rhythms within a small range so the intrinsic rhythms are responsive to sensory signals. They also gate, possibly scan or distribute them. To me it looks rather like the pad-operated traffic signals we have in England which control the traffic with the red and green lights but are also controlled *by* the traffic in some circumstances. There's a rather complex interaction between the controlling system and the vehicle or message signal. When there are no signals or traffic coming, the red and green lights alternate on a time cycle. The moment a vehicle comes in, this cycle is interrupted, as the alpha rhythm is interrupted by a visual signal, but it can also be set by an *afferent* vehicle so that you have a rather complicated system in which traffic and rhythms are continuously interacting.

Stark

We have studied two interesting noise processes in the nervous system; one influencing pupillary aperture which inject in an enormous amount—up to 33 per cent of area fluctuation may be just due to noise modulation. There seems to be no functional role to this noise when studied from a number of different points of view. However, it can be used by the neurophysiologists to understand some of the processes within the brain stem, and we've been able to show that the noise is generated at and about the level of the Edinger-Westphal nucleus. It is non-linear noise and probably has to do with perhaps a multiplicatory relationship between input signals and a neural discharge at this E–W nucleus (25). The second aspect of noise that we've studied has been the irregular impulse discharge that one sees on single nerve fibre trains, and it's very nice that Lord Adrian, who started this work on single unit discharges, is here. One may look in the crayfish at the two signals carried by two single nerve fibres, which are carrying all the information. These interval trains recorded separately and simultaneously are not correlated in detail. This indicates that the detailed interval structure does not in fact carry any signal information that is useful to the animal at the next synapse, at the next signal stage. One can demonstrate this by a variety of experiments in these simple invertebrate animals, and again although it mathematically could be used for, say 100-bits-per-second carrier rates it is *not*, but has a signal capacity of only 3-bits-per-second in Shannon's definition. These seem to be two phenomenon only useful to the neurophysiologist for understanding the physical processes that are generating these impulses (26). On the other hand, a noise process that many people have studied, the micro saccades of the eye, seem to have a function of preventing the adaptation to light. These three kinds of noise processes are in addition to the EEG alpha rhythm, which appears to be a narrow band noise process.

Grey Walter

Now if we could discuss the movie: could you tell me, what is known about subjects' introspective reports while they are controlling their alpha in this way? Is there anything peculiar about their state of mind—I mean do they feel it to be a 'strange' experience at all?

Dewan

Nothing particular I think. One subject, I remember, complained of having a headache for about a week after trying this extensively. In my case, I remember that I once got confused and sent five dots instead of four, but I didn't feel at any time that I was altering my level of attention: rather that I was constantly monitoring something or other—eye position I think. This is another reason to doubt the visual attention hypothesis.

Grey Walter

The reason that I asked this is that when we have used the CNV to control machinery as a technical trick, we often get a very marked diuresis, suggesting that whatever state of mind one gets into has some sort of outflow to the lower centres that control the antidiuretic hormone. Have you come across this?

Dewan

Not at all.

Grey Walter

I was particularly struck by the resemblance in the earlier frames of the movie, when one looked at the girl's eyes, to the eye movement component of *petit mal*—this peculiar upward flicker, and of course one wonders if this has relationship to *petit mal*, functional or otherwise.

Mulholland

Isn't it true that one can suppress *petit mal* by having someone 'pay attention' to visual things?

Grey Walter

Some people can. There are patients who have attacks frequently when they close their eyes. As they close their eyes they generate wave spike trains.

Horn

When the girl's eyes went up, there was a clear oscillation on them. It certainly wasn't 10 Hz but I thought there might have been a 10 Hz component. Has anyone measured the frequency of the oscillation? The point being raised by, I think, Don Lindsley is that when the eyes are rotated upwards there may develop a 10 Hz modulation of the standing potential of the retina which might be responsible for the effect. Would it be possible to impose some other frequency of movement on the eye and see whether the EEG follows that frequency?

Fenwick

This has been done in fact—originally by Adrian with clockwork rods. We have also done it recently with electrical transducers. The vibration of the eye doesn't in any way alter the frequency of the alpha rhythm.

Mulholland

I can voluntarily set my eyes into a sort of tremor which I haven't measured, but it's quite clear that this makes no evident change in the EEG.

Grey Walter

We had a patient with 'jelly nystagmus' at about 16 Hz which he could produce voluntarily. His eyeballs looked as though they had turned into a sort of wobbling jelly, but there was no component at this frequency recorded at the back of his head. Of course there were eye movement potentials at 16 Hz.

REFERENCES

1. Adrian, E. D. and Matthews, B. H. C. (1934) *Brain*, **57**:355
2. Ashby, W. R. (1952) *Design for a Brain.* John Wiley and Sons
3. Breger, L. (In press) *J. abnorm. Psychol.* Monograph
4. Bruce, V. (1960) In Cold Spring Harbour Symposia on Quantitative Biology, 25
5. Dewan, E. M. (1967) *The P (Programming) Hypothesis for REMS.* Presented to APSS Meeting, Santa Monica
6. Dewan, E. M. (1967) *Nature, Lond.* **214**:975
7. Gaarder, K. (1966) *Archs. gen. Psychiat.* **14**:253
8. Goodwin, B. (1963) *Temporal Organization in Cells.* Academic Press:
9. Greenberg, R. and Leiderman, P. (1966) *Comprehensive Psychiat.* **7**:517
10. Lee, Y. W. (1960) *Statistical Theory of Communication.* John Wiley and Sons
11. Mulholland, T. B. and Evans, C. R. (1965) *Nature, Lond.* **207**:36
12. Mulholland, T. B. and Evans, C. R. (1966) *Nature, Lond.* **211**:1278
13. Newman, E. A. and Evans, C. R. (1965) *Nature, Lond.* **206**:534
14. Pittendrigh, C. (1960) In Cold Spring Harbor Symposia on Quantitative Biology, 25
15. Sokolov, Y. N. (1963) *Perception and the Conditioned Reflex*, translated by S. W. Waydenfeld. MacMillan Co., pp. 288–294

16. Tarjan (1962) Proc. of the First International Symposium on Optimizing and Adaptive Control (IFAC)
17. Tsien, H. S. (1954) *Engineering Cybernetics*, McGraw-Hill, p. 253
18. Walter, W. G. (1960) In *Electroencephalography*. Ed. by Hill and Parr. First edition, London, MacDonald and Co., p. 268
19. Walter, W. G. (1962) In *Neural Physiopathology*. (Ed. by Grenell), p. 244
20. Walter, W. G. (1963) *The Living Brain*, Norton
21. Walter, W. G., Cooper, R., Aldridge, V. J. and McCallum, W. C. (1964) *Nature, Lond.* **203**:380
22. Wiener, N. (1958) *Nonlinear Problems in Random Theory*. John Wiley and Sons
23. Wiener, N. (1961) *Cybernetics*, Second Edition. MIT Press
24. Winfree, A. (1967) *J. Th. Biol.* **16**:15

Discussion

25. Stark, L. (1968) Section II: '*Neurological Control Systems*'. 'The Pupil'. *Studies in Bioengineering*. Plenum Press: New York
26. Stark, L. (1968) 'Section I: *Neurological Control Systems*'. The Crayfish. *Studies in Bioengineering*. Plenum Press: New York

Attention, Eye Movements and Motion Detection: Facilitation and Selection in Optokinetic Nystagmus and Railway Nystagmus

Johannes Dichgans and Richard Jung

Neurologische Klinik mit Abteilung für Neurophysiologie der Universität Freiburg i. Br., Germany

Voluntary control of visual attention in man and its relation to eye movements allows one to make quantitative studies of attentive mechanisms by combined subjective and objective methods. The recording of optokinetic nystagmus (OKN) induced by moving stimuli and the scaling of velocity sensation with and without eye movements have disclosed two mechanisms of motion perception which are altered by visual attention: afferent motion detection by moving retinal images and efferent motion detection related to ocular following movements (slow phase of OKN). These two mechanisms are influenced by voluntary attention as, also is optokinetic nystagmus. The selective function of attention can be studied in 'railway nystagmus' during which retinal images moving with different velocity and direction occur in the visual field. Previous observations by one of us (R.J.) on the influence of attention and inattention upon the speed of eye-following in OKN and upon the related alteration of velocity perception (33) were investigated with quantitative methods by Körner and Dichgans using psychophysical scaling of velocity estimation and relating them to eye movements (13, 37).

It is the purpose of this paper to describe some of these experiments on movement detection in relation to visual attention. In the first part changes of OKN are described under various conditions of visual attention, inattention, auditory attention and distraction. A quantification of OKN and motion detection during attentive eye following, inattention and stationary eye fixation follows. These experiments demonstrate that afferent retinal motion detection causes higher velocity sensation than efferent oculomotor motion detection. The second part describes eye movements associated with various modes of attentive selection of visual objects by foveal fixation in railway nystagmus. The results are discussed in relation to the role of slow following movements and quick saccadic eye movements in motion and space perception.

METHOD

In 50 experiments using 40 normal subjects with normal vision, horizontal eye movements were recorded by electronystagmography (33) using bitemporal electrodes with head fixation. Ocular movements were calibrated for visual angles of 15 degrees before and after each experiment to control for significant alterations of the corneo-retinal potential during the test. When alterations occurred, the amplitude and velocity measurements were corrected accordingly during the experiment. Movement stimuli consisted of alternating black and white stripes of equal visual angle (7 degrees), which were projected on a white half-cylindric screen, visual angle 140 degrees, at speeds varying from 5 to 140 degrees per second.

The angular velocity of the slow phases of recorded OKN was measured. Mean values of the quickest slow phases during a short period of 2–5 s of attentive eye following were computed for comparison with subjective scaling. In experiments with visual inattention the average of all slow OKN-phases which occurred during comparative estimation of velocity, was measured. The velocity of retinal images was computed as the difference between the velocity of eye tracking and that of the moving stimulus. Subjective evaluation of the speed of the moving stripes was made according to Stevens' psychophysical methods of magnitude or ratio estimation (53).

Three types of experiment were performed:

1) In the first series 10 subjects were asked to follow the moving stripes with attention, and then to shift to inattention by looking 'through' and trying to ignore the moving pattern. The percentage differences were computed by comparison of perceived speeds in the inattentive *v.* attentive states. Stimulus velocities ranged from 5 to 90 degrees per second in ten steps.

2) In a second series 20 subjects estimated moving stimulus velocities between 10 and 130 degrees per second (7 steps) under conditions of maximal visual attention. Velocity judgments were made relative to a standard or reference of 70 degrees per second, which was assigned an arbitrary value of 10 and presented to the subject before each comparison stimulation.

3) In the third series (10 subjects) velocity estimations were obtained during fixation of a small stationary spot in the moving field. In this experimental condition, movements of retinal images occurred with the same angular speed as the stimulus velocity, since following movements were completely suppressed by fixation of a small stationary spot (the latter was controlled by recording). The explored range was 5–140 degrees per second (17 steps) with the standard of 70 degrees per second repeatedly presented to the subjects between comparison stimulations. Binaural noise precluded unwanted acoustical cues made by the motor running at different speeds.

For railway nystagmus the subjects looked sideways out of a moving train while keeping their heads still. At the same time they attended to different objects at various distances. Moving retinal images in the periphery of foveally fixated objects have different velocities and different directions of movement and these are opposite for background and foreground during middleground fixation. The recording of eye movements was done as the subject viewed a film made from an appropriate position in the moving train, shifting his attention with fixation of various objects in fore-, middle- and background (*Figure 5*).

RESULTS

1. Attention, Distraction, Eye Movements and Motion Perception

Before describing the quantitative evaluation of these experiments some earlier observations on the relation between oculomotor activity, attention, non-attention, distraction and visual imagination should be mentioned. *Figure 1* shows examples of these findings in OKN elicited by stripes moving

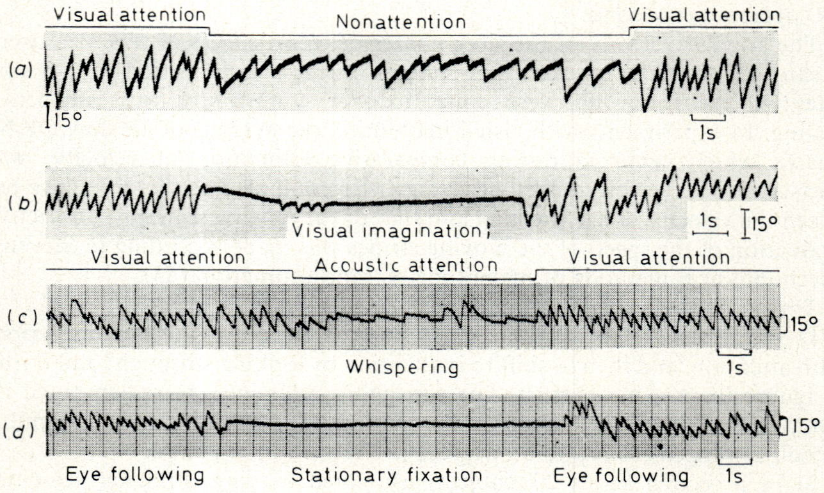

Figure 1. Optokinetic nystagmus (OKN) in normal subjects during visual attention, inattention, imagination, acoustic distraction and fixation

At the beginning and the end of each record OKN is maximal with full attention and adequate eye following to stimulus velocities of 90 degrees per second (*a*, *b*) and 60 degrees per second (*c*, *d*)

(*a*) During inattention the velocity of the slow following phase of OKN drops to 10–30 degrees per second with simultaneous increase of subjective velocity caused by higher speed of moving retinal images (*Figure 2*)

(*b*) OKN is suppressed during visual imagination of a stationary image similar to stationary fixation (*d*)

(*c*) Acoustic distraction by listening to whispering diminishes OKN and nearly suppresses it

(*d*) Complete inhibition of OKN by fixation of a small stationary spot in the moving field

horizontally at 90 degrees per second to the right (*Figure 1a* and *b*) and 60 degrees per second to the left (*Figure 1c* and *d*). All records begin with full visual attention to the moving stimulus thus showing adequate eye tracking which results in a sharp visual image of the contrasting white and black stripes.

During voluntary inattention OKN markedly diminishes, as seen in the velocity of the slow phase which lags behind the stimulus velocity by more than two-thirds (*Figure 1a*, *2*). With 'inattention' all subjects first notice blurring of the moving black-white contrasts, and if they try simultaneously to perceive the velocity of the moving stimulus without following it, an impression of exaggerated velocity occurs. Not one of the 50 subjects tested described slowing during inattention. Some saw the acceleration during inattention but not spontaneously and only after questioning.

In contrast to voluntary visual inattention, an increase of perceived speed was not usually observed during spontaneous slackening of attention. Usually no acceleration was noticed during distraction by auditory and other stimuli, although the slowing of OKN was similar (*Figure 1c*). Obviously it is difficult to get a simultaneous judgment of perceived speed while attending to other stimuli or stationary mental images. The diminution of OKN during such imagery (*Figure 1b*) is often very effective and causes interruption of OKN which may be as complete as during fixation of a *real* stationary object in the moving field (*Figure 1d*).

During stationary fixation it was always possible to get an estimation of the speed of the moving stripes from the subjects. This estimation was invariably in the direction of a higher perceived velocity, reaching values nearly twice the speed seen during the attentive eye following.

(*a*) *Quantification of velocity perception*

With attentive ocular tracking the velocity of the slow phase of OKN approximated the stimulus velocity up to 50–60 degrees per second, but at higher stimulus velocities the slow phase tended to lag behind the velocity of the moving stripes. For comparison with velocity perception the quickest slow phases during attentive eye following were measured. The mean values of ten subjects are marked in *Figure 3* by black dots (. . .). Interspersed between the best or most accurate following movements are irregular periods of slower following. These are associated with periodic diminution of attention, often unnoticed by the subjects themselves. With training, however, several subjects were able to report involuntary slackening of attention which could be overcome with voluntary effort. They reported simultaneous blurring of the stripes and perceived faster motion during periodic diminution of attention.

The limits of maximal (eye following) accuracy varied in different subjects from 50–100 degrees per second. Two subjects could follow up to 110 degrees per second. Mean values of the slow phases of OKN, however, were lower than stimulus velocities exceeding 30 degrees per second, due to the described inattentive periods of slower interspersed phases. These mean values were not included in the final data for attentive states. Magnitude estimation of velocity with a reference level velocity of 70 degrees per second, increased according to stimulus velocities when eye tracking was accurate. At higher speeds, when it lagged behind the moving stimulus, estimation magnitudes markedly exceeded stimulus velocities (*Figures 3* and *4*).

During inattentive states all subjects produced estimations of higher velocity with a larger intra- and interindividual scatter. The resulting curve of velocity estimation during inattention varies with increasing stimulus velocity and shows a progressively steeper slope and increasing distance from the regression line for optimal eye tracking and velocity estimation (*Figure 3*). Ocular following during the inattentive state was always much slower than

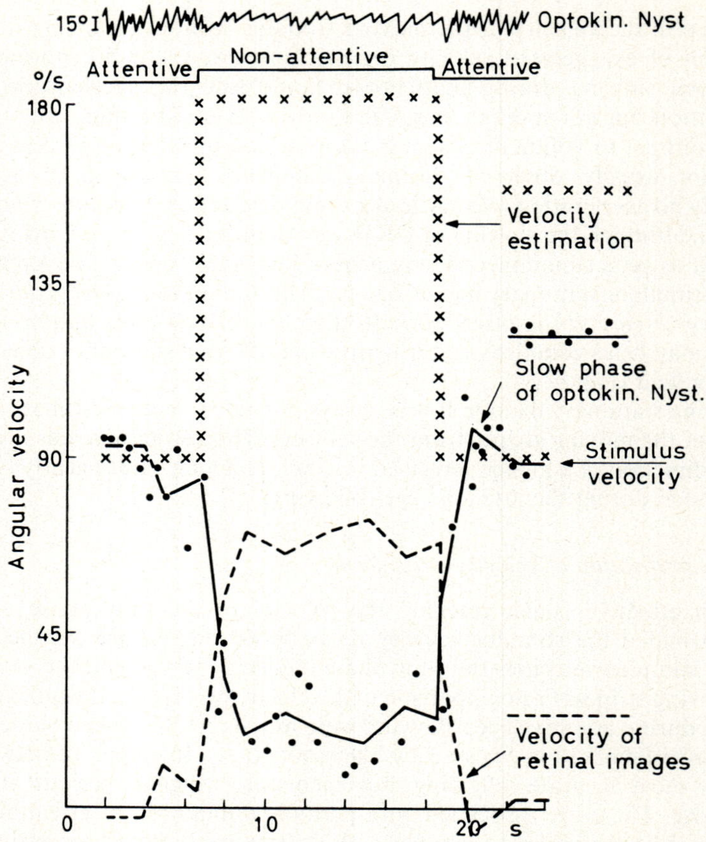

Figure 2. Diminution of optokinetic nystagmus during inattention related to perception of higher velocity

The diminished velocity of 'inattentive' eye following causes a higher velocity of moving retinal images, although stimulus velocity remains constant at 90 degrees per second. The upper record of OKN is the same as in *Figure 1a*. During attention it begins with velocities of the slow phases of OKN (eye following movements) which correspond approximately to stimulus velocity; motion detection is purely efferent by ocular following. During inattention the slow phases lag behind the stimulus velocity. This slower velocity of the 'inattentive' eye following (20–30 degrees per second) causes moving retinal images with velocities of 60–70 degrees per second, computed as the difference between stimulus velocity and eye following. During inattention the velocity was estimated to be approximately double that during attention. This exaggeration of apparent motion disappears with renewed attention

stimulus velocity, commonly below 15 degrees per second, thus lagging further and further behind with higher speeds of the moving stimulus. The magnitude of velocity estimation during inattention increased proportionally with the velocity of moving retinal images, computed as the difference between ocular velocity during eye following and stimulus velocity (*Figure 3*). Some subjects were able to gradually change their visual attention from optimal attentive eye following of 90 degrees per second stimulation, through inter-

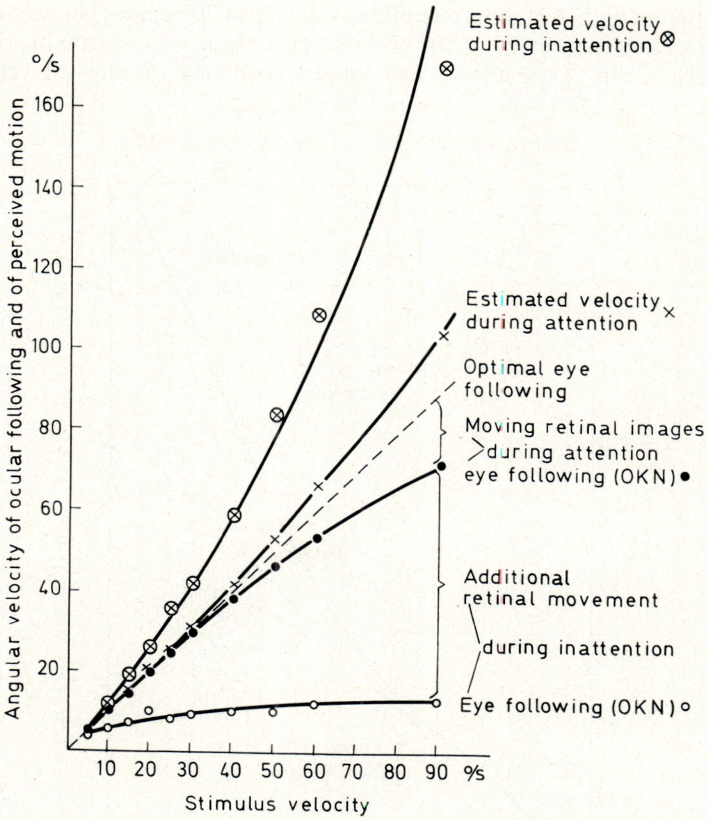

Figure 3. Estimation of various stimulus velocities and OKN during attention and inattention (mean values of 10 *subjects)*

Subjective scalings of velocities and objective measurements of the slow phases of OKN were combined on a common ordinate of angular velocity and an abscissa of increasing stimulus velocities (5–90 degrees per second)

During *attention* the eye following movements and velocity perception are nearly identical with stimulus velocity up 50 degrees per second. Above this, eye following movements lag progressively more behind, resulting in moving retinal images and increased velocity perception at 60–90 degrees per second stimulus velocity

During *inattention* a marked slowing of eye following velocity occurs together with an excess of velocity estimation. Both curves tend to deviate similarly and to form a mirror image from the optimal regression line of eye following and velocity perception

This shows that afferent velocity perception resulting from moving retinal images attains approximately twice the value of efferent velocity perception caused by adequate eye following

mediate velocities of 70–30 degrees per second, to a reduced, inattentive following rate of 10 degrees per second.

(*b*) *Scaling of velocity with attentive eye tracking*

An example of individual results regarding ocular following and velocity scaling is shown in *Figure 4*. It covers stimulus velocities between 10 and 130 degrees per second. A standard stimulus velocity of 70 degrees per second was assigned an arbitrary value of 10. This reference value was repeatedly exposed during the scaling experiment. The subject estimated diminished velocities

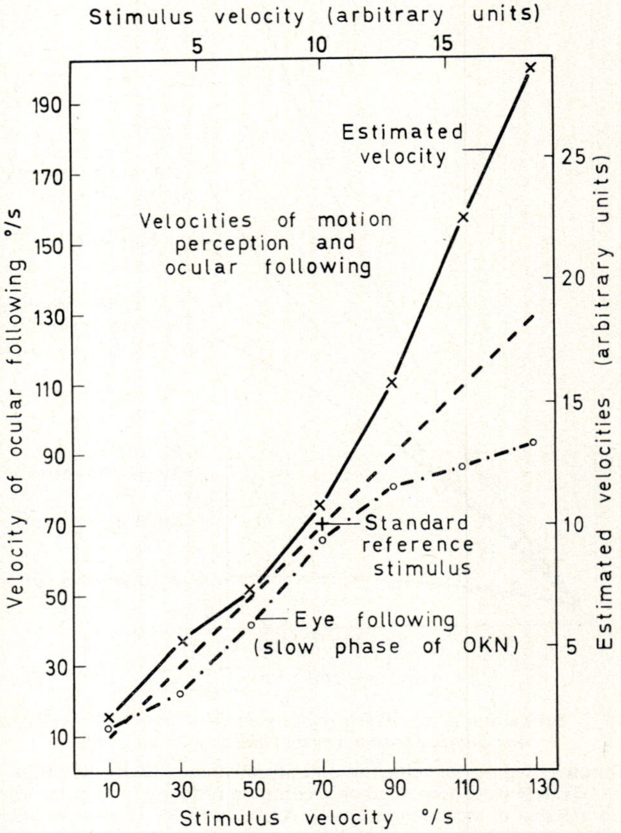

Figure 4. Subjective velocity and attentive ocular following at different stimulus velocities in one subject (♂, 29 years old)

Estimations with full visual attention were made in reference to a standard velocity of 70 degrees per second, set as an arbitrary unit of 10. Eye following is slightly lower than stimulus velocity below 90 degrees per second. Ocular following lags behind stimulus velocity more markedly at 110 and 130 degrees per second. Simultaneously estimated velocity rises in a steeper slope. Thus excessive subjective velocity develops like a mirror image of inadequate eye following, in relation to the optimal regression line. This corresponds to increased velocity of moving retinal images during slower eye following

with numbers below 10 and higher velocities with numbers greater than 10. These estimations approached a value of 30 for real stimulus velocities of approximately 130 degrees per second (i.e. over-estimation of 70–80 degrees per second). The figure shows that despite maximal attention the slow phase of OKN lags behind stimulus velocities above 50 degrees per second and falls well below the optimal regression line of the moving stimulus. Estimated velocity is increased with decrease of ocular following. The reciprocal relation of eye tracking to velocity estimation at higher stimulus speeds is similar to the relation observed for lower speeds and inattention (*Figure 3*). This suggests that increased retinal motion velocity accentuates visual velocity perception. The angular velocity of moving retinal images is multiplied for perceived velocity by a factor of (1·6–1·9). Thus the same rule of higher velocity sensation of afferent motion perception from moving retinal images may be assumed for the higher scaling magnitudes during attentiveness at faster stimulus speeds, as during inattentive following movements illustrated in *Figures 2* and *3*. The common factor in both conditions is that the eye following velocity lags behind the stimulus velocity.

In the subject depicted in *Figure 4*, as in most others, the facilitation of OKN by maximal attention was not strong enough to induce adequate eye following over 90 degrees per second. During visual inattention the slow phases drop behind at much lower velocities and usually do not reach speeds greater than 20 degrees per second (*Figure 1a*, *2*). The perceptual effect of slowed inadequate eye following is an apparently higher velocity of the moving stripes.

In the experiments described the shifts from selective visual attention to visual inattention to the moving stimulus are examples of voluntary alterations of attention. This kind of visual attention is different from general vigilance or alertness, although vigilance is a necessary condition for attention.

(*c*) *Velocity perception without eye movement*

In the third series of experiments stimulus velocities were estimated with controlled suppression of eye following movements by the fixation of a small stationary red spot within the field of moving stripes (*Figure 1d*). Under these conditions the movement of retinal images has the same angular velocity as the stimulus. Over the whole range of stimulus speeds with eyes stationary the subjects over-estimated the real stimulus velocities by a factor of 1·6–1·9 (13, 13a, 37).

(*d*) *Distraction by auditory stimuli and cognitive processes*

During prolonged visual attention the optimal tracking velocity of OKN diminished when distraction stimuli from other sensory modalities occurred. This was studied by applying auditory verbal stimuli in order to distract attention from the visual moving stimulus. Most effective were words and sentences whispered in the ear of the subject. During this distraction the subject ceased to follow the moving visual stimulus with complete attention and simultaneously the velocity of the slow OKN phase dropped (*Figure 1c*).

The reduction of OKN during competition between auditory and visual stimuli shows large intra- and interindividual variations which have not been studied in detail. The lowest ocular velocities during the slow OKN phases occurred when the subject had to listen very attentively and had difficulty in understanding the spoken words. The reduction in the velocity of slow phases of OKN during auditory distraction (*Figure 1c*) was similar to that during voluntary visual inattention when looking 'through' the moving stripes (*Figure 1a*). Although visual imagery of a stationary pattern can suppress the OKN completely, some small quick phases may be interspersed (*Figure 1b*). Distraction (from the moving stimulus) not involving sensory stimuli, e.g. by performing mental arithmetic or concentrated thinking causes a similar reduction of eye tracking speed. In trained individuals this is mostly perceived as blurring or as velocity increase.

The optokinetic stimuli used in the experiments described above were uniform rotating stimuli which cover most of the visual field and move at a constant circular velocity. Changes in stimulus velocity affected all components of the field equally. This stimulus situation is quite different from that producing railway nystagmus.

2. Railway Nystagmus and Visual Selection of Different Motion Velocities

(*a*) *Motion perception*

Looking out of the window of a moving train, one experiences movement of various objects in the outside world apparently streaming backwards at different speeds. The foreground moves rapidly, the middleground more slowly, and background movement is hardly perceptible. Although the train moves at constant speed, there are two variants which determine the velocity of perceived objects relative to the observer: (*1*) their distance from the railway line; (*2*) the relative location and direction of their apparent movement. In contrast to the stimulus situation for OKN (with constant velocity and stimulus distance) the angular velocity of the various objects is dependent upon the distance from the observer. The velocities of objects increase from their first appearance in the anterior periphery as they move to the centre of the field of regard, and decrease similarly as they leave the centre before disappearing in the direction opposite to the moving train. The velocities of the objects vary in relation to their distance because the visual angle is the same for short sections transgressed rapidly in the foreground, and long extensions in the background with very slow movement. One fixates certain objects with foveal vision and the eyes follow the appropriate waxing and waning of speed across the visual field, thus causing the phenomenon known as railway nystagmus.

Motion perception during railway nystagmus principally concerns the fixated object although a general sensation of movement in space may be present. This motion sensation concerns the train moving with the subject, or more often the environment moving in the opposite direction. When one looks 'diffusely' out of the window, movement velocity appears greater than during particular object fixation. However, the speed difference between an object deliberately pursued and one not tracked is less impressive than with motion stimulation of constant angular velocity and OKN. In railway nystagmus during normal daylight viewing conditions the relative velocities of objects at various distances are perceived more clearly than actual speed differences. This is most prominent during the night, when lighted objects seem to stream backwards at very different speeds against a dark background with no clear sensation of depth.

It seems surprising perhaps that fixating of objects in the middleground may cause appropriate movement sensation, even though different opposed retinal movements occur simultaneously in the visual field. The foreground images move quickly in the same direction over the retina and the background images show slow opposed motion when the middleground is fixated visually. Although one might expect marked movement contrast from these different retinal movements this is rarely experienced. It may appear that the most distant background (for example the moon on the horizon) moves slowly in the same direction as the observer while foreground objects are streaming rapidly backwards. These effects appear only when attention is specifically directed to them, and apart from such side effects selective motion sensation is dominated by the pursued objects. Thus a predominance of efferent motion perception during pursuit and a relative suppression of the effects of retinal motion occurs during railway nystagmus.

(*b*) *Railway nystagmus*

Various types of eye movements during railway nystagmus are shown in *Figure 5*. These are more irregular with varying speed of pursuit depending upon the apparent velocities in the fore-, middle- and background (*Figure 5a*). They show less rhythmic sequences than during ordinary OKN (compare *Figures 1d* and *5* from the same subject). When attention is directed to objects entering the fore- or middleground from the direction in which the train moves, a large rapid phase first appears by which the subjects attempt fixation (*Figures 5a* and *b*). The following long pursuit is often interrupted by quick phases in the direction of pursuit, opposite to normal quick phases of OKN and of higher velocity than apparent motion (anticipating gaze, *Figure 5c*). Smaller quick phases of opposite direction to pursuit also occur (*Figure 5b*). Such quick saccades occur more often at the beginning of the long following movements (*Figure 5a*). They seem to represent searching and correcting before an object is properly fixated and accurately tracked. Interruption of pursuit movements by larger rapid phases in the same direction results in square wave forms (*Figures 5a, d* and *e*), larger than the square waved '*Gegenrucke*' of normal fixation. 'Staircase' sequences of small rapid phases may initiate long pursuit phases (*Figure 5b*), and some slow nystagmic lapses of long duration are similar to segments of a sinusoidal curve according to the waxing and waning of apparent velocity. When the subject begins to track an object in the centre, the slow phase has a round form instead of an S-shape (*Figure 5a* and *b*). Rapid phases in the direction of pursuit appear mostly when 'interesting' objects attain relatively high speed in the central part of the field of view. At the end of the long following movements a large quick phase appears again which initiates a new sequence of pursuit. These often start with a few rapid scanning phases. (*Figure 5b*.) The velocities of the tracking movements correspond approximately to that of the fixated object in the middle- or background, but do not properly track the high velocities in the centre of the foreground.

The maximal velocity of eye tracking in railway nystagmus is difficult to measure. Even in the sinusoidal following movements it usually does not exceed 60 degrees per second except in the rare 'mixture' of slow and quick parts as in *Figure 5d*. 90 degrees per second were the highest pursuit velocities in some parts of the slow phases.

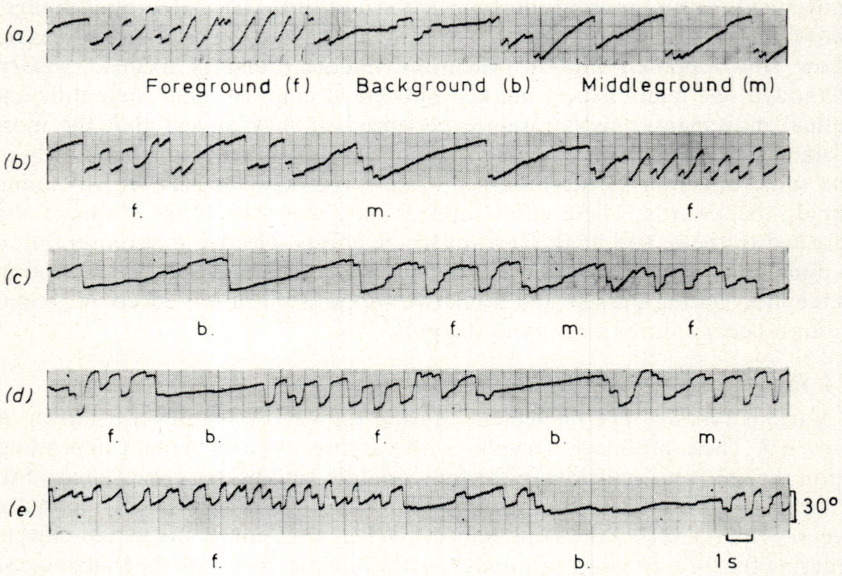

Figure 5. Railway nystagmus and its various forms with visual selection during varying attention to foreground and background

The subject looks at the panorama which appears at the right of the train window showing objects apparently moving from left to right in a film of 50 degree aperture. Bitemporal d.c. recording shows eye movements to the left as a down stroke

(*a*) Attention is wandering from the middle to the foreground. Eye movements try to grasp and fixate selected objects appearing on the left by quick saccadic phases to the left and pursuit movements to the right. Opposite saccades to left and right interrupt the following movement with intermittent fixation of other objects. Rapidly moving objects in the foreground are pursued by slow phases with maximal velocity of 55 degrees per second. Much slower following movements appear when looking to the background with some re-fixations. Long following movements over 50 degrees amplitude occur in pursuit of middleground objects

(*b*) Similar pursuit movements, interrupted by opposite re-fixations, are partly initiated by steplike sequences of rapid phases

(*c*) Long pursuit movements occur when looking to the back and middleground. Pursuits interrupted by small saccades when looking to the foreground result in a step like S-form

(*d*) The first fixation in the foreground on an object with rapid apparent movement results in a deformed pursuit with a quick component which may anticipate the fixation of the rapidly disappearing object. In the next attempts at fixation some rapid phases appear in the direction of pursuit to the right, and these result in large square waves

(*e*) Unsuccessful following movements to very fast objects in the foreground result in square wave forms with rapid components to both sides. Unsteady pursuits with some small rapid phases appear when looking to the background

DISCUSSION

The principal findings are:

1) Attentive facilitation is necessary for correct tracking in OKN.

2) Attentive selection with foveal fixation determines the tracking of a selected object against a moving background during railway nystagmus.

3) Motion perception employs two mechanisms which can be separated by voluntary attention; (*a*) afferent inflow by moving retinal images which

signals movement during inattention, (*b*) efferent outflow with eye following which requires an intention to look.

4) Correct ocular tracking is accompanied by velocity perception which corresponds to that of the real stimulus, whereas moving retinal images exaggerate the velocity information by a factor of up to 1·6–1·9.

5) The slow tracking phase of OKN is the mechanism for correct velocity information by monitoring oculomotor output, controlled by foveal fixation. These results are now discussed together with some neurophysiological and psychological considerations.

Attention and Vigilance as Conditions of Optokinetic Nystagmus

Berlyne (5) has pointed out in this conference the difficulties that even psychologists may have in clearly defining attention, set, attentiveness, and their distinction from vigilance: for this reason we have not tried to develop an unequivocal terminology in the description of our experiments. For simplicity we have employed the terms 'visual attention' and 'inattention'. The term 'selective attention' was used only in reference to the particular case of visual selection in railway nystagmus. Selection, however, is generally a criterion of attention: one of us (R.J.) has compared the selective function of attention to a searchlight of varying direction and breadth and brightness of beam, illuminating the inner and outer world (34). The term 'intention to look' as used here, indicates the consciously directed component of visually attentive OKN. Facilitation of OKN and velocity information are optimal during the 'intention to look' for moving stripes and to follow this stimulus. The term 'visual inattention' does not describe sharply enough the reduced intention to follow the moving stripes, or the change in attitude towards the moving stimulus which results in an altered velocity perception. It may however be sufficient to explain the results of our experiments.

The influence of different components of visual attention and its general precondition, vigilance, may be demonstrated by the effects on optokinetic and vestibular nystagmus. For adequate ocular following, OKN requires foveal fixation and optimal visual attention which is deliberately directed to the moving objects. Vestibular nystagmus is cut down by visual attention and is facilitated by eye closure but requires general wakefulness (39). Vigilance alone is not sufficient for full OKN with adequate eye following. Optimal OKN corresponds to the '*Schau-Nystagmus*' of Ter Braak (54). The slowed eye movements lagging behind stimulus velocity which persist during 'inattentive' OKN may correspond to Ter Braak's '*Stier-Nystagmus*' but its pure subcortical origin in man is doubtful. Only during attentive foveal *Schau-Nystagmus* in man with 'intention to look' at stripes and possibly also for the 'instinctive interest' described by Rademaker and Ter Braak (36, 51) in dogs, are the various criteria of human OKN which were discussed in other papers (33, 39) well developed. These criteria include exact correspondence of eye tracking with stimulus velocity, and the directed shift of the nystagmic beats (*Schlagfeld*) to the side of the quick phases—which would correspond to visual search and foveal fixation of new objects appearing in the visual field. Kornhuber (39) and Hood (31) have shown that after exclusion of foveal vision correct eye tracking becomes much slower and the *Schlagfeld* is converted from the side of the quick phase to the side of the slow phase of OKN (31). Occasionally long pursuit movements occur during ordinary OKN when the subject tries to fixate one contrast border over large parts of the field of view. These voluntary pursuits may be compared to the long pursuit phases of railway nystagmus but their straight line form of constant velocity differs from the rising and falling speed and curved or S-shaped slow phases of railway nystagmus. Only for these rare deliberate long pursuits does the rule fail that OKN beats with its *Schlagfeld* on the side of the quick phase. This *Schlagfeld* to the direction of the quick phase is automatic but may have psychological and biological significance related to the vital importance of novel moving stimuli entering the peripheral visual field, which lose their interest value after they have been fixated and identified. It can be modified by deliberate attentiveness to the right or left side of the field of regard. Conversely to OKN, vestibular nystagmus beats in the *Schlagfeld* of the slow phase.

Vestibular nystagmus is inhibited, in contrast to OKN, by intentional fixation of moving stimuli. In this case OKN prevails and may supplant vestibular nystagmus. Vestibular nystagmus is optimal in the dark and still better with closed eyes suppressing all visual attention, as long as general vigilance is present. Thus vestibular nystagmus is independent of attention, dependent on vigilance and disappears in drowsiness and sleep except in the REM dream stage (35, 52). Optokinetic after-nystagmus is more marked in the dark with open eyes than with eye closure (46) which reduces visual attention and often also induces some diminution of vigilance. Since vigilance is a necessary condition for any attentive state in all modalities its significance need not be discussed here.

Differences between various attentive states and general vigilance may be shown in the experiments involving distraction and concentration. As soon as visual attention is diminished by acoustic attention and calculation (i.e. inner attention) OKN is diminished. In contrast, vestibular nystagmus is facilitated during the same procedure. Calculation increases general vigilance but decreases visual attention. Internalized visual attention accompanied by facilitation of OKN may be demonstrated by Zikmund's (58) experiments on imagining moving visual images and Erickson's (19) hypnotic suggestion of moving stimuli: such conditions may elicit eye movements similar to OKN in the direction of quick phases opposite to the imagined direction of suggested motion. These conditions correspond to special internal attentiveness, although their relation to general vigilance is debatable since rapid eye movements occur in 'dream' sleep. The latter may be related to 'internalization of attention' to the imagery of the dream (1).

When attention is transferred from vision to hearing, this is not restricted to a shift from one afferent sensory modality to another but also involves alteration in the activity of the occipital cortex. Adrian has shown that occipital alpha activation occurs after this transfer of attention from visual to auditory stimuli (2).

An interesting model for the influence of selective visual attention on OKN are observations using visual after-images—which are equivalent to stabilized retinal images. When the after-image is localized outside the foveal region (eccentric after-image) the subject may try in vain to fixate this after-image. Doden and co-workers (15, 16) have shown that attention directed to an eccentric after-image of this kind, causes inversion of OKN and that normal OKN is resumed again when attention returns to the moving stimulus. Kommerell and Mackensen (38) recently have shown that saccadic eye movements appear in the direction of the after-image similar to the quick phases of inverted OKN also without moving stimuli. These correspond in amplitude to a series of attempted saccadic fixations, when the after-image moves further to the periphery with each saccade.

Following Exner's studies of motion perception and his neuronal models of attention (20) it is generally accepted that attentional facilitation acts selectively on certain neuronal systems with simultaneous inhibition of other functions. Hofman postulated that attention transfers retinal motion information into oculomotor effects (29), but the special neural mechanisms mediating facilitation and inhibition of oculomotor functions remain obscure although all studies of OKN agree that visual attention is necessary to elicit adequate OKN (8, 24, 26, 33, 37, 39, 43, 51, 54).

All observations of diminished OKN and movement sensation during inattention, distraction by other sensory stimuli or by mental effort appear to share a common explanation, i.e. diminution of specific visual attention to the moving stimulus. The resultant increase in moving retinal images activates afferent motion detection with its exaggerated velocity information. Accurate velocity perception therefore depends upon attentive ocular following and monitoring of the optimal slow OKN phase, which is controlled by foveal fixation of moving objects.

OKN and Railway Nystagmus, Attention and Visual Selection

The eye movements of railway nystagmus and their role in visual perception were first mentioned in 1866 by Helmholtz (27). Later the term was used by Barany (4) as synonymous with OKN. However, a special study with eye movement recording was lacking. For this reason we have made a detailed description of the sensory phenomenology and eye movements (*see* Results 2) and have postponed quantitative studies of selective attention for later investigation.

Some contrast between OKN elicited in the laboratory and railway nystagmus are explained by difference of the moving stimuli and of the attentional attitudes of the subject in both conditions. In railway nystagmus visual selection and interest determines the object chosen for foveal fixation. Pursuit velocities vary according to the very different apparent velocities of the fixated parts of the panorama. The variant forms of railway nystagmus correspond to an intended pursuit of one selected object in contrast to the regular sequence of slow following phases during experimentally induced OKN. In the latter, diffuse visual interest is directed towards a uniform sequence of stripes moving at constant speed, followed from their first appearance in the peripheral field to the midline region until new contrast stripes appear which are captured with a quick phase and the process repeated. For this reason we generally use the term *following movements* or tracking for the slow phases of OKN and reserve 'pursuit movements', used in the older literature dealing with OKN (17, 18, 24), for visual selection with railway nystagmus.

Since Fox, Couch and Dodge (24) investigated the psychological aspects of OKN it is known that several quick corrective movements (small rapid phases) may disrupt the slow phases of OKN, particularly when the subject is interested in the stimulus objects and when the moving objects are complex in shape. Our records of railway nystagmus show an increased number of quick phases and of quick refixations. These occur mostly in the beginning of long flowing movements as quick searching jerks, directed to both the opposite and pursuit direction (*Figure 5b* and *c*). The unintended '*Schlagfeld*' of OKN on the side from which the moving stimuli appear contrasts with intentional selection in railway nystagmus. Both are directed to moving objects of prominent interest and significance; in OKN, this is to moving stimuli newly entering the peripheral visual field, and in railway nystagmus to selected objects apparently moving with different velocities. OKN occurs only if the moving patterns arouse some attention and 'interest' in the subject. Therefore, moving stripes with a stationary surround are sufficient stimuli for OKN only for men and monkeys. In cats and dogs OKN disappears after a few stripes have passed by, but full OKN is elicited by rotating obviously interesting stimuli, such as rabbits, around a dog (36, 51, 54).

Motion Perception and Eye Movements

Three aspects of these experiments are of particular interest and should be discussed:

(1) *Efferent oculomotor processes during attentive eye following without moving retinal images result in accurate subjective estimates of velocity and correspond most closely to the real velocities of moving stimuli*

The velocities of eye tracking and velocity estimates increase in parallel with increasing speed of the moving stimulus usually up to 50 degrees per second (26, 44, 56). If moving retinal images occur due to diminished attention and foveal fixation during the slow phase of OKN, or with higher stimulus velocities, the correspondence between perceived and real speed is lost and over-estimation occurs.

(2) *Afferent visual signals associated with moving retinal images cause exaggerated estimates of velocity*

These estimations reach 1·6–1·9 times the values of those obtained during adequate eye following of the same stimulus (13, 13a, 37). Following movements are a necessary and normal consequence of retinal image motion in the awake organism, if they are not prevented by stable fixation, by the imagination of stationary images or by reduction of visual attention (*Figure 1*). Thus in our artificial experimental situation the over-estimates of velocity during retinal image motion may be related to the suppression of sensori-motor control of eye following, normally operative during attentive behaviour.

(3) *The selective function of visual attention and ocular pursuit*

This is demonstrated in railway nystagmus when foveal fixation of objects of varying velocity regulates both eye pursuit and movement perception.

The oculomotor effects from different and often oppositely-directed moving retinal images occurring in the periphery of the fixated and pursued pattern are suppressed in favour of this selected object. Only if attention is simultaneously directed to the surround may one observe contrasting directions and velocities.

Early observations on motion perception during eye movements and fixation (briefly described by Fleischl in 1882) mention an approximate doubling of velocity sensation without eye movements (23), and later nystagmographic studies arrived at a similar value (33). Several attempts to distinguish motion detection with and without eye tracking by Aubert (3), Dodge (18), Filhene (22) and Brown (11) offered various hypotheses but no clear explanation. Dodge has observed with moving points of light that 'if they appear to move through equal arcs, the pursued must actually move through about three times the arc of the unpursued' (18). Although these experiments measured amplitude and not velocity they are probably related to the exaggerated afferent velocity perception. Dodge (17, 18) overrates the more powerful moving retinal images as a source of motion perception and understates the motor component of eye following and its 'hypothetical kinaesthetic factor'. Although we agree that proprioceptive signals from the eye muscles are probably of no importance, our results of accurate following motion sensation are a strong argument for the importance of efferent motion detection.

Higher velocity perception without eye movements was often seen and was called the 'Aubert–Fleischl Paradox' by Brown (11). Motion perception was regarded as a very complex phenomenon influenced by many factors. No attempt was made to quantify the different velocities under standard conditions and to compare these with eye movement recordings, although quantitative estimations of sensation have been recognized since Fechner's (21) *Psychophysik* appeared in 1860. Also Stevens (53) did not apply his scaling methods to motion perception. Mashhour made a scaling investigation restricted to slow movements (47).

Mashhour's study on scaling of perceived velocity is the only available work comparable to our experiments. Observations, however, were restricted to stimulus speeds below 5 degrees per second and a few eye movement recordings were made which are difficult to interpret at these slow velocities. We found that eye movements at such low stimulus velocities are automatic, imperative and relatively independent of special visual attention if the subject is looking at the moving stimulus, and if the stimulus area is large. Mashhour found an approximate linear function of velocity scaling with an exponent of 0·8–1·0 of Stevens' power function. This agrees with our results for a larger range of stimulus velocities (5–140 degrees per second). No appreciable difference has been found at these slow velocities below 5 degrees per second between velocity estimation with and without attentive intentional eye following. Only stationary fixation of other objects interrupts eye following and causes a perception of higher velocity at low stimulus speeds. Rachlin (50a) without paying attention to eye movements made scalings of a larger range of small light spot velocities and found an exponent of 0·78.

Sharp vision, eye tracking and blurring in motion perception

Visual inattention of stationary fixation during OKN causes an impressive change of visual perception by blurring or smearing of the moving images. Blur is seen by all subjects as soon as eye tracking lags behind the stimulus velocity. During adequate eye following all moving contrasts, not only the foveally fixated, are seen as sharp black-white boundaries. Since blurred vision of moving retinal images is associated with a sensation of increased speed the question arises as to whether blurring contributes to exaggerated motion sensation. Blurring alone cannot explain the subjective increase of velocity for the following reasons: 1) Initial blurring occurs with refraction anomalies during attentive fixation and correct tracking; 2) The perception of increased speed during inattention is approximately the same with continuous blurred images (myopic vision) as with clear differences between sharp and blurred vision. Although spectacles for refraction anomalies are not necessary in myopia to elicit attentive eye following, it is possible that sharp foveal vision may act as a control mechanism for adequate eye pursuit in normal refraction.

Studies on motion perception thresholds also indicate high sensitivity of the afferent mechanism of retinal motion. Aubert (3) noted that motion thresholds are markedly higher when the eye follows the moving objects than for fixation without eye following. Leibowitz (41) found that velocity thresholds are lowered by 48 per cent when reference lines surround the moving stimulus. We assume that his fixation of the reference lines may have resulted in moving retinal images.

The significance of exaggerated velocity sensation with moving retinal images cannot be discussed here in detail. Elicitation of adequate eye following by this primary stimulus on the retina involves complex cybernetic regulations of optokinetic effects of motion stimulation for which several feedback loop models have been discussed (39, 55). One might construct a model of eye movement control in which the afferent motion information of the retina is doubled by positive feedback. The exaggerated velocity perception of moving retinal images might be the internal aspect of sensorimotor feedback mechanism for optokinetic activation. When eye movement slightly lags behind stimulus velocity, excessive velocity information might sustain OKN. Small movements of retinal images would suffice to trigger adequate oculomotor tracking. A further mechanism of oculomotor regulation would be the faculty of predicting forthcoming movement from only a fragment of information.

Our assumption that the nearly doubled motion information from retinal image movement is part of a control mechanism may be supported by experiments on optokinetic stimulation of immobile eyes. Experiments of Ter Braak in rabbits (54) and recently of Körner and Dichgans in men (37) have shown a motor equivalent to uncompensated moving retinal images. The velocity of the closed non-stimulated mobile eye may reach much higher speeds than the stimulus velocity—in fact up to double the velocity in man and still faster speeds in rabbit (54). This never occurs during visual stimulation of the eye which is following stimulus velocity during optimal attention. Ocular following velocities never exceed stimulus velocities in normal eyes. The presence of an oculomotor equivalent of moving retinal images with its corresponding after-effect during fixation of a small stationary object within the moving field is demonstrated by Mackensen. Although stationary fixation suppresses ocular following a latent OKN is liberated in the dark as a strong optokinetic after-nystagmus (46).

Motion information by slow ocular tracking movements and exclusion of rapid phases of OKN

All data on the speed of eye movements and velocity perception in our experiments have been obtained from the slow tracking movements (slow phases of OKN). The rapid nystagmic phases interrupting the slow phases in the opposite direction were disregarded. It was further assumed that motion information conveyed by moving retinal images was not effective during the quick phases but was functional only during inadequate tracking in slow phases or during stable fixation. Slow following movements are evidently necessary for form vision and for the fixation of moving objects, but it is not evident *a priori* that slow phases and following movements are the sole factors determining efferent velocity detection. What, for example, is the role of the quick phases during nystagmus? Are they only corrective eye movements which initiate the next slow following phases, and is vision suppressed during the quick phases to ensure continuity of perception during the slow phases, as Dodge (17, 18) and many others have assumed? Or can useful information on visual localization be obtained during the quick phase?

Mackensen (45) found that the quick phase of OKN has rather constant high velocities of 200–400 degrees per second, independent of the velocity of the slow phase, but with lower amplitudes slower speeds of the quick phases occur as also in small saccadic movements. Rapid OKN phases resemble eye saccades in many aspects. Rapid nystagmic phases, saccadic eye movements and square wave opposite jerks (*Gegenrucke*) occur only during vigilance, arousal and attention. They disappear during unconsciousness and sleep except when dreams arouse rapid eye movements. The attentiveness causing rapid eye movements need not always be visual because *Gegenrucke* may also occur during eye closure, when the subject is distracted by auditory attention.

Directed gaze movements (*Blickzielbewegungen*) as well as the first quick phases of OKN are mechanisms for the visual capture of interesting objects which attract our attention in the periphery. Good spatial localization of this visual target is necessary to obtain foveal fixation, and this is achieved by rather accurate direction of gaze with a few small corrective movements. Errors of pure oculomotor localization in the dark are relatively small, as Merton (48) has shown. Localization errors, however, are made in the direction of intended gaze after eye muscle paralysis, which prevents eye deviation. This old observation on spatial displacement by Graefe (25) was the first indication of an efferent regulation of space determination by intended eye movement.

The quick nystagmic phase corresponds to the saccadic movements during voluntary gaze. When the eyes are open, rapid shifts of retinal images must occur during the quick eye movements. Latour (40) and Zuber and Stark (59) have demonstrated considerably higher visual thresholds during saccadic eye movements and quick nystagmic phases. This saccadic suppression apparently begins 50–80 ms *before* the saccade and is accompanied by a diminution of the pupillary response (60). Specific saccadic suppression of motion information has not yet been investigated by quantitative methods but some observations imply motion suppression during eye saccades. Vossius has proposed that rapid eye movements have a different control mechanism to the slow following movements (55). In fact it seems probable that the rapid nystagmic phases and saccadic eye movements do not simply cause suppression but represent revaluation of visual information.

Ditchburn (14) has shown that a subject is unable to see simultaneous cathode ray recordings of his rapid eye movements on an oscilloscope. Gaze-directed visual localization in the outer world and efferent motion perception with ocular followings are rather precise and are well integrated to form a stable visual world. This spatial constancy corresponds to Hering's *Raumkonstanz der Sehdinge* (28) and the regulations in Howard and Templeton's 'stability centre' (32). The internal model of the outer world, which we carry within (2, 12) is constantly controlled by sensorimotor information, although the neurophysiological and cybernetic mechanisms of these regulations are not properly understood. Models of eye movement control proposed by Vossius (55), Kornhuber (39) and others cannot yet explain the perceptual constancy (*Raumkonstanz der Sehdinge*), either for visual localization or for motion perception. The *Reafferenz-Prinzip* of von Holst (30) and the matching theory of MacKay (42) are useful general conceptions but still need physiological verification.

Recent investigations by Brindley on electrical stimulation in the visual cortex of conscious blind persons (9) have shown an influence of eye movement on the localized phosphenes arising in the visual area. These phosphenes, caused by local cortical stimulation and seen in certain parts of the visual field are reported as moving with the voluntary movements of the blind eyes in the same direction of gaze as retinal after-images. Since no retinal afference comes from the blind eye, this seems to indicate that eye movements change the model of the visual world represented in the visual cortex.

Electrophysiological findings show that the brain receives and processes some information during rapid ocular movements. Slow potentials probably corresponding to presynaptic inhibition in the cat's lateral geniculate during the REM-phase in sleep (7) and the lambda-waves of the human EEG occurring after saccadic movements and quick nystagmic phases with visual contrast stimuli (6) are signs of information processing and inhibition in the central visual system which are related to rapid eye movements.

A higher threshold for afferent brightness information during eye saccades is not a sufficient control of continuous perception. The quick phase of the saccade is not only unnoticed, but is also bridged over by revaluation of visual information. A central revaluation might include compensation and cancellation of the previous picture. The quick phase of OKN determines not only the rhythmic interruption of the slow phase and the number of refixations but also the spatial location of the object selected and pursued. Further it induces the slow phase of OKN to beat in a certain field as the fast sweep of an oscillograph sets the beginning and end of slow sweep. How efferent motion sensation originates during the slow ocular tracking, how it is facilitated by attention to a precise perception of velocity and at another moment indicates the action of neuronal mechanisms monitoring the oculomotor outflow, is as unknown as the suppressive and revaluation effects of quick phases. Since Helmholtz (27), most authors (10, 49, 57) have agreed that proprioceptive signals from the external eye muscles are not involved.

Our experiments show the close interaction of attention, eye movements and motion perception. The oculomotor function of gaze and nystagmus is dependent upon attention and represents sensorimotor correlates of attentive behaviour. Further investigation might lead to a more precise understanding of the cerebral mechanisms of visual attention and their co-ordination in the cerebral cortex and the reticular formation. Without more precise knowledge about these cerebral mechanisms of eye movement control, however, it seems premature to speculate on the neurophysiological regulation of attentive processes themselves.

SUMMARY

1. The influence of attention on eye movements and motion detection has been investigated by the recording of optokinetic nystagmus (OKN) and the psychophysical scaling of velocity perception.

2. OKN is altered by attention in two ways: (*a*) Facilitation by visual attention to the moving stimuli, which can be adequately followed in this condition by the slow nystagmic phase up to stimulus velocities of 50–90 degrees per second. (*b*) Visual selection with foveal fixation of single moving objects and adequate pursuit in a panorama containing objects with different and varying velocities.

3. Two mechanisms of motion detection were studied by quantitative methods and were separated by voluntary alteration of attention and fixation: afferent motion detection by visual inflow from moving retinal images results in excessively high velocity perception; efferent motion detection by oculomotor outflow from ocular following movements causes adequate velocity perception corresponding to real stimulus velocities. Only the oculomotor mechanism appears dependent upon attentive facilitation.

4. Afferent velocity perception from moving retinal images during diminished attention or fixation of stationary objects is exaggerated, by a factor of 1·6–1·9, when compared with efferent velocity estimation during attentive eye following.

5. The slow phase of OKN shows adequate following at stimulus velocities of 50–90 degrees per second, but only for short periods of optimal attention. Spontaneous slowing of tracking movements occurs together with irregular periodic slackening of attention and blurred vision. More marked slowing is caused by auditory distraction, visual imagination and other modifications of the intention to look at the moving objects.

6. Recordings of railway nystagmus demonstrate the selective function of visual attention, fixation and pursuit. The visually interesting object is selected and the eye movements precisely track its waxing and waning velocity. Foveally fixated objects dominate motion perception, although retinal images

may show opposing movements in the fore- and background of the pursued object. Lengthy ocular pursuit results in adequate reproduction of the changing velocities of the selected object over large parts of the field of regard, and is represented by round and S-shaped slow phases.

7. The results have been discussed in relation to attentive mechanisms of intentional selection, vigilance, distraction and to the constancy of space perception with saccadic suppression during the rapid phase of OKN.

ACKNOWLEDGEMENT

The extensive co-operation of Dr. F. Körner (Tübingen) in the experiments of velocity scaling and the valuable help of Dr. C. Voigt in the experiments on railway nystagmus is gratefully acknowledged. We thank Dr. Barbara Brooks for discussing and correcting the English manuscript.

DISCUSSION ON JUNG

Grey Walter

Dr. Jung's paper reminds me of a number of observations that we've all made at one time or another. One of the effects we should consider again is the famous 'waterfall effect'. Every EEG worker has seen this of course in his early days—the impression that the EEG-paper is running backwards when it stops. But one adapts to this. I can no longer get the waterfall effect with an EEG. The important feature is that only *a part* of the visual field appears to be moving the other way. It is not a nystagmoid movement of the eyes because the effect only operates over part of the field. Do you think that in this complicated process of allowing for apparent movements, and avoiding the horrible effects you get in the cinema with too rapid panning there may be some process related to the waterfall effect, stabilizing moving features with the secondary effect of producing an inverted after-image of movement?

Jung

During a waterfall illusion one may record optokinetic after-effects when the preceding moving stimulus had occupied large parts of the visual field. A neuronal correlate to these after-effects of adaptation for direction-specific motion may be the following observation which Baumgartner made in our laboratory. We could record neurones in the visual cortex responding to moving stimuli which change their tendency to respond to a preferred direction into a tendency to respond to the opposite direction for a period of time. When stimulated by moving patterns in one direction they are strictly direction specific, e.g. responding only to a right-to-left movement. Then suddenly they change, and respond only to movement in the opposite direction—from left to right. There must be some gating effects following motion adaptation—or, if you prefer, after saturation of motion detection—to a certain direction. I think this is the neuronal complement of the waterfall effect in cortical neurones.

Grey Walter

What cortex?

Jung

The visual cortex, and it may be more marked in areas 18 or 19, but Baumgartner could find it also in 17.

Byford

I would like to suggest an alternative explanation for the mechanism of optokinetic nystagmus; but first a definition. Target following is that conscious movement of the eye during which the point of regard is held in the same position on the retina. *Figure 6* shows the eye movements which are produced by a subject seated with a rotating drum, the speed of which is increasing uniformly—small electric lamps are fixed to its inner surface. At the beginning, whilst the drum is rotating at something less than 20 degrees per second, the eye follows a selected target consciously. As the drum velocity increases, optokinetic nystagmus develops and this is a quite different process; it is followed at higher drum speeds by a marked velocity oscillation during which the speed of rotation of the eye alternates wildly. A further increase in drum speed results in a completely inactive eye, and a visual impression of just one straight line of light. *Figure 6* was obtained by recording the corneo-retinal potential; this, whilst adequate for qualitative work, has certain known inaccuracies. If a physical eye movement measuring system is used, the results are less variable and the experimental conclusions are not always in agreement with those derived from corneo-retinal potential recordings.

Figure 7 illustrates the general relationship between target velocity and eye speed for one subject; its shape is common to all subjects, but between-subject variability is considerable. If the eye always moved at the same speed as the target, then all the experimental points would lie along the straight 45 degree line. Points below this line show the eye to be moving more slowly than the targets, and above the line to be moving faster than the targets. The physical measuring system shows the former to be the case. This is very reasonable, because the eye is slowing down the velocity with which targets are moving across the retinal elements, and hence improving resolution; at the same time it is preserving velocity information, even though that information is in absolute terms incorrect. On some occasions, results obtained by the use of the corneo-retinal potential would support the opposite theory (*Figure 8*), i.e. that the eye moves faster than the targets; this is clearly untenable.

The reason for this apparent contradiction would appear to be connected in some way with the calibration procedures, since static calibrations resulting from discrete movements between two targets of known angular separation, and dynamic calibrations resulting from an instruction to follow consciously a slowly moving target, sometimes produce different corneo-retinal potential deflection sensitivities.

Jung

You are quite right to mention the difficulties of calibration. We have tried to overcome these by taking a number of calibrations. The corneo-retinal potential varies with dark adaptation and other conditions, and this is, of course, a cause of error in these experiments with eye movement recordings of long duration. We tried to overcome this by having several calibrations in

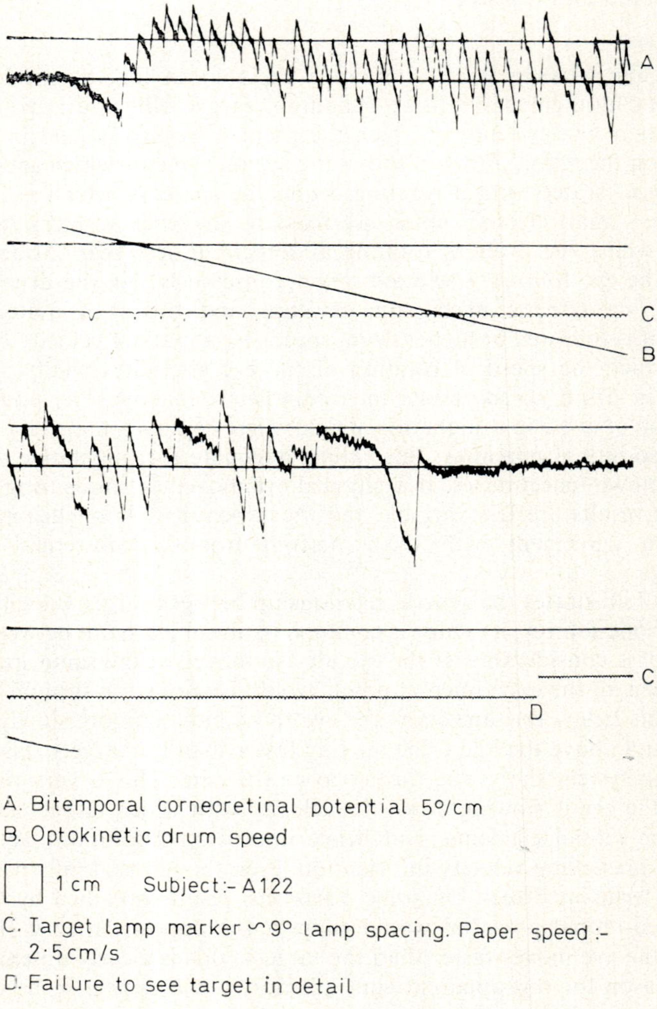

Figure 6. Optokinetic nystagmus resulting from uniformly increasing target speed

between so that one knows if one's corneo-retinal potential has changed. In general your curves are quite similar to ours, but the results differ for maximal eye velocities. Human eye movements usually cannot follow stimulus velocities of more than 90–100 degrees per second. Your decline over 90 may be diminished attention.

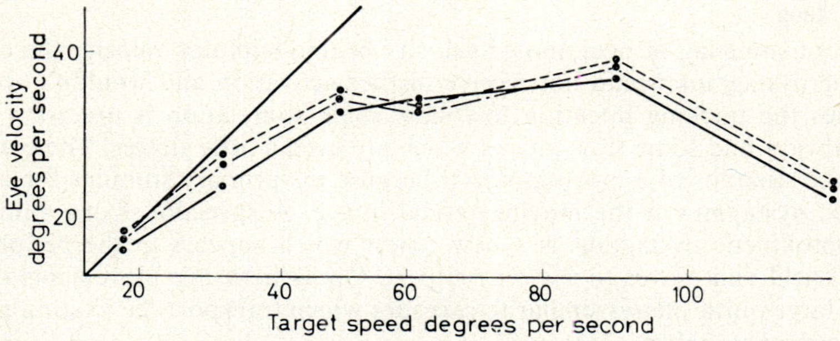

Figure 7. For this subject, both recording methods and either calibration procedure give the same results

------ Corneo-retinal potential—Static calibration
—·—·—Corneo-retinal potential—Dynamic calibration
———Photoelectric recording—Static and dynamic

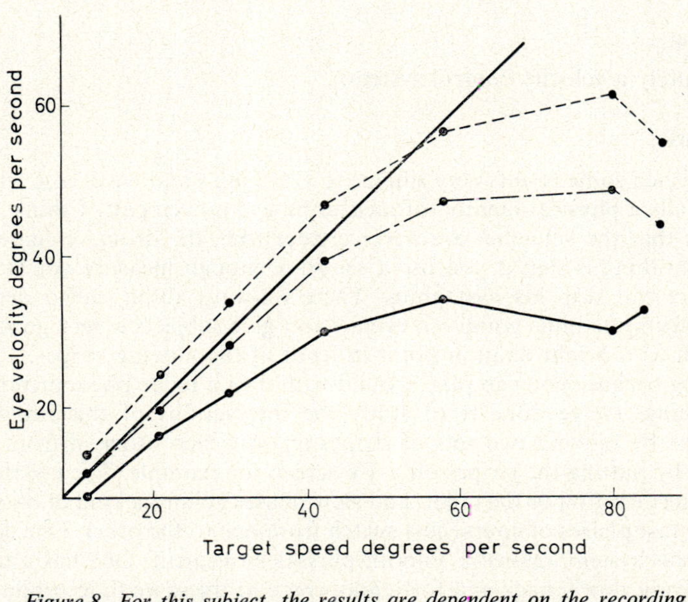

Figure 8. For this subject, the results are dependent on the recording technique and on the calibration procedure

Byford

I was suggesting that except at low velocities this isn't a following process at all, and that there is always a velocity lag between the angular speed of the target and the eye.

Jung

Without some lag of oculomotor velocity behind stimulus velocity the optokinetic nystagmus would not receive further activation and wouldn't go on. Besides the tracking intention to follow some stimulation is necessary and you always find some slow phases which are even a little slower. These again give the stimulus of a moving object, because the primary stimulus for optokinetic nystagmus *is* the moving retinal image. A specially strong stimulus for optokinetic nystagmus is a new object which appears in the peripheral visual field and moves to the central part. Optokinetic eye movements show early large quick phases similar to saccades which transport the fixation point to the moving object.

Creutzfeldt

Is the distance between two targets always the same on the retina?

Stark

No it isn't at all. A constant velocity error for a particular velocity means the distance is changing all the time. So you're thinking of this as a position control and it's a velocity control.

Jung

It's definitely a velocity control system.

Stark

We published some results very similar to Dr. Jung's and we do use what Dr. Byford calls a physical method of measuring eye movements. I think I agree with you that the velocities are never greater than the drum velocities. One interesting thing is that if you use a sensitive enough measure you find that no subject can stop his nystagmus. There's always about half a degree of involuntary nystagmus going on, even if you give subjects a very good reference such as a bright fixation point in front of the moving stripes. It's very interesting because you can play around with it as a reflex phenomenon. One of the things we've done is to study the interactions of this involuntary nystagmus by moving two sets of stripes against each other in front of the subject—by putting the stripes on a TV screen for example. The eye then has to pick one direction or the other, and slow phases go into a kind of oscillation while the fast phases of course just switch from one to the other. One can also interact this system against a smooth pursuit movement. One has a fixation point moving slowly back and forth and which at the same time produces the involuntary nystagmus with stripes. This now behaves quite differently from the case where two sets of stripes are fighting each other. With smooth pursuit movements and nystagmus one gets algebraic addition very interestingly. We were trying to distinguish between two kinds of smooth movements to see whether they were under the control of the same central nervous system mechanism. The most interesting thing I found in these experiments, however, is that I used to think that the slow phase would sort of pull the eye off the fixation point of direct gaze, and then when it got a little too far off, the fast phase would correct this error, bringing the eye forward. But in fact this isn't what happens. As soon as the set of stripes start to move (and one can do this instantaneously with an electronic system rather than with a mechanical one) the eyes move with a saccade in the direction of the moving field, and then they slowly follow in the stripe direction, until back in the primary position, and then saccade again. And if one shuts off the target instantaneously the eyes will slowly drift back to the forward gaze position. These results are quite different from observations of vestibular nystagmus and are contrary to what people have often said about the optokinetic effect (61, 62, 63).

Jung

Well, I described the shift of optokinetic nystagmus to the side of the quick phase about 30 years ago. Biologically it makes sense. What is interesting are the *new* objects that appear in the peripheral field. The objects in the middle field are already known and one pays no attention to them. The compensation is not of velocity but of acceleration. The vestibular receptors are sensitive to acceleration, and that is why I always thought that these speeded up following movements had something to do with the vestibular system and their acceleration information. But now recording railway nystagmus more accurately I am also convinced that the optokinetic system can follow accelerated movements. It is the 'visual grasping of new objects' that makes sense biologically. It is not the fixation and following processes which conserve known objects in the fixation point, but the grasping processes of the quick phase which determine the half field, where optokinetic nystagmus beats.

Byford

In an endeavour to determine the source, I covered the central area of the contact lens and later the outer area. These two conditions are optically similar, but produced quite remarkable results because the character of nystagmus, although still there, changed immensely, it became very much slower, and very much smaller. I have no explanation for this, but it was reproduceable on almost all subjects.

Stark

You mean that peripheral field nystagmus produces slower movements?

Byford

It produces a change in the character of the nystagmus.

Jung

There are some funny variations of this nystagmus from the peripheral field. It's slower and not so regular. Sometimes quick movements to the other side appear also. These come with after-images.

Byford

I have not noticed that except at the higher speeds.

REFERENCES

1. Adey, W. R., Kado, R. T. and Rhodes, J. M. (1963) *Science, N.Y.* **141**:932
2. Adrian, E. D. (1947) *The Physical Background of Perception.* Oxford: Clarendon Press
3. Aubert, H. (1886–1887) 'Die Bewegungsempfindung'. *Pflügers Arch. ges. Physiol.* **39,** 347: **40,** 459
4. Bárány, R. (1922) 'Zur Klinik und Theorie des Eisenbahnnystagmus'. *Acta oto-lar.* **3**:260
5. Berlyne, D. E. 'Development of the Concept of Attention in Psychology'. (This conference)
6. Bickford, R. G. and Klass, D. W. (1964) In (Ed. by Bender): *The Oculomotor System.* New York: Hoeber, p. 293
7. Bizzi, E. (1966) *J. Neurophysiol.* **29**:861
8. Borries, G. V. Th. (1926) *Fixation and Nystagmus.* Copenhagen: Munksgaard
9. Brindley, G. S. and Lewis, W. S. (1968) *J. Physiol.* **196**:479
10. Brindley, G. S. and Merton, P. A. (1960) *J. Physiol.* **153**:127
11. Brown, J. F. (1931) *Psychol. Forsch.* **14**:199
12. Craik, K. J. W. (1943) *The Nature of Explanation.* Cambridge: Univ. Press
13. Dichgans, J., Körner, F., Krejcova, H. and Jung, R. (1967) *Pflügers Arch. ges. Physiol.* **297**:R 84

13a. Dichgans, J., Körner, F. and Voigt, K. (1969) *Psychol. Forsch.* **32**:277

14. Ditchburn, R. W. (1955) *Optica Acta,* **1**:171
15. Doden, W. and Meyer, E. (1961) *Dt. Ophthal. Ges.* **64**:498
16. Doden, W., Meyer, E. and Ritter, G. (1963) *Klin. Mbl. Augenheilk.* **142**:118
17. Dodge, R. (1900) *Psychol. Rev.* **7**:454
18. Dodge, R. (1904) *Psychol. Rev.* **11**:1
19. Erickson, M. H. (1962) *Am. J. clin. Hypnosis.* **4**:181
20. Exner, S. (1894) *Entwurf zu einer physiologischen Erklärung der psychischen Erscheinungen,* 1. Teil. Leipzig & Wien, F. Deuticke
21. Fechner, G. T. (1860) *Elemente der Psychophysik,* Teil 1 & 2. Leipzig: Breitkopf und Härtel
22. Filhene, W. (1922) *Z. Sinnesphysiol.* **53**:134
23. Fleischl, E. v. (1882) Physiologisch-optische Notizen. 2. Mitt. Sitzber. *Akad. Wiss. Wien,* **86,** Abt. 3, 8
24. Fox, J. C., Couch, F. H. and Dodge, R. (1931) *Arch. Neurol. Psychiat.* **26**:23
25. Graefe, A. v. (1854) *Graefes Arch. Ophthal.* **1**:28
26. Grüttner, R. (1939) *Z. Sinnesphysiol.* **68**:1
27. Helmholtz, H. (1896) *Handbuch der physiologischen Optik,* 2. Aufl. Hamburg u. Leipzig: G. Voss
28. Hering, E. (1879) In: *Hermann's Handbuch der Physiologie,* **3** : 343. Leipzig: Vogel
29. Hofmann, F. B. (1915) *Ergebn. Physiol.* **15**:238
30. Holst, E. v. and Mittelstaedt, H. (1950) *Naturwissenschaften,* **37**:464

31. Hood, J. D. (1967) *Acta oto-lar.* **63**:208
32. Howard, I. P. and Templeton, W. B. (1966) *Human Spatial Orientation.* London, New York, Sydney: Wiley & Sons
33. Jung, R. (1953) In: (Ed. by Bergmann, Frey and Schwiegk) *Handbuch d. inneren Medizin*, 4. Aufl., V/1, 1325. Berlin, Göttingen, Heidelberg: Springer
34. Jung, R. (1954) In (Ed. by Delafresnaye, Adrian, Bremer and Jasper): *Brain Mechanisms and Consciousness, Symposium*, 310. Oxford: Blackwell
35. Keser, H. (1947) *Ber. Kongr. Neurol. Psychiat.*, Tübingen, 239
36. Kleyn, A. de and Rademaker, G. G. J. (1928) *Nederl. Tijdschr. Geneesk.* **2**:5529
37. Körner, F. and Dichgans, J. (1967) *Graefe's Arch. klin. expl. Ophthal.* **174**:34
38. Kommerell, G. and Mackensen, G. (1967) *Klin. Mbl. Augenheilk.* **151**:702
39. Kornhuber, H. (1966) In (Ed. by Berendes, Link and Zöllner): *Hals-Nasen-Ohren-Heilkunde*, Bd. III/Teil 3, 2150. Stuttgart: Georg-Thieme-Verlag
40. Latour, P. L. (1962) *Vision Res.* **2**:261
41. Leibowitz, H. (1955) *J. opt. Soc. Am.* **45**:829
42. MacKay, D. M. (1958) *Nature, Lond.* **181**:507
43. Mackensen, G. (1953) *Klin. Mbl. Augenheilk.* **123**:133
44. Mackensen, G. (1954) *Graefe's Arch. Ophthal.* **155**:284
45. Mackensen, G. and Schumacher, J. (1960) *Graefe's Arch. Ophthal.* **162**:400
46. Mackensen, G. and Rudolph, U. (1962) *Graefe's Arch. Ophthal.* **165**:60
47. Mashhour, M. (1964) *Stockholm Studies in Psychol.* **3**. Stockholm: Almquist and Wiksell
48. Merton, P. A. (1961) *J. Physiol.* **156**:555
49. Merton, P. A. (1964) In (Ed. by Bender): *The Oculomotor System*, p. 314. New York: Harper & Row
50. Pompeiano, O. and Morrison, A. R. (1965) *Archs. ital. Biol.* **103**:569
50a. Rachlin, H. C. (1966) *Percept. and Psychophys.* **1**:77
51. Rademaker, G. G. J. and Ter Braak, J. W. G. (1948) *Brain*, **71**:48
52. Reding, G. R. and Fernández, C. (1968) *Electroenceph. clin. Neurophysiol.* **24**:75
53. Stevens, S. S. (1961) In (Ed. by Rosenblith): *Sensory Communication*, 1. New York and London: Wiley & Sons
54. Ter Braak, J. W. G. (1936) *Arch. Neerl. Physiol.* **21**:309
55. Vossius, G. (1960) *Z. Biol.* **112**:27
56. Westheimer, G. (1954) *Arch. Ophthal.* **52**:932
57. Whitteridge, D. (1960) In (Ed. by Field, Magoun and Hall): *Handbook of Physiology, Sect. I, Neurophysiology*, Vol. II, p. 1089. Am. Physiol. Soc.: Washington
58. Zikmund, V. (1966) *Studia Psychol. Slov. Akad. Bratislava*, **8**:254
59. Zuber, B. L. and Stark, L. (1966) *Expl. Neurol.* **16**:65
60. Zuber, B. L., Stark, L. and Lorber, M. (1966) *Expl. Neurol.* **14**:351

Discussion

61. Merril, E. G. and Stark, L. (1963) *Quart. Prog. Rept., Research Lab. Electronics*, M.I.T. **70**:357
62. Merrill, E. G. and Stark, L. (1963) *Quart. Prog. Rept., Research Lab. Electronics*, M.I.T. **71**:268
63. Merrill, E. G. and Stark, L. (1964) *Quart. Prog. Rept., Research Lab. Electronics*, M.I.T. **72**:269

Neurochemical Mediators of Specific and Non-Specific Cortical Activation

Herbert H. Jasper
Canadian Medical Research Council, Research Group in Neurological Sciences, University of Montreal

INTRODUCTION

Lord Adrian, Honorary Chairman of this Conference, was the first to confirm Berger's original observation that arrest of the rhythmic beat of the alpha rhythm, recorded from posterior head regions in man, was related to visual attention. Opening the eyes with attempt to see in a completely darkened room was shown to produce an arrest of the occipital alpha rhythm in the absence of photic stimulation (1, 2, 3). Shifting of attention from visual to auditory stimuli caused an increase in occipital alpha rhythm, suggesting that the selectivity of the attention process might also be reflected in the alpha blocking reaction. It was not necessarily related only to the general level of vigilance.

Recording the electrical activity from the exposed cerebral cortex, during neurosurgical procedures in waking patients carried out with Penfield, we were able to show that 'sensori-motor attention' in the intention or preparation to make a given movement was associated with local blocking of both alpha and beta rhythms of local regions of the pre- and postcentral gyri (27). It would seem, therefore, that local desynchronized activation of the cerebral cortex may be considered, in the waking brain, to be an electrical correlate of the attentive process.

Evoked potentials recorded from specific sensory receiving areas of cerebral cortex in the intact unanaesthetized animal or man do not necessarily change in a systematic manner with attentive behaviour. They may be increased or decreased, depending upon the general state of vigilance or reactivity. In general, they are usually found to be increased, on the average, during drowsiness and slow wave sleep, decreased during alertness, unless they have been previously depressed by habituation. Since they do not vary consistently with desynchronizing cortical activation it may be assumed that they depend upon different synaptic systems, and possibly that they are dependent upon different neurochemical mediators. This would imply that neuronal systems involved in preliminary processing of sensory information may be different from those responsible for attentive awareness (21).

One of the methods of testing the hypothesis that separate neuronal systems may be necessarily involved in attentive behaviour is to analyse the nature of chemical substances which are liberated from the cerebral cortex during desynchronized activation as compared to that of slow wave inactivation of sleep. Comparison between non-specific and specific activation of cortical sensory receiving areas with regard to possible neurochemical mediators can also be carried out. The effect of the administration of these substances upon cortical activity and reactivity, or the effect of specific pharmacological blockade of their action upon cortical electrical activity and behaviour should give information of value concerning possible neurochemical mechanisms of the attentive process. Some preliminary studies along these lines will be reviewed in the present report.

EVOKED POTENTIALS DURING SLEEP, WAKING, AND ATTENTIVE BEHAVIOUR

Even though evoked potentials provide only a very crude and complex indicator of some of the gross synchronized aspects of information processing in cortical sensory receiving areas, their variation with states of sleep, waking, and attentive behaviour would suggest that the blocking or inhibition of transmission and preliminary processing of sensory information at the various synaptic stations to the cerebral cortex is not necessarily involved in the attentive process. Modifications in this processing may occur at any synaptic station from the sense organ to the cerebral cortex, through existing efferent controls and convergent influences. The amplitude of specific evoked potentials, however, is often greater during slow wave sleep than in waking, and may be diminished during alerting and attention, or there may be no significant alteration in amplitude of the primary evoked potential complex.

In experiments originally designed to study habituation of evoked potentials from visual cortex in the cat with repeated photic stimulation, carried out in our laboratories with Roy Steinberg, the average of 50 to 100 successive potentials, summated by the Computer of Average Transients (CAT), failed to show significant attenuation when variations in pupillary diameter were controlled by pupillary dilatation (atropine) and by occluders with permitted constant retinal illumination. During monotonous repetition of the flashes of light, in the absence of other forms of visual stimulation, the animals would often go to sleep as judged by the EEG and behavioural observation. No significant change in amplitude of average visual cortical evoked potentials occurred during states of sleep as compared to wakefulness. The only consistent finding was a shortening of latency during wakefulness, and a modification in the late components of the evoked potential complex. Desynchronization of the EEG was, however, a constant accompaniment of wakefulness and attention (43).

Comparable observations have been made in man during natural sleep and wakefulness, as for example, in the studies of Corletto and associates in Rossi's laboratory in Turin (10). Recording average visual evoked potentials to light flashes in the human subject, with pupils dilated by atropine, they found an increase even in the early components of the average E.P. complex

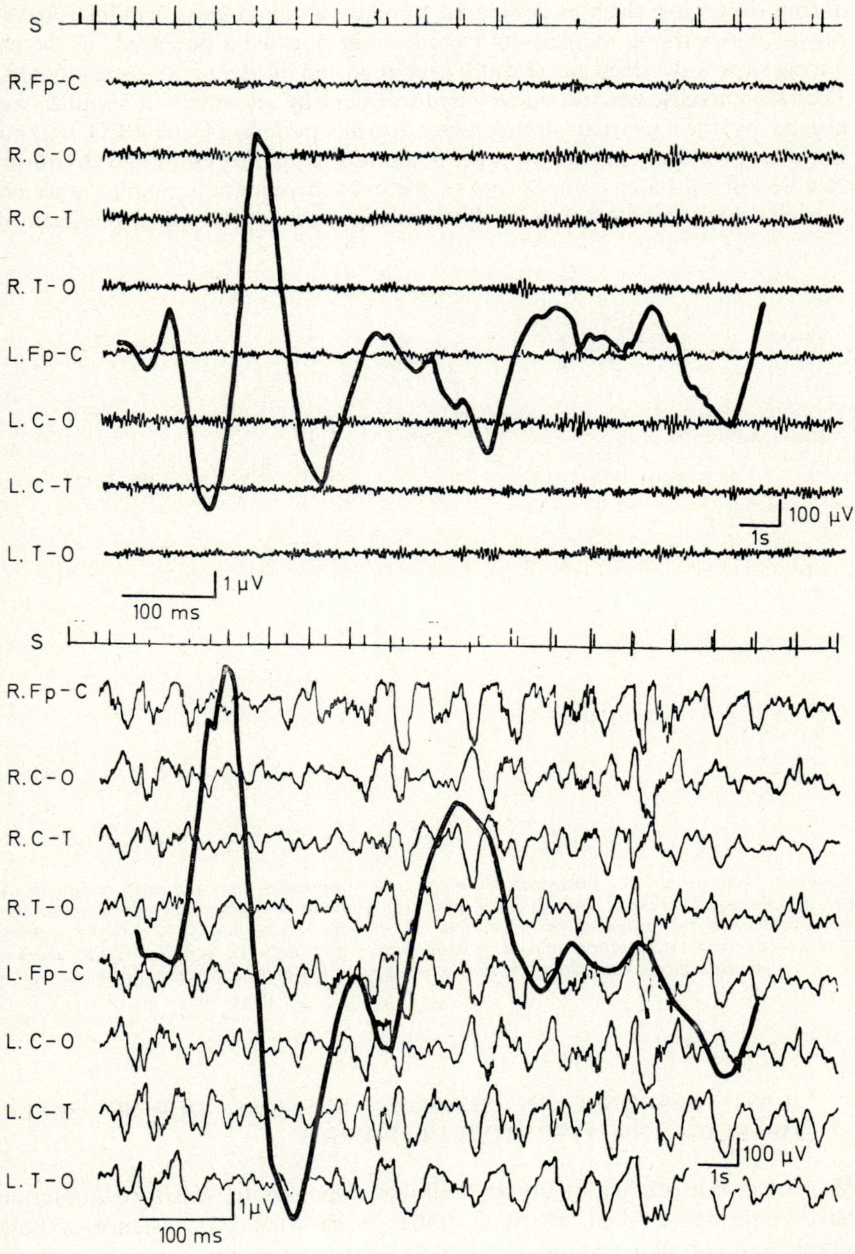

Figure 1. Average of 100 *evoked responses from the parieto-occipital region in a human subject in response to flashes of diffuse light at* 2 s *intervals during relaxed wakefulness* (*above*) *and during slow wave sleep* (*below*). *The pupils were dilated with atropine in both waking and sleeping conditions. Sample EEG records are shown in the background. Note increased amplitude of average potential during sleep*

(From Corletto *et al.*, 1966, by courtesy of Masson & Cie.)

during slow wave sleep as compared to relaxed wakefulness, as illustrated in *Figure 1*. A striking increase in average sensory evoked potentials in the cat during slow wave sleep has recently been reported by Herz and associates (18). That this increase was not due to reinforcement by spontaneous spindles was proven by the fact that stimuli given during periods of the EEG without spindles produced initial average potentials equal to those given during spindle bursts. Later components of the evoked potential complex were enhanced during spindle bursts, as illustrated in *Figure 2*, from their publication.

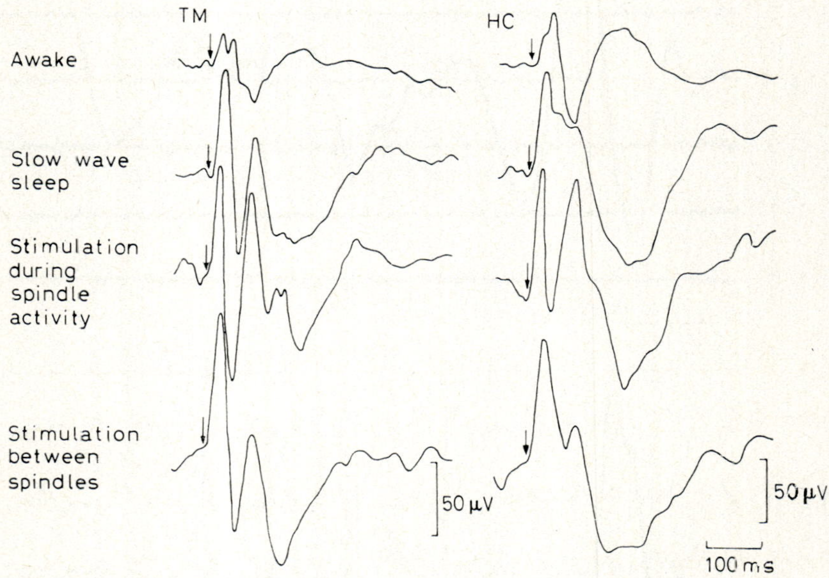

Figure 2. Acoustic evoked potentials from electrodes implanted over auditory cortex in the cat. Responses to 50 *brief tones* (0·3 s *at* 800 Hz), *given at irregular intervals of about* 10 s *were summated by means of the Computer of Average Transients* (*CAT*) *during waking, slow wave sleep, stimulation during spindle burst, and between spindles. Note marked increase in evoked potentials during sleep, and little effect of spindle burst*

(From Herz *et al.*, 1966,
by courtesy of Springer-Verlag)

UNIT RESPONSE IN SENSORY SYSTEMS DURING SLEEP, WAKING, AND ATTENTIVE RESPONSES

Microelectrode studies of unitary cellular responses to sensory stimulation have confirmed evoked potential studies in showing a persistence of both simple and complex responses in cortical sensory receiving areas in animals during natural sleep, or barbiturate anaesthesia. Even the most complex cells in visual association areas were found by Hubel and Wiesel (20) to respond to specific patterns of visual stimulation, or to movement in a certain direction in cats under barbiturate anaesthesia. Latencies may be shortened, patterns of response changed, variability reduced, and unit responses may follow higher frequencies under conditions of alertness, but some cells appear to be inhibited rather than facilitated by wakefulness.

Microelectrode studies of units in the auditory cortex of the unanaesthetized cat have shown that certain cells appear to respond only when the animal pays attention to the auditory stimulus—the 'attention' units of Hubel *et al.* (19). Other cells within the same area of cortex would respond to the auditory stimuli regardless of attention.

In microelectrode studies of the response of single thalamic cells in human subjects, recorded with Bertrand during stereotaxic surgery (22, 23, 24), we have been able to confirm the persistence of regularly repeated responses to tactile stimulation independent of the attention of the patient to the stimulus. Cells in primary sensory nuclei of the ventro-basal complex responded mechanically to each stimulus, even when the patient was not aware of the stimulus due to distraction by attention elsewhere. However, in more mesial and dorsal thalamic regions, outside the primary sensory nuclei, cells were found which were exquisitely sensitive to attention or to the novelty of a given stimulus, failing to respond otherwise. Some of these cells we have called 'novelty detectors'. It is important to note that they were never found intermingled with cells in specific sensory systems, but were organized into a separate system with different properties. This confirms the growing body of evidence to suggest that one should look for separate neuronal systems subserving attentive responses, in contrast to those transmitting and processing sensory information at thalamic as well as at cortical levels of integration.

NEUROCHEMICAL MEDIATORS OF SPECIFIC SENSORY AND NON-SPECIFIC CORTICAL ACTIVATION

Since electrophysiological studies would suggest that separate neuronal systems are involved in the mediation of states of wakefulness and attention, independent, but interacting with those systems involved in the transmission and processing of specific sensory information, it is possible that there may be a neurochemical basis for such differences. Extensive studies of central transmitter substances during recent years should be brought to bear upon this problem (14, 30).

Out of these studies have emerged several chemical substances which appear to be good candidates for the mediation of excitation or inhibition in different central synaptic systems, or possibly for the regulation of excitatory states in assemblies of neurones by means of a form of neuro-humoral action. Recent advances in ultrasensitive, specific and precise analytical techniques for identifying substances synthesized in small quantities (even single cells) of central nervous tissue—including histochemical stains for their precise *in vivo* localization and distribution, the isolation of subcellular fractions, including 'synaptosomes', by fractional centrifugation, and micro-iontophoretic techniques permitting the direction testing, *in vivo*, of the local action of minute amounts of chemical substances liberated from a micropipette in the synaptic regions of single neurones—have provided a wealth of new data of significance to the old problem of chemical mediators of sleep and wakefulness, and synaptic transmissions in various central neuronal pathways.

THE MONOAMINES: NORADRENALINE, SEROTONIN, DOPAMINE

Positive identification and localization of an extensive system of noradrenergic neurones throughout the neuraxis, but principally in the brain stem hypothalamus and limbic system, have been established by the specific fluorescent histochemical studies of Dahlström and co-workers in Sweden (15). The cell bodies showing the specific fluorescent stain for noradrenaline are situated, not within specific sensory motor pathways, but diffusely distributed throughout the medulla, brain stem, hypothalamus, and hippocampus with terminals extending into specific synaptic systems, into the basal ganglia and even reaching the cerebral cortex, particularly in the cingulate region. Descending terminals extend to synaptic areas of ventral horn cells in the spinal cord. The main portion of this adrenergic pathway lies within what would be usually considered the lateral reticular formation of the brain stem, regions usually included in those portions whose activation is responsible for arousal and which produce the classical effects of the reticular activating system with desynchronization of cortical electrical activity. The problem remains, however, as pointed out by Jouvet (28) as to the relation between this adrenergic system and activated or paradoxical sleep as well as to the classical arousal response.

Previous studies by Dell (16), Rothballer (39, 40) and others have shown that electrocortical activation as well as behavioural awakening or arousal can be reproduced by intracarotid injections of very small amounts of noradrenaline. It was thought that this form of activation was due to a portion of the reticular activating system which possessed adrenergic receptors in the hypothalamus and brain stem and that it was not due to a direct action on the cerebral cortex itself. The evidence for this was provided by experiments in which sections of the brain stem at the superior collicular level, the classical *cerveau isolé* preparation, prevented arousal by injection of noradrenaline. Even a hemisection of the brain stem just above the superior colliculus prevented adrenergic arousal in one hemisphere while permitting it to occur from the hemisphere with the intact mid-brain connections. Such experiments led to the conclusion that adrenergic cortical activation was not direct, but that it was mediated by the brain stem activating system.

Some question about the nature of the direct action of noradrenaline has been raised by experiments showing the failure of noradrenaline injected into the blood stream to reach the brain stem except in a few isolated regions as judged by isotope studies (44). This question is not as yet resolved, but it may be that the turnover rate of injected noradrenaline is so rapid in most areas that it fails to be detected by the usual isotope techniques. However, the existence of noradrenaline—containing neurones throughout a portion of the reticular activating system gives positive proof that noradrenaline is an important candidate for a part of the arousal mechanism. There is no evidence, however, that it is the cortical mediator of desynchronized activation accompanying behavioural arousal.

The histochemical studies mentioned above have also revealed a separate system of neurones containing 5 hydroxytryptamine or serotonin. These are located along the midline of the brain stem in a system of neurones, called the Raphe system, quite distinct from the areas containing noradrenaline. Recent studies of Jouvet (28) have suggested that the Raphe system is antagonistic to the lateral reticular formation and may produce sleep and synchronization in its ascending action upon neopallial structures. Local lesions of the midline of the brain stem have been shown by Jouvet to produce prolonged periods of wakefulness in cats and to diminish the serotonin content of forebrain structures.

In summary, there is at present an increasingly definite body of evidence to suggest that noradrenaline does play an important role in the brain stem as well as in hypothalamic and limbic components of the waking or activating system but the means whereby this system mediates its effects to the cerebral cortex is still uncertain. It is necessary, therefore, to search further for the nature of the chemical mediator involved in cortical representation of the activating system with its classical desynchronization of spontaneous cortical electrical activity.

CHOLINERGIC ACTIVATING MECHANISMS

Histochemical methods, particularly those of Shute and Lewis (42), Lewis and Shute (32), using a histochemical stain for cholinesterase, have mapped out an extensive system of ascending cholinergic fibres which they have called 'the ascending cholinergic reticular system' with projection to the neocortex and to subcortical regions, including particularly the striatum and hippocampus. This system of neurones, with cortical terminals present diffusely over the neocortex, also includes portions of the brain stem which have usually been considered a part of the reticular activating system. Of particular interest is the clear selective staining of the intralaminar portions of the thalamus which stand out distinctly from specific thalamic nuclei. This includes the reticular nucleus surrounding the thalamus. Selective histochemical identification of the thalamic recruiting system, with its terminals extending into specific relay nuclei, is the most convincing demonstration of a distinct function for this portion of the thalamus as opposed to specific projection systems themselves. The presence of ascending fibres in the brain stem also is found in regions outside specific sensory or motor conduction pathways.

Acetylcholine has long been considered a good candidate for the cortical desynchronized activation comparable to the arousal response due to the fact that desynchronized activation can be reproduced readily in experimental animals by intracarotid injection of acetylcholine, as first demonstrated by Bonnet and Bremer (4) and confirmed by Rinaldi and Himwich (37, 38). This form of cortical activation occurs even in the *cerveau isolé* preparation and is not dependent upon the brain stem. It is presumed to have a direct cortical action.

Hernández-Peón in this book has presented evidence for the existence of an extensive cholinergic hypnagogic system of neurones mapped out by the injection or placement of high concentrations of acetylcholine in structures which induce physiological sleep in cats. Such results are difficult of interpretation for one cannot be assured that the injection of such high concentrations of acetylcholine actually reproduces the true physiological action of these systems. If such concentrations completely depolarize cholinoceptive cells and cause a blocking of their activity, the system of neurones which apparently produces sleep may well have the opposite action in its normal physiological function. The same question is to be raised about the early experiments of Cordeau *et al.* (11, 12) in which injection of adrenaline and acetylcholine were made into the brain stem. These authors observed in the intact unanaesthetized cat that acetylcholine tended to induce drowsiness and sleep while adrenaline produced wakefulness.

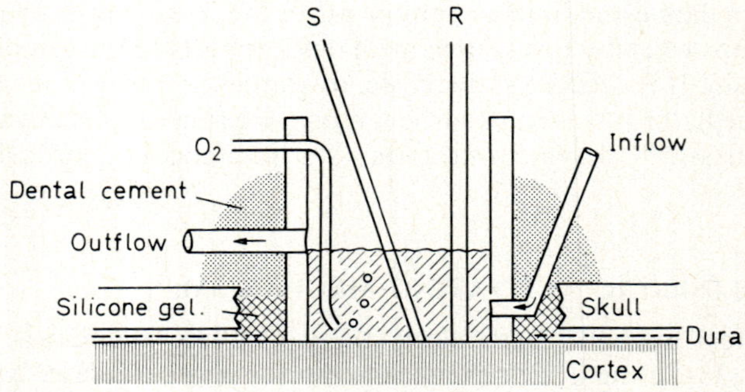

Figure 3. Diagram of cross-section of plastic chamber fixed in skull over exposed cortical surface for collection of superfusate during different physiological conditions in order to analyse for chemical substances which may be liberated as described in text

During recent years further proof that acetylcholine may be involved in desynchronized cortical activation and in the arousal response has been provided by experiments in which the rate of liberation of acetylcholine from the cortical surface has been measured in states of sleep and wakefulness, either following brain stem lesions, anaesthesia or during natural states of sleep and waking.

RATE OF LIBERATION OF ACETYLCHOLINE DURING STATES OF SLEEP AND WAKEFULNESS

By means of a plastic chamber sealed into the skull over the surface of an area of cerebral cortex from which the dura has been removed, sealing the edges with a silicone gel or similar substance, it is possible to collect fluid liberated from the surface of the cerebral cortex for chemical analysis during states of sleep and wakefulness or following various lesions of the brain stem (*Figure 3*). It is necessary, of course, to treat the local area of cortex with an anticholinesterase agent such as eserine or prostigmin before perfusion has begun. Electrical activity can be recorded from within the chamber as well as from other areas of cortex with implanted recording electrodes (6, 7, 8, 9, 33).

The cortex may be perfused with an indwelling push-pull cannula or a surface perfusion chamber.

By the use of these methods it has been found that the rate of liberation of acetylcholine from the surface of the cerebral cortex varies systematically with the state of alertness of the animal (29, 31, 35). Celesia and I have found

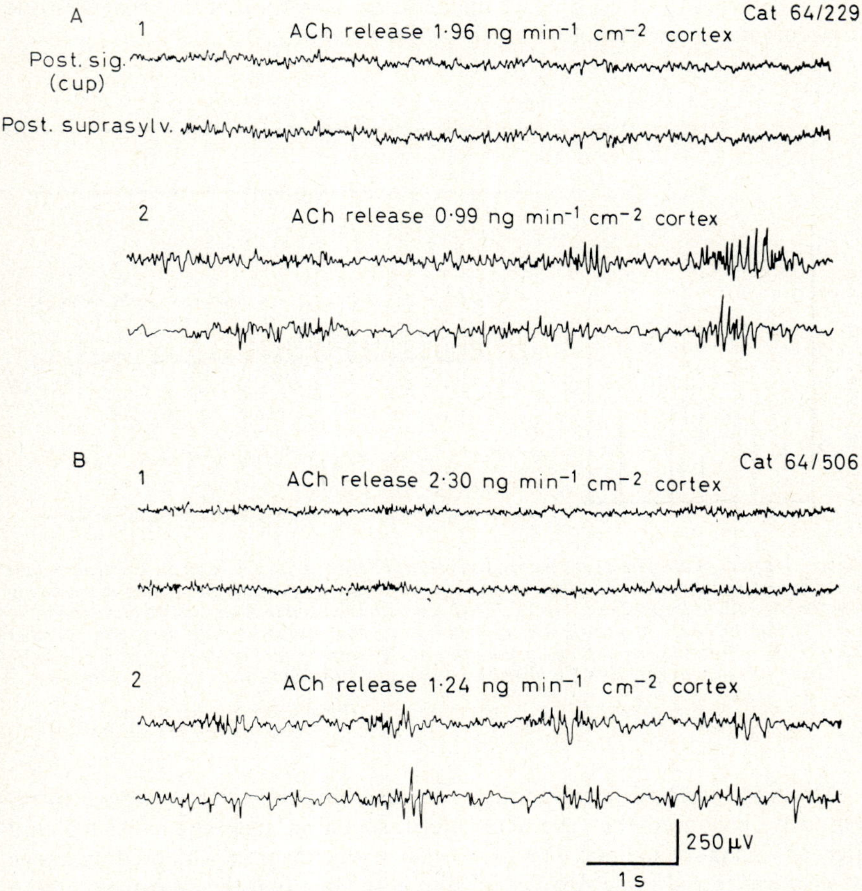

Figure 4. Rate of liberation of acetylcholine from cortical surface of cats during periods of relaxed wakefulness (A-1, 2 ng min^{-1} cm^{-2}), *natural light sleep* (A-2, 1 ng min^{-1} cm^{-2}), *alertness in another animal* (B1, 2·3 ng min^{-1} cm^{-2}) *and during light sleep* (B-2, 1·2 ng min^{-1} cm^{-2})

(From Celesia and Jasper, 1966, by courtesy of Lancet Publications Inc)

that the rate of liberation in the cat from a cortical surface area was approximately 1 ng min^{-1} cm^{-2} when a continuous perfusion method was employed. With natural awakening, as judged by the pupils and desynchronization of the EEG, the rate of liberation of acetylcholine increased to 2 or 3 ng min^{-1}, depending upon the degree of arousal produced (*Figure 4*).

There was a clear relationship between the rate of liberation and the desynchronization of the surface cortical electrical activity in the area of collection. The treatment of the cortex with prostigmin would produce a constant desynchronization of the local area even in the sleeping animal because of the reduction in hydrolysis of naturally liberated acetylcholine in this area. If the rate of liberation was fairly high, within a short time this desynchronized activation would proceed into an epileptiform activation of the local area due to accumulated acetylcholine, as shown in *Figure 5*.

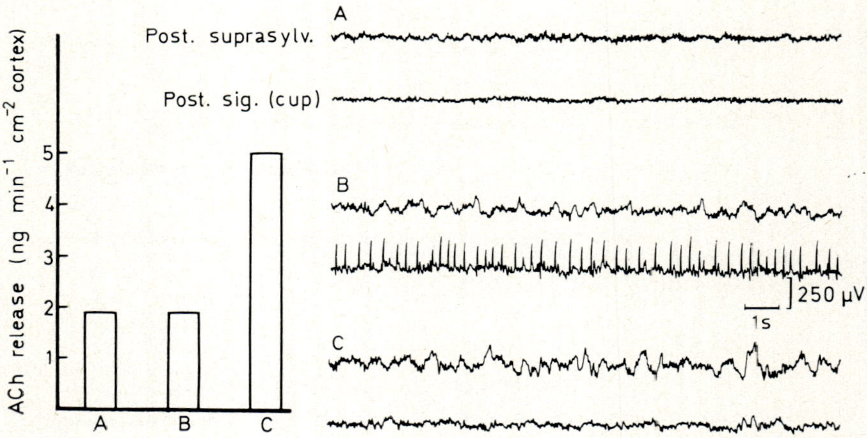

Figure 5. Effect of atropine (1 mg/kg i.v.) *on rate of liberation of ACh from the cortical surface in the intact unanaesthetized cat in relation to the effect of prostigmin applied in perfusing solution directly to local area of cortex. EEG is desynchronized in* A *in waking alert animal—more so due to effect of prostigmin in posterior sigmoid gyrus beneath perfusion chamber (cup). In* B—20 min *later spiking appears in prostigminized area, arrested by atropine in* C. *Note increase in rate of ACh liberation following atropine* (2 to 5 ng min^{-1} cm^{-2})

(From Celesia and Jasper, 1966, by courtesy of Lancet Publications Inc.)

Atropine, injected intravenously, in the amount of 1–2 mg/kg intravenously, produced the 'characteristic dissociation' between the EEG and alert or vigilant behaviour. The EEG became synchronized with predominance of slow waves and some spindles in spite of the fact that the animal reacted to stimuli appropriately and seemed to be behaving as an alert animal. Under these conditions, there has been found a marked increase in the rate of liberation of acetylcholine from the cortical surface. With stimulation of the reticular formation in the atropinized animal, released rates as high as 10 to 12 ng min^{-1} cm^{-1} of cortical surface have been obtained in experiments with Celesia. This is a double dissociation. The EEG continues as a sleeping pattern while the rate of acetylcholine liberated shows a marked increase. This is apparently due to blocking of cholinergic receptors within the cortex by atropine so that more acetylcholine reaches the cortical surface from the sites of liberation though this may not be the only explanation. If the cortex has become activated to the point of epileptic discharge atropine also blocks the epileptic discharge, while at the same time increasing the rate of ACh liberation.

The interpretation of this apparent dissociation between the EEG and behaviour has been a subject of much discussion among neuropharmacologists (5). It has been pointed out that this is not a complete dissociation, for the behaviour of the animal is impaired in other ways, when more complete studies are made of the animal's performance as in the experiments of Sadowski and Longo (41), and Ricci and Zamparo (36), who demonstrated defects in learning ability. There is, therefore, a true dissociation between apparent alertness in behaviour and ability to record experiences and retain them for recall. This raises a question as to whether the desynchronized activation of the cortex can be considered a truly reliable indication of behaviour alertness alone. It is certainly not so in desynchronized sleep for example, and that it may be related to other cerebral functions which are normally linked with alertness, but not necessarily so related under all circumstances.

In chronic experiments carried out by Collier and Mitchell (7, 8, 9) in rats a clear correlation was found between the degree of apparent behavioural alertness of these animals with implanted chronic cannulae, and the rate of acetylcholine liberated from the cerebral cortex. It would seem therefore that in normal waking conditions acetylcholine appears to be involved in a form of cortical activation related to behavioural alertness and possibly to mechanisms of attention. The fact that atropine increases the rate of liberation even in the sleeping animal, with a further increase being produced by arousal or by stimulation of the brain stem reticular formation in the absence of changes in the EEG would still suggest a possible cholinergic mechanism normally involved, even though desynchronizing effects upon cortical electrical activity are not produced.

Specific rapidly conducting afferent pathways to sensory cortex do not appear to be cholinergic, since primary evoked potentials in sensory cortex are not blocked with doses of atropine which block the 'arousal' response. In an attempt to obtain more direct evidence concerning this question, Collier and Mitchell (7) have stimulated the lateral geniculate body (LGB) while collecting acetylcholine from the visual cortex. Then the attempt was made to separate the lateral geniculate from connections with the reticular formation by a section placed mesial and posterior to it in the thalamus and tectal region, They found there was still an increase in ACh output from visual cortex upon LGB stimulation, though reduced in amount. It was concluded that there must be non-specific afferents, as well as specific afferents arising within the LGB, possibly related to the augmenting response of Morison and Dempsey (34).

Repeating the experiments of Collier and Mitchell with Garcia Flores (17) in the cat, we have constructed an oblong perfusion chamber to fit precisely over the visual cortex in the cat without covering other cortical areas. The rate of liberation of acetylcholine during stimulation of the mid-brain reticular formation was compared with that produced by stimulating the lateral geniculate body at different frequencies. In the intact animal there was an increase in ACh liberation following repetitive stimulation of the lateral geniculate body at frequencies below about 60/s, as shown in *Figure 6*. With frequencies of stimulation over 100/s however, there was a further increase in rate of liberation during stimulation of the reticular formation but no increase if the same stimulus was applied to the lateral geniculate body. Thus at higher frequencies, the rate of liberation of ACh due to geniculate stimulation could be dissociated from that due to stimulation of the mid-brain reticular formation as shown in *Figure 6*.

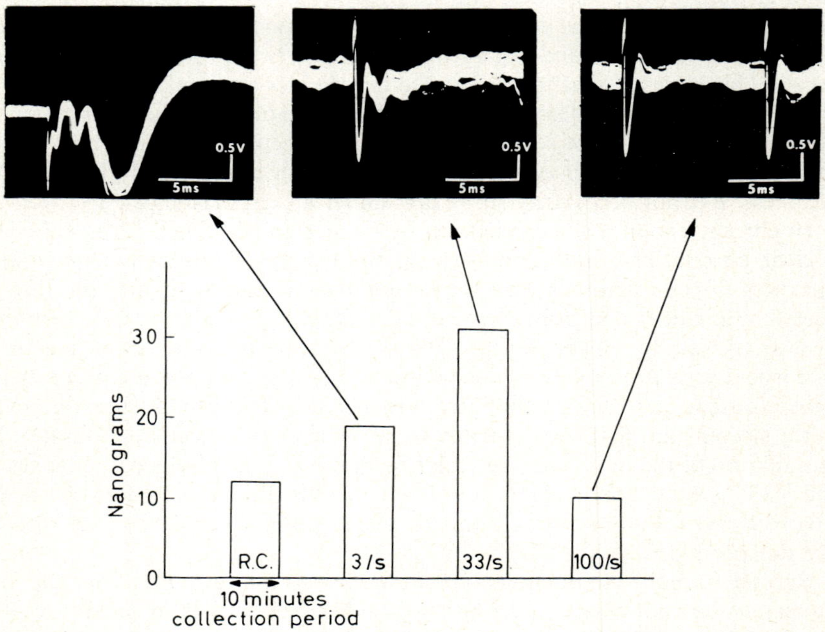

Figure 6. Rate of liberation of ACh from visual cortex in the cat during 10 min *collection periods while stimulating the lateral geniculate body at different frequencies (R.C., resting—no stimulation). Ten superimposed oscilloscope tracings of evoked potentials to geniculate stimulation are shown above for frequencies of* 3, 33, *and* 100 shocks/s. *Note increased ACh output at* 33/s *and decrease at* 100/s

Comparing the rates of liberation of ACh before and after isolation of the geniculate body from the mid-brain and mesial thalamus in the mesial *cerveau isolé* preparation it was found that no significant increase in ACh liberation from the visual cortex could be produced at any frequency of lateral geniculate stimulation in spite of a persistent increase by stimulation of the reticular formation and mesial thalamus in these cases (*Figure 7*). Further controls by cortical undercutting showed that there was no increase by any form of subcortical stimulation in such preparations, ruling out circulatory or other effects which might be independent of the subcortical afferent projections to the cerebral cortex. These results would indicate that when specific visual cortical afferents are isolated from their connections to other brain stem and thalamic structures they do not appear to be cholinergic in visual cortex. This would be consistent with the conclusions reached by other authors that acetylcholine is not the mediator of primary specific afferent activation of the cerebral cortex but rather that it is related to non-specific activation—the mediator for specific afferents being as yet unknown. Synaptic mechanisms for rapid conduction and preliminary processing of sensory information do appear to be based upon different neurochemical mechanisms to those involved in non-specific activation related to states of alertness and attention.

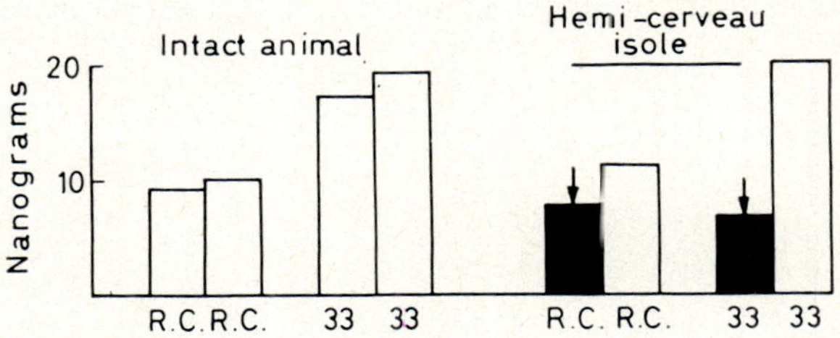

Figure 7. Perfusion chambers were placed over both right and left visual cortical areas in the cat and lateral geniculate bodies were stimulated simultaneously before and after a hemisection of upper mid-brain and thalamus on one side. Note increase in ACh only on intact side

(From unpublished experiments with Garcia Flores (1967))

AMINO ACIDS LIBERATED FROM THE CORTICAL SURFACE DURING THE AROUSAL RESPONSE AND FOLLOWING BRAIN STEM LESIONS

Considerable interest in the physiological action of amino acids has been developed recently by demonstration of the importance of glutamic acid in the activation of cells within the central nervous system by iontophoretic injection (13, 30) and the inhibitory effects of gamma amino-butyric acid (GABA) and related substances. Consequently, it was thought that the rate of liberation of amino acids from the cerebral cortex might be of significance also in our search for cortical activators related to states of sleep and wakefulness.

Using a similar technique of superfusing the cortical surface with a plastic chamber in the skull, the perfusate has been analysed for amino acid content in experiments carried out first with Khan and Elliott (25) and more recently with Koyama (26) using an automatic column chromatographic analyser (Beckman).

In the preliminary experiments with Khan and Elliott (25) we found that there was a significant increase in the rate of liberation of glutamic acid in animals maintained in the alert state (as judged by the EEG) when compared to those in natural sleep or with continuous desynchronization of the EEG following high mid-brain transection in the *cerveau isolé* preparation. Conversely the rate of liberation of gamma amino-butyric acid (GABA) was higher during sleep than during wakefulness, and in fact could not be measured by usual paper chromatographic methods if the animal was maintained alert by intermittent stimulation of the reticular formation.

In more recent studies with Koyama (26) we have found a remarkable change in the pattern of free amino acids liberated from the cortical surface, during states of sleep and wakefulness. These studies are in progress, but an example of a typical finding is shown in *Figure 8* in which a marked increase in glutamic acid was evident following stimulation of the mid-brain reticular formation in the *encephale isolé* preparation. This preparation has been used to avoid changes which might be due to alterations in blood pressure. Also the fact that certain amino acids remained constant rules out the possibility

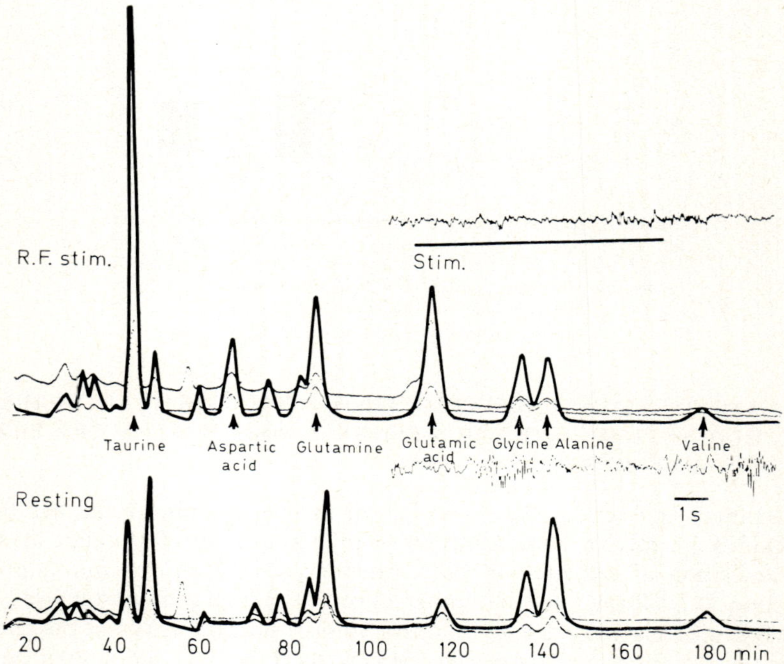

Figure 8. Sample record from Beckman automatic amino acid analyser (Acid series) showing increased rate of liberation of certain excitatory amino acids, especially glutamic acid, aspartic acid, and taurine, as a result of repetitive stimulation of the mid-brain reticular formation in encephale isolé cat preparation with local anaesthesia

(From unpublished experiments with I. Koyama)

that such changes might be due to simple increase in cerebral circulation or in diffusion from the spinal fluid. Other precautions were taken to equalize cerebral spinal fluid pressure with the cortical surface in order to prevent artefacts due to this cause. Under these controlled conditions certain amino acids showed a marked increase during reticular formation stimulation with desynchronization of the cortical EEG while other amino acids remain relatively constant as for example valine and glutamine in *Figure 8*. Significant increases were consistently obtained in glutamic acid and to a lesser extent in aspartic acid while in certain experiments a striking increase was also found in taurine, a sulphonated amino acid as shown also in the example of *Figure 8*. The significance of some of these changes is as yet uncertain, but since glutamic acid is a potent general activator of cortical cells at very low concentrations, as judged by iontophoretic injection methods, it becomes an important candidate for a more diffuse activating process related to sleep and wakefulness, since its rate of liberation in the free form from the cortex is increased during stimulation of the reticular formation with desynchronization of the EEG. We have also been able to confirm clearly the converse relationship with the release of gamma aminobutyric acid being increased during slow-wave sleep (26a).

Table 1. Rate of Release of Amino-Acids from Surface of Cortex in Encephale Isolé Cat, Effect of Reticular Stimulation. Average Values from four Experiments (Nos. 1, 2, 3, 6)

Amino-Acids	*Resting*	*Stimulation**
Taurine	2·88 ± 1·4	7·05 ± 4·0 ns
Aspartic acid	1·01 ± 0·5	3·3 ± 2·6 ns
Glutamine	5·0 ± 0·5	6·8 ± 3·2 ns
Glutamic acid	3·4 ± 1·0	11·8 ± 2·0 p = 0·01
Glycine	3·0 ± 0·3	4·8 ± 1·5 ns
Lysine	1·4 ± 0·28	1·8 ± 0·27 ns
Acetylcholine	0·07 – 0·14	0·21 – 0·27

* 100/s 6V, 1 ms 10 s on and off (Moles × 10^{-10} min^{-1} cm^{-2})

In conclusion, as shown in Table I, it is evident that one must consider changes in cortical amino acid liberation, especially glutamic acid, as well as release of acetylcholine in our search for the chemical correlates of cortical activation related to sleep and wakefulness and possibly to states of arousal or attention. It may be that the longer lasting generalized activation of mid-brain origin is based, in part, upon glutamic acid or other dicarboxylic excitatory amino acids, while the shorter lasting 'phasic' activation may be cholinergic. The possible relationship which may exist between brain stem and hypothalamic noradrenergic activating systems and cortical glutamate metabolism is currently being studied.

DISCUSSION ON JASPER

Adey

I would like to ask Dr. Jasper if he has any data on the distribution of amine substances. The studies by Dahlström and Fuxe and others have shown their anatomical distribution in synaptic terminals according to hierarchical systems that involve the medial thalamic and brain stem structures and also some of the limbic structures. This has raised the question of their possible role as transmitters. Do you have any data on that line?

Jasper

The distribution of the cholinesterase system has been most thoroughly studied by Shute and Lewis at Cambridge. The intralamina nuclei of the thalamus are selectively stained by this method. The hippocampus also gives a very dense cholinesterase stain, and the caudate is most dense. But you were asking about other amines?

Adey

Tryptamine for instance.

Jasper

Work on the catacholamines has just begun, and I haven't enough to speak about that yet. The Swedish work on this subject is the most complete, as you probably know. The differentiation of the catacholamines by the fluorescent staining is very striking. There is a serotone system that runs in the medial brain stem which is separate from the noradrenalin system, and separate from the dopamine system.

Berlyne

Could I ask Dr. Jasper for an elaboration on one point? He was talking about, I gather, these cells in the thalamus that have been connected with attention or attentiveness. He gave an example of their responding to an expectation of being stimulated. I take it that that was only one example. Could I ask what other operations, if any, produced the same effect, and whether this would be related to some ideas that he was expressing the other day about the notion of 'attentiveness'.

Jasper

I used attentiveness, because I thought we might confine attention to the selective component, and vigilance or arousal to the general alertness to all stimuli. This is a semantic problem. Other examples are that visual, auditory or tactile stimulation to certain areas will cause responses in certain cells, either when the patient is not expecting them and ceases to respond on repetition in a very short time, or when his attention is distracted, and in this way you could produce a failure to respond. These multimodal responses are most complicated. Others seem to be behaving in a similar way within a given modality. They are located differently in the thalamus. The first order of complexity is illustrated as those which respond to flexor movements of the fingers on either hand or the toes on either foot—the same unit responding to this movement, bilaterally, hand or foot. These units are also highly susceptible to habituation. At repetition of this movement they cease responding. I am not sure whether they have anything to do with the attentive response, but they are very complex cells and no other form of stimulus will fire them, so they are probably more in the nature of the complex somatic type of unit connected with homologous movements of the four extremities. There are a whole series of complexities here that are very hard to sort out. As a matter of fact, in the operating room we have been unable to find the code that will fire many cells we encounter. Many of them ignore all forms of stimulation or manipulation of the patient. Even to emotional excitement they failed to change their rate of firing. We just have not the right formula for affecting them.

Lilly

I would like to ask Dr. Jasper if he can conceive of any way of examining these chemical reactions to separate out the glial component, the neuronal component, the dendritic, etc. It would seem that the time constants found are long enough that any or all of these systems could be contributing their effects by the present method of analysis. It may be that some fast and microscopic method, possibly optical, could be worked out.

Jasper

This is a difficult problem and impossible with the techniques we are using, and many people have asked us to develop very much faster detection methods which can get the time course of this operation and we are working on this. I think we can make some progress with the *in vivo* isotope counting techniques with specific labels. We hope that this will give us the rapid picture although I am not sure that it will. The content of these substances in fractionations, in centrifuge fractions, gives you some information of value, though limited, I think, as to what it is like in the intact brain. You can lacerate the brain and spin it round and get these various subcellular fractions. I am not really sure whether what we find there is what the normal distribution is in the intact metabolizing tissue. So far therefore we have no real answer to this.

Grey Walter

We should remember that all the electrical changes we record are signs of *chemical* change. We are dealing with chemical batteries and not cat's fur or electrostatic generators. One of the most exciting and important things would be if the work that Dr. Jasper described could be developed in such a way that particular chemical reactions or electron transfers could be associated with particular types of electrical event. This would take us a long way nearer to establishing the real electrochemistry of the brain in relation to the processes of attention or distraction that we are interested in.

Jasper

Of course, our largest programme is in this direction but I will never see the end of the programme, I can assure you! Dr. Cobb, your work interests me very much, in the analysis of the components that were changed in certain circumstances and the effect of atropine which blocks the cholinergic mechanisms specifically is such that it does not affect the early components of the local responses but only the later ones. Have you attempted to dissect your components pharmacologically? It might be worth some of your time.

REFERENCES

1. Adrian, E. D. and Matthews, B. H. C. (1934a) *Brain*, **57**:355
2. Adrian, E. D. and Matthews, B. H. C. (1934b) *J. Physiol.* **81**:440
3. Adrian, E. D. and Yamagiwa, K. (1935) *Brain*, **58**:323
4. Bonnet, V. and Bremer, F. (1937) *C. r. Soc. Biol.* (*Paris*), **126**:1271
5. Bradley, P. B. and Elkes, J. (1953) *J. Physiol.* **120**:13
6. Celesia, G. G. and Jasper, H. H. (1966) *Neurology*, **16**:1053

7. Collier, B. and Mitchell, J. F. (1966a) *J. Physiol.* **184**:239
8. Collier, B. and Mitchell, J. F. (1966b) *Proc. Physiol. Soc., J. Physiol.* **186**:44
9. Collier, B. and Mitchell, J. F. (1967) *J. Physiol.* **188**:83
10. Corletto, F., Gentilomo, A., Rosadini, G., Rossi, G. F. and Zattoni, J. (1966) *Revue Neurologique*, **115**:5
11. Cordeau, J. P. (1962) *Rev. canad. Biol.* **21**:113
12. Cordeau, J. P., Moreau, A., Beaulnes, A. and Laurin, C. (1963) *Archs. ital. Biol.*, **101**:30
13. Crawford, J. M. and Curtis, D. R. (1964) *Br. J. Pharmac. Chemother.* **23**:313
14. Curtis, D. R. and Watkins, J. C. (1965) *Pharmacol. Rev.* **17**:347
15. Dahlström, A., Fuxe, K., Olson, L. and Ungerstedt, U. (1964) *Acta physiol. scand.* **62**:485
16. Dell, P. (1963) In: *Brain Mechanisms* (Ed. byMoruzzi, Fessard and Jasper). Amsterdam: Elsevier
17. Garcia Flores, E. J. (1967) *M.Sc. Thesis*, Department of Physiology: University of Montreal
18. Herz, A., Niedner, I., Fraling, F. and Sommer-Smith, J. (1966) *Expl. Brain Res.* **1**:249
19. Hubel, D. H., Henson, C. O., Rupert, A. and Galambos, R. (1959) *Science, N.Y.* **129**:1279
20. Hubel, D. H. and Wiesel, T. N. (1965) *J. Neurophysiol.* **28**: 227
21. Jasper, H. H. (1966) In: *Brain and Conscious Experience* (Ed. by Eccles). Springer-Verlag, New York Inc.
22. Jasper, H. H. and Bertrand, G. (1964a) *Am. Neurol. Ass. Trans.* **89**:79
23. Jasper, H. H. and Bertrand, G. (1964b) *Physiologist*, **7**:167
24. Jasper, H. H. and Bertrand, G. (1966) In: *The Thalamus* (Ed. by Purpura and Yahr). Columbia University Press: New York
25. Jasper, H. H., Khan, R. T. and Elliott, K. A. C. (1965) *Science, N.Y.* **147**:1448
26. Jasper, H. H. and Koyama, I. (1967) *Fedn. Proc.* **26**:373
26a. Jasper, H. H. and Koyama, I. (1969) *Can. J. Physiol. and pharmacol.* **47**: No. 9
27. Jasper, H. H. and Penfield, W. (1949) *Arch. Psychiat.* **183**:163
28. Jouvet, M. (1967) *Physiol. Rev.* **47**:117
29. Kanai, T. and Szerb, J. C. (1965) *Nature, Lond.* **205**:80
30. Krnjevic, K. (1965) *Proc. Int. Union Physiol. Sciences*, Vol. IV, XXIII, International Congress, Tokyo, 435, *Excerpta Medica*, Series No. 87
31. Krnjevic, K. (1967) *Anesthesiology*, **28**:100
32. Lewis, P. R. and Shute, C. C. D. (1967) *Brain*, **90**:521
33. Mitchell, J. F. (1963) *J. Physiol.* **165**:98
34. Morison, R. S. and Dempsey, E. W. (1942) *Am. J. Physiol.* **135**:281
35. Phillis, J. W. and Chong, G. C. (1965) *Nature, Lond.* **207**:1253
36. Ricci, G. F. and Zamparo, L. (1963) Pharmacology of Conditioning, Learning and Retention, *Proc.* 2nd *Int. Pharmacological Meeting*, Prague
37. Rinaldi, F. and Himwich, H. E. (1955a) *Archs. Neurol. Psychiat.* **73**:387
38. Rinaldi, F. and Himwich, H. E. (1955b) *Archs. Neurol. Psychiat.* **73**:396

39. Rothballer, A. B. (1956) *Electroenceph. clin. Neurophysiol.* **8**:603
40. Rothballer, A. B. (1959) *Pharmac. Rev.* **11**:494
41. Sadowski, B. and Longo, V. G. (1962) *Electroenceph. clin. Neurophysiol.* **14**:465
42. Shute, C. C. D. and Lewis, P. R. (1967) *Brain*, **90**:497
43. Steinberg, R. A. (1965) *Electroenceph. clin. Neurophysiol.* **18**:378
44. Wolfe, D. E., Potter, L. T., Richardson, K. C. and Axelrod, J. (1962) *Science, N.Y.* **138**:440

Electrophysiological Correlates of Selective Attention

Edward C. Beck, Robert E. Dustman and Makoto Sakai
Veterans Administration Hospital and University of Utah, Departments of Neurology and Psychology, Salt Lake City, Utah

We have spent the past several years searching for neural correlates of psychological processes such as attention and learning. Our methods have been those of producing lesions, recording electrical activity, or chemically stimulating different areas of the brain, cortical and subcortical. The areas selected have been those that we and other investigators have presumed to be related to consciousness or mental states. Thus, the work we will report has been very much in the spirit of this conference, the study of the neurophysiology of attention. The results we will report also reflect the spirit of this conference; they have been contradictory and controversial. While often our results have corroborated certain current assumptions about brain function, occasionally they have turned up disquieting exceptions. Like the participants of this conference we have run into repeated conflict in our own laboratory and among ourselves. On occasion we have been prompted to view our contradictions as possible failures and have been inclined to agree with our more pure and critical colleagues, those busy in the spinal cord or comfortably ensconced in the squid axon, that we are perhaps embarked on a hopeless task. The gap between tissue and psyche may be too great a leap to achieve with our limited instrumentation and primitive knowledge of the brain.

Consequently, we would like to spend the first few minutes reviewing earlier work in our laboratory, our itinerary through the mid-brain reticular formation, diffuse and specific thalamic nuclei. As we are now back at the cortex from whence we started we will spend the last part of the presentation with findings involving scalp and cortical recordings, utilizing 'averaged' evoked responses.

We believe the seeming contradiction in the experiments we will describe as well as part of the controversy of this conference can be accounted for as much by errors of logic and wrong assumptions as by contradicting data. Before pursuing this point of view however, allow us to briefly review a few such studies.

ELECTROCORTICAL CORRELATES OF ATTENTION

Early in our experiments we noted that the daily frequency of cortical arousal reactions to a conditioned stimulus increased to a significant level before any motor or behavioural indication of learning could be observed. Cats, permanently implanted with cortical electrodes, were drugged with small dosages of bulbocapnine or made somnolent by standing them in warm water overnight and their EEGs recorded continuously during the formation of a conditioned response. The drug bulbocapnine had no readily detectable effect on the rate of conditioning (3) and usually produced such slow EEG patterns that electrocortical arousal reactions were well demarcated. The frequency of LVF (low-voltage-fast) reactions increased daily and when daily frequency of LVF was plotted over several days the resulting curves appeared similar to learning or performance curves.

Such EEG changes we interpreted as evidence of cortical activation through brain stem and thalamic reticular systems as described by Moruzzi and Magoun (26). In point of fact, the experiments of others (28, 34, 40) making simultaneous recordings from these structures supported this assumption.

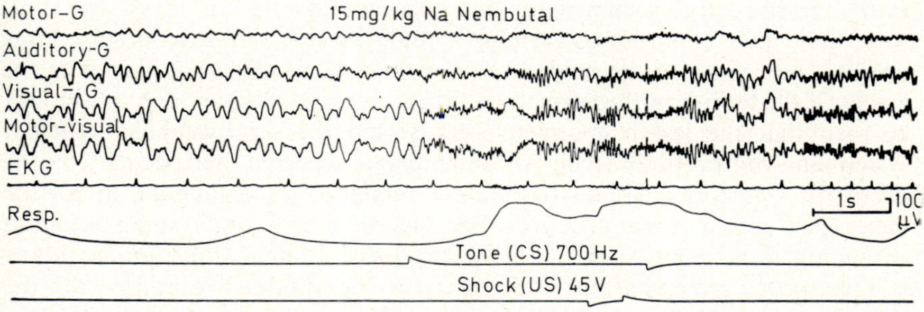

Figure 1. A LVF reaction and a change in respiratory rate to a CS in an anaesthetized cat. The animal was trained under light anaesthesia

We found, however, that with overlearning there was a tendency for the EEG arousal response to diminish and there were instances where conditioned responses occurred during relatively slow electrocortical activity, suggesting a distinction should be made between the original acquisition of a response and its continued performance after learning had occurred. That is, that increased attention is required early in the learning process hence LVF electrical activity at cortex, hence reticular system participation. LVF electrical activity at cortex and reticular system involvement is apparently no longer essential after correct responses are well established (2, 3, 8, 36).

To test these assumptions further we trained cats under barbiturate anaesthesia. During light anaesthesia, 15–20 mg/kg sodium nembutal, the cats slumbered in a hammock while a low intensity, low pitch (700 Hz) tone preceded a foot shock by 3 s. The animal aroused to the shock but care was taken to present the tone during periods of obvious somnolence. With repeated pairings, upwards of 10 sessions of 50 trials each, cats gave EEG and respiratory CRs as illustrated in *Figure 1*. When tested in the alert state, conditioned flexions were observed in 3 of 5 animals so trained. In these instances it will be noted that LVF activity was never seen during the training, except to the

US. It suddenly emerged as part of the conditioned response as may be seen in *Figure 1.* Note the LVF change follows the CS signal by about $\frac{1}{2}$ s and precedes the US shock by almost 2 s. These findings are contrary to our previous notion that LVF and increased attention are essential to the conditioning process (4).

CRITICAL BRAIN AREAS

We next explored the possibility that certain central brain stem areas might be critical for the performance of conditioned reflexes. The areas we had in mind were 1) the mesencephalon including the reticular formation, and 2) certain midline thalamic nuclei such as centre median and dorsal medial nucleus. A number of animals were conditioned to a tone using an unavoidable foot shock as the unconditioned stimulus. After a conditioning criterion was met, electrolytic lesions were placed stereotaxically in the area of the medial mesencephalon and medial thalamus. Seven animals survived this rather drastic surgery sufficiently for behavioural and EEG evaluation. It was found that conditioned responses could be re-established despite large debilitating lesions of the medial brain stem systems. *Figure 2* illustrates the extent of the lesions in the animal with the most extensive damage, and it may be seen that the lesion completely transected the mesencephalic reticular system and central brain stem. Although death occurred unexpectedly on the twenty-first postoperative day the animal remained in a cataleptic stupor the major part of this time, and yet when placed in the conditioning situation struggling took place. The CS alone produced both a transition to lower voltage, faster brain rhythms and respiratory conditioned responses. On the 18th postoperative day, feeble, but discrete flexions were seen to the CS in 6 of 45 presentations.

Other cats with less extensive lesions in the mesencephalon showed a progression from complete stupor during the first postoperative weeks to ultimate spontaneous walking and eating 8–12 weeks later. Each demonstrated high level conditioned responses with re-training.

In the animal with the smallest lesions of the series, only centre median and the dorsal medial nucleus of the thalamus approached total destruction. However, Zona incerta and field H_1 of Forel were heavily damaged. As with animals with lesions in the mesencephalon a profound cataleptic stupor prevailed throughout the first postoperative week. On the 26th day, however, conditioned responses had returned to a high level with retraining, and the animal could stand, walk slowly, eat spontaneously, and reacted promptly by turning to the appropriate side when touched.

Others have reported similar results. Adametz (1) comparing cats with large bilateral reticular lesions done in one stage with equally large lesions done in two or more stages reported that with the multiple stage lesions the animals were alert shortly after surgery, exhibiting seemingly normal sleep-wake cycles and behaviour characterized by walking and eating unaided, despite total destruction of the brain stem reticular formation. Chow *et al.* (9) showed that conditioned cortical potentials, i.e. repetitive electrocortical responses resulting from pairing a tone with flickering light, could be elicited after

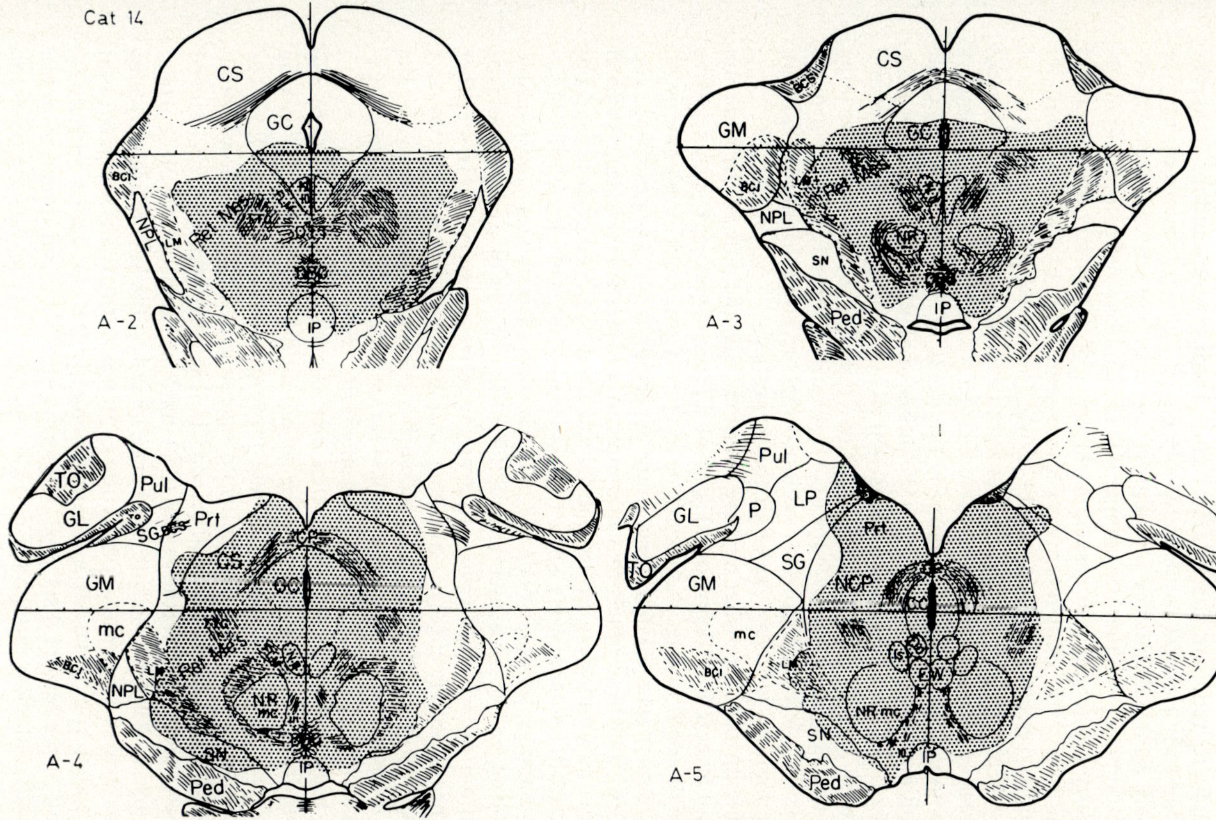

Figure 2. Lesions transecting the mesencephalic reticular system and central brain stem lesions are projected on to the stereotaxic planes of the histological atlas of Jasper and Ajmone-Marsan. In this figure only areas of total destruction are included and the lesions run roughly 1 mm *posteriorly and anteriorly beyond the levels shown. These lesions were the largest of the series of animals*

(From Doty, Beck and Kooi, 1959, by courtesy of Academic Press)

lesions in rostral, midline and anterior thalamus, as well as complete destruction of the mid brain reticular formation.

Lately, Bignall (5) has demonstrated that subcortical structures such as the mesencephalic reticular formation, centre median and midline thalamic nuclei, currently thought to mediate polysensory cortical responses, may be completely removed, but the polysensory cortical evoked responses will recover and interact in a manner comparable to that observed prior to ablation. Polysensory convergence in associational areas of cortex has been regarded for some time with great interest by those searching for the 'anatomy' of conditioning, selective attention, and the like. Interaction and convergence of evoked potentials initiated by different stimuli in polysensory cortical areas are suggestive of neurophysiological substrate of such processes as sensory conditioning. Critical to this notion is the point of view that the reticular formation and its extension through the diffuse thalamic system may provide generalized and localized activation as well as states of hyperexcitability and inhibition in various cortical areas. This has been one hypothesis describing how the brain could selectively tune in or out various stimuli as well as provide the necessary interconnections for such processes as conditioning (15, 27, 33). Bignall's findings seriously challenge this hypothesis.

It would appear from what we have presented that the bulk of the midbrain core is not essential for behavioural arousal, sleep and waking cycles, selective attention or conditioning. But permit us to muddy the waters even more with contrary data from our laboratory. In the experiments we have described, and those cited, it will be noted that common to all was a debilitating surgical procedure. Not only were critical subcortical areas damaged or destroyed but in all probability there were such things as post surgical shock, diaschisis, vascular damage, cerebral oedema, etc. Also, in all instances, considerable time evolved before an evaluation of the effects of the procedure could be made. In our experiments weeks of retraining were necessary and in all other studies the immediate effect of damaging or eliminating a brain area could not be evaluated as the methods precluded such a possibility. What would be the effect if in the midst of purposeful behaviour or a learned response the reticular formation was immediately but momentarily eliminated?

It has been known for some time that brief electrical pulses applied to any of several areas of cortex will produce behavioural arrest in which a waking animal momentarily ceases all ongoing movement (11, 32). This effect has been presumed to be due to corticifugal projections of impulses to lower brain centres. In the experiments we will describe (13) seizure discharge was induced by minute injection of acetylcholine (0·01 ml 5 per cent) in various cortical sensory areas. When the seizure discharge spread to the subcortical reticular system by way of corticifugal projection of impulses there was a behavioural arrest, an abolition of the conditioned response. In the absence of corticifugal projection of the seizure discharge to lower centres the conditioned response was not affected, even though the cortical area of discharge was the receiving area for the CS, such as the auditory area for a series of clicks.

In *Figure 3a* note that subcortical recording is confined to channels 9 through 12. No seizure discharge may be seen in these channels but may be clearly seen in channels 2, 3 and 5 which localizes the discharge to left lateral

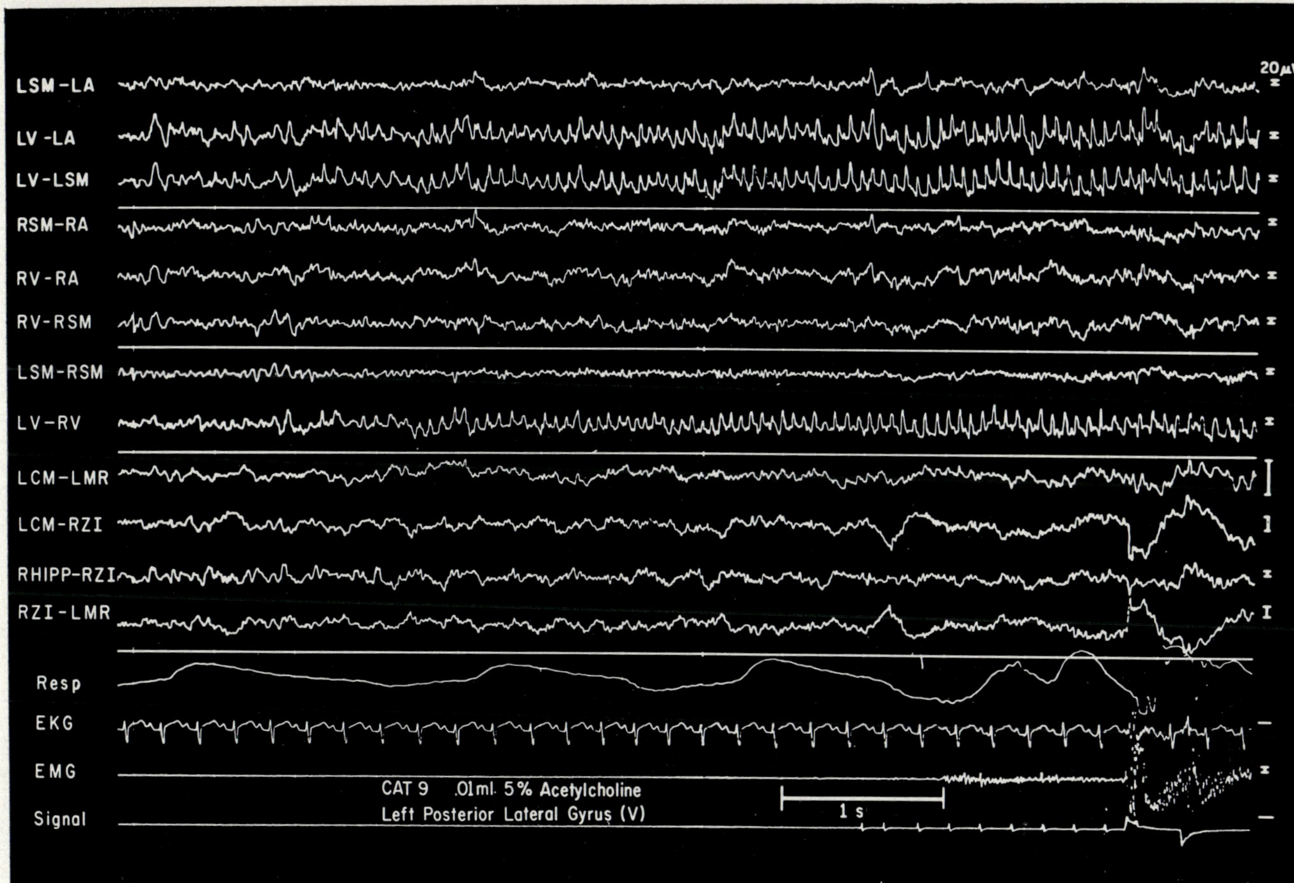

Figure 3a. Responses approximately 3 min following the induction of seizure discharge by injection of acetylcholine in the region of the left posterior lateral gyrus (visual area) in a conditioned cat. The animal produced good flexion response despite local spiking in the lateral gyrus. Note that the seizure discharge is confined to cortex

(From Fleming, 1963, by courtesy of the Editor, *E.E.G. Journal*)

gyrus. Despite the discharge the cat responds appropriately with a CR. In *Figure 3b* the seizure is clearly seen to spread to subcortical sites with a concomitant behavioural arrest and abolition of the CR. Thus, if lower centres are involved in the seizure discharge, the animal does not respond, if they are not involved, behaviour continues in a seemingly appropriate manner.

The difference between this result and those previously cited is interesting. In animals with lesions, appropriate responses, inferring some level of attentiveness, were elicited despite massive destruction of the mid brain core; while in the present experiment a temporary interruption of function, a mere transient 'loading of the lines' of corticifugal projections to the mid brain, resulted in a complete behavioural arrest and in what might be termed a state of confusion in the animal. How do we reconcile these different results? It may be that some small part of the reticular formation was spared in the lesion experiments. Thompson (35), in an excellent review of such studies, pointed out that, in the majority of experiments reported, a very small medial portion of the reticular formation survived. While this may be true, we feel that there is a more parsimonious explanation for the differing results. It seems unlikely that a small remaining ribbon of mid brain could account for the almost complete reconstitution of the damaged animal's behaviour. We suggest the differences can be reconciled by errors of logic rather than of technique.

As investigators in the area of brain research we repeatedly refer to the 'dynamic' nature of the brain and the nervous system, yet we design, and we interpret, our experiments on completely different assumptions. Our assumptions and our arguments are mainly Aristotelian. That is, we really don't see brain function as dynamic at all, we see it as a static thing, a thing that can be taken apart one piece at a time. We tend to regard a structure as doing a certain thing, regardless of the conditions we impose upon it, and regardless of its relationship to other parts and to the whole.

Other sciences, such as Physics and Chemistry have found Aristotelian concepts are only partly true and have substituted Galilean logic or the notion of relativity in their scientific thinking. From their point of view an entity does not exist by itself. In determining an entity, the *situation* assumes as much importance as does the object. The object's identity and function may change with the situation.

If we apply this type of logic to the studies we've described, the conflicting results are understandable. Possibly differences expressed at this conference could be similarly explained. In our experimental design we imposed vastly different situations on the organism so we got different results. An animal bereft of a large area of its brain may be forced to use other systems ordinarily not involved in whatever behaviour is being studied. If we dam up the usual course of a stream, it will flow in another direction. When it is present and healthy the reticular formation most probably plays a critical role in attention and consciousness. However, the activation process involved in alerting and attending, whatever it might be, is surely a function of the interrelationship of a great number of structures *which are not constant in their relationship and change with each situation.*

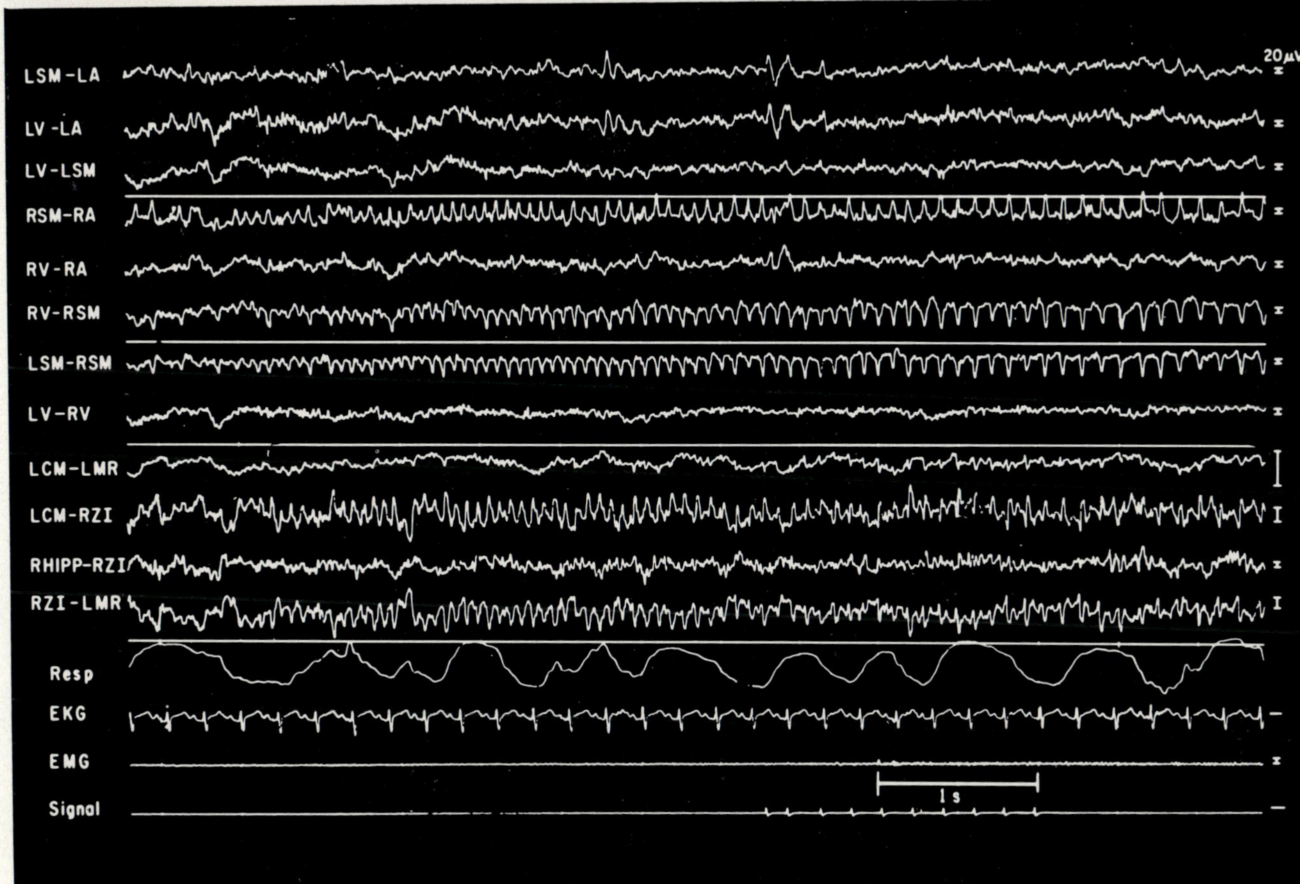

Figure 3b. Response recorded approximately 3 min *following the injection of acetylcholine into the cerebrospinal fluid overlying the right coronal gyrus (somatosensory area) in the same animal as in Figure 3a. Note the absence of a flexion CR (EMG response) in the presence of the downstream projection of cortical seizure discharge. Arrow points to what might be the initiation of a respiratory response*

(From Fleming, 1963,
by courtesy of the Editor, *E.E.G. Journal*)

CEREBRAL EVOKED POTENTIALS AND CONDITIONING

Lately, we have been studying cerebral evoked potentials, recorded from cortex in cats and from scalp in humans. A number of investigators believe that attending, learning and other mental processes may be reflected in the size and wave form of the cerebral evoked response. During inattention or during habituation, the cerebral evoked response has been reported to diminish; while during attention it has been observed to increase in size. For example, attenuation of the evoked response in the visual system to a repeated presentation of a light pulse (habituation) has been reported all along the visual pathway from the optic tract to primary visual cortex implying some inhibitory influence at all levels of that system (7, 9, 25, 29). Similarly, attenuation of the auditory evoked response to a repeated click has been reported in the auditory pathway from the dorsal cochlear nucleus to auditory cortex (14, 19, 38).

Likewise, with increased attention to an auditory stimulus (a conditioned response to a click) enhancement of the auditory response has been reported at the dorsal cochlear nucleus (14, 19) and at the medial geniculate body and auditory cortex (18). Summed evoked responses were also reported to be enhanced in human subjects when they were attending to a click (10, 16) or to a flash (17). In the major portion of these experiments it was the late component or secondary waves that were affected most strongly.

When two stimuli were presented simultaneously and subjects were asked to attend to one and ignore the other, the amplitudes of the late wave components of summed evoked responses to the ignored stimulus were attenuated or reduced while the opposite effect was noted with the response from the attended stimulus (17, 31). This suggests that selective attention may be reflected in the electrical activity of the brain.

Despite these studies, the notion that the size of the evoked response along the sensory pathways changes concomitantly with the 'information' qualities of the stimulus has been seriously challenged. Some argue the concept is an oversimplification. More seriously, a number of contradictory observations have been reported. No attenuation of the evoked response to a repeated light flash was obtained along the visual pathways and at visual cortex in cats with their eyes atropinized (12) nor did a repeated click result in any consistent attenuation of the evoked response in the auditory pathways or at the dorsal cochlear nucleus when the position of the cat's head was carefully controlled (39). It is interesting that Marsh had previously reported such changes in 1961 but evidently repeating some of his experiments with Worden had changed his mind. Recording from the posterior suprasylvian gyrus—presumed to be an associational area in the cat—there were no consistent changes noted in size or form of the evoked response during habituation on repeated experiments with a number of animals (37).

Reports of modification of the evoked response during increased attention are also contradictory in several respects. Attenuation rather than potentiation of the primary as well as late wave components of cortical evoked responses recorded from primary visual cortex has been reported when a cat was attentive to a mouse (21). Enhancement as well as attenuation of both primary and late components of the cortical evoked response recorded from the primary sensory receiving areas to a click and to a flash was observed by Jane, Smirnov and Jasper, when the animal was distracted by a rat or the squeal of a rat (22). Similarly, a reduction rather than an increase in amplitude of the cortical evoked response in the posterior supersylvian gyrus to click, flash or electric shock to the forepaw was found when the animal was distracted by either auditory, visual or tactile stimuli (37).

While EEG tracings roughly correlate with psychological states, at least up to the time of this conference, there is, as yet, no consistent agreement regarding the occurrence of habituation or attention as reflected by evoked responses in sensory systems of the waking brain. These inconsistencies forewarn any simple interpretation of either facilitatory or inhibitory sensory transmission in terms of the amplitude of evoked response.

To account for reported discrepancies, possibly we must abandon the general notion that activation as maintained by the brain stem reticular formation and diffuse thalamic system is a relatively undifferentiated, general energizing process such as turning up a 'volume control' to increase or decrease evoked response amplitudes. A suggested neurophysiological substrate of attention is a constantly changing process wherein there is continuous interplay with the background brain activity and signal. This relationship may change in a variety of ways during any period of attending. One way attention may be reflected neurophysiologically is by selective stability of the ongoing electrical activity in certain areas of the brain, thereby reducing the signal to noise ratio in that area or to that sense modality. That is, certain structures that modulate the brain in some fashion, possibly the reticular formation or the anterior thalamic nuclei, converge on certain systems selectively by sustained activation, thus producing a stable level of cerebral activity and consequently stability of wave form of the evoked response. If this is true, then the signal to noise ratio, that is, the relationship between components of the evoked response and background brain activity would be more constant in that part of the brain or in that system. Evoked responses during such an epoch would be much less variable and, consequently, if averaging methods were used would appear to increase in amplitude during attentive periods by the nature of the time-locked summation employed in most averaging computers. Such an explanation might resolve some conflicts in the literature. Investigators using computers such as the CAT and recording from human scalp, generally report enhancement of summed evoked response during attentive periods. On the other hand, those using cathode ray oscilloscopes and recording from the dura of experimental animals have often reported attenuation or increased stability during attentive states or when the animals were alert.

To investigate these possibilities, experiments were designed towards answering the following questions. To what degree does the evoked response stabilize during periods of attention and will such stability result in what appears to be increased amplitude when computer averaging methods are used? Will this stability be restricted to certain systems or areas such as the visual area when the animal's attention is to vision? Will associational areas of the cortex which are polysensory show differential effects to one modality when that modality has been conditioned?

Five cats, chronically implanted with electrodes on the cortical surface of the marginal, posterior suprasylvian and middle sylvian gyri, were trained to make a discriminative avoidance response (hind-leg flexion) to a train of flashes or a train of clicks while these were presented randomly during the same session. All animals were trained to respond to flash and click respectively with a CR. Thus each animal had two training situations, one in which flash was the CS, and one in which click was the CS. Summed evoked responses

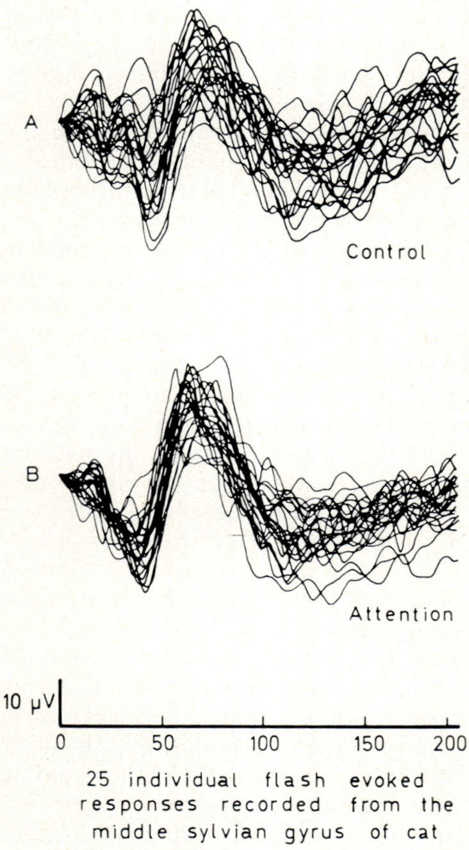

Figure 4. Individual evoked responses to a series of light pulses separated by 1 s. *In one instance* (*A*), *the animal is indifferent to the stimulus, in* (*B*) *the cat has been trained to associate a foot shock with the flash.* (*Upward deflection indicates negativity*)

initiated by each sense modality, vision and audition, were recorded from the three gyri. These were then compared during periods when the stimulus had no significance (control), when the stimulus became a conditioned stimulus (attention), and when the stimulus occurred while the animal was responding to the other stimulus with a conditioned response (distraction). Variability was investigated by replicating these situations three times.

Figure 4 shows 25 individual flash evoked responses recorded from middle sylvian gyrus during a period of indifference to the light in A, and during a period of attention, B. It may be seen in B of the figure that the 25 responses are markedly more stable during the attentive period.

Figure 5 illustrates changes in amplitude of the visually evoked response during levels of attention. Amplitude of components of the summed evoked response to flash recorded from marginal and posterior suprasylvian gyri did not increase with increased levels of attention. Those recorded from middle sylvian gyrus, however, did. It will be noted that a secondary wave component

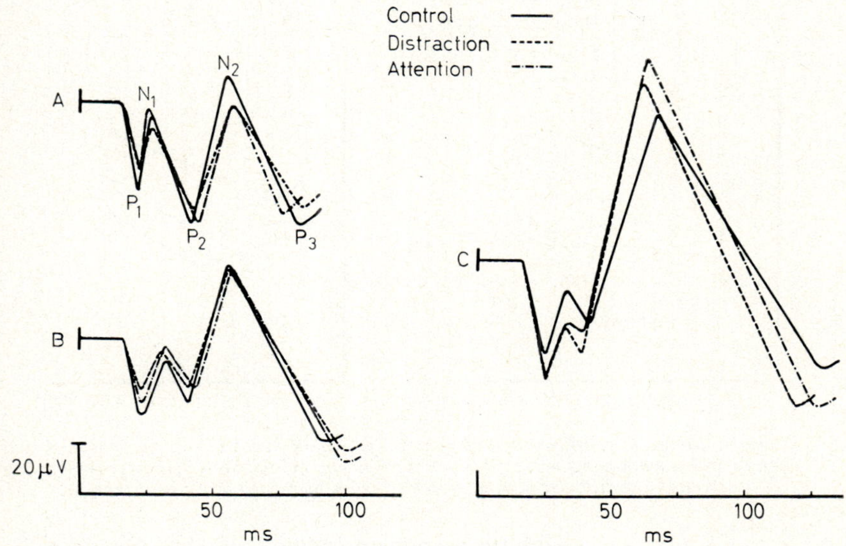

Figure 5. Composite summed evoked potentials from marginal (A), posterior suprasylvian (B) and middle sylvian (C) gyri during control, distraction and attentive periods. These responses were constructed by plotting the mean peak delay and mean peak to trough amplitude of the various waves in the response of five animals and drawing connecting lines between these points. The evoked response of each of the five animals was the sum of 75 responses. Thus the composite response is the sum of 375 responses. In this and subsequent figures, upwards deflection indicates negativity. Amplitude calibration is the same for all responses

with a peak delay of 60–70 ms potentiated reliably during periods of attention and distraction. It should also be noted that the greatest increase was during the presumably greatest attentive period, i.e. when the flash was the CS. With click there was the same trend with a similar secondary component at about 70 ms increasing concomitantly with levels of attention.

Comparison of the stability of the wave form of evoked responses by coefficients of correlation is illustrated in *Figure 6*. Here again a consistent trend may be seen and this is again only evident in recordings from the middle sylvian gyrus.

To return to our questions, we found that the evoked response did indeed stabilize during periods of attention and that this stability was restricted to a certain area, namely the middle sylvian gyrus, for both sense modalities, vision and audition. Contrary to our notion that stability of wave form alone would account for potentiation of evoked response components during periods of attending, we found that single evoked responses also increased reliably in amplitude. The marked potentiation of summed evoked responses during conditioning, as extracted by our CAT computer, we found to be the combined product of an increased amplitude of single evoked responses and a higher stability of the wave form of evoked responses over time.

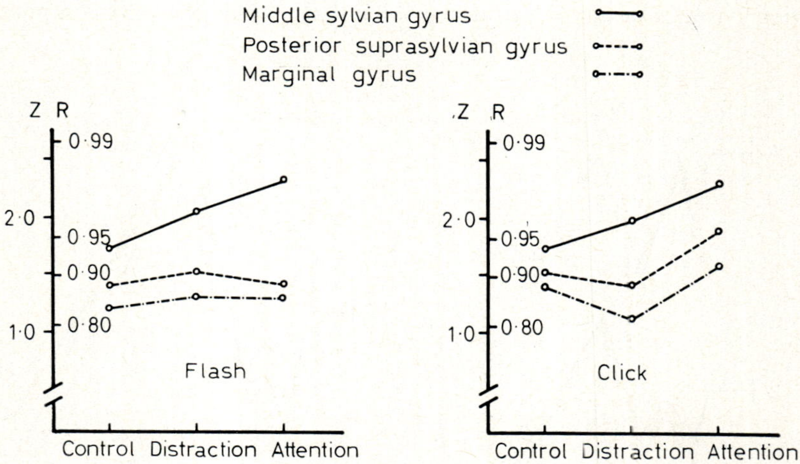

Figure 6. Means of coefficients of correlation of summed evoked responses to flash and click recorded from marginal, posterior suprasylvian and middle sylvian gyrus, showing increased stability of evoked responses with increased levels of attention

We are, of course, stuck with the surprising finding that what appears to be a polysensory response, extremely sensitive to changes in attentive states is best recorded from the sylvian gyrus, an area generally identified as an auditory projection area. We had anticipated that the posterior suprasylvian gyrus, often referred to as predominantly an association area, would be more inclined to change with a psychological state. In these experiments (*see Figure 5*) it did not change at all.

Considering the marked responsiveness of the middle sylvian gyrus to a flash in the present study, there are two possibilities. The first is that it is indeed an associational area in addition to its auditory functions. The second possibility is, because of its location, i.e. in close proximity to the orbital and insular cortex, both well known to have intimate functional relationship with the brain stem reticular formation and diffuse thalamic nuclei (23), there may be some reflection of active or passive volume conduction from these areas to the recording site in the middle sylvian gyrus. This latter possibility remains to be tested.

POTENTIATION AND ASYMMETRY OF LATE COMPONENTS OF THE EVOKED RESPONSE

We have found similar changes in evoked responses recorded from human beings during changes in attentiveness, but first a few words about our methods.

We have repeatedly observed eight distinct wave components within the first 300 ms of the 'averaged' evoked response of most subjects. *Figure 7* illustrates the eight wave components for (*A*) one individual, (*B*) a composite of ten individuals, and (*C*) a second composite of ten different individuals.

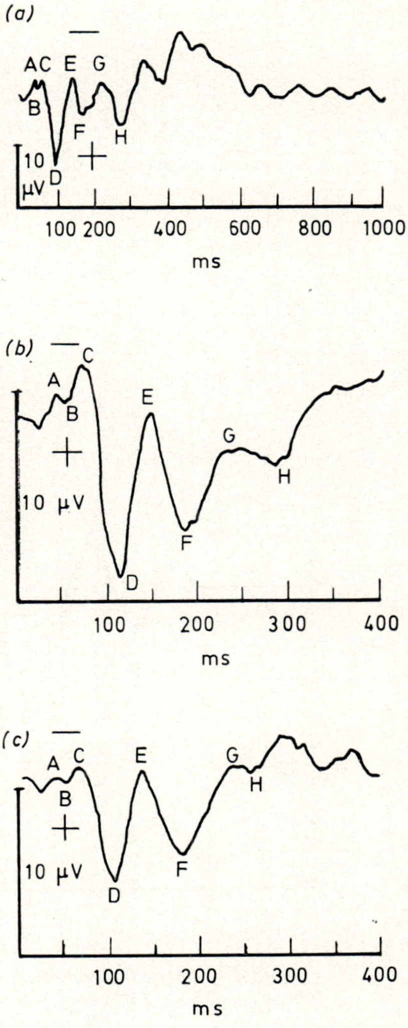

Figure 7. Composites of summed ('averaged') visually evoked potentials illustrating the recurrence of eight distinct waves in the first 300 ms *of the response. The initial positive, seen at about* 25 ms, *is not observed in all individuals so is not included among the eight waves.* (*a*) *One person's summed evoked response from* 240 *flashes,* (*b*) *a composite summed evoked response of ten individuals. This composite was constructed from* 2400 *responses,* (*c*) *a composite of ten other individuals constructed from* 1000 *responses,* 100 *from each subject*

(From Dustman and Beck, 1965, by courtesy of the Editor, *E.E.G. Journal*)

Composite evoked responses were constructed by taking the recorded visually evoked response of each individual of a given group and adding them in a computer. The purpose of the figure is 1) to emphasize that eight components are seen repeatedly, although the last two, G and H in *Figure 7*, are so variable that we have not included them in the experiments we will report, and 2) to illustrate composite evoked responses. While composite evoked responses will probably obscure some components, if there are subtle but consistent changes common to several members of the group, as may occur with a change in the level of attentiveness, these changes should be enhanced by such a procedure and seen more clearly in a composite response.

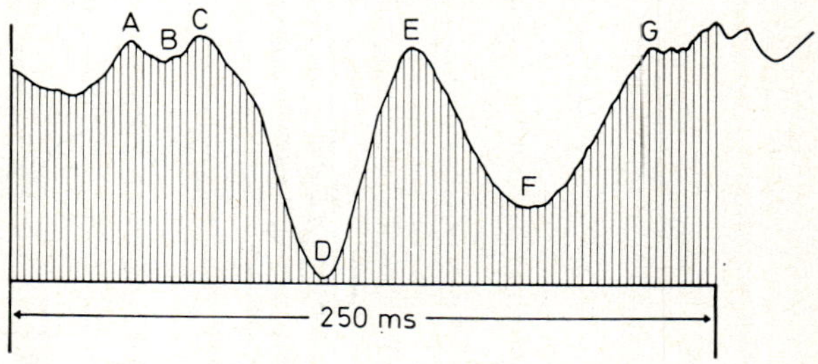

Figure 8. A figure illustrating a method for measuring amplitudes of phase changes as explained in the text

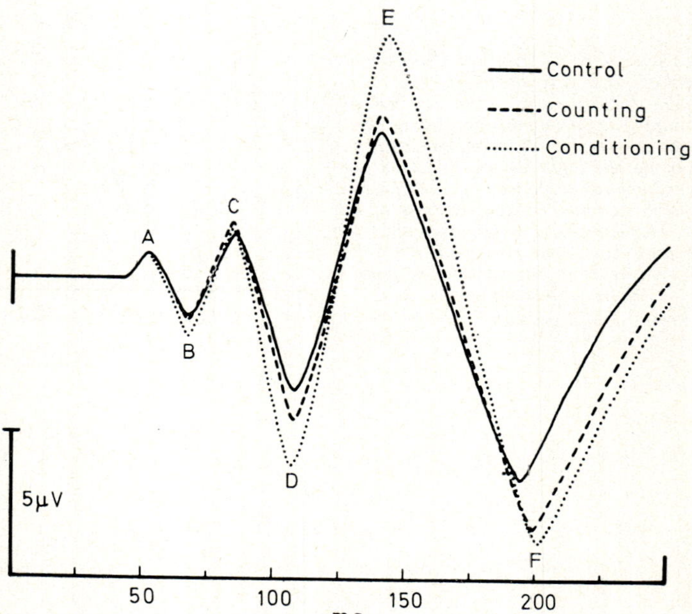

Figure 9. Composite summated evoked responses of 10 *individuals during three conditions. Each composite is constructed of* 7,500 *responses*, 750 *responses from each individual*

(From Sakai, Gindy and Dustman, 1966, by courtesy of the American Psychological Association)

Incidentally we have elsewhere shown the stability of the evoked response over weeks of time. This is important for a measurable degree of stability is essential if we are to study changes in the response due to certain imposed conditions.

As the wave components of the evoked response are determined by phase changes across time, the phase or voltage change of two or more evoked responses can be measured and compared by coefficients of correlation. In *Figure 8* a baseline is established immediately below the largest positive

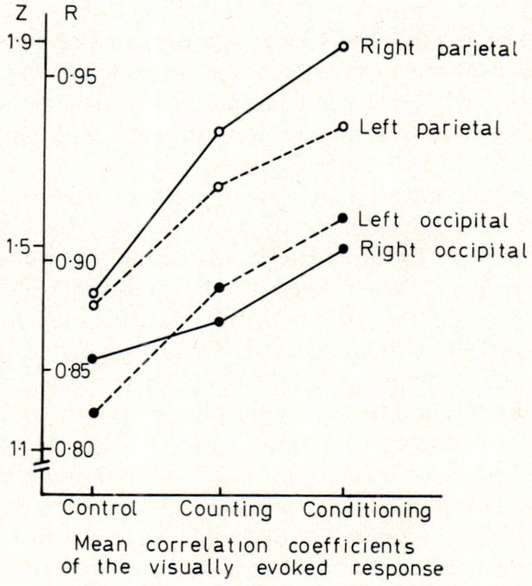

Figure 10. Means of coefficients of correlation of summed evoked responses showing increased stability of evoked responses with increased levels of attention. This is particularly striking with those responses recorded from right parietal scalp

deflection and 100 equally spaced ordinates (2·5 ms apart) are erected from this baseline to intersect the scribed wave form. Amplitude measure can be made of the distance from this baseline to each of the intersections. These 100 measures describe the rise and fall of the wave pattern and can be correlated or compared with 100 similarly obtained measures from another evoked response.

We studied change in the visually evoked responses of 10 young male adults during a number of conditions. The conditions imposed were the following: (*a*) two control sessions, one at the beginning of each experiment and one randomly interspersed among other conditions; (*b*) a period in which the subject counted flashes by a previously instructed ratio so that a certain level of attention was maintained; (*c*) a period in which the subject was fitted with a finger cuff and shocked if he did not flex his finger to each flash. There were three replications, separated by weeks, for each of the described conditions.

Figures 9 and *10* illustrate the results. The amplitude of four wave components, B, D, E, and F, were significantly greater ($p < 0{\cdot}05$) in evoked responses recorded during the conditioning period than the other two periods. The amplitudes of waves D, E, and F recorded during the 'counting' period tended to be reliably greater than during the control periods.

The stability of the evoked responses showed the same trend. Evoked responses were reliably more stable (higher correlations) during conditioning than responses recorded during counting. Those recorded during counting were significantly more stable than those during the control periods. This trend in increased stability with increased levels of attention is illustrated in *Figure 10*. It should be stressed that the greatest changes in amplitude with increased attention were always seen in those responses recorded from parietal scalp and were most accentuated in the right parietal leads. The same trend was also evident in the increased stability of the response with increased levels of attention as can clearly be seen in *Figure 10*.

Why should electrocortical changes with levels of attention be most evident in the minor hemisphere? We honestly do not know. However, recent experiments in our laboratory have repeatedly implicated the right parietal area in findings that may be closely related to the neurophysiology of attention. Rhodes (30) comparing the visual evoked responses of 20 bright children (group mean IQ of 130) with an age matched group of 20 dull children (mean IQ of 79) found one of the main differences in their evoked responses was the tendency of the bright child to have hemispheric asymmetry with much larger evoked responses recorded from right parietal scalp. In the dull child no hemispheric differences were apparent. Bigum (1966) doing his dissertation in our laboratory with mongoloid children has reported the same situation when the mongoloid child is compared with an age matched normal child. All but two of 24 normal children demonstrated hemispheric asymmetries in their evoked responses with the larger response occurring in the right hemisphere, particularly noticeable in the parietal area. The mongoloid children showed no hemispheric differences.

The effects of alcohol ingestion on the amplitude of the visual and somatosensory evoked response (24) shows trends peculiarly similar to what is encountered in the retarded and mongoloid subjects. With alcohol ingestion of moderate amounts, 1–3 ounces, there is a general attenuation of the amplitude of late components of the evoked response in parietal areas. However, the attenuation is most marked in the right parietal area. The asymmetry of left and right parietal responses, seen in most normal subjects, disappears with mild intoxication; both hemispheres become equal in amplitude.

We feel a common denominator of these later studies may be the level of attentiveness, namely, normal and bright children are by definition more attentive than dull or mongoloid youngsters; alcohol is a depressant and level of attention is probably reduced after ingestion; during counting or conditioning the level of attention is most probably increased. In all these instances there is a marked change in amplitude and stability of the evoked response recorded from right parietal scalp in the same direction as the level of attention. The findings are clear and reliable but the explanation is obscure.

DISCUSSION ON BECK

Horn

Dr. Lindsley thought that counting might be distracting, but you have used counting as an attentive task.

Beck

Well, in our counting procedure, subjects don't just count numbers serially, they are instructed to count to a certain amount, like 1, 2, 3, 4, 5, 6, 7 and then to start over again and count to, say 13. At this point they are instructed to press a key. Thus we can monitor attention. We found something interesting in another procedure. We had some engineers working with us interested in our methods. To keep ourselves alert we often square numbers at each flash, counting thus 2, 4, 8, 16 and so forth. This isn't easy and the background EEG is always LVF. With the engineers this wasn't the case. They could square numbers without much effort, and there were occasional epochs of alpha. Squaring numbers to them was probably like humming a tune to a musician. So with the counting procedure for the children we tried to make it a little complicated. You see it was not just simple counting.

Horn

Did you look at the EEG at the time the evoked potentials were being studied? During visual attentiveness the EEG is usually desynchronized, while during inattentiveness it is usually synchronized. The size and shape of evoked potentials are determined to a very large extent by the state of the 'background' EEG.

Beck

Yes, we're well aware of that. First of all, the EEG is monitored. We didn't do frequency analysis which we wanted to do. There was never a response given when there was a slow pattern, and when we took the responses off the tape afterwards, we discarded any that would be questionable. Bear in mind that the session in which we were recording the evoked response, is only a matter of a few minutes, say 3 or 4 min, for a total of 100 flashes—5 min at the outside. If it's a long session we get the person out of the room, walk them round and so forth. We make a great effort to control the background activity.

Horn

Was there any correlation between the state of the background EEG and the size or shape of the evoked potential?

Beck

We just had fast activity all the time.

Horn

I see. They weren't related then?

Beck

Well, we couldn't say. We just made an effort to keep the people alert. We controlled for habituation by always running in about 20 or 30 flashes. We think it's a rapid process, habituation occurs to the first few flashes, and then we start our experiment. This late habituation, say after 100 responses, well, we didn't see this.

Mulholland

Well, since evoked potential seems to be replacing the alpha rhythm as an index of attention, I think some of the questions about the relationship between background rhythms and the evoked potential really need to be worked out. It would be nice to somehow sample the EEG in such a way that you really randomly presented the stimulus that was going to evoke a potential—at least with respect to the dominant frequency in the EEG.

Beck

I think this is generally done. Parenthetically, we have tried to correlate the alpha frequency, as others have done, and we find none of the earlier waves or components of the evoked response correlate with this, nor the background frequency of the mongoloid child we reported on, correlate with their late components. But of course very late components, those sometimes called the 'ringing' response correlate very highly with alpha and may represent the return of alpha following the stimulus. Interestingly, some people without visible alpha show a 'ringing' response. I rarely show alpha in my record, yet occasionally I have a prominent after-discharge in my evoked response. The earlier components I don't think are correlated. Admittedly some levels of consciousness will change amplitude of components but this does not mean they are related to alpha phase.

Mulholland

I'm looking at the more elementary level of method, that is that background can add in and increase amplitude of the evoked potential. This may occur if you're using attentive tasks which produce an increase or decrease in the background.

Grey Walter

I agree with Dr. Mulholland that we should try to relate these evoked responses, CNV and so on to intrinsic rhythms as much as one can. It's a very puzzling field because in general there's almost no relationship between CNV for example, and presence or absence of alpha either from time to time or from person to person. However, sometimes there is, as Dr. Beck has said what we call the 'ringing' effect, when an alpha component becomes coherent following a visual stimulus. If a conditional stimulus is a click there will be a CNV but no interaction with the alpha rhythms, but if in the same subject the conditional stimulus is a flash then in some people an alpha component will become coherent and may grow up to the end of the CNV and involve anterior as well as posterior regions. If the CNV is attenuated by equivocation then the coherent rhythm will also decline. It is almost as though there were a sort of timing rhythm set off, following the flash, counting 1, 2, 3, 4, 5, 6, 7, 10, Go! This is quite unusual but it appears in normal people. The other point is the triggering of stimuli by, or in synchrony with an intrinsic rhythm. We started doing this over 20 years ago or so, as a sort of clinical test, in order to evoke seizure patterns, but we left it alone because we didn't have the time to develop it. It was taken up later by the Russians a good deal. Bechtereva in Leningrad uses this as a clinical test. She has found that if an intrinsic rhythm is a 'normal' one, an alpha rhythm for example (by normal I mean it has some physiological function) then varying the phase of a stimulus, particularly of course a visual stimulus, in relation to the alpha rhythm will change the wave form and general character of the response, because the intrinsic rhythm is involved in some way with the stimulus. However, if it is a pathological rhythm even if it is still within the normal frequency range, it will be unaffected by the stimulus. This suggests that the cortex is cut off in some way—say by a tumour—from the afferent system and therefore changing the phase of the stimulus with respect to the rhythm will not change the wave-form. They use this method of identifying the pathognomonic characters of intrinsic rhythms as opposed to their physiological characters. This is a quite interesting application of this effect and helps to indicate whether a rhythm is involved in the process of perception and attention.

REFERENCES

1. Adametz, J. H. (1959) *J. Neurosurg.* **16**:85
2. Babiyan, S. M. (1961) *Bull. exp. biol. Med.* **50**:881
3. Beck, E. C., Doty, R. W. and Kooi, K. A. (1958) *Electroenceph. clin. Neurophysiol.* **10**:279
4. Beck, E. C. and Kanai, Tatsui (1958) *Am. Psychol.* **13**:531 (abstract)
5. Bignall, K. E. (1967) *Expl. Neurol.* **18**:56
6. Bigum, H. B. (1968) *Ph.D. Thesis.* Univ. Utah
7. Bogacz, J., Vazulli, A., Hondler, P. and García-Austt, E. (1960) *Acta neurol. latinoam.* **6**:353
8. Chow, K. L. (1960) In *Recent Advances in Biological Psychiatry* (Ed. by Wortis). Grune and Stratton: New York
9. Chow, K. L., Randall, W. and Morrell, F. (1966) *Electroenceph. clin. Neurophysiol.* **20**:357
10. Davis, H. (1964) *Science, N.Y.* **145**:182

10a. Doty, R. W., Beck, E. C. and Kooi, K. A. (1959) *Exp. Neurol.* **1**:360

10b. Dustman, R. E. and Beck, E. C. (1965) *Electroenceph. clin. Neurophysiol.* **18**:433

11. Fangel, C. and Kaada, B. R. (1960) *Electroenceph. clin. Neurophysiol.* **12**:575
12. Fernandez-Guardiola, A., Roedon, E., Fanjul, L. and Costella, C. (1961) *Electroneceph. clin. Neurophysiol.* **13**:564
13. Fleming, D. E. (1963) *Electroenceph. clin. Neurophysiol.* **15**:84
14. Galambos, R. (1956) *J. Neurophysiol.* **19**:424
15. Gastaut, H. (1958) In *Ciba Foundation Symposium on the Neurological Basis of Behavior* (Ed. by Wolstenholme and O'Connor). Little, Brown and Co.: Boston
16. Gross, M. H., Begleiter, H., Tobin, M. and Kissin, B. (1965) *Electroenceph. clin. Neurophysiol.* **18**:451
17. Haider, M., Sprong, P. and Lindsley, D. B. (1964) *Science, N.Y.* **145**:180
18. Hearst, E., Beer, B., Sheatz, G. C. and Galambos, R. (1960) *Electroenceph. clin. Neurophysiol.* **12**:137

19. Hernández-Peón, R. (1961) In *Sensory Communication* (Ed. by Rosenblith). M.I.T., Wiley: New York
20. Hernández-Peón, R., Scherrer, H. and Valasco, M. (1956) *Acta neurol. latinoam.* **2**:8
21. Horn, G. (1960) *Brain*, **83**:57
22. Jane, J. A., Smirnov, G. D. and Jasper, H. H. (1962) *Electroenceph. clin. Neurophysiol.* **14**:344
23. Kaada, B. R. (1960) In *Handbook of Physiology*, Sec. II (Ed. by Field, Magoun and Hall). Washington, D.C., American Physiological Society
24. Lewis, E. G. (1968) *Ph.D. Thesis*, Univ. of Utah
25. Macadar, O., Ginés, A., Bove, L. C. and García-Austt, E. (1963) *Acta neurol. latinoam.* **9**:315
26. Moruzzi, G. and Magoun, H. W. (1949) *Electroenceph. clin. Neurophysiol.* **1**:455
27. Morrell, F. (1958) In *Reticular Formation of the Brain* (Ed. by Jasper *et al.*). Little, Brown and Co.: Boston, 545
28. Morrell, F. (1960) *Electroenceph. clin. Neurophysiol.*, Suppl. **13**:65
29. Palestini, M. and Leifschitz, W. (1961) In *Brain Mechanisms and Learning* (Ed. by Fessard *et al.*). Blackwell Sci. Publ.: Oxford, 413
30. Rhodes, L. E. (1967) *Ph.D. Thesis.* Univ. of Utah
30a. Sakai, M., Gindy, K. and Dustman, R. A. (1966) *Proc. Am. Psychol. Assoc.* **2**:139
31. Satterfield, J. H. (1965) *Electroenceph. clin. Neurophysiol.* **19**:470
32. Segundo, J. P., Arana, R. and French, J. D. (1955) *J. Neurosurg.* **12**:601
33. Sharpless, S. and Jasper, H. (1956) *Brain*, **79**:655
34. Shumilina, A. I. (1959) *Sechenov. physiol. J.*, U.S.S.R. **45**:5
35. Thompson, R. F. (1967) In *Foundations of Physiological Psychology*. New York: Harper and Row
36. Thompson, L. W. and Obrist, W. P. (1964) *Electroenceph. clin. Neurophysiol.* **16**:332
37. Thompson, R. F. and Shaw, J. A. (1965) *J. comp. physiol. Psychol.* **60**:329
38. Webster, W. R., Dunlap, C. W., Simons, L. A. and Aitkin, L. M. (1965) *Science, N.Y.* **148**:654
39. Worden, F. G. and Marsh, J. T. (1963) *Electroenceph. clin. Neurophysiol.* **15**:866
40. Yoshii, N. (1957) *Electroenceph. clin. Neurophysiol.*, Suppl. **6**:75

A Neurophysiological and Evolutionary Model of Attention

Raúl Hernández-Peón*
Instituto de Investigaciones Cerebrales, Mexico City

Throughout life from early infancy, in humans attention is a process of paramount importance first for the perception of the external world, then for ego integration, and later for intellectual activities including social interaction. Following the definition of attention characterized by active selective receptivity with simultaneous exclusion of irrelevant information, it can be stated that the newborn infant presents unspecific alerting reactions but is unable to attend selectively to particular stimuli. It is only later on, when maturation of the brain reaches a certain stage, that attention begins to develop in the child who is gradually acquiring maximal voluntary control.

Granting the widely accepted Lamarckian principle according to which ontogenetic development represents a recapitulation of phylogenetic evolution, the question arises as to at which level of the animal scale did 'attention' evolve. A correlation between evolutionary anatomical changes and the appearance of new physiological activities helps to predict the anatomical substrate of those functions. As far as behaviour and their psychological counterparts are concerned, I firmly believe that the combination of basic neurological science and ethological studies into a new neuro-ethological discipline devoted to the investigation of the evolution of neural mechanisms underlying the various aspects of behaviour, will shed considerable light on the understanding of the highly evolved and complex human brain and mind relationships.

With these ideas in mind, and completely aware of the unsurmountable limits of this endeavour, we have initially approached the problem of evolution of the neurophysiological mechanisms of wakefulness and sleep by comparing the EEG and other electrophysiological indices in freely moving mammals, reptiles, and birds, with implanted cerebral electrodes and cannulae.

Since attention refers to the waking state in which selective conscious experiences are sharply defined, any description of the neurophysiological mechanisms involved must start by localizing the anatomical substrate of wakefulness and by considering the conscious experience continuum.

In this paper, I will attempt to reach an integrative conceptualization of

*Dr. Hernández-Peón was killed in a motor accident on his return to Mexico shortly after presenting this paper. Ed.

unpublished results obtained in our Institute in collaboration with several colleagues. Accepting total responsibility for speculative ideas, benevolence for criticism might be granted some day if it succeeds in fostering research along new avenues with a co-ordinated multidisciplinary approach.

A CATECHOLAMINOCEPTIVE VIGILANCE SYSTEM

Ideas as to which parts of the brain underly wakefulness have been considerably clarified in the last two decades following the pioneer work of Magoun (18) and his associates. Three lines of evidence have led me to a definition of what may be designated the Vigilance System:

1) From experiments utilizing high-frequency electrical stimulation in lightly anaesthetized cats, Moruzzi and Magoun (19) localized an arousing region in the central core of the brain stem extending from the medulla oblongata up to the ventromedial part of the thalamus which they called the Ascending Reticular Activating System (ARAS). Very soon, it was found by electrophysiological methods that the rostral mesodiencephalic structures of the ARAS receive corticofugal projections from certain cortical areas (3) which share arousing properties with the disclosed brain stem region as demonstrated by high-frequency electrical stimulation in spontaneously sleeping monkeys (20). Arquistriatal projections to the mid-brain reticular formation are probably involved in the alerting responses also induced in freely moving cats by electrical stimulation of the amygdala (16).

2) Lesion and freezing experiments have shown that the essential region for wakefulness extends from the rostral pontine tegmentum to the caudal diencephalon at the level of the posterior hypothalamus and the neighbouring subthalamic area. The caudal pontine tegmentum and the rostral part of the bulbar reticular formation contain hypogenic neurones with dominant tonic activity. Indeed, wakefulness is considerably prolonged immediately after making localized lesions involving nucleus reticularis pontis caudalis (14), or nucleus reticularis gigantocellularis (12) or else following a midpontine transection (1). Since total neocortical ablation in dogs transforms the monophasic wakefulness-sleep rhythm into a polyphasic rhythm (17), it may also be inferred that certain tonic activity not indispensable for wakefulness is present in the cortical arousing areas projecting to the rostral part of the brain stem. Thus, whereas the latter is concerned with 'obligatory wakefulness', the former plays a role in 'facultative wakefulness'.

3) An extensive overlapping of arousing and hypnogenic neurones in many brain regions has become evident from experiments in which arousal or sleep have been induced by high or low frequency electrical stimulation of the same point. Based on the idea that central synaptic transmission is chemically mediated, and that post-synaptic membranes specifically react to the corresponding transmitters, we have been able to functionally dissect hypnogenic and arousing pathways in the CNS by local application of microcrystals of substances considered as likely transmitter candidates through cannulae permanently implanted (6). When acetylcholine was applied in a highly circumscribed anatomical pathway, a state of sleep, as behaviourally and electrophysiologically defined, ensued. By the same token, all the be-

havioural and several electrophysiological manifestations of arousal and alertness were produced by local application of nor-epinephrine in discrete regions such as the mid-brain reticular formation, medial preoptic region, ventromedial hypothalamus, and part of the grey substance in the spinal cord. Chemical specificity of the arousal response is supported by the finding that epinephrine only produced arousal when deposited in the spinal grey substance but not in the mid-brain reticular formation. All the experimental evidence cited above, lends support to the concept of a catecholaminoceptive Vigilance System extending throughout all the levels of the neuroaxis. The essential part, which is located at the rostropontinemesodiencephalic level, receives descending corticofugal and amygdalofugal projections as well as ascending spinobulbopontine anatomical connections. It is likely that in their descending trajectory, corticofugal impulses pass both through unspecific thalamic nuclei and through the preoptic region in the forebrain. On the other hand, the arousing bulbopontine reticular neurones may be intermingled with the hypnogenic neurones located in the same anatomical area. A more extensive systematic exploration with localized chemical stimulation is necessary to elucidate those questions.

AN EVOLUTIONARY HYPOTHESIS OF THE VIGILANCE SYSTEM

The existence of catecholaminoceptive vigilance neurones in the mammalian spinal grey substance support an evolutionary hypothesis of wakefulness in parallel with the evolutionary hypothesis of sleep postulated by the author (7). According to this view, the phylogenetically old Vigilance System had a representation in the primitive neural tube, and therefore, remnant parts still remain in the spinal cord. Along with the process of encephalization and differentiation of the brain stem, the most important group of vigilance neurones became localized in its rostral segment. And finally, as the forebrain and cortex developed in mammals, corticofugal projections appeared to establish a more refined control of the subcortical neurones, resulting in more adaptive vigilance states such as 'facultative wakefulness' and 'voluntary attention'. Evolutionary considerations on the phylogenetic origin of noradrenoceptive vigilance neurones leads one to wonder about their presence in the mid-brain of submammalian forms with a primitive cortex like reptiles. Indeed, in experiments carried out in collaboration with Rojas-Ramírez and Tauber (21), it was found that local application of microcrystals of norepinephrine in the mid-brain reticular formation of the reptile *Ctenosaura pectinata* elicited periodic bursts of alertness every 3 or 4 min characterized by an increased voltage and frequency of the electrical activity in the forebrain and the mid-brain, a hypertonic reaction of the nuccal muscles, and abundant eye movements. Similar arousal reactions were also produced in the sleeping iguanid by high-frequency electrical stimulation of the mid-brain reticular formation. Association of the fact that the batracian brain contains epinephrine but not nor-epinephrine, with the finding that epinephrine induces arousal when applied locally in the mammalian central nervous system only from the spinal grey substance and from the bulbar reticular formation (2), suggests that epinephrine might have been the primitive synaptic excitatory transmitter of the Vigilance System. It also would seem to imply that demethylation into nor-epinephrine occurred with the appearance of reptiles on Earth approximately 300 million years ago. Now if this conjecture is correct, local application of epinephrine into the batracian mid-brain reticular formation should produce an arousal reaction which is absent in the mammalian brain.

CATECHOLAMINE PRODUCTION IN THE VIGILANCE SYSTEM

If the vigilance neurones located in the rostral part of the brain stem receive nor-adrenergic presynaptic terminals—as their chemical specificity to this substance suggests—it should be possible to detect changes of nor-epinephrine concentration in that region in parallel with changes of vigilance reticular neuronal activity. For instance, overactivation of the Vigilance System might be associated with a partial exhaustion of nor-epinephrine production. As predicted by the hypothesis, in experiments carried out in collaboration with Serrano, Drucker, Chavez-Lara and Ramírez del Angel, when cats were deprived of sleep for 10–12 days, a significant diminution of nor-epinephrine concentration in the mid-brain tegmentum co-existing with a remarkable increase of dopamine and a smaller increase of serotonin were found. Since it is known that dopamine represents the chemical substrate for nor-epinephrine synthesis, it seems logical that overactivation of the metabolic synthesizing machine will result not only in partial exhaustion of the final product, but also in an increase of its precursor derived in part from nor-epinephrine overproduction, and serving as a depot to meet higher requirements of nor-epinephrine synthesis. If this interpretation is correct, blocking nor-epinephrine synthesis by alpha methyl -DOPA should considerably reduce the survival period of the experimental animals deprived of sleep.

MAMMALIAN CORTICAL REACTIVITY DURING THE WAKEFULNESS-SLEEP CONTINUUM

Contrary to the traditional assumption according to which the cortex becomes activated during wakefulness and inhibited during sleep, abundant experimental evidence accumulated in the last few years indicates that cortical reactivity is greater during the periods of deep sleep than during alertness (13). This seems to be a general principle applicable both to the neocortical specific analysers and to the paleocortex more concerned with the storage of recent memory tracings. Because of the possible implications for a neurophysiological interpretation of dreams and hallucinations we have studied (in collaboration with Vera-Trueba) cats with electrodes permanently implanted in the reactivity of the entorhinal cortex, which is the homologue of the hippocampal gyrus in the primate brain. The cortical after-discharge produced by local repetitive electrical stimulation of the entorhinal cortex (at a frequency of 100 Hz over 3 s) increases progressively in duration from alertness to relaxed wakefulness, slow sleep, and rapid sleep, and reaching maximal amplitude in this state. Accordingly, local potentials in the entorhinal cortex evoked by applying single pulses either to a point 1 mm away from the recording site, or to the neighbouring pyriform cortex also increased in amplitude because the wakefulness-sleep continuum had its minimum during alertness and its maximum during the periods of rapid sleep with rapid eye movements. Such an increase of cortical entorhinal evoked potentials can be shown to follow a linear function suggesting that the discharging neuronal pool is smaller during alertness, and largest during rapid sleep. If reduction of cortical reactivity during alertness is the result of active inhibition (whereas larger evoked potentials are the consequence of physiological disinhibition), the above mentioned gradient of cortical reactivity should disappear by local application of substances known to interfere with the processes of pre- and post-synaptic inhibition. Indeed, we have found that after local application of picrotoxin through cannulae electrodes permanently implanted in the entorhinal cortex, the evoked potentials were enhanced during alertness, reaching the same amplitude as during rapid sleep, and remaining extremely stable throughout all the stages of the wakefulness-sleep continuum. If cortical inhibition during wakefulness is related to the activity of reticular vigilance neurones, sustained activation of the latter—as that produced by sleep deprivation—should lead to refractoriness of the cortical inhibitory interneurones linked with the corresponding reticular axons, thus resulting in cortical hyper-reactivity. Current experiments carried out in collaboration with Drucker and Satinoff indicate that this is indeed the case. Sleep deprivation of from 1 to 7 days elicited an enhancement of the entorhinal cortical evoked potentials which recovered their initial amplitude after the animals were allowed to sleep one night. Such an increase of cortical reactivity can account for the thought disorders and hallucinations observed in humans after several days of sleep deprivation (23).

Tonic reticulo-cortical inhibition during wakefulness seems to provide the necessary background for the logical association of memory tracings. If the excitability of cortical mnesic neurones were not restricted by inhibitory processes, a constant flow of memories and thoughts, as occurring in certain psychotic patients, would be an inevitable outcome. With the available experimental data, a correlation can be attempted between the degrees of cortical reactivity and the main types of mnesic association characteristic of the various stages of the wakefulness-sleep continuum. Decreased mnesic association as occurring during emotional excitement would result from maximal cortical inhibition. Moderate diminution of reticulo-cortical inhibition allows logical thinking during relaxed wakefulness and slow sleep. And finally, a practically complete release of reticulo-cortical inhibition during rapid sleep explains the illogical association of memory tracings forming the manifest content of dreams. Co-existing with extensive reticulo-cortical inhibitory influences during wakefulness, selective reticulo-cortical facilitation during attention must obviously be present. From a teleological point of view, reticulo-cortical inhibition would serve the purposes of enhancing the signal-to-noise ratio in cortical circuits, and of preventing a chaotic bombardment of the subcortical neural system involved in the integration of conscious experiences by a great amount of irrelevant stored information.

REPTILIAN CORTICAL REACTIVITY DURING THE WAKEFULNESS-SLEEP CONTINUUM

As important as cortical comparative neuroanatomy, and perhaps even more, is the understanding of the evolution of cortical functioning. As an initial approach to this problem, Tauber, Rojas-Ramírez and Hernández-Peón (21) have made electrical recordings from the forebrain and mid-brain of the iguanid *Ctenosaura pectinata* during all the stages of wakefulness and sleep. The EEG arousal reaction in both regions consisted in an increased electrical activity shown in amplitude and frequency. Accordingly, the amplitude of the EEG diminished throughout sleep including the periods accompanied by REMs. These results suggest that during arousal a diffuse cortical activation occurs in the reptilian brain. Further experiments carried out in collaboration with Rojas-Ramírez support this interpretation. Indeed, photic and local evoked potentials from the forebrain were enhanced during the transition from relaxed wakefulness to alertness. Therefore, it seems that in the reptilian brain, reticulo-cortical excitatory influences prevail over rudimentary reticulo-cortical inhibition.

AVIAN CORTICAL REACTIVITY DURING THE WAKEFULNESS-SLEEP CONTINUUM

Much less is known about EEG and other electrophysiological indexes of wakefulness and sleep in birds than in reptilian or mammalian species. Appropriate selection of species for studies concerning phylogenetic evolution of specific functions is a necessary requisite for avoiding misleading interpretations and conclusions. For instance, Jouvet (15) has concluded that rapid sleep is a phylogenetically recent state appearing in mammalian forms because he and his colleagues were unable to record eye movements during sleep in turtles, and only 0·1 per cent in chickens. However, when recordings are made from foveate reptiles as the *Ctenosaura pectinata*, and from birds with binocular vision and high visual acuity, clear cut periods of rapid eye movements during sleep are found in these two submammalian classes. In collaboration with Rojas-Ramírez we have made electrophysiological recordings (EEG, EMG of the neck muscles and eye movements) through electrodes permanently implanted in the hawk *Buteo Jamaicensis arborealis*. From these studies it became evident that the avian EEG more closely resembles the mammalian than the reptilian EEG. There are both slow (4 –7 Hz) and rapid low voltage waves during wakefulness. However, a certain dominance of slow waves prevails during 90 per cent of sleep whereas low voltage fast activity is dominant during arousal and during brief periods of sleep accompanied by bursts of REMs. The total duration of the periods of desynchronization with bursts of rapid eye movements represents 8–10 per cent of the total sleep time. It is likely that reticulo-cortical inhibition in highly evolved birds is better developed than in their reptilian ancestors. Recordings of evoked potentials and of cortical single units with microelectrodes are necessary to establish this important point.

SENSORY FILTERING IN REPTILES

From a long series of studies, it is now well established that transmission of sensory signals is modified at their entrance to the CNS, and in some cases at the receptor itself according to the significance of the stimulus, and therefore, to the state of attention or inattention of the animal. Sensory filtering rapidly fluctuates among different modalities and even within the same modality. A prevailing tonic inhibitory influence at the first sensory synapse during wakefulness partially blocks the entrance of a great amount of irrelevant signals to the CNS, thus preventing its chaotic activation. Tonic descending sensory inhibition is as important as ascending reticulo-cortical inhibition for achieving selective facilitation of sensory channels. The question arises as to the phylogenetic origin of sensory filtering. In an attempt to answer this question, in collaboration with Rojas-Ramírez we have recorded from the first trigeminal sensory relay in the reptile *Ctenosaura pectinata* the potentials evoked by stimulating the infra-orbital nerve. Here we found that the amplitude of the trigeminal evoked potentials significantly diminished during the EEG arousal reaction provoked by the presence of the experimenter or by touching the body of the animal. Therefore, the presence of sensory inhibition at the first synapse of the reptilian brain when in a state of alertness may be taken as indicative of a more primitive origin for this mechanism than that of reticulo-cortical inhibition.

RETICULAR ORIGIN OF SENSORY FILTERING

From abundant experimentation it is evident that electrical stimulation of the mid-brain reticular formation in cats inhibits transmission at the first synapse of various specific afferent pathways, and that lesions in the same tegmental area enhances sensory transmission (5, 10). Because the vigilance neurones essential for wakefulness are located in the same anatomical area, it was initially thought that sensory filtering during attention was related to the reticular vigilance neurones (4). However, more recent experiments in which tactile evoked potentials have been recorded from the spinal trigeminal sensory nucleus in cats during sleep (11) have disclosed that sensory inhibition at the first synapse occurs not only during attention but also during the periods of rapid sleep with rapid eye movements. Since the arousal threshold is considerably raised during these periods, the above mentioned results support the conclusion that sensory inhibition is not related to the vigilance neurones being maximally inhibited during the periods of rapid sleep. The additional finding that mid-brain tegmental lesions eliminate sensory inhibition both during wakefulness and during rapid sleep still supports the idea of a brain stem origin for such inhibition. The question remains as to which is the function of the reticular neurones related to sensory inhibition. Since the only common function occurring during attention and the periods of rapid sleep usually associated with dreaming in humans is that of integration of the conscious experiences known as sensory perception and dreaming, the author has postulated the existence of a Conscious Experience System functionally independent of the Vigilance System although overlapping in the same anatomical region (8, 9). If this interpretation is correct, it should be possible to dissect these systems functionally and detect independent electrophysiological correlates.

POSSIBLE ELECTROPHYSIOLOGICAL CORRELATES OF THE CONSCIOUS EXPERIENCE SYSTEM

There is common agreement by introspective analysis that the degree or intensity of our conscious experiences does not follow a linear function throughout the wakefulness-sleep continuum. For instance, our conscious sensory perceptions are more sharply defined during attention than during relaxed wakefulness, and occasionally dreams may be even more 'vivid' than a waking experience. Electrophysiological monitoring of sleep in humans has revealed that mental activity is present throughout all the sleep period although certain characteristics usually differentiate that present during slow sleep and that occurring during rapid sleep. A type of mental activity very similar to logical waking thought is usually present during slow sleep as revealed by correlating verbal reports on disruption of sleep. When the subject is awakened in the first of those periods, and asked to report his dreams, not infrequently he denies having been asleep because his subjective experience was that of a type of thinking which may be continuous with the terminal waking thinking before the onset of sleep. A search for objective signs of activation of conscious experience neurones in experimental animals is undoubtedly a difficult task. If the reticular neurones concerned with conscious experience are intermingled with those underlying wakefulness, how can we selectively activate the former? Obviously sensory evoked potentials recorded from the mid-brain reticular formation represent simultaneous activation of both neuronal pools receiving sensory inflow. In addition to sensory information, however, the Conscious Experience System receives information related to past experiences which has no functional relationship with the Vigilance System. As the hippocampal formation seems to represent an important locus for the storage of recent memories, efferent fibres of the fornix probably represent part of the conducting pathways between the Recent Mnesic System and the Conscious Experience System. If this assumption is correct, stimulation of the fornix should selectively activate the mid-brain reticular neurones concerned with the integration of conscious experiences. As there is no way of inquiring about the subjective feelings of an animal, our assumption can only be tested indirectly by correlating the amplitude of the reticular evoked potentials with our own subjective experiences in similar physiological or pharmacologically induced states. The results obtained from experiments carried out in collaboration with Rojas-Ramírez and Goldberg fall in line with our theoretical assumptions. Indeed, the amplitude of the mid-brain reticular potentials evoked by fornix stimulation is largest during the periods of rapid sleep with bursts of rapid eye movements, is smallest during relaxed wakefulness, and intermediate during selective attention. Furthermore, the amplitude of those potentials became reduced during unspecific alertness. Since psychedelic drugs like LSD (also called 'conscious expanding drugs') produce an immense intensification of conscious experiences, we decided to test the effects of LSD on these reticular evoked potentials. Again, as predicted by the hypothesis, LSD significantly increased the amplitude of the fornix-reticular evoked potentials. Preliminary experiments carried out in collaboration with Satinoff and Drucker lend further support to the idea of an independent origin of reticulo-cortical and reticulofugal inhibition at the first sensory synapse. In cats deprived of sleep for several days, while local entorhinal cortical evoked potentials became enhanced, the trigeminal evoked potentials at the spinal *V* sensory nucleus were remarkably reduced or disappeared. Since recovery in both brain locations occurred after allowance of one night of sleep, the results can be interpreted as indicating greater resistance to fatigue for the inhibitory neurones related to sensory filtering and conscious experience than for the inhibitory elements of the reticulo-cortical pathways concerned with wakefulness and arousal. Furthermore, the apparent hyperactivation of the conscious experience neurones may be partially accounted for by an excessive discharge of the disinhibited cortical neurones.

NEUROPHYSIOLOGICAL MODELS OF WAKEFULNESS IN REPTILES AND MAMMALS

From the material previously discussed, a series of inferences may be integrated into an evolutionary model of attention. The absence of selective attentive behavioural responses in reptiles coincides with the diffuse cortical activation observed in these animals. On the other hand, attentive behaviour in birds with binocular vision is co-existent with the appearance of EEG desynchronization in which a process of reticulo-cortical inhibition probably plays a role—as it also does in mammals. Since sensory filtering is a seemingly primitive mechanism already present in reptiles, it may be concluded that, although necessary, such a mechanism is not *sufficient* for attention, which requires cortical processes not yet present in the primitive cortex of the reptilian brain. It may be concluded that attention is a feature of suprareptilian origin first evolving in birds and reaching higher development in mammals. Attention requires reticulo-cortical inhibitory influences related to the Vigilance System as well as corticofugal excitatory influences originated in the neocortex and controlling the brain stem vigilance neurones. In all probability, the reticulofugal inhibitory and facilitatory influences constituting sensory filtering at the first synapse have a more subtle functional organization in higher mammals because of better developed connections between neocortical analysers and the brain stem neurones involved in conscious integration.

A logical intermediate event between subcortical conscious integration of cortical information and triggering of arousal or attention is the activation of reticulo-cortical pathways between the brain stem neurones of the Conscious Experience System and the cortical representation of the Vigilance System. The emergence of paraspecific descending efferent pathways capable of achieving independent unimodal control of sensory transmission represents a new phylogenetic contribution to the mechanisms underlying finer sensory discriminations. Accepting Thorpe's conclusion that 'consciousness at one grade or another is a widespread feature of animal life' (22), we may add that attention is a privilege of suprareptilian vertebrates having its phylogenetic origin in birds and reaching in *Homo sapiens* a glorified culmination.

ACKNOWLEDGEMENT

The investigation reported in this paper was supported by the National Institute of Mental Health, USPHS, under Grant MH 10003-04 and by the Foundation's Fund for Research in Psychiatry under Grant #66-340.

DISCUSSION ON HERNÁNDEZ-PEÓN

Beck

Assuming that the iguana is like the reptile that I am familiar with, with reduced temperature they should sleep. With increased temperature they should be very alert. If you live in the West and note the behaviour of snakes you're very familiar with this phenomenon. Did you manipulate temperature?

Hernández-Peón

In most experiments we kept the temperature strictly constant all the time, but in some experiments we raised the temperature during the night. When we did this the animals nevertheless went to sleep. Normally, of course, they are always awake during the day in sunshine and asleep at nights—there is not a strict relationship between temperature and the state of wakefulness.

Grey Walter

Dr. Rechtschaffen in Chicago has an alligator which he keeps in a thermostatic cupboard and it is flashed every few seconds day and night. He's had it now for 6 months and its brain goes on responding to the flashes week after week. It never sleeps at its constant temperature and there's no sign of habituation. The flash responses just go on and on and it turns out rolls and rolls of slow-speed paper; this alligator is sitting there doing nothing except being fed flies and flashes. It tends to go torpid when it's cold, but it never sleeps.

Hernández-Peón

Well, the iguana has a fixed cycle of sleep. Actually they start to sleep not earlier than 9 o'clock in the evening, usually about 10 or 11 o'clock, and the periods of deeper sleep—which we base on the arousal reaction produced by electrical stimulation of the mid-brain—are between 5 and 7 o'clock in the morning.

Jung

Did I rightly understand that after sleep deprivation you found an increase of dopamine in the reticular formation? Wouldn't you also expect Serotonin alterations? Dr. Cramer in our laboratory tried to find different incorporations of amino acids in the brain stem and the reticular formation during sleep deprivation. There were some differences, but they cannot yet be correlated with the catecholamine content. Since Jouvet found adrenergic neurones and neurones containing serotonin in the pontine nuclei to be regulators of paradoxic and slow wave sleep, serotonin is now expected to influence certain slow sleep mechanisms. Therefore the serotonin alterations would be most interesting.

Hernández-Peón

Yes, there is a definite increase in serotonin.

Jung

In which region of the reticular formation?

Hernández-Peón

This is one gramme of mid-brain tegmentum; in other words I cut off the colliculi, the lateral lemnisci and take one gramme of what remains, which is principally mid-brain reticular formation.

Horn

Dr. Hernández-Peón has covered a tremendous area, and there are a lot of questions I would like to ask him. Some of your conclusions are based on the assumption that the recordings were made from homologous structures—the 'cerebral cortex' of reptiles, birds and mammals. To what extent is this assumption justified? There is much controversy as to what forebrain structure in these three classes of animals are homologous.

Hernández-Peón

Actually the reptile has only hippocampal cortex and pyriform cortex. The so-called germinal cortex is a very small area which has been a source of controversy amongst comparative neuroanatomists. Certainly the reptile has no neo-cortex. The bird has a little more of this so-called germinal cortex and, of course, the neocortex fully appears in the mammal.

Horn

Aren't there many difficulties in inferring functional identity to structures in the brains of these three different animals? If there is some change in the electrical activity of a structure, how does one set about interpreting this change across the animal classes?

Hernández-Peón

My interpretation is that in the entorhinal cortex of the mammal there is a dominance of inhibition during alertness, and wakefulness in general, and disinhibition during rapid sleep. In the reptile, however, there is marked facilitation during alertness. We have not made recordings during the period of rapid eye movements, but certainly, in comparing these states of alertness in the reptile with those in the mammal, the changes in evoked potentials go in opposite directions. Where there is a facilitation in the reptile there is a reduction in the mammal.

Horn

Dr. Lindsley, Dr. Beck and Dr. García-Ausst showed that there is a remarkable stability in the very early components of the evoked responses in human subjects. In so far as these components can be correlated with activity in any particular part of the brain they might be correlated with activity in the primary visual areas. This stability persisted even when the subjects were attending to or were distracted from the stimulus. Dr. Jasper showed that the responses of single cells in the somatic sensory nuclei of the thalamus are reproducible on all occasions, irrespective of the state of attention. You, however, report in animals (where it is operationally more difficult to control attentive behaviour), that there is a very considerable difference in size of the evoked response between that seen in the relaxed animal and that observed in what you call specific attentiveness. There appears to be a disparity between the results found in man and those found in animal. How is this discrepancy to be resolved?

You said that the changes in the response observed at the trigeminal nucleus were due to selective attention or specific alertness. Could you say how you distinguish this state from non-specific alertness?

Hernández-Peón

Showing a mouse or rat to a cat is an example of specific alertness.

Horn

But it is not all that simple. Presenting a mouse to a cat elicits all sorts of effects, the cat looks at the mouse, its ears are pricked up, there is piloerection, the cat may flick its tail and there is probably an increase in the output of adrenaline, to mention only a few. Some of these responses may be a necessary part of the selective response to the mouse, but others clearly are not. How does one distinguish between the necessary and the accessory: the specific and the non-specific?

Hernández-Peón

Non-specific alertness would be, for instance, if the experimenter entered the room or if the cat was alert but not focusing on any particular stimulus.

In the first trigeminal synapse, certainly, we are recording the presynaptic and the postsynaptic components of the response. We can test that by measuring latencies and by using other procedures such as hypoxia. With hypoxia the second postsynaptic wave is reduced whereas at the beginning, the presynaptic wave is enhanced and it is only after several minutes that the presynaptic component is reduced. That is why we have concluded that there is both pre- and postsynaptic inhibition. We are more sure of that in the trigeminal pathway than in the olfactory pathway for in the olfactory bulb we cannot, of course, record a presynaptic component. This is because, probably, the first slow wave is the result of the dendritic activation of the glomerular layer and the repetitive discharge is probably due to fields of activity reverberating in sequence.

Horn

What is the explanation for the reduction in the pre- and postsynaptic potentials in the trigeminal nucleus?

Hernández-Peón

I am not clear what you mean.

Horn

You observe a reduction in the pre- and postsynaptic responses in this nucleus when the animal is in a state of alertness and conclude from this that the signals are being blocked by descending impulses. Are there not other possible explanations? If one group of first order neurones is stimulated this causes presynaptic inhibition of other first order neurones (24). Surely it is possible that when an animal is aroused—sits up or whatever—there might be pilo-erection, which might stimulate cutaneous afferent fibres and cause effective presynaptic inhibition of other fibres around them. This would probably give some at least of the effects that you have found.

Hernández-Peón

We don't think that interpretation follows because if it did the same phenomenon should occur after making a lesion in the mid-brain tegmentum. What actually happens is that for 2 or 3 days after removal of the mid-brain tegementum the animals are in a comatose state. After the animal has recovered we find that if the room is darkened the animal goes to sleep: the instant we turn on the light the animal is aroused, and when we turn off the light the animal goes to sleep again. By arousing the animal and alerting him by clinking a key-ring the potential does not change at all. It remains just the same.

Horn

One explanation for that might be that the lesion damaged the sympathetic outflow controlling the pilomotor muscles.

Hernández-Peón

That is, if you are thinking of the reflex levels inside the mid-brain, but we know that there are sympathetic reflexes at the pontomedullary level and at the spinal level.

Horn

Yes, but there's also one at the hypothalamic level which would have been knocked out by the lesion.

Kennedy

I am now going to take the Chairman's prerogative and announce that this session is now over, but before breaking it up I would like to ask Professor Jasper if he has something to say to us.

Jasper

Thank you Dr. Kennedy for the privilege to express what I am sure all of us are thinking—how much we are grateful to the organizers of this Symposium, Dr. Mulholland and Dr. Evans, for the really splendid job they have done. I must say it's probably more successful than they had anticipated at the beginning. I don't remember any symposium that I have attended where the level of presentation and discussion has been so consistently high. I think that one proof of that is that our Guest of Honour is still with us.

We have other people to thank, the various foundations who have supplied the financial support and I am sure we want the organizers to thank them for themselves, as well as us, for making it possible. Also to the NPL, the Director, Dr. J. V. Dunworth, and the staff behind the scenes, and by no means least, Miss Rosenmann who has worked so hard to make this such a pleasant and efficient meeting. We are very grateful to you all. I think we all want to say again how delighted we are to have had you with us, Lord Adrian, and particularly for not being a typical Guest of Honour, i.e. the kind who doesn't often come but sends a deputy, and then only for the first five minutes. Guest of Honour is probably not the best description, for you have really been an active participant. I must really say this, that we shall always remember this an an epoch-making symposium. Some symposiums are, but most are not, and this occurred at a particularly opportune time for this particular subject, and I believe that it is going to be a real landmark in the field.

Kennedy

Lord Adrian, I think it would be most fitting if you had the last word at the end of these three particularly interesting days.

Adrian

Thank you Dr. Kennedy. I am really glad to have the opportunity to say a few words, if you will all forgive me for going on at the end of a busy and stimulating day. I had been intending to hold my peace and let you do all the talking, but some of the topics raised have been so interesting and some of the discussions so challenging that I have been quite unable to avoid putting in a word here and there. If you had been able to insert electrodes into my optic thalamus, or somewhere else in my brain stem, I think you would have found that during the past three days I have been suffering from acute spasms of envy. I really do envy those who are working now in this exciting field on the problem of attention, and equipped with continually developing techniques to back them up. Until the last war I should say that the main advances were on the clinical side, and many of these came from Hans Berger, although we were not always aware of this. Now I should say that the great advance has been in the techniques. I remember years ago going to Washington, and seeing a prototype of what must have been one of the earliest analogue computers—it looked like nothing more than a roomful of dining tables covered with wooden discs, and I remember wondering whether the time would come when we should need computers to help us in our work on the brain. And today computers do much of the work, and even make some of your mistakes for you. They have made an enormous difference to the investigation of the very complex problems which we face in the central nervous system. And there have been other great technical advances since the war, electron microscopy applied to biological material, chromatography, and so on. In spite of these some of our problems seem to remain unsolvable, and some will only be solved when we have evolved new systems of philosophy, but when I think of the great uncharted areas which are left for you to explore, I envy you even more for the adventures you will have in mapping—or trying to map them. The only development in the reverse direction, the only real drawback, is the enormous amount of literature which all this is producing. Nevertheless this symposium makes it quite clear how really useful a meeting of this kind can be, not only to the scientists who attend it and who benefit from the interchange of ideas, but even to those who can only see it in print, when some of the ideas, we hope, will already be out of date.

So I shall end by saying how glad I have been to attend this symposium, to have listened to expert discussion, and to be more than ever convinced that research on the brain is certain to lead us to new understanding.

REFERENCES

1. Battini, C., Moruzzi, G., Palestini, M., Rossi, G. F. and Zanchetti, A. (1958) *Science, N.Y.* **128**:30
2. Cordeau, J. P., Moreau, A., Beaulnes, A. and Laurin, C. (1963) *Archs. ital. Biol.* **101**:30
3. French, J. D., Hernández-Peón, R. and Livingston, R. B. (1955) *J. Neurophysiol.* **18**:74
4. Hernández-Peón, R. (1959) *Acta neurol. latinoam.* **5**:279
5. Hernández-Peón, R. (1961) *Sensory Communication*, p. 497. The M.I.T. and John Wiley & Sons, Inc., New York: London
6. Hernández-Peón, R. (1965a) In (Ed. by Akert, Bally and Schade) *Sleep Mechanisms: Progress in Brain Research.* Elsevier Publ. Co., Amsterdam, **18**:96
7. Hernández-Peón, R. (1965b) *J. nerv. ment. Dis.* **141**:623
8. Hernández-Peón, R. (1965c) *Excerpta Med.* **93**:123
9. Hernández-Peón, R. (1967a) *Proc. Int. Cong. Neurol. Surg.* Copenhagen, 1965, p. 429
10. Hernández-Peón, R. (1967b) In (Ed. by Vinken and Bruyn) *Handbook of clin. Neurol.* Elsevier Publishing Co.
11. Hernández-Peón, R., O'Flaherty, J. J. and Mazuchelli-O'Flaherty, A. L. (1965) *Expl. Neurol.* **13**:40
12. Hernández-Peón, R., O'Flaherty, J. J. and Mazzuchelli-O'Flaherty, A. L. (1967) *Brain Research*, **4**:243
13. Hernández-Peón, R. and Sterman, M. B. (1966) *Ann. Rev. Psychol.* **17**:376
14. Jouvet, M. (1963) In (Ed. by Hernández-Peón) The Physiological Basis of Mental Activity. *Electroenceph. clin. Neurophysiol. Suppl.* **24**:133
15. Jouvet, M. (1967) *Scient. Am.* **216**:62

16. Kaada, B. R., Anderson, P. and Jansen, J., Jr. (1954) *Neurology*. **4**:48
17. Kleitman, N. and Camille, N. (1932) *Am. J. Physiol.* **22**:474
18. Magoun, H. W. (1963) *The Waking Brain*. Thomas, Springfield, Ill.
19. Moruzzi, G. and Magoun, H. W. (1949) *Electroenceph. clin. Neurophysiol.* **1**:455
20. Segundo, J. P., Arana-Iñiguez, R. and French, J. D. (1955) *J. Neurosurg.* **12**:601
21. Tauber, E. S., Rojas-Ramírez, J. A. and Hernández-Peón, R. *Electroenceph. clin. Neurophysiol.*, in press
22. Thorpe, W. H. (1966) Ethology and Consciousness. In (Ed. by Eccles, J. C.) *Brain and Conscious Experience*. Springer-Verlag, Berlin, 470
23. Williams, H. L., Morris, G. O. and Lubin, A. (1962) In (Ed. by West, L. J.) *Hallucinations*. Grune & Stratton, New York and London, p. 158

Discussion

24. Eccles, Eccles and Magni (1961) *J. Physiol., Lond.* **159**:147

List of Participants

Guest of Honour:

The Rt. Hon. Lord Adrian, O.M., F.R.S., Trinity College, Cambridge, England

Chairmen:

Professor H. H. Jasper, University of Montreal, Canada
Dr. W. Grey Walter, Burden Neurological Institute, Bristol, England
Dr. J. L. Kennedy, Institute for Educational Development, New York, U.S.A.

Participants:

Dr. W. R. Adey
University of California
Los Angeles, Calif., U.S.A.

Dr. E. C. Beck
Veterans Administration Hospital
Salt Lake City, Utah, U.S.A.

Professor M. B. Bender
Mount Sinai School of Medicine
New York, U.S.A.

Professor D. E. Berlyne
University of Toronto, Canada

Dr. G. H. Byford
R.A.F. Institute of Aviation Medicine
Farnborough, England

Dr. W. A. Cobb
The National Hospital
London, England

Dr. O. Creutzfeldt
Max Planck Institut für Psychiatrie
Munich, Germany

Dr. E. M. Dewan
Air Force Cambridge Research Lab.
Bedford, Mass., U.S.A.

Dr. C. R. Evans
National Physical Laboratory
Teddington, England

Dr. P. B. C. Fenwick
The Maudsley Hospital
London, England

Dr. E. García-Austt
Institute of Neurology
Montevideo, Uruguay

Dr. W. R. Goff
Veterans Administration Hospital
West Haven, Conn., U.S.A.

Dr. R. Hernández-Peón
Institute of Cerebral Studies
Mexico City, Mexico

Dr. G. Horn
University of Cambridge, England

LIST OF PARTICIPANTS

Professor R. Jung
Neurologische Universitätsklinik
Freiburg, Germany

Professor H. P. Koepchen
Physiologisches Institut der Universität
Munich, Germany

Dr. J. C. Lilly
Maryland Psychiatric Research Centre
Baltimore, Md., U.S.A.

Professor D. B. Lindsley
University of California
Los Angeles, Calif., U.S.A.

Dr. C. T. Lombroso
Children's Hospital Medical Centre
Boston, Mass., U.S.A.

Dr. W. C. McCallum
Burden Neurological Institute
Bristol, England

Professor A. F. Mirsky
Boston University School of Medicine
Mass., U.S.A.

Dr. T. B. Mulholland
Veterans Administration Hospital
Bedford, Mass., U.S.A.

Dr. I. Ruttkay-Nedecky
Institute of Normal and Pathological Physiology
Bratislava, Czechoslovakia

Professor L. Stark
University of California
Berkeley, Calif., U.S.A.

Dr. V. Zikmund
Institute of Normal and Pathological Physiology
Bratislava, Czechoslovakia

Index